NANOTECHNOLOGY SCIENCE AND TECHNOLOGY

INNOVATIONS IN NANOMATERIALS

NANOTECHNOLOGY SCIENCE AND TECHNOLOGY

NANOTECHNOLOGY SCIENCE AND TECHNOLOGY

INNOVATIONS IN NANOMATERIALS

AL-NAKIB CHOWDHURY
JOE SHAPTER
AND
ABU BIN IMRAN
EDITORS

nova publishers

New York

Library of Congress Cataloging-in-Publication Data

Innovations in nanomaterials / editors, Al-Nakib Chowdhury, Joe Shapter, and Abu Bin Imran (Department of Chemistry, Bangladesh University of Engineering and Technology, Dhaka, Bangladesh, and others).
    pages cm. -- (Nanotechnology science and technology)
 Includes bibliographical references and index.
 ISBN 978-1-63483-548-0 (hardcover)
 1. Nanostructured materials--Industrial applications. 2. Bioengineering--Materials. 3. Electronic apparatus and appliances--Materials. I. Chowdhury, Al-Nakib, editor. II. Shapter, Joe, editor. III. Imran, Abu Bin, editor.
 TA418.9.N35I555 2014
 620.1'15--dc23

                    2015028217

Published by Nova Science Publishers, Inc. † New York

CONTENTS

PREFACE

The problems facing mankind currently and those it will face in the future are and will be incredibly complex. As such, novel and innovative solutions to these problems will have to be explored. Problems centred, for example, in supplying enough energy or water in the future will have to be solved via sustainable solutions, as it is now clear that the continuation of current practice for an ever-increasing load will mean irreparable damage to the planet. Indeed, there has already been considerable damage to the Earth's environment which must be repaired and steps will have to be taken to ensure future activities cause no further environmental damage. As the population increases, the challenge of feeding everyone will be an increasingly challenging issue. The increasing population of the planet will also put pressure on society to take care of people. The actions in this area will have to not only involve greater monitoring of things, such as air and water quality, but will also require new therapeutic paradigms to help cure the ill more quickly and less invasively than the current case.

The solutions to these challenges will have to be based on new approaches. These novel approaches will have their foundations in the development of new materials and protocols. The rise of nanotechnology over the last fifteen years or so now provides the opportunity for development of innovative and ground breaking approaches to society's pressing problems. This book provides a snapshot of the current development of new materials, ranging from nanoparticles to nanotubes to graphene to various hybrid materials and composites. These materials will without a doubt be the cornerstones of the novel approaches used to tackle mankind's most pressing problems in the 21st century.

Nanoparticles represent an exciting new class of materials which have very different properties than those of the bulk material. For example, while bulk gold is a relatively inert noble metal, gold nanoparticles have considerable catalytic activity. Many nanoparticles show very promising results in therapeutic applications such as in the treatment of cancer. There is also great promise for the use of nanoparticles as antibacterial agents. Silver is the most common element explored in the area but other materials are now started to be reported.

Carbon based materials such as graphene and nanotubes have been investigated for some time and hold great promise. The strength, flexibility and conductivity of these materials among other exciting properties offer tremendous opportunities in electrode fabrication, energy production and storage as well as the potential to make membranes for applications in both water purification and gas sensing.

The array of nano-sized materials at researchers' disposal and their unprecedented ability to control and manipulate the properties of the materials offers the opportunity for items that are currently in use to be made with novel properties and functions that can be tuned to the exact application. For example, water desalination membranes can be constructed with precise control over pore diameters ensuring a more efficient separation which will provide not only higher quality water but also use less energy in the process. Similar approaches have been used to improve gas separations which will have far reaching consequences in controlling and improving air quality in the future.

Nanomaterials such as nanoparticles or various titania dioxide structures have also been shown to have tremendous photocatalytic activity. These materials can reduce pollutants back to fundamental species such as carbon dioxide and water. Such approaches promise to be able to clean up effluent from factories as well as improve water quality through removal of bacteria.

In the future, it will be critical to monitor the environment in an effort to control harmful agents and hence prevent problems before they begin. Nanomaterials hold the promise of both more sensitive detection of a wide range of species and a wide range of sensor deployment. The small size of the sensors will mean sampling at multiple locations with very regular frequency will be possible providing highly detailed information on things such as air or water quality. Such sensing networks will make the control of pathogens easier and make response times to outbreaks faster. This will improve the general health of all mankind.

Finally, it is important to appreciate that the materials discussed in this book and indeed being developed in many labs around the world are novel with unprecedented properties. As such, it is absolutely critical that full testing of the materials in terms of toxicology and environmental impact is undertaken. It would be of little value to solve one issue only to create another. The challenge of such testing should not in any way be underestimated. It is important to realise that unlike a bulk material, the properties of nanomaterials can vary depending on many factors such as size, exact structure or surface chemistry among others. This means the testing regime is much bigger than any similar challenges faced in the past and hence this work most have a high priority in labs developing applications for nanomaterials.

This book explores a variety of nanomaterials, their applications and their impact. We hope the reader finds it a useful resource in the summary of the current state of research in this area.

Professor Al-Nakib Chowdhury
Professor Joe Shapter
Dr. Abu Bin Imran

In: Innovations in Nanomaterials   ISBN: 978-1-63483-548-0
Editors: Al-N. Chowdhury, J. Shapter, A. B. Imran   © 2015 Nova Science Publishers, Inc.

Chapter 1

METAL OXIDE NANOPARTICLES: TOXICOLOGICAL IMPACT ON BACTERIA

Gitashree Darabdhara[1], Manash R. Das[1,*]
and Rabah Boukherroub[2,†]

[1]Materials Science Division, CSIR-North East Institute
of Science and Technology, Jorhat, Assam, India
[2]Institut d'Electronique, de Microélectronique et de Nanotechnologie,
Villeneuve d'Ascq, France

ABSTRACT

Nanomaterials in the size range < 100 nm are finding numerous applications in the field of medicine, biosciences, information technology, etc. Along with benefits, nanomaterials owing to their very small size, their large surface area and capability of producing reactive oxygen species pose toxic threats. Metal oxides (MO) such as aluminium oxide (Al_2O_3), titanium dioxide (TiO_2) and zinc oxide (ZnO) are some of the common industrial additives in various applications such as abrasives, opacifiers in paints, paper or plastics and in semiconductor technology. Metal oxide nanoparticles (MO NPs) display enhanced activity in contrast to their corresponding bulk counterpart. Antimicrobial activity of MO NPs has been a subject of research in the last few years. Several studies revealed the toxicity of MO NPs towards mammalian cell lines, bacteria, plants and shellfishes. For example, it was found that ZnO and TiO_2 NPs reduce microbial biomass and bacterial diversity and composition. Similarly, Al_2O_3, Fe_2O_3, Fe_3O_4 and CuO NPs have toxic effects against several bacterial strains. Adhesion of the NPs to bacteria have been established by several instrumental techniques such as Fourier transform infrared (FTIR) spectroscopy, UV-visible spectrophotometry, Raman spectroscopy, zeta potential measurements, scanning electron microscopy (SEM), transmission electron microscopy (TEM), ion beam microscopy, etc.

---

[*] Manash R. Das: Materials Science Division, CSIR-North East Institute of Science and Technology, Jorhat 785006, Assam, India (E-mail: mnshrdas@yahoo.com).

[†] Rabah Boukherroub: Institut d'Electronique, de Microélectronique et de Nanotechnologie, UMR CNRS 8520, Université Lille 1, Avenue Poincaré-BP 60069, 59652 Villeneuve d'Ascq, France (E-mail: rabah.boukherroub@iemn.univ-lille1.fr).

This chapter is focused on the antibacterial activity of different MO NPs in comparison with their bulk counterparts, the mechanism of adhesion of NPs to different bacterial cell wall surface and their impact on environment.

INTRODUCTION

In recent years, with the development of nanotechnology, a lot of novel nanomaterials are prepared, their novel properties are being gradually discovered, and their applications in various fields have also advanced greatly. It is well established that when materials are made in the form of very small size (nm), there is an extensive change in their physical and chemical properties and these changes lead to establishment of new phenomena [1]. The significant differences between nanomaterials and their bulk counterparts have led to rapid development in the field of nanotechnology [2, 3]. The field of nanotechnology is related to the production and application of nanostructured materials and considered as one of the fastest developing fields by many scholars [4-6]. Increasing attention has been paid to nanomaterials due to their large surface area as well as high reactivity [7-10]. The use of nano materials (NPs < 100 nm) is finding numerous applications in the field of materials science, nanomedicine, biosciences and information technology due to their unique physical, chemical, electronic, electrical, mechanical, magnetic, thermal, dielectric, optical and biological properties [11-16]. In the recent years, metallic nanoparticles (NPs) that represent an intermediate aspect between bulk materials and atoms/molecules have been a matter of investigation [17]. Among metal NPs, metal oxide (MO) NPs have been widely used for both industrial and household applications [7]. They are of economic importance owing to their unique physicochemical properties like dissolution properties, electronic charge, small size and large surface to mass ratio. MO NPs have been commonly used in pharmaceutical products, in daily usable products like sunscreen lotions and other cosmetic products, in electronic industries for making semiconductors, etc. ZnO and TiO_2 NPs are common elements in cosmetics, skin care products, UV detectors and semiconductors [18]. Also ZnO NPs are used as drug carriers, as fillings in medical materials, biosensors, biogenerators, bioelectrodes, electroluminescent devices and ultraviolet laser diodes [19].

Owing to their magnetic properties, Fe_2O_3 and Fe_3O_4 NPs are of particular interest [6]. They are applied in a large number of areas such as gas sensors, ion exchangers, catalysis, magnetic recording devices, magnetic resonance imaging, waste water treatment, toners and inks for magnetic data storage xerography, etc. [20]. CuO NPs are successfully used in catalysis, gas sensors, high temperature superconductors, solar energy conversions and field emission emitters [21-25]. Al_2O_3 NPs are widely applicable in the field of catalysis, heat transfer fluids, "polymer modification," etc. [26]. All the above mentioned MO NPs are also found to exhibit excellent antimicrobial activities.

Nanomaterials due to their size effects and large surface to volume ratio apart from displaying unique physical and chemical properties also exhibit distinctive biological properties and antimicrobial agents based on nanomaterials have received much attention in the recent years. Toxicity caused by nanomaterials is a major concern in the rapidly developing field of nanotechnology. MO NPs are produced in large amounts by a number of industries and as such find their way for discharge into the environment *via* manufacturing sewages or through leakage during transportation processes.

MO NPs are mainly released in air, water and soil and hence it is mandatory to estimate the harmful effects of these released NPs towards human health and environment.

Microorganisms are usually used to predict the toxicity caused by MO NPs [18]. An effective way is to observe the consequence on bacterial cells that are exposed to these MO NPs. Bacteria represent a very important class of microorganisms and play a vital role in the functioning of ecosystem. Study of the interaction of bacteria with NPs is thus an important subject matter to understand the influence of NPs on bacteria. Furthermore, being single celled organisms, bacteria represent an interesting model to study the influence of NPs on functioning of cell/organism [4]. As a result of the advancement of biotechnology, many resistant bacterial strains have been generated, which developed resistance to different germicides and antibiotics. So, formulation of new bactericidal agents which are non-toxic with no irritation is highly desirable. Recent literature focused on the use of metal NPs (Ag, Au, etc.) as effective lethal agents towards bacterial cells and MO NPs (Al_2O_3, Fe_2O_3, Fe_3O_4, CuO, ZnO, TiO_2, etc.) in this regard exhibit excellent biocidal activity against both Gram positive and Gram negative bacterial strains [1].

MO NPs have been found to exhibit admirable bactericidal properties. Therefore, it is expected that these particles may serve to formulate new generation of antibacterial materials. Moreover, antimicrobial agents based on inorganic substances display better stability than organic based antimicrobial agents. So researchers have concentrated their efforts on the synthesis of antimicrobial agents based on metal and MO NPs [14, 27].

This chapter consists of a comprehensive review of the current research work on antibacterial activity and adhesion of different MO NPs (Al_2O_3, Fe_2O_3, Fe_3O_4, CuO, ZnO, TiO_2) to bacterial cell walls, their characterization and their impact on environment.

BACTERIA AND THEIR SURFACE PROPERTIES

Bacteria are an essential link in the food web and play an important role in the ecosystem [28]. Bacteria, single celled microorganisms, are found in abundance in near surface of geological systems and are involved in various biogeochemical processes like transport of contaminants and their degradation, sorption of metals from minerals, dissolution of minerals and their precipitation, and in various other activities [29].

Soil and water contamination in the rural areas is a major threat to environment and human health, so it is very important to recognize the phenomenon of bacterial adhesion to mineral surfaces for the better understanding of subjects related to bacterial cell transport and their fate [30]. The interfacial chemistry between bacteria and water is very important as it plays a vital role in various bacterial reactions. As such it is very important to understand bacterial surface chemistry. The deviation of bacterial cell structure and composition with different environmental variables like pH is responsible for different bacterial interactions [29]. Bacterial adhesion to soil and aquatic systems are influenced by the chemical properties of bacteria and abiotic environmental surfaces.

Gram positive bacterial cell wall consists of 40-80% peptidoglycan (a polymer of *N*-acetyl glucosamine and *N*-acetyl muramic acid), techoic acid (polymer of glycopyranosyl glycerol phosphate) and teichuronic acid (similar to techoic acid but replaces phosphate with carboxyl groups) [30].

In Gram negative bacteria, there is an outer membrane in addition to peptidoglycan layer that contains phospholipids, lipoproteins, lipopolysaccharides and proteins. Unlike Gram positive bacteria, Gram negative bacteria lacks of techoic or teichuronic acid. With the aid of several macroscopic and molecular tools, it was possible to identify major proton active functional groups on bacterial surfaces. Bacterial cell wall consists of numerous polymers and macromolecules possessing carboxyl, hydroxyl, phosphate and carbohydrate related moieties, which can possibly form coordinative bonds with functional groups present on the mineral surfaces. Carboxylic, phosphate and carbohydrate groups are subtle to variations in solution pH and deprotonation of carboxylic and phosphate groups leads to a net negative charge of the bacterial cell surface in the pH range of 4-9 [29]. Chemical properties of bacteria and abiotic environmental surfaces influence the interaction between them. Long-range electrostatic interactions and short range interactions are two different interactions that exist between them. Short range interactions are controlled by (1) chemical (covalent, ionic, hydrogen) bonding, (2) van der Waals forces, and (3) hydrophobic effects. Furthermore, steric effects also influence microbial adhesion, which is important for overlapping regions of polymer segments.

Cytotoxicity of oxide NPs towards microbes has been a matter of research from the past decades. Locating the site of NPs toxicity is a key point for understanding toxicity mechanism. So adhesion process of NPs to bacterial surface is an important argument of NPs bacterial cytotoxicity [28].

ANTIBACTERIAL ACTIVITY OF MO NPS AND THEIR CHARACTERIZATION

Antimicrobial activity of MO NPs is very significant. Various studies have revealed the toxicity of MO NPs towards mammalian cell lines, bacteria, plants, shellfishes, etc. The toxicity of MO NPs towards bacteria has drawn much attention and this interaction provides us with evidence regarding the influence of NPs once they are released to the ecosystem. Moreover, bacteria as a single celled organism proved to be a good example to study the impact of MO NPs and to have an understanding of how NPs can affect cell or organism functioning. Also, the toxicity of these MO NPs at cellular levels must be undertaken in media with ionic strength close to natural fresh waters, which provides us with information whether NPs are harmful to ecosystem, particularly water system [4]. Various instrumental techniques such as FTIR, TEM, SEM, selected area electron diffraction (SAED), zeta potential measurements, atomic force microscopy (AFM), etc. are employed for characterizing the effect of MO NPs on bacterial cells. A brief overview will be given on different MO NPs and their antibacterial activities against different bacteria strains.

1. Al_2O_3 NPs

Al_2O_3 is widely used as an abrasive agent or insulator because of its good dielectric and abrasive properties [4]. Al_2O_3 NPs have an extensive use in industry as well as personal care products.

These oxide NPs are found to exhibit antibacterial activities against different bacteria strains [1]. Jiang et al., studied the effect of Al_2O_3 NPs as well as micron ranged Al_2O_3 particles on three different bacteria strains: *B. subtilis*, *E. coli* and *P. fluorescens* [4]. For the experiment, bacteria were grown in 50 mL Tryptic Soy (TS) medium for 1 day and the temperature was maintained at 30ºC in an incubation shaker. The bacteria were then separated from the medium by centrifugation and washed with NaCl solution. Finally, the bacteria were suspended in NaCl aqueous solution and the bacterial suspension diluted to 2×10^8 cells/mL for toxicity measurements. 10 mL of the bacterial solution was taken in a test tube to which was added 100 µL of Al_2O_3 NPs solution to obtain an exposure concentration of 20 mg/L. The bacterial solution and Al_2O_3 NPs were mixed in a vortex and incubated for 2 h at 30ºC. The toxicity of the Al_2O_3 NPs was then measured by colony count method by comparing with the control after 24 h of incubation. Furthermore, the toxicity characterization was performed using zeta potential measurements and TEM imaging. It was found that Al_2O_3 NPs exhibited substantial toxicity to the three bacteria strains. The death rate was 57% for *B. subtilis*, 36% for *E. coli* and 70% for *P. fluorescens*. However, Al_2O_3 bulk particles (BPs) did not exhibit any negative effect on any of the three bacteria strains. Thus, it was concluded that Al_2O_3 NPs showed good antibacterial activity than their bulk counterpart.

In another study by Jiang et al., on the effect of Al_2O_3 NPs against *B. subtilis*, *E. coli* and *P. fluorescens*, toxicity assessment was performed by FTIR spectroscopic technique [28]. It was observed that exposure of bacteria to Al_2O_3 NPs changed the IR spectra of bacterial cells. The NPs caused protein structural damage and phospholipid molecular damage. Zhang et al., studied the interaction of corundum (α-Al_2O_3) immobilized on surfaces with *E. coli* in an aqueous environment using AFM with the aim to better understand surface properties and interaction forces [31]. They found that the adhesion force strength was significantly influenced by the NPs size. The dependence of adhesion force between *E. coli* and α-Al_2O_3 NPs of different sizes arose from effective contact area. Smaller NPs had better adhesion to bacterial cell surface. Gokulakrishnan et al., investigated the effect of Al_2O_3 NPs (size < 50 nm) on five bacteria strains namely *P. aeruginosa*, *Klebsiella sp.*, *S. pneumonia*, *S. aureus* and *Streptococcus species* [32]. The antibacterial activity of all these four bacteria strains was tested by well diffusion method. Bacteria were grown in 20 mL of sterile molten Mueller Hinton agar media and 10^8 cells were grown on sterile petri plates containing the medium with overnight culture. Solid medium was then gently pierced with a cork borer to make a well. Finally, Al_2O_3 NPs were added to the well (50 µg/mL) and incubated for 24 h at 37 ± 2ºC and antibacterial study was performed in terms of zone of inhibition (in mm). Al_2O_3 NPs displayed excellent antibacterial activity against *P. aeruginosa* (11.00 ± 0.15) and *S. pneumonia* (11.00 ± 0.41). However, the Al_2O_3 NPs did not exhibit any sensitivity against *Klebsiella sp, S. aureus* and *Streptococcus species*. Similarly, Ravikumar et al., studied the effect of Al_2O_3 NPs against poultry pathogens *Klebsiella, E. coli, Staphylococcus* and *Salmonella species* by well diffusion method [33]. The diameter (in mm) for the zone of inhibition against all the four bacteria strains was found to be: *Klebsiella species* (6 ± 0.39), *E. coli* (10 ± 0.41), *Staphylococcus* (7 ± 0.26) and *Salmonella* (8 ± 0.32). The results confirmed the activity of Al_2O_3 NPs against all the four bacterial species. A number of similar studies were performed by different research groups on antibacterial activity of Al_2O_3 NPs and found positive results in their investigations, which proved good antibacterial properties of Al_2O_3 NPs [34-36]. The mechanism of action of these MO NPs on bacterial cells will be discussed later in this chapter.

2. CuO NPs

Cu, an important trace element in living organisms, plays a dynamic role in protein functioning. With rapid growth of nanotechnology in the global market, the use of engineered NPs is also increasing. CuO NPs are extensively employed in a variety of applications such as pigment in ceramics, as dispersed NPs to increase thermal conductivity, gas sensors, as catalyst and also to replace Au and Ag in ink-jet printable electronics [18, 37, 38]. The antimicrobial activity of CuO NPs is also worth mentioning. The antimicrobial activity of CuO NPs has been examined by using different microorganisms like *E. coli, B. subtilis, S. aureus,* etc.

Azam et al., assessed the toxicity of CuO NPs against human pathogenic bacteria of both Gram positive and Gram negative like *E. coli, P. aeruginosa, S. aureus* and *B. subtilis* by well diffusion method [39]. In a typical method, bacterial cells were cultured in Muller Hinton broth at 35 ± 2°C on a rotary shaker at 160 rpm. Bacterial cells were grown by spreading 100 μL of fresh culture having 10^6 colony forming units (CFU)/mL on agar plates. The plates were then left standing for 10 min. 8 mm wells were prepared in the agar plates for test of nanomaterial antibacterial activity. To prevent the wells from leakage, they were sealed with a drop of molten agar (0.8% agar). To the wells, 100 μL (50 μg) of CuO NPs sample was poured and incubated overnight at 35 ± 2°C for antibacterial activity studies. It was observed that CuO NPs displayed admirable action against both types of bacterial cells. CuO NPs exhibited a zone of inhibition of about 21 mm against *B. subtilis* and 15 mm in case of *E. coli*. Similar patterns were observed for the other two bacteria strains. It was seen that Gram negative bacterial strains of *E. coli* and *P. aeruginosa* had zone sizes that were reduced by 28% and 33% than those of Gram positive bacterial species, suggesting that Gram negative species exhibited better resistance to this nanomaterial in comparison to the Gram positive bacterial strains.

Theivasanthi et al., synthesized CuO NPs *via* two methods viz. electrolysis route and chemical reduction approach and tested their antibacterial action against different bacterial strains [40]. CuO NPs synthesized using electrolysis method exhibited a surface to volume ratio of 0.25, which was found to be higher than that of CuO NPs synthesized using chemical reduction method (0.05). Antibacterial effect of the CuO NPs was tested by using standard zone of inhibition microbiology assay. The CuO NPs synthesized using electrolysis method displayed diameter of zone of inhibition against *E. coli* and *B. megaterium* of 15 mm and 5 mm, respectively, while CuO NPs prepared using chemical reduction method exhibited a zone of inhibition of about 8-10 mm for *E. coli* cells. The better antibacterial activity of CuO NPs synthesized by electrolysis method was attributed to greater surface to volume ratio of the NPs. Dasari et al., performed similar type of antibacterial study of CuO NPs against *E. coli* cells by spread plate counting method and found that CuO NPs exhibited excellent antibacterial activity against the selected bacterial strains [18]. Dimkpa et al., investigated the toxicity of CuO NPs against beneficial rhizosphere isolate *Pseudomonas chlororaphis 06* and compared it to that of ZnO NPs [41]. The toxicity of the commercially available CuO NPs was found to be higher than that of ZnO NPs. Neither NPs released alkaline phosphatase from the cells' periplasm, indicating minimal outer membrane damage. Their study attributed the toxicity of CuO NPs to accumulation of intracellular reactive oxygen species. In a similar manner, CuO NPs were found to be the most toxic as compared to NiO, ZnO and Sb_2O_3 NPs against the bacterial strains *E. coli, B. subtilis* and *S. aureus* [42]. Toxicity assessment was carried out *via* colony count method. Bacterial culture was prepared using Luria-Bertani broth at 37°C overnight and all tests were performed using LB agar plates as a function of NPs concentration. All the MO NPs displayed lethal effects against all the three bacteria strains. CuO NPs turned to be the most toxic followed by ZnO, NiO and Sb_2O_3 NPs. Effects of metals ions dissolved from MO NPs were found to be negligible. A number of studies were reported by other researchers to explore the antibacterial properties of CuO NPs [43-51]. CuO NPs synthesized by different methodologies exhibited excellent antimicrobial activity against both Gram positive and Gram negative bacterial strains, antifungal activity as well as anticancer activity and generated knowledge for their use in chemotherapy.

3. ZnO NPs

ZnO is widely used as a semiconductor because of its wide-band-gap of 3.3 eV and its non-toxicity [5]. ZnO NPs are also extensively used in industrial and consumer products like in sunscreens to scatter and reflect ultraviolet radiation in sunlight, as fillings in medicinal materials and as drug carriers [52, 53]. ZnO NPs as a versatile nanomaterial was also investigated for various applications in biosensors, biogenerators, bioelectrodes, electroluminescent devices and ultraviolet laser diodes [19]. Although ZnO NPs have several positive applications yet the harmful impacts of this nanomaterial on living cells cannot be ignored [52]. A number of reports are available on the toxicological impact of ZnO NPs on different microbial strains [51, 54-63]. Huang et al., studied the effect of ZnO NPs against *S. agalactiae* and *S. aureus,* two etiological agents of several infective diseases in humans [19]. The ZnO NPs investigated in the study were synthesized in ethylene glycol medium by hydrolysis of zinc acetate. The bacterial strains were grown in ordinary broth medium containing beef extract, peptone and NaCl. Bacteriological tests were realized in solid agar plates with varied concentrations of ZnO NPs. The bactericidal tests were performed by well diffusion method as well as colony count method and characterized by different instrumental techniques such as TEM and SAED. TEM images demonstrated morphological changes in bacterial cells in contact with ZnO NPs (poly vinyl alcohol-coated ZnO having narrow size distribution) as shown in Figure 1. TEM images revealed the ability of the NPs to penetrate inside the bacterial cells thus damaging the cell membrane. The study exposed the fact that low concentrations of ZnO NPs did not cause any cellular destruction. In Figure 1 (a), NPs are seen to penetrate inside the cell of *S. agalactiae* and illustrates cellular internalization of ZnO NPs and cell wall disorganization. In Figure 1 (b), cell membrane damage of some *S. agalactiae* is clearly observed. The cells of *S. aureus* are also seen to attach to ZnO NPs; partial cell membrane damage as well as cellular division is also represented in Figure 1 (c, d). Thus, TEM images showed association of ZnO NPs with bacterial cell membrane and the cells internalize some of the particles.

Furthermore, SAED pattern appears as a series of spots, each spot corresponding to particular diffraction from a plane in the crystal structure. SAED patterns of ZnO NPs inside and outside the cell of *S. agalactiae* are displayed in Figure 2. From Figure 2 (a), it is clear that in the cellular matrix, the crystal wurtzite structure of ZnO NPs is retained. The NPs grow along [0001] and its side surfaces are demarcated by {0110}.

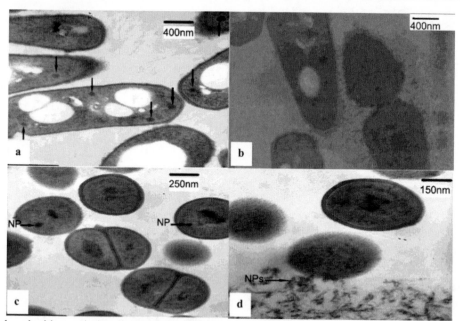

Figure 1. TEM images of bacterial cell section upon interaction with ZnO NPs: (a, b) *S. agalactiae* and (c, d) *S. aureus*.

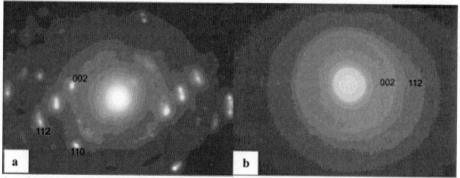

Figure 2. SAED patterns of ZnO NPs (a) outside the cell and (b) inside the cell of *S. Agalactiae*.

However, in the SAED patterns of ZnO NPs inside the bacterial cell, the crystal structure of ZnO NPs is no longer retained, but changes into an amorphous structure as a result of dipping the cell sap into ZnO NPs and partially damaging the NPs crystal structure as illustrated by the diffused SAED pattern in Figure 2 (b). The SAED pattern of ZnO NPs inside the cell is dark, although the [002] and [112] faces can be easily differentiated. From the SAED results it can be concluded that ZnO NPs display biocidal properties towards selected bacterial strain and also the crystal structure of ZnO NPs undergoes changes inside the cell.

Jiang et al., examined the bacterial toxicity of ZnO NPs along with Al_2O_3, SiO_2 and TiO_2 NPs and their respective bulk counterparts with the aim to understand the size and/or chemical composition effects by evaluating the contribution of dissolved metal ions to the overall toxicity [4]. In addition, the interaction of the NPs with bacteria and their surface properties were investigated using zeta potential measurements and TEM imaging. Bactericidal effects were tested against *B. subtilis*, *E. coli* and *P. fluorescens* using bacterial colony count method. ZnO NPs were found to be the most toxic among all the NPs investigated with a death rate of 100%. ZnO Bulk Particles (BPs) also exhibited toxicity towards *E. coli* and *P. fluorescens,* but were less toxic than ZnO NPs. ZnO NPs completely destroyed *B. subtilis* bacterial cells. Although ZnO NPs displayed much higher toxicity than ZnO BPs, both released similar amounts of Zn^{2+} ions and so the toxicity difference between ZnO NPs and BPs was likely due to higher toxicity of ZnO NPs. It is to be noted that under similar experimental conditions, metal ions from Al_2O_3, SiO_2 and TiO_2 (NPs and BPs) were not detected in the liquid phase. Al_2O_3 NPs induced a death rate of 57% against *B. subtilis*, 36% towards *E. coli* and 70% against *P. fluorescens* whereas SiO_2 caused a mortality rate of 40% against *B. subtilis*, 58% towards *E. coli* and 70% towards *P. fluorescens*. However, TiO_2 NPs as well as BPs did not induce any toxicity against the two bacterial strains.

The MO NPs attachment to the surface of bacteria was examined by TEM. All the NPs showed similar behaviour for the three bacteria strains. Figure 3 displays TEM images of the different MO NPs in interaction with *P. fluorescens*. The results clearly indicate that Al_2O_3, SiO_2 and ZnO NPs coat the whole bacterial cells. Interestingly, Al_2O_3 and SiO_2 NPs were present only on the bacterial cells, suggesting that these NPs attach preferentially to cell walls rather than aggregate together (Figure 3 A and B). ZnO NPs interaction with *P. fluorescens* is different from that of Al_2O_3 and SiO_2 NPs. Indeed, ZnO NPs covered the bacterial cell, but also scattered everywhere on the TEM grid (Figure 3D), suggesting that their interaction with the bacterial cells is weaker although these particles were more toxic than Al_2O_3 and SiO_2 NPs. On the other hand, TiO_2 NPs formed large aggregates of several hundred nanometers (Figure 3C). Zeta potential measurements are usually performed in bacterial studies to understand the influence of surface charge on bacteria and NPs and thus the mechanism of NPs toxicity (Figure 4). From the zeta potential measurements made by Jiang et al., the surface charge present on different MO NPs as well as bacterial cells was identified [4]. For the study, the NPs stock solution was diluted in NaCl solution and divided into a number of aliquots of different pH. Samples were then stored for 2 h for achievement of equilibrium and then subjected to ultrasonication for dispersion. The zeta potential measurements of the bacterial cells were performed by dissolving 2×10^8 cells/mL in NaCl. From the measurements, the surface charge of *B. subtilis* was determined to be –41.3 mV, *E. coli* (–7.20 mV) and *P. fluorescens* (–32.3 mV) at pH 6.5. At this pH value, Al_2O_3 (+30 mV) and SiO_2 (+35 mV) were positively charged, while TiO_2 (–21 mV) and ZnO (–5 mV) were negatively charged. For the highly positively charged NPs (Al_2O_3 and SiO_2), strong electrostatic interactions with negatively charged bacteria would contribute to NPs' adhesion to the cell walls. Similarly, it is expected that electrostatic force would play an important role in TiO_2 repulsion from the bacterial cell wall. Although electrostatic interactions are believed to be dominant in the adhesion of positively charged NPs, it is not excluded that for ZnO NPs, receptor-ligand interactions are most important.

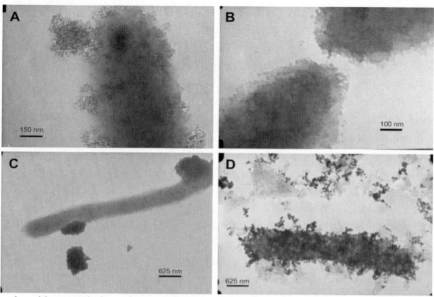

Reproduced with permission from reference [4], Copyright© 2009, Elsevier Publication, Environmental Pollution, 2009, Vol. 157, Pages 1619-1625.

Figure 3. TEM images of NPs adhesion to the surface of *P. fluorescens*: (A) Al_2O_3, (B) SiO_2, (C) TiO_2, (D) ZnO NPs.

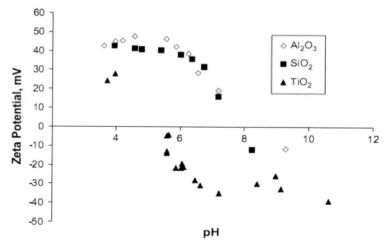

Reproduced with permission from reference [4], Copyright© 2009, Elsevier Publication, Environmental Pollution, 2009, Vol. 157, Page 1619-1625.

Figure 4. Zeta potential of MO NPs measured in 1g/L NaCl.

Carboxyl, amide, phosphate, hydroxyl groups and carbohydrate-related moieties in the bacterial cell wall may provide sites for the molecular-scale interactions with the oxide NPs [64, 65]. In a separate report, Jiang et al., studied the effects of Al_2O_3, TiO_2 and ZnO NPs on the bacteria cells (*B. subtilis*, *E. coli* and *P. fluorescens*) and bacterial surface biomolecules using Attenuated Total Reflectance (ATR)-FTIR [28]. FTIR is a powerful characterization tool for studying surface functional groups present on bacterial cell and their interaction with MO NPs. The NPs toxicity is connected with their adhesion onto bacteria, damage to the structure and change in the physico-chemical properties of surface biomolecules [66]. The major biomolecules on bacterial cell, which mainly interacts with MO NPs when the NPs attach to the cell surface, includes lipopolysaccharides (LPS), lipoteichoic acid (LTA), proteins and phospholipids.

Figure 5 displays the FTIR spectra of bacteria cells before and after exposure to MO NPs. All the three bacteria investigated in this work exhibited similar FTIR spectra [28]. The bands at 1638 and 1540 cm^{-1} are assigned to amide I and amide II stretching, respectively. The amide I band is associated with the carbonyl stretching vibration $v(C = O)$ in the peptide bond groups, while the amide II band is due to the bending vibration $\delta(N\text{-}H)$ and $v(C\text{–}N)$. Upon interaction with MO NPs, a slight shift and shape change in amide I band were observed, suggesting changes in protein structure. A small peak at 1745 cm^{-1} arising from ester stretching $v(C = O)$ in the spectrum of *B. subtilis* was absent in the spectra of *E. coli* and *P. fluorescens*. This peak decreased after interaction with ZnO NPs (Figure 5a). For Gram positive bacteria *B. subtilis*, a band at 1059 cm^{-1} due to a combination of $v(C\text{–}O, C\text{–}O\text{–}C, C\text{–}O\text{–}P)$ was observed. For Gram negative bacteria *E. coli* and *P. fluorescens*, a similar stretching band was observed at 1079 cm^{-1}. The stretching vibration at 1079 cm^{-1} is assigned to sugar vibration of intact bacterial cells and hence it can be inferred that the sugar content of Gram negative bacteria is high. In case of *B. subtilis*, the band at 1059 cm^{-1} shifted to lower wavenumbers upon interaction with Al$_2$O$_3$ NPs and ZnO NPs.

Because this IR absorbance is related to hydroxyl, carbonyl, carboxylic acid and phosphate groups, its shift towards lower wavenumbers is likely due to binding of functional groups on bacterial cells to MO NPs. However since bacterial cells are very complex systems, it is very difficult to assign which bacterial surface biomolecule or functional group on bacteria interacts with NPs in reality. To further gain insight on the effects of MO NPs on the cell wall, the changes in amide I band of bacteria along with those of bovine serum albumin (BSA) and protein G in terms of their second derivative (to enhance their spectral resolution) are displayed in Figure 6. BSA has high α-helix content with an amide I band at 1653 cm^{-1}, while protein G contains both α-helices and β-sheets with the latter being dominant.

The vibration bands at 1653 and 1636 cm^{-1} correspond to α-helix and β-sheets, while the band at 1645 cm^{-1} is due to unordered segments. Protein BSA contains only 54-68% α-helices and very little β-sheets and thus exhibited only α-helix bands in the second derivative spectra. In the second derivative spectra of all the three bacteria, the stretching intensity of the β-sheet region was higher than the α-helix region, which may be attributed to the high content of β-sheet structure in the cell protein (Figure 6 (a)). Interaction with MO NPs caused a shift of the bands of β-sheets of bacteria and protein towards lower frequency thus generating new bands at 1630 and 1627 cm^{-1}. The band at 1636 cm^{-1} is assigned to antiparallel β-sheet.

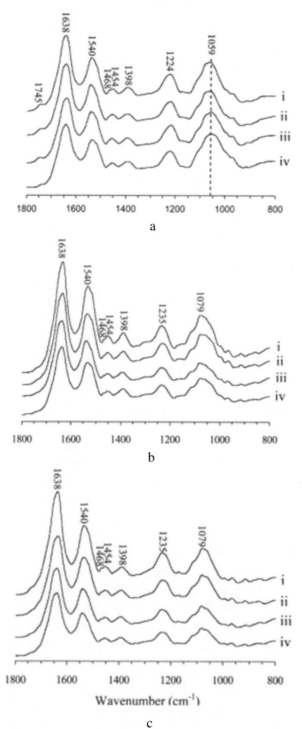

c

Figure 5. ATR-FTIR spectra of Gram positive *B. subtilis* (a), Gram negative bacteria *E. coli* (b), and *P. fluorescens* bacteria (c). The different spectra are for (i) bacterial cells alone and bacteria cells with (ii) TiO_2 NPs, (iii) Al_2O_3 NPs, (iv) ZnO NPs.

The vibration bands at 1653 and 1636 cm^{-1} correspond to α-helix and β-sheets, while the band at 1645 cm^{-1} is due to unordered segments. Protein BSA contains only 54-68% α-helices and very little β-sheets and thus exhibited only α-helix bands in the second derivative spectra. In the second derivative spectra of all the three bacteria, the stretching intensity of the β-sheet region was higher than the α-helix region, which may be attributed to the high content of β-sheet structure in the cell protein (Figure 6 (a)). Interaction with MO NPs caused a shift of the bands of β-sheets of bacteria and protein towards lower frequency thus generating new bands at 1630 and 1627 cm^{-1}. The band at 1636 cm^{-1} is assigned to antiparallel β-sheet.

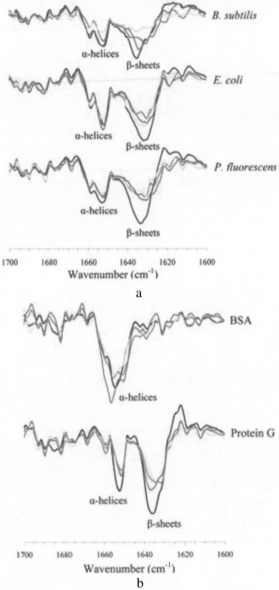

Figure 6. Second derivative of amide I band of bacteria (a) and protein (b); (dark blue) bacteria or protein alone, (cyan) with TiO$_2$ NPs, (magenta) with Al$_2$O$_3$, (yellow) with ZnO NPs.

Appearance of new peaks upon exposure to NPs suggests changes in hydrogen bonding strength and interaction between β-strands. A strong hydrogen bond resulting from intermolecular or intramolecular bonding may be the reason for low-frequency shift. The α-helix bands at 1653 and 1659 cm^{-1} are not affected (no shape or position change) upon NP exposure to bacteria and protein G. But when BSA was subjected to NPs, a new band at 1650 cm^{-1} appeared.

From these studies, it is clear that exposure to Al_2O_3, TiO_2 and ZnO NPs caused changes of the IR spectra of bacterial cells, LTS, LPS, proteins and PE due to biomolecule adsorption on the NPs surface. Bacterial biomolecules adsorption on the NPs surface during NPs exposure causes several changes inside the bacterial cell, which is responsible for bacterial toxicity. ZnO NPs, the most toxic among the investigated NPs induced the most serious molecular damage to PE [28].

Jin et al., reported antimicrobial activity of ZnO NPs against *L. monocytogenes, S. enteritidis* and *E. coli* [67]. In their investigation, they utilized ZnO quantum dots (ZnO-QD) that were bound to polystyrene film (ZnO-PS) or polyvinylprolidone gel (ZnO-PVP) and found that both ZnO-QD and ZnO-PVP exhibited antimicrobial activity against all three bacteria strains whereas ZnO-PS exhibited no such activity. Their study suggested the use of ZnO NPs in food systems for inhibition of bacteria. Li et al., examined antibacterial activity of ZnO NPs coated onto poly (vinyl chloride) based films against *E. coli* and *S. aureus* [68]. ZnO coated films displayed better bactericidal effects against *S. aureus*. However, such activity of the coated films was absent against fungi species *Aspergillus flavus* and *Penicillium citrinum* probably due to the complexity of fungal cell wall. Their findings suggest that ZnO NPs can be used for coating plastic films to prepare antimicrobial packages against bacteria like *E. coli* and *Staphylococcus*.

Sinha et al., investigated the bactericidal effects of ZnO NPs against mesophilic and halophilic bacteria like *Enterobacter sp., Marinobacter sp.*, and *B. subtilis* [69]. Toxicity was more noticeable on Gram negative bacteria than Gram positive bacteria due to the presence of thicker peptidoglycan layer on Gram positive bacteria. ZnO NPs induced 50% reduction in the growth of *Enterobacter sp.*, and 80% for *Marinobacter sp.* However, bulk ZnO exerted minimal growth reduction. The presence of negatively charged cardiolipins on the cell surface of halophiles may favour electrostatic interaction between ZnO NPs and cell surface responsible for attachment of ZnO NPs on bacterial cell surface, leading to nanotoxicity. Similarly, Azam et al., reported the toxic effect of ZnO NPs against human pathogenic bacteria *E. coli* and *P. aeruginosa* (Gram negative), *S. aureus* and *B. subtilis* (Gram positive) by well diffusion method [39]. ZnO NPs demonstrated admirable effect against all the bacteria with zone of inhibition of 25 mm for *B. subtilis* and 19 mm for *E. coli*. Similar results were observed in the case of the other two bacterial strains.

You et al., studied the effect of ZnO NPs on *E. coli* and bacteriophages MS2 [70]. Toxicity assessment was carried out by standard double agar layer (DAL) method and a turbidimetric microtiter assay. No bacteriophage inactivation was observed in presence of the highest concentration of ZnO NPs (20 mg/L). However, in binary bacteria-phages system in which *E. coli* was exposed to MS2 and ZnO NPs at once, the dynamic changes of active bacteria and MS2 phages during incubation showed that exposure to ZnO NPs increased the number of phages by 2-6 orders of extent. All these findings suggested that bacterial viruses like MS2 can infect host *E. coli* when exposed to NPs simultaneously. Desari et al., evaluated the toxicity effect of several MO NPs like ZnO, CuO, Co_3O_4 and TiO_2 on *E. coli* strain under light and dark conditions using spread plate counting method and LC_{50} value determination method [18]. Their results revealed that ZnO NPs were the most toxic among all the MO NPs against the selected bacterial strain. The mechanism of MO NPs toxicity was determined by oxidative stress, reduced glutathione (GSH), lipid peroxidation (LPO) and metal ions.

Dutta et al., investigated the effect of transition metal doped and matrix embedded ZnO NPs on *E. coli* bacterial strain [52].

The studies were carried out both in Luria-Bertani medium and on solid agar medium in the presence and absence of light. Toxicological assessment was quantified in terms of minimum inhibitory concentration, minimum bactericidal concentration, colony forming unit counts and by growth curves and disk diffusion tests. The study revealed a significant change in bactericidal properties at various concentrations of ZnO NPs, but also a dependence on the structural and chemical properties of the ZnO NPs. The size of the ZnO NPs, the presence of oxygen defects in ZnO NPs, the presence of activator atoms in the host lattice, and the surface properties play a critical role in imparting toxicity behaviour of ZnO NPs. The results indicated that toxicity of the NPs varies with structural and chemical properties and can be modified for use as bactericidal agents. Dijaz et al., reviewed the enhancement of antibacterial property through preparation of metal ion doped NPs in which it was reported that ZnO doped with Ti exhibited antibacterial property against *E. coli* and *S. aureus* [51]. The antibacterial property was related to reduction in particle size and crystallinity.

Hsu et al., modified the surface of two ZnO NPs samples (labelled A-ZnO and B-ZnO) with five different surface modifying molecules (1-dodecanethiol (DT), 3-methylthiophene (MT), trichlorododecylsilane (SA), (3-aminopropyl) trimethoxylsilane (SB) and 1,8-octanedithiol (OT)) and examined their effects against *E. coli* adopting colony count method [71]. Surface modifying molecules are present as a thin layer on the ZnO NPs and affect aggregation size, zeta potential and release of zinc ion from the NPs. The magnitude of effects on toxicity of the two ZnO samples depends on the surface modifying agent and the same modifying agent can also have opposite effects on the two ZnO samples. Significant change in the survival rate of *E. coli* is observed for samples modified with SA and DT. In case of SA and DT, toxicity change is observed in the opposite direction. For SA, significant decrease in toxicity of A-ZnO is observed whereas significant decrease in toxicity of B-ZnO in case of DT is evident. Pre-existing conditions of the ZnO NPs are responsible for the effects of surface modifier molecules. The opposite effect of the same surface modifying agent on A-ZnO and B-ZnO indicates that toxicity lowering for a certain modifying agent cannot be considered as universal for all ZnO NPs.

4. TiO₂ NPs

TiO_2 is widely present in different industrial products and consumer goods; it is a white pigment and very broadly used because of its brightness and high refractive index [72]. TiO_2 is extensively used in paints, papers, plastics, inks, medicine, pharmaceuticals, food products, toothpaste, etc. TiO_2 NPs are present in cosmetics and skin care products and widely used in heterogeneous photocatalysis [18]. TiO_2 NPs have been reported to be toxic against different aquatic organisms and other microorganisms. Antimicrobial activity of TiO_2 NPs is another aspect that is widely explored by the researchers [73-79]. Shah et al., showed that anatase phase TiO_2 NPs can behave as antimicrobial agent upon UV light activation and are highly capable of inhibiting growth of methicillin-resistant *S. aureus* [80]. Similarly, Erdural et al., synthesized TiO_2 NPs *via* hydrothermal method and demonstrated their efficient antibacterial activity against *E. coli* cells [81]. Desari et al., evaluated the toxicity effect of several MO NPs like ZnO, CuO, Co_3O_4 and TiO_2 on *E. coli* strain under light and dark conditions using spread plate counting method and LC_{50} value determination method [18]. In their results, TiO_2 NPs was found to be toxic against the selected bacterial strain.

Ahmad et al., explored the bactericidal effect of TiO_2 NPs against *E. coli* using colony count and disc diffusion assay [82]. For their study, stock TiO_2 NPs with a concentration of 1 mg/mL in distilled water was subjected to sonication for obtaining a homogeneous suspension. *E. coli* was cultured overnight and added to Luria broth containing TiO_2 NPs of varied concentration and was incubated for 16 h at 37°C. In the disk diffusion method, 0.2 mL of fresh bacterial culture was inoculated into 5 mL of sterile Luria broth and incubated for 3-5 h. 100 µL of the culture were added to an agar medium and divided into three plates separately. 50 µL of TiO_2 NPs sample were added to 8 mm diameter wells and incubated at 37°C for 18 h. It was observed that upon increasing the NPs concentration, the diameter of zone of inhibition increased with a maximum of 17 mm for 100 µg of TiO_2 NPs. Thus, TiO_2 NPs exhibited potent bacterial effect as examined in their investigation. In a similar type of study, Haghi et al., demonstrated the bactericidal effect of TiO_2 NPs on *E. coli* [83]. Desai et al., studied the antimicrobial effect of TiO_2 NPs against *E. coli*, *K. pneumonia* and *S. aureus* under visible light illumination [84]. TiO_2 NPs were produced *via* a sol-gel synthesis route. Bactericidal activity of the TiO_2 was examined by adding 0.1 M TiO_2 NPs to bacterial suspension and subjecting them to sunlight. After regular intervals of time, samples were withdrawn and the number of surviving cells was determined *via* viable count method. TiO_2 NPs displayed excellent antibacterial activity against all the three bacterial species. Roy et al., studied the effects of nano-TiO_2 with different antibiotics against Methicillin-Resistant *S. aureus* (MRSA) [85]. They adopted the disk diffusion/well diffusion method to study antibacterial activity against MRSA culture. The presence of 20 µg/disc of TiO_2 NPs increased the antibacterial activities of all antibiotics against test strain with minimum 2 mm to maximum 10 mm of well size. An inhibition zone of 10 mm was noted for MRSA against penicillin G and amikacin whereas TiO_2 exhibited a synergic effect on the antibacterial activity of nalidixic acid when tested against MRSA. Thus the results revealed the importance of TiO_2 NPs in the antimicrobial action of beta lactums, cephalosporins, aminoglycosides, glycopeptides, macrolids, lincosamides and tetracycline against MRSA. Moreover, in presence of TiO_2 NPs, MRSA displayed decreased antimicrobial resistance against various antibiotics. Dijaz et al., also reviewed the photochemical properties of TiO_2 NPs which enhance antibacterial properties of TiO_2 NPs [51]. TiO_2 NPs under UV light generate reactive oxygen species (ROS) and under UV, TiO_2 induces genetic damage in human cells and tissues. However, doping of TiO_2 NPs with metal ions helps to overcome this problem and improves the antibacterial and photocatalytic properties of TiO_2 NPs.

5. Iron Oxide NPs

Iron oxide is an important transition MO having wide technological importance in present day world. Till present, sixteen pure phases of iron oxide are known. Iron oxide finds applications in several fields such as catalysis, pigments, flocculants, sorbents, gas sensors, ion exchangers and in several other areas. Iron oxide NPs like magnetite (Fe_3O_4) and haematite (Fe_2O_3) are intensively applied in magnetic recording, magnetic data storage devices, toners and inks for xerography, magnetic resonance imaging, waste water treatment, etc. [20]. Because of their magnetic properties, an optimal drug delivery system can be achieved by use of an external magnetic field [86].

Fe_2O_3/Fe_3O_4 NPs were also reported to exhibit antimicrobial activities [87-94]. Schwegmann et al., examined the effect of Fe_3O_4 on *S. cerevisiae* and *E. coli* cells [6]. Fe_3O_4 NPs were synthesized and allowed to react with the microorganism cells. The interaction and toxicological assessment were characterized by zeta potential measurements and SEM imaging technique. From zeta potential measurements, it was seen that at pH 4 the dominating forces between the NPs and microorganisms were electrostatic forces whereas at pH 10 the dominating force was repulsive. No much adverse effect on the survival of microorganisms at selected pH was observed. At pH 4, high bactericidal effect on *E. coli* cells was seen, whereas at pH 10 only a very low bactericidal effect was highlighted. The toxicological effect on *E. coli* appeared to depend on the extent of sorption, which in turn depends on electrostatic forces. Sorption isotherms fitted the Langmuir model, indicating formation of a monolayer. When attractive force between MO NPs and surface of bacteria is high, most of the bacteria surface is covered with MO NPs. SEM imaging technique was used to study the surface morphology of the MO NPs and also their adhesion to bacterial cell surface. The bacterial cells at pH 4 were prepared with MO NPs concentration of 24 mg/L for SEM analysis and the resulting images are shown in Figure 7. The results reveal that a high density of Fe_3O_4 NPs is attached to *E. coli* and *S. cerevisiae*.

Zhang et al., investigated the size effect of α-Fe_2O_3 NPs adsorption on *E. coli* bacterial strain [2]. Studying the size effect of NPs is very important to understand the toxicity imparted by these nano-range materials towards living cells. The adsorption rates were found to be faster for small α-Fe_2O_3 NPs than those for large NPs, which are expressed as a number of adsorbed α-Fe_2O_3 NPs per unit cell surface area. Their study tried to explain the kinetic behaviour of α-Fe_2O_3 NPs interacting with microbial cells. Large NPs in the size range of 76–98 nm reached adsorption equilibrium faster (30–40 min) than smaller NPs (60–90 min). Adsorption rates (rate of concentration drop) in mg of Fe per unit litre per time (Fe/(L.s)) followed the order 98 nm > 76 nm > 53 nm > 26 nm. However, the rate of adsorption when calculated as the number of adsorbed α-Fe_2O_3 NPs per unit cell surface area per unit time ($\#/(m^2.s)$) was found to be faster for smaller NPs than for large NPs. Theoretical adsorption study was carried out using the Extended Derjaguin-Landau-Verwey-Overbeek (ELDVO) combined with interfacial force boundary layer theory (IFBL) and was in good agreement with the experimental data.

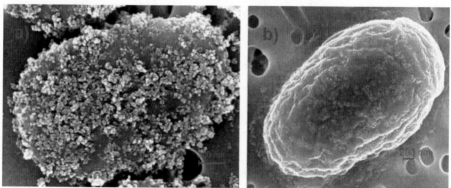

Figure 7. SEM images of Fe_3O_4 NPs attachment onto (a) *E. coli* and (b) *S. cerevisiae.*

Fast mobility and lower energy barrier in total interaction energy were responsible for faster kinetics of small NPs. Their study thus provided an understanding of NPs' adsorption. In a separate study, Zhang et al., explored the effect of α-Fe_2O_3 on *E. coli* cells [31]. α-Fe_2O_3 NPs were selected in their study because of their stability and distinct size distribution and were used as reference particles in colloid studies. The interaction between the MO NPs and bacteria was studied by AFM and SEM imaging. The group conclusively remarked that the size dependence of adhesion force between MO NPs and *E. coli* cells depends on effective contact area and these interactions are responsible for toxic effects of MO NPs.

Behera et al., reported the synthesis of Fe_3O_4 NPs by co-precipitation method and their use as active antibacterial agent against human pathogenic bacteria [17]. Gram positive as well as Gram negative bacteria were selected namely *S. aureus, S. flexneri, B. licheniformis, B. brevis, V. cholera, P. aeruginosa, S. aureus, S. epidermidis, B. subtilis* and *E. coli* for the study. Bactericidal assessment was carried out by zone inhibition method. The Fe_3O_4 NPs had maximum bactericidal effect against *B. licheniformis* with zone of inhibition of about 22 ± 0.70 mm and exhibited no effect on *S. flexneri* and *P. aeruginosa.*

A number of factors are responsible for antibacterial effect of Fe_3O_4 NPs. Frenk et al., explored the effect of Fe_3O_4 NPs on soil bacteria present in two different soil types: a sandy loam and a sandy clay loam [95]. This study established that Fe_3O_4 NPs lowered the soil microbial activities and also affected bacterial community in sandy loam type soil, which was low in organic matter and soil clay fractions. However, the activity was low in case of sandy clay loam type soil, which possesses higher organic matter content as well as higher clay fractions. This study revealed the toxic effect of MO NPs against soil bacterium and also it was inferred that the presence of clay fractions and organic matter reduced the toxic effect of Fe_3O_4 NPs. Liu et al., investigated the effect of adhesion of α-Fe_2O_3 and γ-Al_2O_3 NPs on *E. coli* bacterial cells [96]. Bacteria adhesion on NPs was achieved by simple chemical reaction and bactericidal effect was demonstrated by zeta potential measurements and FTIR spectroscopy. The adhesion of bacterial cells to α-Fe_2O_3 NPs was found to be less in comparison to γ-Al_2O_3 NPs owing to less positive charge on α-Fe_2O_3 NPs compared to γ-Al_2O_3 NPs. FTIR investigation also revealed that chemical reactions between functional groups on *E. coli* and α-Fe_2O_3 NPs are responsible for toxic effect on bacterial cells.

MECHANISM OF BACTERIAL ADHESION

Recent advances in nanotechnology suggest that MO NPs have a wide range of advanced applications in several fields as already discussed in the previous sections. With increasing use of these MO NPs, their release into the environment is inevitable and the toxicity of these particles has gained the attention of researchers and environmentalists all over the globe. However, there are a lot of controversies and uncertainties regarding the toxicity mechanism of these MO NPs. Different scientists have opined different mechanisms for MO NPs toxicity on bacterial cells depending on reaction conditions, type, size and shape of MO NPs, etc. The exact toxicity mechanism of MO NPs is not understood completely [97]. In human cells, the toxicity may be due to their intervention with varied biochemical processes and to ion leakage, leading to increased intracellular ion concentration which results in interference with free metal ion homeostasis [98].

Chang et al., attempted to review the toxicity effects and mechanism of ZnO and CuO NPs on bacterial cells in which mention was made that the damaging mechanism by MO NPs includes oxidative stress, coordination effects, non-homeostasis, geneotoxicity, etc. [7]. The toxicity of ZnO and CuO NPs depends on interaction between MO NPs and biomolecules. It mainly involves protein unfolding, fibrillation, thiol cross-linking and enzymatic activity loss.

MO NPs diffuse across the cell membrane due to their small size; ion channels and transporter proteins permit MO NPs to cross the plasma membrane as shown in Figure 8 (a). Certain MO NPs also enter *via* endocytosis. Cu^{2+} and Zn^{2+} ions from MO NPs enter cells by transport and ion/voltage-gated channels. MO NPs then directly interact with mitochondria and redox active proteins, which stimulate ROS production inside the cell [7, 99]. ROS are the natural by-products of the metabolism of respiring organisms [100]. ROS then affect DNA strand breaking and also gene expression. ROS effect of Cu^{2+} and Zn^{2+} is displayed in Figure 8 (b). Furthermore, Cu^{2+} and Zn^{2+} also have the capacity of forming chelates with biomolecules and can dislodge metal ions in specific metalloproteins, resulting in protein inactivation as illustrated in Figure 8 (c). MO NPs release Cu^{2+} and Zn^{2+} ions, which dislocate cellular metal cation homeostasis which in turn is responsible for toxicity as shown in Figure 8 (d). In a different study carried out by Li et al., on the antibacterial activities of different MO NPs, a mechanism based on photogenerated ROS was proposed [101].

According to their results, NP systems generate ROS, which persuade oxidative stress responsible for antibacterial activity [102]. This is based on the principle that when MO NPs are irradiated by light with photo-energy greater than their band gap, an electron from the valence band is transferred to the conduction band, which produces a hole (h^+) in the valence band. Conduction band electrons and valence band holes possess high reducing and oxidizing power, respectively.

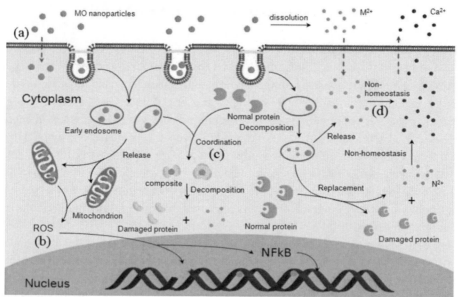

Figure 8. Different paths of cellular toxicity of ZnO and CuO NPs: (a) NPs' entry inside cell, (b) oxidative stress effect, (c) coordination effect, (d) non-homeostatis effect of Zn^{2+} and Cu^{2+} ions.

Through a reductive process, electrons generate superoxide anion ($O_2^{\cdot-}$) by reaction with molecular oxygen. Also the holes present in the valence band react with adsorbed water and/ or hydroxyl ions to generate hydroxyl radical (OH) *via* an oxidative process.

Aqueous reactions of $O_2^{\cdot-}$ mostly produce singlet oxygen ($^1O_2$), while. OH is a strong and non-selective oxidant which can damage biomolecules including carbohydrates, nucleic acids, lipids, proteins, DNA and amino acids. $^1O_2$ can irreversibly damage the treated tissues and causes biomembrane oxidation and degradation. $O_2^{\cdot-}$ is the major precursor of the three types of ROS ($O_2^{\cdot-}$, $\cdot$OH and $^1O_2$), which contribute towards major oxidative stress in biological systems [103]. The antibacterial study performed by Jiang et al., attributed the toxicity to dissolved metal ions from the oxides [4]. However, the actual mechanism was not clearly understood. Most of the studies pointed to photosensitivity and production of ROS under specific wavelength high-intensity light. Desari et al., investigated the cytotoxicity mechanism imparted by MO NPs (ZnO, CuO, Co_3O_4, TiO_2) against *E. coli* under light and dark conditions [18]. The toxicity mechanism was determined *via* measurements of oxidative stress, GSH, LPO and released metal ions. In the presence of light, ZnO and TiO_2 produce ROS as they are photoreactive. Production of ROS was higher in TiO_2 treated *E. coli* but ZnO was more toxic.

In *E. coli* treated with ZnO and TiO_2 the production of ROS is concentration dependent, which approves the toxicity relation of MO NPs and ROS production in *E. coli*. However, the toxicity of CuO, Co_3O_4 NPs does not account for ROS production. The production of ROS was in the order $TiO_2 > ZnO > CuO > Co_3O_4$. In dark conditions, the toxicity may be attributed to release of metal ions. GSH plays an important role in antioxidant defence and depletion of GSH leads to increased levels of ROS, nitrogen species, dysfunction of mitochondria and depletion of ATP. Exposure to MO NPs leads to decreased level of GSH.

The study detected highest decrease of GSH for ZnO NPs and hence stated to be the most toxic among the studied MO NPs. LPO is a pointer of oxidative stress, which is defined as oxidative degradation of membrane lipids, leading to cell damage. Increase in the level of LPO indicates increased toxicity. Exposure to MO NPs causes increase of LPO level, which confirms that the toxicity of MO NPs is based on oxidative stress. To sum up, the toxicity of MO NPs was more significant under light than in dark conditions.

Likewise, a number of studies have been performed by various research groups to highlight the mechanism of NPs toxicity towards bacterial cells and many have opined different mechanisms [104]. The exact mode of mechanism is not clearly understood. Further studies are required to assert the mechanism governing the antibacterial activity of MO NPs.

CONCLUSION

MO NPs display prospective applications in optoelectronics, nanodevices, nanoelectronics, nanosensors, information storage, catalysis, and biotechnological fields, as stated. Toxicological impact on bacteria has attracted much attention. NPs release from industrial effluents into the surrounding environment causes contamination of both soil and water and threaten human and environmental health. So understanding the NPs adhesion mechanism onto bacterial cell is of utmost importance for addressing environmental complications.

Over the years, a number of studies have been carried out on antibacterial activities of several MO NPs on various bacterial strains. However, none of these studies established exactly the toxicity mechanism of the MO NPs. So, extensive studies in this field are required to understand the toxicological mechanism of MO NPs. Moreover, amalgamation of nanotechnology and biology provides scope for the development of new materials with better use to scientific community as well as mankind.

ACKNOWLEDGMENTS

The authors are thankful to the Council of Scientific and Industrial Research for the financial support (CSIR-NEIST Project title: MLP 6000(WP2)) and Director, CSIR−NEIST, Jorhat for his interest to carry out the work. GD acknowledges DST, New Delhi, India for DST-INSPIRE Fellowship grant. RB acknowledges financial support from the Centre National de la Recherche Scientifique (CNRS), Lille1 University and Nord Pas-de-Calais region.

REFERENCES

[1]   Rana, S.; Kalaichelvan, P. T. *Adv. Biotechnol.* 2011, 11, 21-23.
[2]   Zhang, W.; Rittmann, B.; Chen, Y. *Environ. Sci. Technol.* 2011, 45, 2172-2178.
[3]   Baun, A.; Hartmann, N. B.; Grieger, K.; Kusk, K. O. *Ecotoxicol.* 2008, 17, 387–395.
[4]   Jiang, W.; Mashayekhi, H.; Xing, B. *Environ. Pollut.* 2009, 157, 1619-1625.

[5]   Chitra, K.; Annadurai, G. *Int. Food Res. J.* 2013, 20, 59-64.

[6]   Schwegmann, H.; Feitz, A. J.; Frimmel, F. H. *J. Colloid. Interface Sci.* 2010, 347, 43-48.

[7]   Chang, Y. N.; Zhang, M.; Xia, L.; Zhang, J.; Xing, G. *Materials* 2012, 5, 2850-2871.

[8]   Dehner, C. A.; Barton, L.; Maurice, P. A.; Dubois, J. L. *Environ. Sci. Technol.* 2011, 45, 977-983.

[9]   Amelia, M.; Lincheneau, C.; Silvi, S.; Credi, A. *Chem. Soc. Rev.* 2012, 41, 5728-5743.

[10]   Joh, D. Y.; Kinder, J.; Herman, L. H.; Ju, S. Y.; Segal, M. A.; Johnson, J. N.; ChanGarnet, K. L.; Park, J. *Nature Nanotechnol.* 2011, 6, 51-56.

[11]   Sun, Y.; Xia, Y. *Science* 2002, 298, 2176-2179.

[12]   Hodes, G. *Adv. Mater.* 2007, 19, 639-655.

[13]   Wang, D.; Kou, R.; Choi, D.; Yang, Z.; Nie, Z.; Li, Z.; Saraf, L. V.; Hu, D.; Zhang, J.; Graff, G. L.; Liu, J.; Pope, M. A.; Aksay, I. A. *ACS Nano* 2010, 4, 1587-1595.

[14]   Balusamy, B.; Kandhasamy, Y. G.; Senthamizan, A.; Chandrasekaran, G.; Subramanian, M. S.; Tirukalikundram, K. *J. Rare Earths* 2012, 30, 1298-1302.

[15]   Abboud, Y.; Saffaj, T.; Chagraoui, A.; Bouari, A. E.; Brouzi, K.; Tanane, O.; Ihssane, B. *Appl. Nanosci.* 2013, 4, 571-576.

[16]   Gunalan, S.; Sivaraj, R.; Rajendran, V. *Prog. Natural Sci.: Mater. Int.* 2012, 22, 693-700.

[17]   Behera, S. S.; Patra, J. K.; Pramanik, K.; Panda, N.; Thatoi, H. *World J. Nano Sci. Eng.* 2012, 2, 196-200.

[18]   Dasari, T. P.; Pathakoti, K.; Hwang, H. M. *J. Environ. Sci.* 2013, 25, 882-888.

[19]   Huang, Z.; Zheng, X.; Yan, D.; Yin, G.; Liao, X.; Kang, Y.; Yao, Y.; Huang, D.; Hao, B. *Langmuir*, 2008, 24, 4140-4144.

[20]   Mohapatra, M.; Anand, S. *Int. J. Eng. Sci. Technol.* 2010, 2, 127-146.

[21]   Chowdhuri, A.; Gupta, V.; Sreenivas, K.; Kumar, R.; Mozumdar, S.; Patanjali, P. K. *Appl. Phys. Lett.* 2004, 84, 1180–1182.

[22]   Jammi, S.; Sakthivel, S.; Rout, L.; Mukherjee, T.; Mandal, S.; Mitra, R.; Saha, P.; Punniyamurthy, T. *J. Org. Chem.* 2009, 74, 1971–1976.

[23]   Zhang, D. W.; Yi, T. H; Chen, C. H. *Nanotechnology* 2005, 16, 2338–2341.

[24]   Dar, M. A.; Kim, Y. S.; Kim, W. B.; Sohn, J. M.; Shin, H. S. *Appl. Surf. Sci.* 2008, 254, 7477–7481.

[25]   Yin, M.; Wu, C. K.; Lou, Y.; Burda, C.; Koberstein, J. T.; Zhu, Y.; O'Brien, S. *J. Am. Chem. Soc.* 2005, 127, 9506–9511.

[26]   Pakrashi, S.; Dalai, S.; Prathna, T. C.; Trivedi, S.; Myneni, R.; Raichur, A. M.; Chandrasekaran, N.; Mukherjee, A. *Aquatic Toxicol.* 2013, 132-133, 34-45.

[27]   Kon, K.; Rai, M. *J. Comp. Clin. Pathol. Res.* 2013, 2, 160-174.

[28]   Jiang, W.; Yang, K.; Vachet, R. W.; Xing, B. *Langmuir* 2010, 26, 18071-18077.

[29]   Jiang, W.; Saxena, A.; Song, B.; Ward, B. B.; Beveridge, T. J.; Myneni, S. C. B. *Langmuir* 2004, 20, 11433-11442.

[30]   Parikh, S. J.; Chorover, J. *Langmuir* 2006, 22, 8492-8500.

[31]   Zhang, W.; Stack, A. G.; Chen, Y. *Colloids Surf. B* 2011, 82, 316-324.

[32]   Gokulakrishnan, R.; Ravikumar, S.; Raj, J. A. *Asian Pac. J. Trop. Dis.* 2012, 2, 411-413.

[33]   Ravikumar, S.; Gokulakrishnan, R. *Int. J. Pharm. Sci. Drug Res.* 2012, 4, 157-159.

[34] Ansari, M. A.; Khan, H. M.; Khan, A. A.; Cameotra, S. S.; Saquib, Q.; Musarrat, J. *J. Appl. Microbiol.* 2014, 116, 772-783.

[35] Ansari, M. A.; Khan, H. M.; Khan, A. A.; Pal, R.; Cameotra, S. S. *J. Nanopart. Res.* 2013, 15, 1-12.

[36] Fajardo, C.; Sacca, M. L.; Costa, G.; Nande, M.; Martin, M. *Sci. Total Environ.* 2014, 473-474, 254-261.

[37] Karlsson, H. L.; Cronholm, P.; Hedberg, Y.; Tornberg, M.; Battice, L. D.; Svedhem, S.; Wallinder, I. O. *Toxicol.* 2013, 313, 59-69.

[38] Ahamed, M.; Alhadlaq, H. A.; Majeed Khan, M. A.; Karuppiah, P.; Al-Dhabi, N. A. *J. Nanomater.* 2014, 2014, 1-4.

[39] Azam, A.; Ahmed, A. S.; Oves, M.; Khan, M. S.; Habib, S. S.; Memic, A. *Int. J. Nanomed.* 2012, 7, 6003-6009.

[40] Theivasanthi, T.; Alagar, M. *Ann. Biol. Res.* 2011, 2, 368-373.

[41] Dimkpa, C. O.; Calder, A.; Britt, D. W.; McLean, J. E.; Anderson, A. *J. Environ. Pollution* 2011, 159, 1749–1756.

[42] Baek, Y. W.; An, Y. J. *Sci. Total Environ.* 2011, 409, 1603–1608.

[43] Dinesh, R.; Anandaraj, M.; Srinivasan, V.; Hamza, S. *Geoderma* 2012, 173-174, 19-27.

[44] Padil, V. V. T; Cernik, M. *Int. J. Nanomed.* 2013, 8, 889-898.

[45] Ren, G.; Hu, D.; Cheng, E. W. C.; Vargas-Reus, M. A.; Reip, P.; Allaker, R. P. *Int. J. Antimicrob. Agents* 2009, 33, 587-590.

[46] Jadhav, S.; Gaikwad, S.; Nimse, M.; Rajbhoj, A. *J. Cluster Sci.* 2011, 22, 121-129.

[47] Pandiyarajan, T.; Udayabhaskar, R.; Vignesh, S.; James, R. A.; Karthikeyan, B. *Mater. Sci. Eng. C* 2013, 33, 2020-2024.

[48] Shaffiey, S. F.; Shaffiey, S. R.; Ahmadi, M.; Azari, F. *Nanomed. J.* 2013, 1, 198-204.

[49] Sivaraj, R.; Rahman, K. S. M.; Rajiv, P.; Narendhran, S.; Venckatesh, R. *Spectrochim. Acta* 2014, 129, 255-258.

[50] Germi, K. G.; Shabani, F.; Khodayari, A.; Kalandaragh, Y. A. *Metal-Org. Nano-Metal Chem.* 2014, 44, 1286-1290.

[51] Dizaj, S. M; Lotfipour, F.; Barzegar-Jalali, M.; Zarrintan, M. H.; Adibkia, K. *Mater. Sci. Eng. C* 2014, 44, 278-284.

[52] Dutta, R. K.; Sharma, P. K.; Bhargava, R.; Kumar, N.; Pandey, A. C. *J. Phys. Chem. B* 2010, 114, 5594-5599.

[53] Li, S.; Song, W.; Gao, M. *Procedia Environ. Sci.* 2013, 18, 100-105.

[54] Raj, M. S.; Roselin, P. *Int. J. Pharm. Biosci.* 2012, 3, 267-276.

[55] Jacob, S. J. B.; Bharathkumar, R.; Ashwathram. G. *World J. Pharm. Res.* 2014, 3, 3044-3054.

[56] Xie, Y.; He, Y.; Irwin, P. L.; Jin, T.; Shi, X. *Appl. Environ. Microbiol.* 2011, 77, 2325-2331.

[57] Azizi, S.; Ahmad, M.; Mahdavi, M.; Abdolmohammadi, S. *Bioresources* 2013, 8, 1841-1851.

[58] Divya, M. J.; Sowmia, C.; Joona, K.; Dhanya, K. P. *Res. J. Pharm., Biol. Chem. Sci.* 2013, 4, 1137-1142.

[59] Firouzabadi, F. B.; Noori, M.; Edalatpanah, Y.; Mirhosseini, M. *Food Control* 2014, 42, 310-314.

[60] Tayel, A. A.; El-Tras, W. F.; Moussa, S.; El-Baz, A. F.; Mahrous, H.; Salem, M. F.; Brimer, L. *J. Food Safety* 2011, 31, 211-218.

[61] Kim, A. R.; Ahmed, F. R.; Jung, G. Y.; Cho, S. W.; Kim, D. I.; Um, S. H. *J. Biomed. Nanotechnol*. 2013, 9, 926-929.

[62] Premanathan, M.; Karthikeyan, K.; Jeyasubramanian, K.; Manivannan, G. *Nanomedicine* 2011, 7, 184-192.

[63] Joshi, P.; Chakraborti, S.; Chakrabarti, P.; Singh, S. P.; Ansari, Z. A.; Husain, M.; Shanker, V. *Sci. Adv. Mater*. 2012, 4, 173-178.

[64] Omoik, A.; Chorover, J. *Biomacromolecules* 2004, 5, 1219-1230.

[65] Leone, L.; Ferri, D.; Manfredi, C.; Persson, P.; Shchukarev, A.; Sjoberg, S.; Loring, J. *Environ. Sci. Technol*. 2007, 41, 6465-6471.

[66] Omoike, A.; Chorover, J.; Kwon, K. D.; Kubicki, J. D. *Langmuir* 2004, 20, 11108-11114.

[67] Jin, T.; Sun. D.; Su, J. Y.; Zhang, H.; Sue, H.-J. *J. Food Sci*. 2009, 74, M46-M52.

[68] Li, X.; Xing, Y.; Jiang, Y.; Ding, Y.; Li, W. *Int. J. Food Sci. Technol*. 2009, 44, 2161-2168.

[69] Sinha, R.; Karan, R.; Sinha, A.; Khare, S. K. *Bioresource Technol*. 2011, 102, 1516-1520.

[70] You, J.; Zhang, Y.; Hu, Z. *Colloids Surf. B* 2011, 85, 161-167.

[71] Hsu, A.; Liu, F.; Leung, F. C. C.; Chan, W. K.; Lee, H. K. *Nanoscale* 2014, 6, 10323-10331.

[72] Shi, H.; Magaye, R.; Castranova, V.; Zhao, J. *Part. Fibre Toxicol*. 2013, 10, 1-33.

[73] Kumar, S.; Bhanjana, G.; Kumar, R.; Dilbaghi, N. *Mater. Focus* 2013, 2, 475-481.

[74] Massard, C.; Bourdeaux, D.; Raspal, V.; Feschet-Chassot, E.; Sibaud, Y.; Caudron, E.; Devers, T.; Awitor, K. O. *Adv. Nanopart*. 2012, 1, 86-94.

[75] Haghi, M.; Hekmatafshar, M.; Janipour, M. B.; Gholizadeh, S. S.; Faraz, M. K.; Sayyadifar, F.; Ghaedi, M. *Int. J. Adv. Biotechnol*. 2012, 3, 621-624.

[76] Gao, Y.; Truong, Y. B.; Zhu, Y.; Kyratzis, I. L. *J. Appl. Polym. Sci*. 2014, 131, 40797-40810.

[77] Yadav, H. M.; Otari, S. V.; Koli, V. B.; Mali, S. S.; Hong, C. K.; Pawar, S. H.; Delekar, S. D. *J. Photochem. Photobiol. A* 2014, 280, 32-38.

[78] Jayaseelan, C.; Rahuman, A. A.; Roopan, S. M.; Kirthi, A. V.; Venkatesan, J.; Kim, S. K.; Iyappan, M.; Siva, C. *Spectrochim. Acta A* 2013, 107, 82-89.

[79] Malarkodi, C.; Chitra, K.; Rajeshkumar, S.; Gnanajobitha, G.; Paulkumar, K.; Vanaja, M.; Annadurai, G. *Pharm. Sinica* 2013, 4, 59-66.

[80] Shah, R. R.; S. Kaewgun, B. I.; Lee, T. R.; Tzeng, J. *J. Biomed. Nanotechnol*. 2008, 4, 339-348.

[81] Erdural, B. K.; Yurum, A.; Bakir, U.; Karakas, G. *Nato Sci. Peace Security* 2008, 2, 409-414.

[82] Ahmad, R.; Sardar, M. *Int. J. Innovative Res. Sci. Eng. Technol*. 2013, 2, 3569-3574.

[83] Haghi, M.; Hekmatafshar, M.; Janipour, M. B.; Gholizadeh, S. S.; Faraz, M. K.; Sayyadifar, F.; Ghaedi, M. *Int. J. Adv. Biotechnol. Res*. 2012, 3, 621-624.

[84] Desai, V. S.; Kowshik, M. *Res. J. Microbiol*. 2009, 4, 97-103.

[85] Roy, A. S.; Parveen, A.; Koppalkar, A. R.; Ambika, P. M. V. N. *J. Biomater. Nanobiotechnol*. 2010, 1, 37-41.

[86] Tran, N.; Mir, A.; Mallik, D.; Sinha, A.; Nayar, S.; Webster, T. J. *Int. J. Nanomedicine*, 2010, 5, 277-283.

[87] Zhang, W.; Shi, X.; Huang, J.; Zhang, Y.; Wu, Z.; Xian, Y. *ChemPhysChem* 2012, 13, 3388-3396.

[88] Taylor, E. N.; Kummer, K. M.; Durmus, N. G.; Leuba, K.; Tarquinio, K. M.; Webster, T. J. *Small* 2012, 8, 3016-3027.

[89] Arokiyaraj, S.; Saravanan, M.; Udaya Prakash, N. K.; Valan Arasu, M.; Vijayakumar, B.; Vincent, S. *Mater. Res. Bull.* 2013, 48, 3323-3327.

[90] Bardajee, G. R.; Hooshyar, Z.; Rastgo, F. *Colloid Polym. Sci.* 2013, 291, 2791-2803.

[91] Ebrahiminezhad, A.; Davaran, S.; Rasoul-Amini, S.; Barar, J.; Moghadam, M.; Ghasemi, Y. *Curr. Nanosci.* 2012, 8, 868-874.

[92] Mukherje, M. *Int. J. Mater. Mech. Manufact.* 2014, 2, 64-66.

[93] Deng, C. H.; Gong, J. L.; Zeng, G. M.; Niu, C. G.; Niu, Q. Y.; Zhang, W.; Liu, H. Y. *J. Hazardous Mater.* 2014, 276, 66-76.

[94] Behera, S. S.; Patra, J. K.; Pramanik, K.; Panda, N.; Thatoi, H. *World J. Nanosci. Eng.* 2012, 2, 196-200.

[95] Frenk, S.; Moshe, T. B.; Dror, I.; Berkowitz, B.; Minz, D. *Plos One* 2013, 8, 1-12.

[96] Liu, Z. D.; Li, J. Y.; Jiang, J.; Hong, Z. N.; Xu, R. K. *Colloids Surf. B* 2013, 110, 289-295.

[97] Hajipour, M. J.; Fromm, K. M.; Ashkarran, A. A.; Aberasturi, D. J.; Larramendi, I. R.; Rojo, T.; Serpooshan, V.; Parak, W. J.; Mahmoudi, M. *Trends Biotechnol.* 2012, 30, 499-511.

[98] Llop. J.; Lopis, I. E.; Ziolo, R. F.; Gonzalez, A.; Fleddermann, J.; Dorn, M.; Vallejo, V. G.; Vazquez, R. S.; Donath, E.; Mao, Z.; Gao, C.; Moya, S. E. *Part. Part. Syst. Charact.* 2014, 31, 24-35.

[99] Karlsson, H. L.; Gustafsson, J.; Cronholm, P.; Moller, L. *Toxicol. Lett.* 2009, 188, 112-118.

[100] Jones, C. M.; Hoek, E. M. V. *J. Nanopart. Res.* 2010, 12, 1531-1551.

[101] Li, Y.; Zhang, W.; Niu, J.; Chen, Y. *ACS Nano* 2012, 6, 5164-5173.

[102] Vani, C.; Sergin, G. K.; Annamalai, A. *Int. J. Pharma Bio Sci.* 2011, 2, 326-335.

[103] Pathakoti, K.; Huang, M. J.; Watts, J. D.; He, X.; Hwang, H. M. *J. Photochem. Photobiol. B* 2014, 130, 234-240.

[104] Kim, A. R.; Ahmed, F. R.; Jung, G. Y.; Cho, S. W.; Kim, D. I.; Um, S. H. *J. Biomed. Nanotechnol.* 2013, 9, 926-929.

In: Innovations in Nanomaterials
Editors: Al-N. Chowdhury, J. Shapter, A. B. Imran

ISBN: 978-1-63483-548-0
© 2015 Nova Science Publishers, Inc.

Chapter 2

ADVANCEMENTS IN NANOPARTICLE DETECTION AND CHARACTERIZATION IN PLANT TISSUES

Xingmao Ma[1,] and Min Chen[2]*

[1]Zachry Department of Civil Engineering, Texas A&M University,
College Station, Texas, US
[2]Dan F. Smith Department of Chemical Engineering, Lamar University,
Beaumont, Texas, US

ABSTRACT

Engineered nanoparticles (ENPs) are increasingly detected in the environment and in biological tissues. Once in the biological matrix, ENPs may undergo complex physical and chemical reactions, leading to the formation of nano-entities with highly different sizes, shapes and chemical properties than the primary nanoparticles. Determination of the physical and chemical characteristics of these nano-entities in biological tissues holds significant importance for the toxicity, fate and transport of ENPs in biological systems. Previous research on the nano-bio interface has mostly focused on the characterization of ENPs before exposure. Detailed study on the characterization of ENPs in biological tissues following their uptake and accumulation was rare. The paucity of knowledge on this important aspect of bio-nano interactions is, at least partially, due to the lack of viable technologies to obtain desirable information on nanoparticles after they enter into biological tissues. Fortunately, recent developments in respect of analytical technologies demonstrated significant promises for ENP detection and characterization in biological matrix. This chapter aims to provide a broad review on a suite of microscopic and spectroscopic technologies which have been employed to generate important information about ENPs in in plant tissues.

* Corresponding Author address. Email: xma@civil.tamu.edu. Tel: (1)979-862-1772; Fax: (1)979-584-1542;

INTRODUCTION

The market share of nanotechnology-enhanced commercial products reached \$174 billion in 2007 and is expected to grow to \$2.5 trillion by 2015 [1]. As a result, increased release of engineered nanoparticles (ENPs) into the environment is highly likely. ENPs primarily refer to man-made materials with at least two dimensions between 1 nm ~ 100 nm [2]. ENPs at this size range fall into a transitional zone between individual molecules and the corresponding bulk materials and therefore possess unique properties distinctively different from their bulk counterparts [3, 4]. For examples, ENPs generally possess much larger specific surface area (SSA) and high reactivity than the bulk materials with the same composition. ENPs also differ from their natural counterparts in that they are generally more homogenous (monodispersed), have a more uniform conformation and are frequently designed to have more specific surface properties. Consequently, their environmental impact and behavior may be drastically different from the natural nanoparticles and these differences have attracted significant attention and concerns.

US Environmental Protection Agency categorize ENPs into four main groups based on their composition: carbonaceous NPs, metallic NPs, dendrimers and composites. Carbonaceous NPs are the most abundant ENPs and primarily consist of fullerenes and nanotubes. Fullerenes are enclosed cage-like structures comprised of twelve 5-member carbon and unspecified 6-member rings in defect-free form. Structures with fewer hexagons have more reactive carbon sites and higher strain energies. The most representative fullerene is nC_{60}, an icosahedrally symmetrical structure supporting an equivalent electronic state and bonding geometry for each carbon atom [5]. Carbon nanotubes (CNT) can be defined as carbon whiskers, which are perfectly straight tubules of nanometer dimensions with properties close to that of an ideal graphite fiber. Generally, two distinct types of CNTs exist, depending on whether the tubes are made of more than one graphene sheet (multi walled carbon nanotube, MWCNT) or only one graphene sheet (single walled carbon nanotube, SWCNT).

Metallic ENPs include both elemental metallic ENPs (e.g., AuNPs, AgNPs) and metal oxide NPs (e.g., TiO_2 NPs, CeO_2 NPs, ZnO NPs.). Metallic ENPs have found applications in a range of industries such as sunscreen and information industry [6]. Also included in this category are Quantum dots (QDs) which are artificial "droplets" of charge that can contain anything from a single electron to several thousand electrons [7]. QDs are typically in the dimension of a few nanometers to micrometers. QDs display many similar quantum phenomena as real atoms and nuclei and therefore are often used as models of different atoms by simply changing its size and shape. Dendrimers are spheroid or globular nanostructures used in different industries deriving mainly from its large voids inside and versatile functional groups on the surface. The size, shape, topology, flexibility and molecular weight can be controlled precisely by generation (shell), chemical composition of the core, internal branching and surface functionalities.

Nanocomposites are combinations of NPs with other NPs or larger, bulk-type materials. These materials such as nanosized clays are added to auto parts and packaging materials to enhance their mechanical and thermal properties. According to US EPA, about half of the newly produced ENPs are multi-element, with increasing complexity and functionality. For example, the ferromagnetic ENPs in the form of $M_xFe_3\text{-}xO_4$ (M = Co, Zn, Mn or Ni) have been the focus of many research works recently. The unique features displayed by these ENPs are important for a number of industrial applications such as nanomaterial-based catalysts, biomedicine, magnetic resonance imaging, data storage, environmental remediation, nanofluids and optical filters.

Once ENPs are released into the environment, they interact with all components of the ecosystem. For example, ENPs can be adsorbed on the surface of soil particles, plant root surface or taken up by plant roots and transported up to plant shoots, depending on the physicochemical properties of ENPs and plant species. Several important processes concerning the fate and effect of ENPs in agroecosystems are illustrated in Figure 1. Plants are at the core of the agroecosystem and display intimate interactions with ENPs. The presence of ENPs can be enhancive, inconsequential or phytotoxic to plant growth and several great reviews have been published recently on this important topic [8-10]. Plants can also take up ENPs from the environment and accumulate them in different tissues. Plant uptake represents an important pathway for human exposure to ENPs and therefore the investigation of plant uptake and accumulation of ENPs, their distribution among different plant tissues and subcellular localization has been an intensive research area in the plant nanotoxicology field. The advancement holds great importance for human health and sustainable development of nanotechnology.

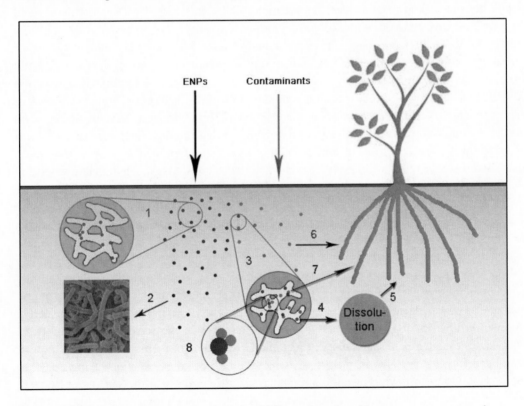

Figure 1. Important processes that may occur after ENPs are introduced into an agroecosystem. 1. sorption/aggregation, 2. microbial toxicity, 3. interactions with co-existing contaminants and adsorption onto soil particles, 4. nanoparticle dissolution, 5. plant uptake of dissolved ions, 6. plant uptake of environmental contaminants, 7 plant uptake of nanoparticles or nanoparticle-contaminant complexes, 8. higher resolution of nanoparticles-contaminant complex.

Zhu and colleagues [11] provided the first evidence that ENPs can be taken up by plants. They showed that iron oxide NPs (Fe_3O_4) was taken up by pumpkin plant roots (*Cucurbita maxima*) and transported to plant leaves. Lin and colleagues [12] investigated the uptake and translocation of carbonaceous nanoparticles by rice plants (Oryza sativa) and found that fullerene C_{70} could be easily taken up by plant roots and transported to plant shoots. Their study also demonstrated that C_{70} could be potentially transported downward from leaves to roots through phloem if C_{70} enters into plants through aerial uptake. Similar observation was not seen for multi-walled carbon nanotubes (MWCNTs). With a two-photon excitation microscope, another group of researchers demonstrated that MWCNTs primarily adsorbed on root surface as individuals and aggregated CNTs even though one or both ends can pierce through root cap cell walls [13]. However, recent research using radioactive labels and other more sensitive techniques indicate that CNTs can be taken up by plants and transported from roots to shoots, even though the transport potential from roots to shoots is extremely small [14, 15].

In contrast to MWCNTs, most metallic ENPs can be taken up by plants and accumulated in plant tissues. Au, Ag, CeO_2, TiO_2, and even some nanostructures such as NaYF4:Yb,Er nanocrystals could be taken up by different plant species and sequestered in plant tissues [16-20]. Once ENPs are in plant tissues, ENPs may undergo different physical, chemical and biological processes in plant tissues. For example, ENPs may aggregate in plant tissues and our previous investigation with 20 nm citrate-coated silver NPs showed that the NPs aggregated to much large sizes and were retained mostly in the intercellular spaces of Arabidopsis roots [21]. After CeO_2 was exposed to cucumber for three weeks, soft x-ray scanning transmission microscopy (STXM) and near edge x-ray absorption fine structure (XANES) analysis showed that measurable amount of CeO_2 was biotransformed to $CePO_4$ in plant roots and to cerium carboxylates in plant shoots [22]. In a separate study with CeO_2 and four agricultural crops, another research group showed that the oxidation state of CeO_2 was unaltered in the root tissues of these four species as determined by x-ray absorption spectroscopy (XAS) [23]. Similarly, Servin and colleagues investigated that uptake and tanslocation of TiO_2 NPs by cucumber and noticed the transport of Ti from roots to leaves. In particular, they observed high concentrations of Ti in leaf trichomes, but found no biotransformation of Ti in plant tissues [24]. Notably, they reported that anatase TiO_2 NPs were primarily associated with root tissues and rutile TiO_2 were more concentrated in the leaf tissues and tricome structure, suggesting that rutile TiO_2 might have a higher potential to transport from roots to shoots than anatase TiO_2. In a subsequent study, the authors verified the accumulation of TiO_2 in cucumber fruits for the first time for both crystal phases, but no transformation was detected [25]. The discrepancies on ENP biotransformation may derive from the differences of ENPs, plant species or differences in the sensitivities of analytical technologies used by the researchers. For metallic ENPs, some of them may dissolute in plant tissues and for others, ENPs in plant tissues may come from the reduction of metallic ions inside plant tissues following metallic ion uptake [26]. Regardless of the sources of ENPs in plant tissues, plant uptake and accumulation of ENPs by agricultural crops, especially the edible tissues, is a great concern and the determination of ENP characteristics, such as their concentration, aggregation, chemical stability and so on in plant tissues is of critical importance to understand the health and safety impacts of ENPs. Current technologies are inadequate to provide all the information concerning the state of ENPs in plant tissues. For example, most microscopic imaging techniques only allow qualitative investigation on tiny fraction of plant tissues, which are often not representative of the whole plant tissues. In addition, some techniques have high detection limits, requiring the exposure of plants to unrealistically high concentrations. While tissue extraction techniques such as Inductively Coupled Plasma Mass Spectroscopy (ICP-MS) or Inductively Coupled Plasma Optical Emission Spectrometry (ICP-OES) allow quantitative determination of metallic ENPs, detailed information on the state and localization of ENPs in plant tissues are lost after the acid digestion. Nonetheless, these technologies are important for the investigation of ENPs in plant tissues and this chapter will review several most commonly used technologies in studies of ENP plant interactions. For each technology, a brief introduction on the principles of the technology will be followed by a discussion on their advantages and limitations in their applications for ENP plant interaction studies.

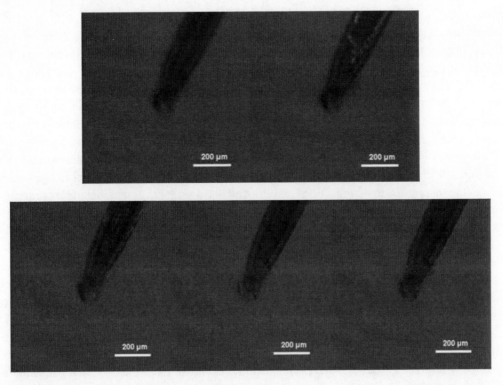

Figure 2. Confocal microscopic images collected at different optical sections of radish fine roots treated with CeO$_2$ NPs. The distance between two optical planes is 23.5 μm. yellow arrows indicate the vascular tissues in radish roots. Xy: xylem.

NANOPARTICLE DETECTION AND CHARACTERIZATION TECHNIQUES IN PLANT TISSUES

Confocol Microscope

Confocal microscope is a light microscope which can be operated at both fluorescence mode and transmitted or reflected mode. In the latter mode, confocal microscope can generate better quality images with more details than conventional light microscope images for two important reasons: point by point illumination of samples and installation of pinhole systems to reject out of focus light [27, 28]. As a consequence of the point by point illumination, longer imaging time is necessary. However, this operation allows the confocal microscope to focus on only a thin focal plane and virtually section and image the specimen layer by layer. After the completion of the imaging at different planes called "optical sections", the images can be stacked up and reassembled to obtain 3D images. Abundant information is available in the literature concerning the principles of confocal microscope and will not be repeated here. For the imaging of biological samples, confocal microscope is mostly operated in fluorescence mode. In this mode, an artificial fluorophore is typically introduced into cells but the fluorophore is not needed in the imaging of ENPs in plant tissues because ENPs themselves can function as fluorophores. What is important is to find the right wavelength for different ENP materials to fluorescence. We have successfully detected silver (Ag) ENPs and cerium oxide (CeO_2) ENPs in different plant root tissues [8, 21]. Figure 2 lists a set of confocal microscopic images acquired at different optical sections of radish roots following their exposure to CeO_2 ENPs for three weeks. As indicated, CeO_2 ENPs adsorption on the surface is insignificant, but CeO_2 accumulation on the root tip was observed. The image in Figure 2 C shows the vascular tissues and that signals were detected in the position of the vascular tissues, suggesting that CeO_2 might have entered the vascular tissues and transported along the xylem. The observation is consistent with some other studies employed with different analytical techniques. The advantage of confocal microscope for the detection of ENPs in plants is that confocal microscope can process relatively large samples so that the distribution of ENPs inside plant tissues can be determined. However, confocal images only indicate the approximate localizations as illustrated in Figure 2 and cannot reveal the subcellular localization of ENPs due to its resolution. Detailed subcellular localization is typically obtained with higher resolution microscopes such as electron microscopy which will be discussed in the following section. However, recent developments on super-resolution optical microscopes improved the resolution by over an order of magnitude [29] and can result in broader applications of the fluorescence microscopes in detecting ENPs in biological tissues in the future. These super-resolution optical microscopes apply varying principles to circumvent the light diffraction limit, but the approaches generally fall into two categories. The first approach is to confine the fluorescence to a much smaller region by the use of spatial sequence readout and the second approach involves the localization of a single emitter and fit it with the stochaistic functional techniques [30]. Examples of super-resolution microscopes include spectral precision distance microscope (SPMD), stochastic optical reconstruction microscope (STORM) and spatially modulated illumination microscopy (SMI) and so on. These super-resolution microscopies normally involve some software processing of the images and have not been used in ENP detection in plants even though potentials are enormous. At present, conventional confocal microscopes are still being used and another limitation of the current confocal microscopy for detecting ENPs in plants is that plant leaves naturally fluoresce which interferes with the detection of ENP fluorescence, therefore, confocal microscope is not popular for the detection of ENPs in plant shoot tissues for now. New development in the computing line may help separate the signals in the future and enhance the applications of confocal microscope in the plant nanotoxicology field. Finally, plant root tissues contain many porous structures which are filled with air. Air can occasionally generate some background noises and the imaging of ENP plant interactions therefore must be accompanied with control root comparisons.

Transmission Electron Microscopy (TEM)

Transmission electron microscopy (TEM) is common technique to monitor ENPs in plant tissues. TEM operates similarly as conventional optical light microscope to a great extent. The TEM has three major components: the illumination source, the objective lens and the imaging system. The goal of the illumination source is to provide monochromatic (same wave length) electron beam. The beam can be operated in two modes: parallel beam and convergent beam. Parallel beam can be used to form TEM images while the convergent beam can be used for scanning images in scanning transmission electron microscope (STEM). The objective lens and the sample holder is the place where electron beams interact with samples to generate images or diffraction patterns, which will be focused and amplified in the imaging system. The replacement of light with electron significantly improves its resolution power because the resolution is directly related to the wavelength as defined by the Rayleigh criterion and the operation of TEM at high energy levels will yield better resolution images, but this will also increase electron aberrations. In TEM imaging, the mass contrast constitutes the primary mechanism for image formation. When electrons hit the specimen, the elastic scattering of electrons is related to the number of atoms, with thick samples scattering more electrons. Electron scattering is also associated with atomic numbers of elements, with high atomic number elements scattering more electrons than lower atomic number elements. This is how ENPs are detected in biological tissues with TEM. Due to the increased resolution of electron microscope, the subcellular localization of ENPs can be determined with TEM. In our previous studies, we have successfully applied this technology to determine silver nanoparticles in Arabidopsis roots and first reported that silver nanoparticles may be internalized through plasmadesmata [21]. However, the use of TEM for ENP detection in plant tissues is mostly qualitative for biological tissue due to the small sample sizes the TEM can handle. The result is therefore not representative of the whole plant tissues. Most previous research focused on root tissues (more accurately, root tip region) when TEM was employed in their research because root is the most likely tissue to trap ENPs when root exposure was adopted [31]. Another drawback of using TEM in ENP detection in plant tissues is the lengthy and tedious sample pretreatment process. The tissues must be fixed in glutaraldehyde (mostly commonly applied solvent) and then sliced into thin pieces by microtone. Such requirements limit the more popular application of TEM. Also, TEM only provides 2D image, and the interpretation of the image must be performed carefully. Due to the collision of electrons with air particles, TEM must be operated in high vacuum. Contamination by air may leave dark spots on the specimen and these spots may be confused as ENPs. In this case, the shape and size of ENPs will help distinguish ENPs with contamination spots, but only the compositional analysis will provide the most accurate confirmation on the nature of the spots in biological samples. As with other microscopic technologies, TEM has made big strides and high resolution TEM (HRTEM) becomes increasingly popular in studying the phenomena at nanoscale [32]. The clustering of HRTEM with other spectroscopic techniques such as x-ray diffraction or high-spatial resolution electron energy loss spectroscopy (EELS) could potentially provide both the physical characteristics and chemical valence state for ENPs in plant tissues. The promises are yet to be realized.

Scanning Electron Microscope (SEM)

SEM is another commonly used imaging method to investigate ENP in plant tissues. However, SEM is generally used as a surface imaging technology and in the context of ENP and plants, SEM is primarily used to detect the adsorption or attachment of ENPs on plant root or shoot surfaces. SEM can be operated in different modes by operating the facility at different energy levels. When the images are generated by the electrons dislodged from the specimen surface (secondary electrons), the electrons will form points with different brightness on the electron detector according to their distances to the detector. Therefore, secondary electron mode can generate topographical information on sample specimens. When images are produced from the electrons reflected back by the specimen (backscattered electrons), the resolution is reduced and topographical information is lost. However, backscattered electrons can provide compositional contrast because the brightness of signals of backscattered electrons depends on the atomic number and the metallic ENPs all have higher atomic numbers than carbon and therefore can be detected by backscattered electrons, Figure 3. For ENP detection in plant tissues, backscattered electrons are more indicative than secondary electrons. The higher energy used in SEM operation can cause ionization of atoms and generate X-ray which could give compositional information of the atoms. SEM-EDS (energy dispersive X-ray spectroscopy) is a popular technique to determine the composition of interested spots when ENPs are suspected in an SEM image. Conventional SEM operates in high vacuum and the specimen needs to be electrically conductive, therefore, they are typically coated with a conductive layer (e.g., carbon layer). For plant ENP interactions, the high energy used in SEM operation results in the damage of plant tissues after SEM and repeated monitoring is unlikely.

A new type of SEM which allows the equipment to operate at much lower vacuum and no pretreatment are developed. This relatively new type of SEM is called environmental scanning electron microscope (ESEM). Non conductive specimens such as plant tissues can be directly viewed. The ESEM mode even enables the investigation of wet samples (e.g., fresh tissues in a sample holder with liquid) directly [33]. The tissues can be placed in the commercial QuantomiX (WETSEM®) and directly imaged under SEM. The key component of the capsule is a thin membranous partition which separates a fully hydrated sample from the high vacuum of SEM chamber. The thickness and the materials of the membrane are designed to make it practically transparent for 10-30 keV electrons of the beam, and yet tough enough to withstand the pressure differences. Recently, WETSEM has been enabled to conduct microanalysis with EDS, offering a rare opportunity to conduct compositional analysis on the native samples [34]. WETSEM has shown great potential for ENP detections in plant tissues. However, it should also be mentioned that this technology was developed to image cells in their native state. Plant tissues are significantly larger than a single cell and the loading of plant tissues in the QuantomiX capsule can be tricky in order to avoid the attachment of plant tissues on the membrane.

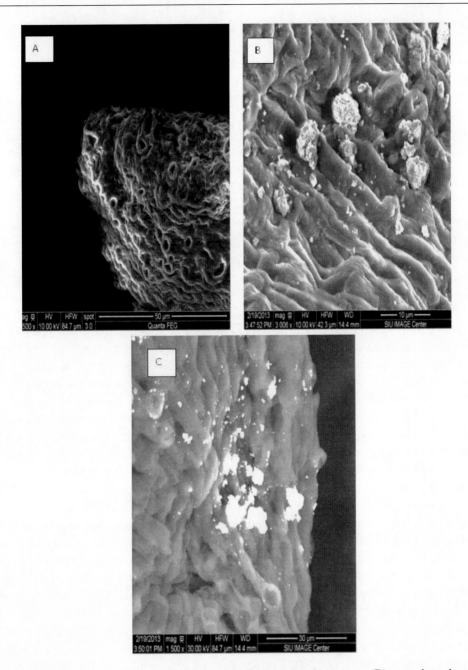

Figure 3. Scanning electron microscopic images of (A) control soybean roots, (B) secondary electron images of soybean roots exposed to silver ENPs and (C) backscattered electron images of soybean roots exposed to silver ENPs.

Inductively Coupled Plasma-Mass Spectrometry (ICP-MS)

ICP-MS is a highly sensitive and selective method for elemental analysis and has played significant roles in many research and industry fields. Liquid samples introduced into the instrument go through a series of processes including atomization and ionization through the hot plasma torch. Before the elements reach the mass spectrometer, most of them are converted into singly charged ions. Elements with low ionization efficiency such as P, C. N will be difficult to be analyzed by ICP-MS, but the majority of elements in the periodic table can be analyzed by ICP-MS. The application of ICP-MS in ENP detection in biological tissues including plant tissues is very common [35]. However, to determine ENPs in plant tissues with conventional ICP-MS, plant tissues are typically digested with strong acids which convert ENPs into ions and therefore all the features of nanoparticles in plant tissues are lost. As a result, the conventional ICP-MS only provides quantitative concentration of the element. However, to accurately gauge the toxicity of ENPs, the size and size distribution, the biotransformation of ENPs in plant tissues are also essential.

Most recently, single particle ICP-MS (sp-ICP-MS) method has been developed and is being used for both qualitative and quantitative analyses of ENPs [35]. sp-ICP-MS is a cutting edge technique capable of detecting the concentrations and masses of metal elements in each particle, thus determining the size of metallic ENPs at environmentally relevant concentrations [36, 37]. The principles of sp-ICP-MS method for ENPs detection in aqueous samples have been well presented by Degueldre et al.[38] Briefly, ENPs are first suspended homogenously in a solution and the suspension is then introduced into plasma of ICP-MS by a nebulizer and spray chamber. The elements are then ionized in the plasma to generate a plume of ions rather than a continuous stream of ions. The generated plume of ions are transported to mass spectrometer and detected as a pulse (transient signal). The frequency of the pulse is proportional to the concentration in the sample and the intensity of the pulse is proportional to the size. If dissolved metal elements are also present in the suspension sample, they are detected as "background". In this approach, dissolved metal concentration, particle concentration, and particle size distribution can be obtained simultaneously. The dwelling time of the facility is probably the most important parameter in sp-ICP-MS operation to ensure that the signal comes from one single NP at each sampling measurement. The technique is mostly used in liquid samples, but it is gradually adopted for the analysis of ENPs in more complex biological and environmental samples. A recent study applied sp-ICP-MS to determine the silver ENPs in chicken meat and found that these silver ENPs became slightly smaller after two days of equilibrium in chicken meat, possibly due to dissolution [39]. These authors also found that some silver ENPs were biotransformed into silver sulfide in chicken meat. The successful application of sp-ICP-MS in ENP detection and characterization in animal tissues is encouraging for similar applications of ENP detection in plant tissues. However, plant tissues differ from animal tissues because each plant cell is surrounded by a cell wall and plant cells are "glued" together by lignin. The unique structure of plants makes the extraction of ENPs more challenging than animal tissues. Therefore, the primary challenge to use sp-ICP-MS in ENP detection in plant tissues probably lies in the development of satisfactory strategy to extract ENPs from plant tissues without compromising the properties of ENPs. In spite of its unprecedented potentials, it should be mentioned that sp-ICP-MS, like any other technologies, has a detection limit concerning the particle size and particle concentration in plant samples. For the particle size, it is associated with the ionization efficiency of the element and therefore is element dependent. A recent study reported the lower bound of 40 different types of ENPs which can be reliably analyzed by sp-ICP-MS [40].

CONCLUSION

ENP detection in the environment is expected to increase with rapid development of nanotechnology and the uptake and accumulation of ENPs by plants, by agricultural crops in particular, can be an enormous food safety and security concern. Detection and characterization of ENPs in plant is therefore, of critical importance. Even though many available analytical technologies can be harnessed for this purpose, each facility has its own limitations. Fortunately, many new advancements in the analytical technologies offer exciting opportunities to rapidly and more accurately detect and characterize ENPs in plants. With this great need, the author is optimistic that some of the technologies will be mature in a few years and become more broadly used, some of them (e.g., sp-ICP-MS) may even become a routine technique for ENP detection and characterization in plant tissues and other biological matrices.

ACKNOWLEDGEMENT

Dr. Ma acknowledge the financial support of the USDA-AFRI (#2011-67006-30181) and (#2012-67005-19585).

REFERENCES

[1]   Rejeski, D. Managing the molecular economy. *Environ. Forum*. 2010, 27, 36-41.
[2]   American Society for Testing and Materials. Standard Terminology Relating to Nanotechnology. E2456-06. 2006, West Conshohocken, PA.
[3]   Taylor, R and Walton, D. R.M. The chemistry of fullerenes. *Nature*, 1993, 685-693.
[4]   Christian, P., Kammer, Von der. F., Baalousha, M., Hoffmann, Th. Nanoparticles, structure, properties, preparation and behavior in environmental media. *Ecotoxicol*, 2008, 17, 326-343.
[5]   Johnson, R. D., Meijer, G., Bethune, D. S. C-60 has icosahedral symmetry. *J. Am. Chem. Soc.* 1990, 112, 8983-8984.
[6]   Klaine, S J., Alvarez, PJ. J., Batley, G. E., Fernandes, T. F., Handy, R. D., Lyon, D. Y., Mahendra, S. Y., McLaughlin, M. J. and Lead, J. R. Nanomaterials in the environment: behavior, fate, bioavailability and effects. *Environ. Toxicol. Chem.* 2008, 27, 1825-1851
[7]   Kouwenhoven, L and Marcus, C. Quantum dots. *Physics World*. 1998, June, 35-39.
[8]   Ma, X., Geisler-Lee, J., Deng, Y., Kolmakov, A. Interactions between Engineered Nanoparticles and Plants: Phytotoxicity, Uptake and Accumulation. *Sci. Total Environ.* 2010, 408, 3053-3061.
[9]   Dietz, K., Herth, S. Plant nanotoxicology. Trends in Plant Science. 2011, 16, 582-589.
[10] Toxicity, uptake and translocation of engineered nanomaterials in vascular plants. *Environ. Sci. Technol.*, 2012, 46, 9224-9239.
[11] Zhu H, Han J, Xiao J., Jin Y. Uptake, translocation and accumulation of manufactured iron oxide nanoparticles by pumpkin plants. *J. of Environ. Monit*, 2008, 10: 713-717.

[12] Lin S, Reppert J, Hu Q, Hunson JS, Reid ML, Ratnikova T, Rap AM, Luo H and Ke PC Uptake, translocation and transmission of carbon nanomaterials in rice plants. *Small*, 2009, 1128-1132.

[13] Wild, E and Jones, K.C. Novel method for the direct visualization of in vivo nanometerials and chemical interactions in plants. *Environ. Sci. Technol.* 2009, 43, 5290-5294.

[14] Khodakovskaya, M. V., Silva, K.,Nedosekin, D. A., Dervishi, E., Biris, A. S., Shashkoe, E. V., Galanzha, E. I., Zharov, V. P. Complex genetic, photothermal and photoacoustic analysis of nanoparticle-plant interactions. *PNAS*, 2011, 108, 1028-1033.

[15] Larue, C., Pinault, M., Czarny, B., Georgin, D., Jaillard, D., Bebdiab, N., Mayne-LHermite, M., Taran, F., Dive, V., Carriere, M. Quantitative evaluation of multi-walled carbon nanotube uptake in wheat and rapeseed. *J. Hazard Mater.* 2012, 227-228, 155-163.

[16] Hischemoller, A.; Nordmann, J.; Ptacek, P.; Mummenhoff, K.; Hasse, M. In-vivo imaging of the uptake of unconversion nanoparticles by plant roots. *J. of Biomed. Nanotechnol.* 2009, 5, 278-284.

[17] Lopez-Moreno, M., Rosa, G. D. L., Hernandez-Viezcas, J. A., Peralta-Videa, J.R.,, Gardea-Torresdey, J. L. X-ray absorption spectroscopy (XAS) corroboration of the uptake and storage of CeO$_2$ nanoparticles and assessment of their differential toxicity in four edible plant species. *J. Agr. Food and Chem.* 2010, 58, 3689-3693.

[18] Lee, W., An, Y., Yoon, H and Kweon, H., Toxicity and bioavailability of copper nanoparticles to the terrestrial plants mung bean (Phaseolus radiatus) and wheat (Triticum awstivum): plant uptake for water insoluble nanoparticles. *Environ. Toxicol. Chem.*, 2008, 27 (9), 1915-1921.

[19] Ma, Y.; Kuang, L.; He, X.; Bai, W.; Ding, Y.; Zhang, Z.; Zhao, Y.; Chai, Z. Effects of rare earth oxide nanoparticles on root elongation of plants. *Chemosphere* 2010, 78, (3), 273-279.

[20] Larue, C., Laurette, J., Herlin-Biome, N., Khodja, H., Fayard, B., Flank, A. M., Brisset, F., Carriere, M. Accumulation, translocation and impact of TiO2 nanoparticles in wheat: influence of diameter and crystal phase. *Sci. Total Environ.* 2012, 431, 197-208.

[21] Geisler-Lee, J, Wang, Q, Yao, Y, Zhang, W, Geisler, M, Li, K, Huang, Y, Chen, Y, Kolmakov, A, Ma, X. Phytotoxicity, accumulation and transport of silver nanoparticles by Arabidopsis thaliana. *Nanotoxicology* 2013, 7, 323-337

[22] Zhang, P.; Ma, Y.; Zhang, Z.; He, X.; Zhang, J.; Guo, Z.; Tai, R.; Zhao, Y.; Chai, Z. Biotransformation of ceria nanoparticles in cucumber plants. *ACS nano.* 2012, 6, 9943-9950.

[23] Lopez-Moreno, M. L.; de la Rosa, G.; Hernandez-Viezcas, J. A.; Peralta-Videa, J. R.; Gardea-Torresdey, J. L. X-ray absorption spectroscopy (XAS) corroboration of the uptake and storage of CeO$_2$ nanoparticles and assessment of their differential toxicity in four edible plant species. *J. Agric. Food Chem.* 2010, 58, 3689-3693.

[24] Servin, A. D., Castillo-Michel, H., Hernandez-Viezcas, J.A., Corral, D. B., Peralta-Videa, J. R., Bardea-Torresdey, J. L. Synchrotron micro-XRF and mcro-XANES confirmation of the uptake and translocation of TiO2 nanoparticles in cucumber (Cucumis sativus) plants. *Environ. Sci. Technol.* 2012, 46, 7637-7643.

[25] Servin, A. D., Morales, M. I., Castillo-Michel, H., Hernandez-Viezcas, J.A., Munoz, B., Zhao, L., Nunez, J. E., Peralta-Videa, J. R., Bardea-Torresdey, J. L. Synchrotron

verification of TiO2 accumulation in cucumber fruit: a possible pathway of TiO2 nanoparticle transfer from soil into the food chain. *Environ. Sci. Technol.* 2013, 47, 11592-11598.

[26] Harris, A. T., Bali, R. On the formation and extent of uptake of silver nanoparticles by live plants. *J. Nanoparticle Res.* 2008, 10 (4), 691-695.

[27] Sheppard, C. J. R., Shotton, D. M. Introduction. In Confocal Laser Scanning Microscopy; Springer-Verlag New York Inc.: New York, 1997, 1-13.

[28] Prasad, V., Semwogerere, D., Weeks, E. R. Confocal microscopy of colloids, *J. Phys.: Condens. Matter*, 2007, 19, 113102-113127.

[29] Galbraith, C. G. and Galbraith, J. A. Super-resolution microscopy at a glance. *J. Cell Sci.*, 2011, 124, 1607-1611.

[30] Vogelsang, J., Steinhauer, C., Forthmann, C., Stein, I. H., Person-Skegro, B., Cordes, T., Tinnefeld, P. Make them blink: probes for super-resolution microscopy. *ChemPhysChem*, 2010, 11, 2475-2490.

[31] Taylor, A., Rylott, E. L., Anderson, C. W. N., Bruce, N. C. Investigating the toxicity, uptake, nanoparticle formation and genetic response of plants to gold. *PLOS one*, DOI: 10.1371/journal.pone.0093793.

[32] Wang. Z. L. New developments in transmission electron microscopy for nanotechnology. *Adv. Matar.*, 2003, 15, 1497-1514.

[33] Donald, A. M. The use of environmental scanning electron microscopy for imaging wet and insulating materials. *Nat. Mater.* 2003, 2, 511-516.

[34] Tiede, K., Tear, S. P., David, H., Boxall, A. B. Imaging of engineered nanoparticles and their aggregates under fully liquid conditions in environmental matrices. *Water Res.*, 2009, 43, 3335-3343.

[35] Pokhrel, L. R., Dubey, B. Evaluation of developmental responses of two crop plants exposed to silver and zinc oxide nanoparticles. *Sci. Tot. Environ.*, 2013, 452-453, 321-332.

[36] Mitrano, D. M., Ranville, J. F., Bednar, A., Kazor, K., Hering, A. S. Higgins, C. P. Tracking dissolution of silver nanoparticles at environmentally relevant concentrations in laboratory, natural and processed waters using single particle ICP-MS (spICP-MS). *Enviorn. Sci.: Nano.*, 2014, 1, 248-259.

[37] Laborda, F., Bolea, E., Jimenez-Lamana, J. Single particle inductively coupled plasma mass spectrometry: a powerful tool for nanoanalysis. *Analytical Chem.* 2014, 86, 2270-2278.

[38] Degueldre, C., Favarger, P. Y., Wold, S. Gold colloid analysis by inductively coupled plasma-mass spectrometry in a single particle mode. *Analytical Chimica Acta.*, 2006, 555, 263-268.

[39] Peters, R. J. B., Rivera, Z. H., Bemmel, G. V., Marvin, H. J. P., Weigel, S., Bouwmesster, H. Development and validation of single particle ICP-MS for sizing and quantitative determination of nano-silver in chicken meat. *Anal Bioanal Chem.*, 2014, 406, 3875-3885.

[40] Lee, S., Bi,X., Reed, R. B., Ranville, J. F., Herckes, P., Westerhoff, P. Nanoparticle size detection limits by single particle ICP-MS for 40 elements. *Environ. Sci. Technol.*, 2014, 48, 10291-10300.

In: Innovations in Nanomaterials   ISBN: 978-1-63483-548-0
Editors: Al-N. Chowdhury, J. Shapter, A. B. Imran   © 2015 Nova Science Publishers, Inc.

Chapter 3

CONDUCTING POLYMERS: THEIR ROUTE TO NANOBIONICS APPLICATIONS VIA ATOMIC FORCE MICROSCOPY

Michael J. Higgins[*], *Gordon G. Wallace,* *Paul J. Molino, Amy Gelmi and Hongrui Zhang*

ARC Centre of Excellence for Electromaterials Science,
Intelligent Polymer Research Institute, AIIM Facility, Innovation Campus,
University of Wollongong, Wollongong, NSW, Australia

ABSTRACT

In bionic applications, where electronics meets biology, conducting polymers have played an important role in the development of electrode-tissue interfaces. Fundamental to these advances are the enabling tools which help us to understand, and manipulate, the cell-material interface at the nanoscale. Therefore, in this review, we firstly provide an introduction to bionics and highlight the attributes of conducting polymers in the field. A focus is then given on the role of Atomic Force Microscopy in elucidating nanoscale and molecular level properties, which is critical to gaining precise control over the conducting polymer-cell interface. We provide an extensive review on several AFM approaches used to study a range of properties and interactions, including polymer structure, mechanical properties, actuation, electrical properties, through to interaction forces, protein adhesion and the probing of living cell surfaces at the single molecule level.

[*] Corresponding Author Address: Assoc. Prof. Michael Higgins, ARC Australian Research Fellow, ARC Centre of Excellence for Electromaterials Science, Intelligent Polymer Research Institute/AIIM Faculty, Innovation Campus, Squires Way, University of Wollongong NSW 2522. T +61 2 4221 3989; Email: mhiggins@uow.edu.au.

INTRODUCTION

The combination of biological and electronic systems is described as 'bionic' and applicable for control of electrically excitable tissues such as nerve or muscle tissue within the body [1]. Bionic devices such as the cochlear implant and bionic eye require biocompatible electrode interfaces that are vital for communication between the device and living tissue. The electrodes must be capable of supplying high-density electrical charge [2], should not provoke an inflammatory response, and have low impedance [3]. Low impedance is vital to ensure electrical stimulation is applied at thresholds that avoid cell damage [4]. Safe stimulation levels for neural prostheses vary between 0.5 to 4000 μC cm^{-2}, depending on the placement and application of the electrode [5]. Low impedance is also important for efficient charge transfer at the electrode-tissue interface and decreases the energy required for stimulation (ideal for bionic devices that require a battery). A high charge storage capacity is desirable as the electrode is able to store a relatively large charge without undergoing irreversible, and possibly cytotoxic, Faradaic reactions [6]. Bionic device electrodes currently use conventional metallic materials such as gold, platinum, platinum alloys, and iridium oxide to deliver stimulation [3] [7]. These metals have excellent conductivity, are stable and functional for long-term implants, and do not chemically react with the surrounding tissue [3]. For example, platinum is used in cochlear implant electrodes as it is chemically inert, non-toxic, and has low impedance and long-term stability during electrical stimulation [8]. However, the physical properties of metallic surfaces, particularly the Young's Modulus, can have negative effects on surrounding tissues. For example, hard metals can provoke an inflammatory response during insertion of the electrode or after surgery due to chronic movement of the electrode [9] [10].

ORGANIC CONDUCTING POLYMERS

Organic Conducting Polymers (OCP) as electrodes and electrode coatings are an alternative to metallic electrodes in bionic devices [11]. OCPs have organic aromatic backbones and conduct electricity due to delocalized electrons in the conjugated p-orbitals. OCPs act as semiconductors and exhibit both electronic and ionic conductivity and have been investigated as conductive materials in many bionic applications [1]. For example, poly(3,4-ethylenedioxythiophene) (PEDOT) has a high charge injection limit (15 mC cm^{-2}) and wide potential limit window compared to metallic materials (see Table 1) and has been explored as coatings for neural microelectrodes [5].

Table 1. Charge-injection limits of electrode materials for stimulation in the central nervous system. Adapted from [4]

Material	Mechanism	Maximum Q_{inj} (mC cm^{-2})	Potential Limits vs Ag|AgCl (V)
Pt and PtIr alloys	Faradaic/capacitive	0.05 – 0.15	-0.6-0.8
Iridium oxide	Faradaic	1 – 5	-0.6-0.8
Titanium nitride	Capacitive	~1	-0.9 to 0.9
PEDOT	Faradaic	15	-0.9 to 0.6

It is the electrical properties of OCPs that are extremely interesting for applications in bionic devices. The conductivity of these materials is within the semi- conductor range (0.1-1000 S/cm), which is acceptable for electrode applications [2]. Due to 3-D microtopography and porosity, the surface area of OCPs is much greater than conventional metal electrodes and thus leads to a higher charge density and lower impedance [12]. The charge injection mechanism for OCP electrode materials is more advantageous for biological applications compared to metals; redox reactions occurring within OCP results in electronic current being converted to ionic current [13]. This electronic-to-ionic conversion of current is seemingly more compatible with living cells that utilize ionic currents.

The physical properties of OCPs are more advantageous than their metallic counterparts; they are pliable, flexible and lightweight compared to metals, in addition to being inexpensive [14]. The softer surface of these polymers provides inherent compatibility with biological systems, thus affording them superior biocompatibility compared to conventional metallic electrode materials. A supplementary advantage of OCPs is the incorporation of dopants into the polymer structure. A dopant is a molecule that is incorporated into the polymer during synthesis to enable conductivity. The nature of the dopant (such as size, charge and chemical structure) will modify the properties of the polymers, specifically physical properties [15-17], surface chemistry [18, 19], and electrical conductivity [20-24]. The biocompatibility of OCPs can be further enhanced through the incorporation of biological dopants without compromising the conductivity of the polymers [25, 26].

The properties of OCPs, physical, chemical, and electrical, have a direct influence on the proliferation, growth and differentiation of living cells. Cells respond to surface properties through several mechanisms and hence the surface properties of OCPs need to be carefully considered. The presence of a dopant changes their nanotopography and surface chemistry, a feature that can be used to enhance cell growth or control cell differentiation. Finally, their electrical properties play a very important role; the conductive properties dictate their ability to deliver charge to cells [20, 24], control the release of dopants (e.g., drug molecules) [23], or mechanically stimulate through electro-actuation processes [27].

Therefore, OCP electrodes have the unique ability to control cell interactions through various mechanisms, including:

- Nanoscale physical properties;
- Incorporation of biomolecular dopants;
- Electrical stimulation;
- Controlled drug release; and
- Mechanical (actuation) stimulation.

BIOLOGICAL DOPANTS

The use of biological dopants (biodopants) is at the forefront of synthesizing biocompatible OCPs. Biological dopants have included proteins [28], peptides [29, 30] and extracellular matrix (ECM) components [16, 31-34]. The inclusion of biomolecules enables specific interactions with the cell membrane to influence cell function. For example, laminin peptide fragments incorporated into the conducting polymer, Polypyrrole (PPy), promotes specific cell adhesion to the polymer and enhances the density of adhered cells compared to non-bioactive dopants [30].

Glycosaminoglycans (GAGs), biological molecules produced by the ECM, are increasingly utilized as OCP biodopants. GAGs make good candidates for doping as they are negatively charged and soluble in aqueous solutions. They are a major component of the ECM and cellular structure within the central nervous system and other tissues, and are also found on cell surfaces. They play important roles in cellular functions, such as growth factor signalling, cell division, wound healing, homeostasis and tissue morphogenesis [35], and interact with specific moieties on both the cell membrane surface and ECM proteins [36]. Cell adhesion to surrounding tissue is mediated by the presence of cell surface integrin receptors, which bind to ECM components such as fibronectin, vitronectin, laminin and collagen [37]. In turn, these ECM components bind to GAGs [38], forming a critical continuum for mediating cell adhesion.

ELECTRICAL STIMULATION OF CELLS

Electrical stimulation is shown to enhance the growth of excitable cells such as skeletal muscle and neural cells [16, 20-22, 24, 33, 39]. A difference in the oxidation state of the polymer can control the growth of endothelial cells. For example, PPy/ dodecylbenzene sulfonate (DBSA) shows a significant reduction in cell density on polymer substrates in their reduced state compared to the oxidized (as-grown) state [40]. Similarly, mammalian cells display behaviour dependent on the oxidation state of PPy whereby reducing the polymer leads to prevention of cell spreading and causes cells to 'round-up' [33]. Electrical stimulation of PEDOT/polystyrene sulfonate (PSS) films prior to cell seeding shows that oxidation of the polymer affects cellular interactions [41]. Reduction of PEDOT/PSS films promotes cellular adhesion and proliferation (Figure 1A, left), however, oxidation of the polymer results in cell detachment and death of epithelial MDCK cells (Figure 1A, right). The proposed mechanisms involving the interaction of the ECM protein, fibronectin (FN), is shown in Figure 1B. However, no change in cell adhesion occurs when the polymer is stimulated 24h after cells have been on the surface, indicating that once cell adhesion is established switching the redox state does not affect the cell viability.

A potential gradient along the polymer and its relationship to FN protein adsorption has been used to investigate the influence of oxidation state on cell behaviour [42]. 3T3-L1 fibroblast-adipose cells were deposited on a PEDOT/tosylate (TOS) polymer with a potential bias of -1V to +1V applied across it before the seeding of cells. The fluorescent imaging showed there was a distinct gradient of cells distributed across the oppositely biased polymer, where the cells prefer the oxidized side of the film.

It is implicit from the above studies that the effect of electrical stimulation, or switching of the polymer's redox surface properties, is an important aspect in controlling cell adhesion and growth. Furthermore, in order to understand these cell-material interactions, characterization tools must be at least as sensitive as the cells themselves; the tools must probe at the nanoscale and molecular level in relevant physiological environments. Similarly, the polymers and their properties must be assembled in such a way that control over the distribution of material composition is possible with nanometre resolution. This is where Scanning Probe Microscopy techniques such as Biological-Atomic Force Microscopy (Bio-AFM) are coming of age in nanoscale characterization and multidisciplinary research and henceforth the focus of this chapter.

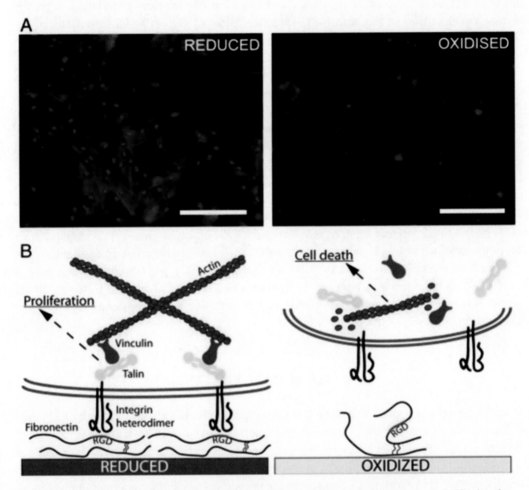

Figure 1. (A) On reduced films, fibronectin (FN) in an extended conformation presents RGD sites for cell binding and growth. On oxidized films, FN in a more compact conformation conceals RGD sites and prohibits cell attachment and growth. (B) Schematic drawing of the proposed mechanism for cell interaction via FN on reduced and oxidized PEDOT. Adapted from [41].

ATOMIC FORCE MICROSCOPY (AFM)

The inception of AFM represents a major landmark event along the timeline of nanotechnology [43]. More recently over the last two decades, the use of AFM in the biological sciences has increased rapidly due to its ability to operate under aqueous physiological conditions [44], measure across different length-scales ranging from single living cells through to individual proteins [45, 46], and easily integrate with optical/fluorescence microscopy and electrical techniques to enable simultaneous acquisition of a wide range of physical, chemical, and electrical information [47]. The advent of a broader range of available commercial AFM systems and different modes of AFM provides new opportunities for advanced research and experimental flexibility. For example, AFM systems and designs (e.g., petri dish holders and heaters) that integrate well with optical techniques and facilitate live cells studies, including electrochemical cell configurations that enable observations of cellular dynamics as a function of electrical stimulation, represent exciting developments in this area [48]. These attributes of Bio-AFM fundamentally makes for an ideal characterization technique when one merges nanotechnology and electromaterials with biology in the field of Nanobionics.

AFM is a versatile tool and particularly applicable to studying a wide-range of highly sought after nanoscale properties of OCPs, whether in the form of thin films coatings, fibers, or nanostructures. AFM can be performed immediately on as-deposited polymer samples, without additional sample preparation, in ambient conditions or liquids, providing a quick and practical approach. A range of different AFM modes can be applied, notably for conducting polymers these include standard Contact or Tapping Mode imaging, Kelvin Probe Force-AFM (surface potential imaging), Conductive-AFM, and Force Measurements. In most cases, the topographical image is acquired simultaneously during acquisition of electrical signals to enable correlation of structural-electrical properties. AFM reveals lateral nanoscale properties across macro-sized polymer samples but also enables the ability to image individual polymer nanostructures such as single nanotubes or nanowires. Furthermore, single nanostructures can be located by imaging, and then the AFM tip can be precisely positioned over a region of interest (e.g., single nanowire) and probed using a range of AFM modes. Here, we firstly present some of the various AFM modes and commonly used approaches to probe the nanoscale properties of OCPs.

AFM IMAGING

AFM imaging is applied extensively in current research and remains a popular technique of choice to determine the structure and morphology of OCP substrates, films or nanostructures. Imaging is often performed *ex-situ*, typically after polymer growth or different treatments such as heating or chemical modification. In-situ experiments provide an attractive option for revealing dynamic changes in morphology during polymer deposition or in response to external stimuli. The flexibility of AFM to operate in gaseous and liquid environments, such as those mimicking conditions required for vapour-phase or electrochemical polymerization, allows easy implementation of in-situ experiments. Most commercial AFM systems are equipped with the necessary hardware, such as environmentally-controlled liquid or electrochemical cells, to undertake biological and electrochemical experiments. In general, conducting polymers are sufficiently stable, or mechanically robust, to enable straightforward imaging. However, softer, lower modulus polymers in liquids may require the use of intermittent contact mode to reduce lateral forces that cause damage to the surface during imaging.

NANOSCALE MORPHOLOGY AND GROWTH

In studies on chemical polymerization of polyaniline (PA) coatings, AFM imaging shows that the film thickness increases in the first 5-7 min of growth and saturates at 30 nm [49]. Early time-point films are smooth, consisting of tightly packed particles, and only attain a maximum roughness of 3 nm, suggesting that film growth originates from nucleation of particles where the lateral growth rate is greater compared to the height growth. Ex-situ AFM imaging of PPy doped with chloride (Cl), sulfate (SO_4^-), perchlorate (ClO_4^-) and dodecylsulfate (DDS) reveals that the size of individual particles increases linearly with the logarithmic of film thickness [50, 51]. However, there is an abrupt increase in particle size once the thickness has reached ≈ 1 μm for Cl and ClO_4^- doped films. Ex-situ AFM imaging of PSS doped PPy films prepared via electrochemical polymerization with growth times ranging from 10 – 600 sec shows a linear increase in thickness with polymerization charge [52]. Thickness – charge conversion coefficients of 3.9 μm·C/cm^2 are obtained. The roughness, ranging from 10-30 nm, shows a semi-logarithmic increase with polymerization charge [52]. The effects of many electrochemical parameters such as dopant type and growth conditions (e.g., monomer/dopant concentration, electrolyte, electrode substrate) are understandably of interest in related studies using AFM. Higher temperatures cause an increase in surface roughness [53], likely due to increases in migration rate of the monomer, while lower pH produces thicker, rougher films [54]. The use of a diverse range of dopants in PPy, including biological molecules like GAGs and even another conducting polymer, poly(2-methoxyaniline-5-sulfonic acid) (PMAS), does not dramatically alter the characteristic 'cauliflower' polymer morphology of OCPs but leads to considerable variations in surface roughness [55] (Figure 2A-E). It should be noted that AFM imaging does not effectively convey the degree of porosity, as the tip can only penetrate a limited depth. For example, PPy doped with dextran sulfate (DS) shows a highly porous structure in SEM images yet AFM images give what appears to be a continuous surface [56]. Thus, AFM and SEM imaging techniques should be utilized in a complementary fashion when assessing film morphology.

Ex-situ and especially in-situ AFM imaging provide unique insight into the fundamental mechanisms of polymerization growth. The imaging can be done simultaneously during the growth process to capture the very early events of polymer deposition on the electrode. The polymer growth, involving nucleation of the particles, follows either instantaneous and/or progressive growth which can proceed in 2D or 3D [57]. Instantaneous nucleation involves growth with a constant number of particles (without new particles forming), while progressive nucleation involves the continual growth of particles, as well as the formation of new ones. In 2D growth, the size of the particles expands more rapidly in the lateral direction across the electrode, while 3D growth occurs equally in all directions. AFM images of PPy growth on gold/highly orientated pyrolytic graphite electrodes indicate growth as being instantaneous 2D prior to individual particles interconnecting upon which the growth converts to progressive 3D [57]. Once individual particles have reached a certain size, they cease to grow and are afterward accompanied by newly formed particles. This nucleation process shows instantaneous followed by progressive growth, though variations in growth occur depending on the type of dopant [58]. AFM studies have referred to the presence of 'active' nuclei in the early stages of deposition [59], with their initial growth observed as progressive

3D. Observations of more complex growth structures, such as thin networks or islands, are less easily described by instantaneous/progressive growth mechanisms [60].

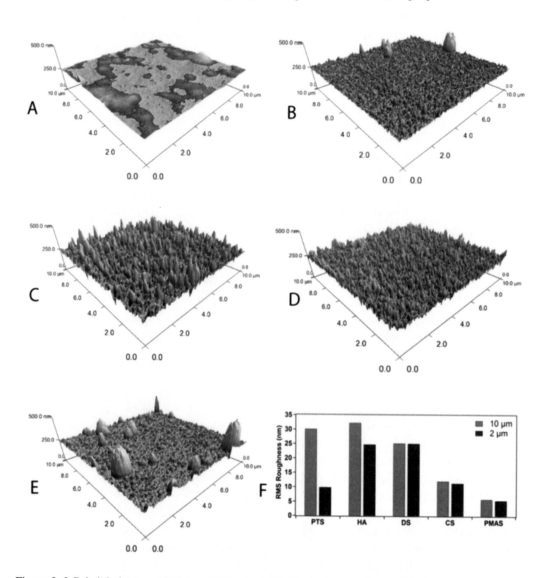

Figure 2. 3-D height images of 10 μm AFM scans. (A) PPy/poly(2-methoxyaniline-5-sulfonic acid), (B) PPy/chondroitin sulfate, (C) PPy/hyaluronic acid, (D) PPy/dextran sulfate, (E) PPy/para-toluene sulfonate, and (F) Histogram showing the RMS roughness values for each film over scan areas of 2 μm and 10 μm. Reproduced from [55].

Following insightful studies on growth mechanisms, understanding the properties after growth and particularly during changes in redox cycling is ongoing research. In-situ AFM imaging combined with applying cyclic voltammetry or stepped potential measurements enables researchers to correlate the polymer structure, along with surfaces roughness, to changes in oxidation state, polymer-dopant interactions and intercalation of ions as they exchange at the polymer-electrolyte interface. Dynamic changes in morphology of PPy/ClO$_4^-$ in response to redox cycling show that films prior to oxidation/reduction, and uptake of solvent/ions and swelling, have an aligned fibril structure that becomes random and granular as the films undergoes cycling [61]. The resolution of the AFM is sufficient to distinguish the morphologies of short-packed fibrils of $\approx$ 20 nm diameter, associated with short oligopolypyrrole units, while the granular structure is believed to originate from longer polymer chains.

NANOMECHANICAL PROPERTIES

The measurement of mechanical properties using AFM typically falls under two main interests; firstly, one involves recording the amount of indentation for a given applied force and then fitting of the data to mechanical contact models (e.g., Hertz theory) to calculate the Young's Modulus. Phase imaging also resolves lateral surface varations in stiffness, though this approach is generally non-quantitative. A second main interest is the study of mechanical responses, or actuation, of films as they undergo reversible swelling and contraction due to the uptake/release of ions during oxidation and reduction. In this case, dynamic in-situ measurements are achieved by placing the tip in contact with the film and the polymer actuation recorded as changes in the z-height over time. AFM imaging can also be performed, typically across a single line, during the polymer actuation to provide additional information on changes in surface morphology.

THIN FILMS COATINGS

One type of AFM-based nanoindentation technique uses a set of reference samples of known modulus (independently measured using depth sensing indentation (DSI) measurements) to reliably quantify hardness and modulus of bulk OCP films in air [62]. Modulus values of 1.3 GPa for PEDOT/PSS films are comparable to those values obtained in micro-tensile tests [63]. Similar analysis of AFM force versus indentation curves shows that polyaniline has a higher modulus in the range of 2.4-4.8 GPa [62]. The modulus of OCPs typically drops by 50% in liquid [64]. Specifically, the variation in stiffness of differently doped PPy films immersed in phosphate buffer saline (PBS) is illustrated by comparing force versus indentation curves [55] (Figure 3A). Greater indentation indicates more compliant films, with the curves fitted to extract Young's modulus. In these measurements undertaken in PBS, PPy/ poly(2-methoxyaniline-5-sulfonic acid) (PMAS) has a significantly lower modulus of 30 ± 2.0 MPa compared to PPy/chondroitin sulfate (CS) with a value of 293 ± 31 MPa (Figure 3B) [55]. Similar values for PPy/dextran sulfate (DS) (706 ± 44 MPa) and PPy/hyaluronic acid (HA) (660 ± 49 MPa) are significantly higher than PPy/CS and PPy/PMAS. PPy/para-toluene sulfonate (pTS) has the highest value of 1000 ± 87 MPa compared to all other films [55] (Figure 3B). The modulus of PPy/pTS (1000 MPa) is in the range of that expected for polymers with no plasticizer, indicating that no or very little amount of water is adsorbed. PPy/PMAS is known to have hydrogel-like properties and its high water content of > 90% expected to contribute to the low modulus. A significant increase in modulus for the other four doped film (pTS, HA, DS, CS) is suggested to relate to their reduced water content. This is somewhat unexpected for the high molecular weight HA, which has a high-water binding capacity, suggesting that an effect of dopant-polymer interactions on overall film properties (e.g., porosity) plays a major role in water uptake. Using high molecular weight HA as the dopant can produce brittle PPy films [31], however AFM studies using force curve mapping shows softer, more compliant films when lower molecular weight HA is incorporated as the dopant [65]. In the latter work, these films are perhaps the softest OCP ever recorded, with modulus values in the range of 2MPa. More recently, the use of a relatively new AFM technique based on quantitative nanomechanical imaging shows that the modulus of PEDOT/PSS films decrease from 200-250 MPa in air to 23 MPa in deionized water [66]. Hydrated PEDOT/PSS films have significantly lower modulus than those doped with pTS (139 MPa) and ClO_4^- (68 MPa), suggesting that excess sulfonate groups of the PSS significantly increase the ability to imbibe water and hydrate the films. However, the possibility that the AFM indents a compressible, brush-like layer of only excess PSS at the polymer surface may contribute to the lower modulus values.

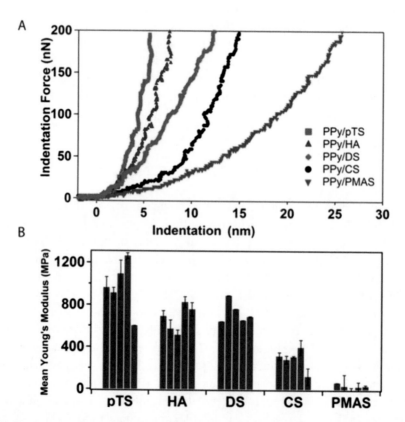

Figure 3. (A) Force-indentation curves for differently doped Ppy films. (B) Histogram of mean Young's modulus, each column represents the average value on one spot on the surface. Error bars are standard error. Reproduced from [55].

Based on the above studies, the modulus of hydrated OCPs ranges widely between 1GPa down to a few MPa depending on the dopant. This range is perceived as an attractive intermediate of mechanical properties for an OCP coating bridging the interface between hard metals (100-200 GPa) and much softer gels, cells and tissues ($< \approx 10$ kPa), enabling a more stable and less invasive electrode-tissue interface. AFM measurements have made a significant contribution in this area, particularly with the ability to probe the mechanical properties of thin films completely immersed in biologically relevant fluids. The unique ability to quantify highly localized and lateral heterogeneities in surface modulus, which cells may effectively sense on the nanoscale, will continue to play an important role in the nanoscale characterization of OCPs, especially as their interfacial surface chemistries and composite forms (e.g., gel composites) become evermore complex for bionic applications.

NANOWIRES AND NANOTUBES

The ability of AFM to pinpoint highly localized force measurements on individual nanostructures of <100 nm in diameter enables research on quantifying the mechanical properties of individual nanowires and nanotubes that are either directly on a substrate, or freely suspended over channels. AFM force curves are typically performed at several points along the nanowire or nanotube and their mechanical properties, modulus and hardness, determined using either mechanical contact models or 3-point bending tests. These approaches have been applied to various OCP nanostructures, including nanowires, nanotubes, nanofibers, and nanoribbons that have potential for interfacing as nanostructured electrodes in bionics. PEDOT/PSS nanowires suspended over silica microchannels have a modulus of 3-10 GPa [67]. The values for the single nanowires are significantly higher than those of bulk PEDOT/PSS (2-3 GPa) and PPy/PSS (1-2GPa) films and attributed to several reasons, including lower number of stress defects, increased pi-conjugation and polymer chain length/cross-linking. AFM measurements of single PPy nanotubes (NT) and helical polyacetylene (HPA) nanofibers lying directly on silica or freely suspended over microchannels show modulus values of 1GPa and 0.5 GPa, respectively [68]. Force versus indentation curves of the single PPy/NT and HPA nanofibers on the substrate are linear and suitably analyzed using contact mechanical models. Suspended HPA nanofibers also give a linear mechanical response, however, PPy/NT initially show a non-linear curve below z-piezo displacements of approximately of 60 nm, followed by a linear response at higher forces. The linear component can be fitted to a clamped beam model to extract the modulus, though reasons for an initial non-linear mechanical response of PPy/NT, which is not observed for HBA nanofibers, are not clear. AFM studies have also found a dependence of elastic tensile modulus on the size of individual PPy/NT [69, 70]. Force measurements on single NT, ranging from 35-160 nm outer diameter, suspended over the pores of polyethylene terephthalate (PET) membranes show the modulus significantly increases with a decrease in thickness or outer diameter [69]. Dramatic increases in modulus are particularly evident below an outer diameter of 50-60 nm. As modulus is related to the material structure perfection, this dependence is suggested to be due to an increase in the degree of structural ordering, or reduced number of defects, as the nanotubes decrease in size. At present, it appears that AFM-based studies on the mechanical properties of individual nanowires/nanotubes have only been done in air, not liquid environments. Such liquid measurements are a logical next step for understanding their nanomechanical interactions with biological systems.

NANOSCALE ACTUATION

As mentioned above, another major interest in nanomechanical properties involves the study of mechanical actuation of OCP thin films as they undergo electrochemically-driven swelling and contraction. The approach involves in-situ Electrochemical-AFM to enable measurement of film height whilst electrochemically switching between oxidized and reduced states. This ability to directly visualize dynamic changes in height on the order of a few nanometres up to microns in real-time has been appealing to researchers. Effects of film thickness and dopant on the actuation mechanisms and performance, including parameters such as % strain and strain-rate are often investigated. In early efforts, PPy films doped with either ClO_4^- or pTS were subject to sequential oxidation and reduction by applying potential square wave pulses between 0.2 and -0.8 V [71]. For ClO_4^- initial reduction induces a 50% increase in height but soon after the film contracts and further cycles do not cause regular changes in height (or volume). pTS doped films show smaller increases in height (10%), though the reversibility of actuation is more consistent. Due to increases in height during initial reduction, it is suggested that charge compensation in the early stages of redox reactions is mostly driven by cation transport rather than anion expulsion from the polymer. Micro-patterned films, $\approx$ 30μm wide and 1.5 high, of PPy doped with DBSA provide a clever approach to enable AFM imaging of the entire PPy structure during actuation [72, 73]. The actuation shows a highly anisotropic, reversible 30-40% change in height, or swelling, between the oxidized and reduced states in the direction perpendicular to the substrate but much less (2%) in the parallel direction. These previous studies have led to similar work on the effect of film thickness on actuation during cyclic voltammetry at different scan rates [52]. The extent of actuation in thicker, micrometer-scale films is highly dependent on scan rate. Significantly, the actuation height for thicker films shows a strong nonlinear decay with increasing scan rate, which becomes more linear and less scan rate dependent as the film thickness decreases [52]. As a consequence, there is an 18-fold difference in the actuation height between the thickest (3.17 μm) and thinnest films (97 nm) at 10 mV/s, which decreases significantly to an 8-fold difference at a scan rate of 200 mV/s. This nonlinear decay in actuation has previously been related to diffusion limited processes (e.g., semi-infinite planar diffusion), where the extent of actuation is limited by the rate and distance of ion transport into and out of the polymer. In contrast, the transition to a linear relationship for the nanometer thick films suggests that the majority of fast ion diffusion into the polymer is primarily occurring at the solid-liquid interface, thus shifting the electrochemical actuation process from a diffusion limiting to a current limiting system. A 1.5-2 times improvement in strain is seen for the thinnest film (97 nm) over the next highest thickness value (652 nm), revealing that strain performance increases as the thickness of the supported film enters the <100 nm regime [52]. However, the increases in strain/strain rate for nanoscale thick films are significantly less than those predicted by simple models, indicating that rate limiting mechanisms other than the absolute film thickness (e.g., modulus, porosity) are also at play.

Nanoactuation of a unique kind has been studied by combining force indentation curves and in-situ electrochemical-AFM [74]. Firstly, force indentation curves on thin polybithiophene (PBT) films show plastic deformation, resulting in permanent indentations with depths of 2-15 nm at applied forces > 300 nN [74]. The aim is then to use in-situ electrochemical-AFM imaging to investigate the effect of electrochemically switching the polymer redox state on modulating the topography of the nanoindentations. It is proposed that by oxidizing the films, their swelling due to the uptake of the dopant anions from the electrolyte should 'erase' the nanoindentations and recover the polymer (permanent shape) due to smoothing out of the topography. Furthermore, it is expected that upon electrochemical switching to a reduction potential, if the polymer retains the 'memory' of the nanoindentations (temporary shape), then induced ion expulsion and subsequent contraction of the film should result in the recovery of the nanoindentations. To assess whether the films retain the 'memory' of the nanoindentations, grid arrays of nanoindentation are prepared and their topographies imaged after the application of -1 V and then after switching back to 1 V [74]. Nanoindentations are clearly visible prior to electrochemical switching (Figure 4A), and the polymer recovery subsequently occurs after the application of -1 V (Figure 4B), though in some cases only partially (see nanoindentations (3, 4, 5, 6, 9). Importantly, the nanoindentations completely reappear after switching back to 1 V (Figure 4C), thus supporting a reversible recovery process shifting from a temporary ↔ permanent ↔ temporary shape without reprogramming. Height profiles taken across each of the nanoindentations before electrochemical switching (red trace) and after polymer recovery (black trace) and indentation recall (blue trace) show the extent to which the dimensions of the nanoindentations change during the switching process (Figure 4D). Unlike classical thermoset and thermoplastic shape memory polymers, this shape memory process is unique in that it utilizes electrochemical control of the polymer redox state to conceal, and temporarily store, preformed nanoscale surface patterns, which can later be recalled.

AFM studies have been performed to gain a better understanding of the role of mechanical actuation under conditions that are more relevant to the intended biological application. For example, [55] studied the effect of different biological dopants on mechanical actuation induced by electrical stimulation waveforms, specifically biphasic ± 100 mV pulse ranging from 0.01-10Hz, considered to be applicable to clinical stimulation applications. It was of interest to determine if mechanical actuation is evident given the existence of rate-limited ion diffusion at high frequency, biphasic stimulation conditions. For this work, Figure 5A (i) shows that the Z-piezo signal plateaus during the constant potential region of the pulse at the slowest stimulation of 0.1 Hz, indicating that the film reaches maximum expansion/contraction during the stimulation cycle. At 1 Hz (Figure 5A, ii), the actuation decreases and shows a triangular profile, while no actuation is discernable at 10 Hz (Figure 5A, iii). Figure. 5B shows the absolute change in film height varies from ≈ 2 nm for 1Hz and 5-10 nm for 0.1Hz. Figure. 5C shows the maximum strain calculated using the mean thickness value of each film. PPy/CS (4.7%) and PPy/PMAS (4.6%) show significantly higher strains at 0.1 Hz compared to PPy/HA (3.3%), PPy/pTS (2.2%) and PPy/DS (1.6%). In this study, there was also an attempt to relate the actuation to roughness and modulus. It was found that irrespective of whether the dopant is biologically derived, the physical properties tend to group together with films having either a low roughness, low modulus and high strain, or vice versa. When investigated as substrates for supporting the growth and differentiation of skeletal muscle cells, these two groupings of the properties correlate with the differing ability of the PPy polymer to support the cells [16].

PPy incorporating the antipsychotic drug, risperdone, can function as an electrically controlled in vitro drug release system [75]. Actuation of these films is observed when the potential is alternated between ±0.6V at a frequency of 0.5 Hz. Freshly prepared PPy films demonstrate cation-driven actuation; the polymer swells on reduction to its neutral state as an influx of mobile cations balances out the excess of pTS anions. However, after 6 days of aging, mixed ion driven actuation is observed where the polymer is seen to shrink and expand on both oxidation and reduction. The amplitude of actuation in aged films is reduced to around half that of the fresh films. It is found that the rate of drug release is dependent on the number of actuation cycles experienced by the polymer.

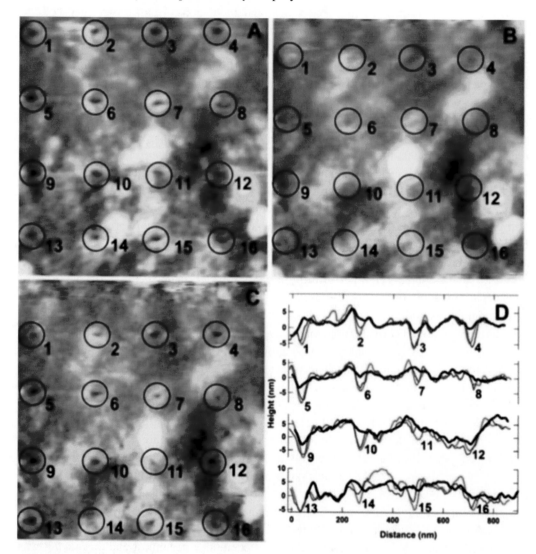

Figure 4. (a) AFM height images showing an array of nanoindentations before electrochemical switching (A), after polymer recovery at -1V (B) and after indentation recall at 1V (C). (D) Height profiles taken across each row of nanoindentations in the array before electrochemical switching (red trace), after polymer recovery at -1V (black trace) and after indentation recall at 1V (blue trace). Adapted with permission from [74].

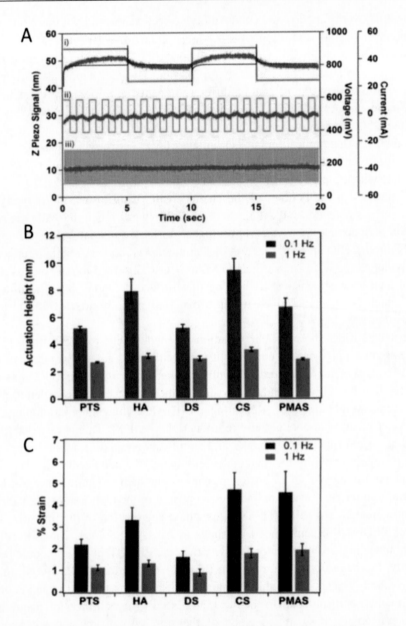

Figure 5. (A) PPy/CS data for biphasic waveform stimulation and actuation with frequencies. Red represents the voltage signal, green represents the Z piezo signal and black represents the current signal; (i) 0.01 Hz (ii) 0.1 Hz (iii) 1 Hz. (B) Histogram of mean actuation height for 0.1 and 1 Hz stimulation. (C) Histogram of mean % strain for 0.1 and 1 Hz stimulation. All error bars are standard error. Reproduced from [55].

NANOSCALE ELECTRICAL PROPERTIES

Kelvin Probe Force Microscopy (KPFM), or surface potential AFM imaging, and Conductive-AFM (C-AFM) are the two main techniques applied to study electrical characteristics of OCPs. These techniques enable highly localized nanoscale measurements of surface potential, work function and conductivity, as well as mapping of these properties with lateral nanometer spatial resolution over an area of up to ≈ 100 μm. Measurements are typically restricted to air environments (dry state). Initial studies endeavored to understand structure-property relations of conductivity, particularly the nanoscale origin of bulk conductivity, followed by the effects of different dopants and polymer treatments. C-AFM provides a means to study electrical properties of individual nanowires, nanotubes, and other nanostructures that are otherwise difficult to quantify using micro- and bulk techniques. C-AFM also involves the application current-voltage (I-V) curves, which can be pinpointed to specific positions on individual nanostructures. Bandgaps obtained from I-V curves of C-AFM correlate with those of adsorption spectra studies.

KPFM shows that the nodules of the characteristic 'cauliflower' morphology of OCPs have lower surface potential, indicating areas of higher work function, while the opposite is the case for peripheral regions that have higher surface potential values or lower work function (Figure 6). Higher work function of the nodules is related to higher conductivity, or more highly doped regions. The surface potential is also observed to vary across the surface over lateral nanoscale dimensions, indicating that surface potential is not well-defined and that of E_o which is usually ascribed to these materials actually arises from an average of distributed values.

KPFM and C-AFM show similar variation in conductivity for PBT films [77, 78]. In undoped, as-grown polymer the conductivity is confined to nodule structures, which are suggested to consist of more ordered, crystalline higher molecular weight domains compared to peripheries that have lower conductivity due to the presence of less-ordered, amorphous lower molecular weight polymer. Repeating the C-AFM imaging after doping, however, shows a greater increase in conductivity in the peripheries [78]. This is explained by doping that should be accompanied by an uptake of a significant amount of counter-ions, as well as solvent. Due to their more highly ordered, dense polymer, the nodule structures are less likely to accommodate this uptake of ions/solvent and sustain such structural changes. Therefore, doping is more preferential in the peripheries compared to the nodules despite their expected higher redox potential. For PEDOT/PSS films, phase segregation of the polymer is believed to consist of PEDOT/PSS surrounded by shells of excess PSS [79]. Therefore, the nodule centres have much higher intrinsic conductivity than the PEDOT-depleted nodule peripheries. Consequently, the main obstacle for current flow is between the PEDOT-rich grains, while the current is easily transported within the nodules. It is shown that the vertical conductivity in OCP films, i.e., perpendicular to the substrate, is up to three orders of magnitude lower than the lateral conductivity in the plane of the film [79]. Combined with conductivity measurements, this is elucidated by AFM imaging of films in cross-section that show pancake-shaped PEDOT-rich islands separated by lamellas of PSS [79]. EFM has been used to resolve the issue of length-scales that govern conductivity in OCP [80]. This work explains conductivity in terms of 'metallic islands' and inquires whether these metallic islands are confined to single polymer strands, or are in the form of large semi-crystalline areas, or if they exist at all. Measurements were performed on monolayers of polybithiophene whilst applying a small directional current to the film [80]. Sharp potential drops are observed across the monolayer, which are interpreted as dependencies of conductivity on directionality of the polymer backbone. When the backbone of the individual polymer chains are aligned with the direction of the current, the measured conductivity is high, while perpendicular alignment of the backbones significantly lowers the conductivity. This is supported in EFM images showing drops in potentials at boundaries of differently orientated backbones of up to 100-500nm in domain size.

Understandably there is much interest in the effect of dopants and growth conditions on improving conductivity. C-AFM has been applied in several studies for this purpose. Based on C-AFM data, [81] explains that sodium dodecyl sulfate (SDS), which forms large micelles in aqueous media, assists in the formation of large particles but these are not well electrically connected. In contrast, ClO_4^- is homogeneously dispersed in the polymer matrix, most likely due to its lowest basicity, and provides higher conductivity. pTS doped PPy films have the highest conductive state ascribed to its relatively high basicity in comparison with that of ClO_4^- and high polarizability of the pTS. It is stated that high basicity allows the radical cations, i.e., polarons, of PPy to easily associate with the dopants, stabilizing the conductive states. The AFM observations of morphological and electrical properties of the PPy films are credited for helping provide the above explanations. Overall, KPFM and C-AFM are notable in studying the effect of various parameters on thin conducting polymer films and show that the conductivity generally increases with an increase in growth time, dopant concentration, and use of non-aqueous solvents as the electrolyte. Through the use of electrical AFM techniques, we now appreciate a generally accepted picture of an inhomogeneous conductivity that occurs over nanoscale length-scales, and that this spatial pattern of conductivity is inextricably linked to their nanoscale morphology.

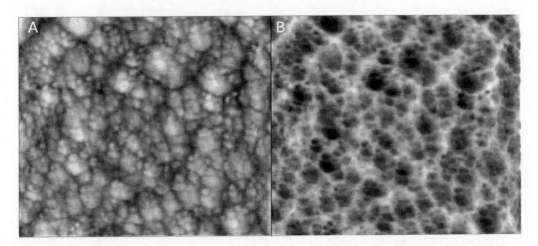

Figure 6. (A) Topography image and (B) KPFM image of surface potential of a PPy/pTS film grown potentiostatically at 0.6 V for 5 min, thickness 150nm. Reproduced with permission from [76].

ELECTRICAL PROPERTIES OF INDIVIDUAL NANOMATERIALS

Electrical AFM modes lend themselves to probing intriguing electrical properties that arise as OCPs in the form of individual nanowires and nanotubes are scaled down to nano-sized dimensions. The ability to pinpoint localized measurements along individual nanostructures opens a new paradigm of electrical characterization under vacuum, ambient, or especially liquid environments. C-AFM is used, typically in the I-V spectroscopy mode, to demonstrate I-V linear characteristics of OCP nanomaterials. Whilst some type of extraordinary electrical properties may be expected, several studies have quantified conductivity values that are only comparable to bulk films. PPy nanotubes prepared using pores of track-etched polycarbonate, ranging from 50-200 nm in diameter, give conductivities of $\approx$ 1 S/cm that are consistent with bulk films [82]. PEDOT/PSS nanowires, < 10 nm in diameter, give similar values of 0.06 – 0.6 S/cm [83]. Studies show that size does matter; at smaller dimensions a more highly-ordered polymer structure starts to dominate the conductivity. For example, AFM conductance measurements on single PPy nanobelts give constant values of $\approx$ 2-2.5 S/cm for widths > 100 nm, though this value increases linearly to 4.5 as the width approaches 50-60 nm [84]. However, other studies using C-AFM attain appreciable conductivities. The conductivity of PPy nanotubes is shown to be 5000 S/cm when the wall thickness reaches 15 nm [85], while others report values of 3200 S/cm for PPy nanobelts [86]. It is noted in the literature that some caution must be taken when comparing values, as both the loading force and applied bias voltage during the C-AFM imaging or I-V curve can influence the measured current signal [84]. Lastly, C-AFM of single gold and silver nanoparticles adhered to OCP films provides a means to study the properties of electrical nanoscale contacts between the polymers and metals. Sufficient electrical contact could be made between noble metal nanoparticles and either PPy, poly(3-methylthiophene) (P3MET) and PEDOT [87]. It turns out that PEDOT is the most conductive at the sites where the particles sit on the surface, suggesting that PEDOT better aligns sulfur atoms on the surface to enable strong bonding of gold and silver nanoparticles.

SURFACE CHEMICAL PROPERTIES

AFM is primarily a physical measurement tool and does not have inherent capabilities for analysis of chemical composition, as achieved by spectroscopy or x-ray techniques. AFM is better known for probing local surface chemistry using chemically functionalized tips. Various strategies for surface modification of the tips include the use of self-assemble monolayers, covalent coupling via 1-ethyl-3-(3- dimethylaminopropyl)carbodiimide (EDC) and N-hydroxysuccinimide (NHS), biotinylated-BSA, histadine tags and polyethylene glycol (PEG) linkers [88-91]. Whilst this provides an in-direct approach for detecting chemical surface groups via binding forces, in principle, it can be a very powerful tool for chemical and nanoscale spatial recognition at the level of single molecules. This type of measurement is typically classified under measurements involving force spectroscopy [88-90], which are presented in more detail below..

FORCE MEASUREMENTS: INTERACTION FORCES OF OCP

Thermodynamic expressions and force laws provide a means to calculate energies and forces based on knowledge of the interacting surfaces yet the ability to make comparisons with direct measurements using force measurement techniques is invaluable. The direct measurement of intermolecular and surface forces is becoming an increasingly explored area for OCP, as there is growing interest in the effect of electrical fields on binding kinetics and biomolecular interactions [92]. The principle of AFM force measurements are shown in Figure 7A. They involve measuring the change in deflection of a flexible cantilever with sharp tip whilst bringing the tip into contact and then withdrawing it from a surface. Tip-surface interaction forces, F, acting on the cantilever are easily measured using simple Hooke's Law, $F = kd$, where k is the cantilever stiffness (spring constant) and, d is the cantilever deflection. These AFM measurements, or force curves, can be applied to measure double-layer forces. The AFM tip can also be functionalized with ligands (e.g., protein) to recognize and detect the binding forces of complementary surface molecules [88].

AFM force measurements reveal that there can be complete charge compensation during applied potentials in P3MET and PPy films [93]. This relates to an interesting observation that no diffuse double layer interaction at the polymer-solution interface is evident in the force curves [93], a situation that is very different from metal or semi-conductor electrodes. For these measurements, a silica colloidal probe, negatively charged at pH > 3 in 0.001M potassium perchlorate ($KClO_4^-$), shows no indication of electrostatic attraction or repulsion over a potential range of +0.4 to -0.8V vs Ag/AgCl, suggesting the charge in the film is fully compensated by anions. This behaviour is distinctly different to that of a gold electrode, which has similar surface roughness. The main finding is that ions mainly reside within the polymer film and that any electrode capacitance that could be ascribed to the polymer film would be an inner or compact layer capacitance. It is also possible that the force curves, with no evident double layer interaction, results from an ionic concentration that is so high within the polymer pores that all of the compensation occurs in a compact and extremely thin, few angstroms, diffuse layer that is not detected.

Despite the above study, repulsive forces have been detected on OCP as a function of the applied potential. AFM force measurements between a silicon tip and sulfonated polyaniline (SPANI) in 1mM KCl at pH 2.5 and 25°C show interaction forces that are dependent on the applied potential [94]. At low pH, an interaction between a slightly negatively charged tip and SPANI film with negatively applied bias produces a repulsive force that extends out to 20 nm (Figure 7B, i). As the applied potential is increased towards positive values the repulsive force diminishes eventually to the point where a net attractive force and tip-polymer 'pull-off' adhesion of 2.0–2.5 nN is present (Figure 7B, ii).

Force measurements can be performed where a known potential is also applied to the AFM tip and polymer (Figure 7C). The interaction between a gold coated tip with applied -200 mV bias and PPy/HA film in 0.005mM NaCl at neutral pH and room temperature similarly produces interactions that are dependent on prior charging of the polymer [65]. The interaction however becomes more complex as a function of the lateral position of the tip across the surface. For uncharged (as-grown) films, a purely repulsive interaction occurs on nodules of the characteristic 'cauliflower' morphology (Figure 7C, i) but additional short range attractive forces and 'pull-off' adhesion of 0.5 nN appear within the peripheries [65] (Figure 7C, ii). When the polymer is charged prior to the measurements with +200mV, the repulsive force significantly diminishes and the attractive force and pull-off adhesion are again present, as in Figure 7C (ii), yet on this occasion the interaction is not dependent on the lateral position of the tip (i.e., nodules versus peripheries). An interesting aspect of this research is that charging of these polymers relates to the ability of adipose stem cells to adhere to the surface. Because the charging does not significantly change the topography and modulus of the polymer, it is apparent that electrostatic forces play a primary role in the binding and bioactivity of surface adsorbed extracellular proteins that promote cell adhesion.

Assessment of the 'pull-off' adhesion force between a silicon tip and SPANI films show that it tracks the electrochemically induced charge at the polymer surface [94]. An adhesion versus electrode potential curve exhibits a titration-like curve response (Figure 8) where the minimum and maximum of the adhesion represents a surface that is saturated with positive or negative charge. Least squares fitting to these curves exhibit an inflection point that corresponds to the potential where the electrostatic force transforms from repulsive to attractive. For this silicon tip-SPANI interaction, a measured inflection point of -125 mV is approximately equal to the half-way potential in cyclic voltammetry measurements.

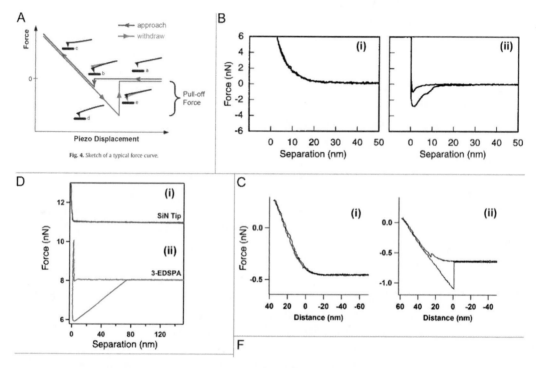

Figure 7. (A) Schematic of AFM force curve. a) tip approaches; b) attractive force and tip-surface contact; c) repulsive contact force; d) tip-sample adhesion; e) 'pull-off' adhesion and tip withdrawal. (B) Force curves for interaction between silicon tip and SPANi-coated electrode in 1 mM KCl at 25°C and pH 2.5 at applied potentials of (i) -350 mV and (ii) +250mV(vs AgQRE). Reproduced from [94]. (C) Force curves for interaction between gold coated tip (biased at -200mV) uncharged PPy-HA film in 0.005 M NaCl(aq) electrolyte at the location of (i) nodule structures and (ii) peripheries of nodules structures of the polymer morphology. Reproduced from [65]. (D) AFM force curves for interaction between (i) plasma treated AFM silicon nitride tip (SiN) and (ii) aminosilinized tip (3-EDSPA) and PPy/CS in phosphate buffer saline. Adapted from [95].

CHEMICAL MAPPING OF POLYMER SURFACE

Chemical modification of AFM tips and surfaces enables the interactions of functional groups (e.g., –COOH, -NH$_2$, -OH and –CH$_3$). This approach is done extensively using model surfaces (e.g., self-assembled monolayers), though only recently pursued for OCPs. For example, a series of functional groups are introduced onto the AFM silicon tip and force measurements are performed after each functionalization step to assess their involvement in the interaction with the OCP [95]. Plasma treated silicon nitride tip (SiN$_3$) bearing –OH groups, which are hydrophilic and negatively charged at neutral pH, show a small repulsive force and no adhesion to as-grown PPy/CS films with no applied potential (Figure 7D, i). In contrast, 3-ethoxydimethylsilylamine propyl (3-EDSPA) treated tips terminated with protonated NH_3^+ groups at neutral pH show an attractive force during approach followed by a 'pull-of' adhesion of 2.0 nN (Figure 7D, ii). This net attractive force and adhesion between the positively charged tip and negatively charged polymer surface indicates the presence of anionic sulfate groups of CS at the polymer surface. Gluteraldehyde (GAH) functionalized tips bearing carbonyl groups are reactive for primary amines to enable protein crosslinking. Due to the presence of carbonyl groups these tips are negatively charged and show a small repulsive force but unlike the silica tips a 'pull-off' adhesion of 0.5 nN is present (force curves not shown). These tips could potentially undergo a Shiff's base reaction to couple with –NH groups of the polymer, however, the magnitude of the adhesion forces does not suggest the formation of covalent bonds.

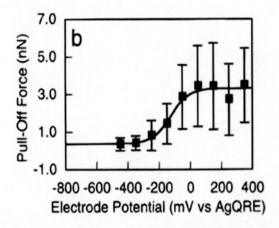

Figure 8. Pull-off adhesion between silicon tip and SPANI electrode as a function of the electrochemical potential applied to the substrate in 1 mM KCl at 25°C and pH 2.5. The solid lines represent least squares fits to a titration curve. Each data point represents the average of 256 measurements and the error bar is the standard deviation. These curves give $E°$ values of -100 mV. Adapted with permission from [94].

AFM force mapping involves collecting a grid-array of force curves across the sample [96]. Each pixel in the force map represents a single force curve from which parameters such as modulus and adhesion can be extracted and displayed as a 2-dimensional image by plotting the values as function of the X-Y position. To correlate material properties with topography, the latter can be simultaneously acquired by determining z-piezo height where the tip makes contact with the surface relative to the starting piezo height during a force curve. Recently, this approach was used to detect functional groups, initially incorporated onto the monomer, at the surface of electrochemically polymerized films of PEDOT/hydroxy (OH), PEDOT/ phosphorylcholine (PC), and mixtures of the two [97]. Various tip functionalities were employed, with dendron-coated probes consisting of anthracene moieties to promote pi-pi interactions with the OCP backbone. The anthracene modified probes show the ability to discern differences between homopolymers of PEDOT/OH and PEDOT/PC based on the degree of polymer chain stretching, or interaction length, during the binding. Thus, using polymer stretching as a mechanical 'marker', AFM force mapping is able to detect nanoscale domains of PEDOT/OH and PEDOT/PC in mixed films.

BIO-AFM AND BIOMOLECULAR INTERACTIONS

Through a variety of strategies to chemically functionalize the AFM tip, the attachment of many different proteins in active and desired conformations has made it possible to directly measure the forces between a single protein and substrate of interest. AFM force spectroscopy has shown it is possible to deconvolute the effects of the intrinsic surface heterogeneity of OCP, as any given single x-y location of a force curve can actually represent a specific single protein–surface interaction bearing its own characteristic fingerprint. Our group has implemented this approach to understand the interactions between fibronectin and PPy films doped with different biomolecules, including CS and HA [95, 98]. The principle behind this type of force spectroscopy experiment is depicted in Figure 9A and highlights the ability to probe protein-OCP interactions both at the molecular and nanometer scale under electrical stimulation.

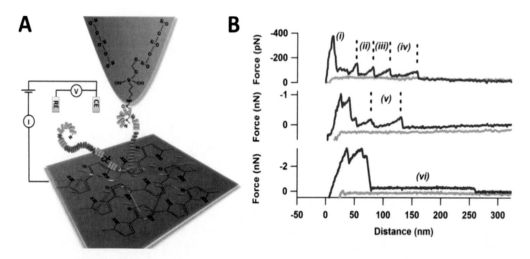

Figure 9. (A) Schematic diagram of AFM tip functionalized with FN interacting with a conducting polymer electrode (blue), polypyrrole (PPy), comprising a conjugated backbone (black chemical structure) with entrapped dopants (green) such as chondroitin sulphate (CS), hyaluronic acid (HA) or dextran sulfate (DS). Force spectroscopy measurements were performed as a function of an applied voltage using electrochemical–AFM. The PPy-electrode operated as the working electrode in 3-electrode electrochemical cell. (B) AFM force curves for the interaction of FN with non-electrically stimulated PPy/CS showing the trace (orange) and retrace (blue) curves. The peak at (i) corresponds to initial detachment of the tip and fibronectin molecules from the surface. The two subsequent peaks (1st and 2nd dashed lines) and their spacing of 27.1 nm (ii) and 28.5 nm (iii) correspond to the sequential unfolding of FNIII modules (~ 75 amino acid residues). A peak spacing of 47.8 nm at point (iv) is greater than that for FN unfolding and correlates with multiple detachment of FN-polymer binding sites. Adapted with permission from [98].

A direct approach is to covalently functionalize the AFM tip with a protein of interest. The tip is brought into contact usually for a delayed period to initiate binding to the OCP surface Figure 9A. AFM tips functionalized with fibronectin (FN), a well-known ECM protein that facilitates cell adhesion, show 'pull-off' adhesion on PPy doped with GAGs (Figure 9B, i) [95, 98]. This adhesion is due to bulk interactions (e.g., multiple proteins) of the protein functionalized tip. After 'pull-off' adhesion, other types of adhesion events occur, including sawtooth-forces due to sequential unfolding of folded FN domains as the tethered molecules are stretched (Figure 9B, ii, iii, iv) or plateau forces that involve the desorption or 'peeling' off of FN molecules from the polymer surface (Figure 9B, vi) [95]. The former has a very characteristic profile during the extension of single proteins in that the forces to unfold individual domains (peaks) are $\approx$ 100-200 pN and spacing between peaks is equal to the fully extended length of unfolded domain (i.e., 28 nm = $\approx$ 75 amino acid residues) (Figure 9B, ii, iii, iv) [98]. Multiple sawtooths also arise when there are multiple binding sites along the length of the protein (Figure 9B, vi). The interaction of FN with PPy doped CS, DS, HA and pTS show sawtooths due to multiple binding sites with peak spacings of $\approx$ 60 nm that correlate with the distance between heparin-binding domains of FN, suggesting their involvement in the interaction [98]. Binding of individual domains gives average forces ranging from 100-150 pN and the FN freely interacts along its length, allowing binding at heparin domains, and is also able to extend up to its contour length ($\approx$ 175 nm) under tensile forces. Repeating experiments using FN functionalized AFM tips on more hydrophobic, polythiophene (PT) films show that the extension of the FN is greatly reduced to distances of $\approx$ 25 nm which correlates with dimensions of FN in its folded conformation, suggesting that this conformation is retained, e.g., hydrogen bonds are not disrupted during interactions with these low surface energy polymers [99].

To further assess differences in adhesion between films doped with GAGs, force-volume mapping was performed with the FN functionalized tips to spatially map the dependence of adhesion on topography of PPy/CS and PPy/HA films [95]. The adhesion maps of PPy/CS and PPy/HA show a difference in both the strength and lateral distribution of the protein adhesion. PPy/CS shows a higher strength of adhesion compared to PPy/HA. Each adhesion event (red pixels) from an adhesion map can be overlaid onto the topography image (Figure 10A and B). Due to a significantly higher occurrence of adhesion across the surface for PP/CS (77% surface coverage), there is no correlation between adhesion and topography to the point where the high distribution of adhesion effectively masks the topography (Figure 10A). Conversely for PPy/HA, the adhesion is more distributed along the nodule regions, as opposed to peripheries, and thus occurs at much lower density (24% surface coverage) (Figure 10B). C-AFM scans were taken on PPy/CS and PPy/HA to further assess the correlation between the FN adhesion, topography and conductivity of the films. PPy/HA shows regions of conductivity that are confined to the nodule regions of the polymer, while the peripheries show little or no measurable current. The conductivity of PPy/CS is however more uniformly distributed across the surface with no clear correlation with the topography, indicating more homogenous doping of the polymer. The dependence of the adhesion on the topography is further depicted by showing the protein interaction with both PPy/HA (Figure 10D) and PPy/CS (Figure 10C) using actual 3-D height images overlaid with the conductivity, where areas of black indicate low or no conductivity, while areas in green indicate higher conductivity. The density of the adhesion acquired in the force maps closely correlates with the distribution of conductive, or more CS or HA doped, regions across the film, confirming that FN adhesion to the polymer is mediated by specific interactions with the dopants.

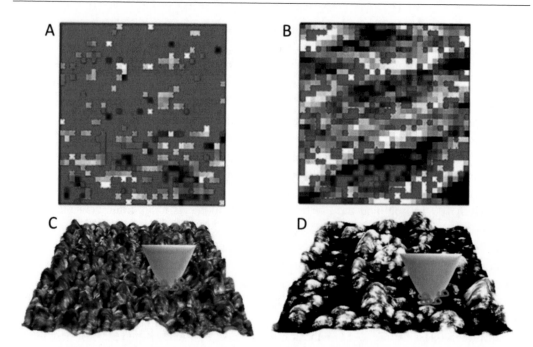

Figure 10. Adhesion image (red pixels) overlaid on topography images of PPy/CS (A) and PPy/HA (B). Below - Schematic of protein-surface interaction using actual 3-dimensional topography with overlay of corresponding conductivity (green is conductive, black is non-conductive, Z scale 4 nA). Scan area is 1 μm and AFM tip is drawn roughly to scale (≈ 30 nm tip radius). (C) CS is more homogenous giving a higher probability of FN adhesion. (D) HA displays inhomogenous conductivity resulting in lower probability of adhesion. The protein (red) may not adhere to non-conductive areas (black) of the polymer. Adapted from [95].

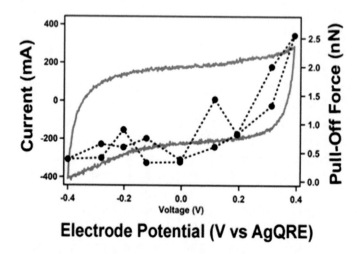

Figure 11. Pull-off adhesion force (dashed curve) versus voltage and corresponding cyclic voltammograms (solid curve) for PPy/CS films. Adhesion values (black circles) represent an average from individual force curves collected at each time point during 3 cyclic voltammogram cycles performed at a scan rate of 50 mV/s. No changes in pull-off adhesion were observed for slower scan rates of 5mV/sec (data not shown). Adapted with permission from [98].

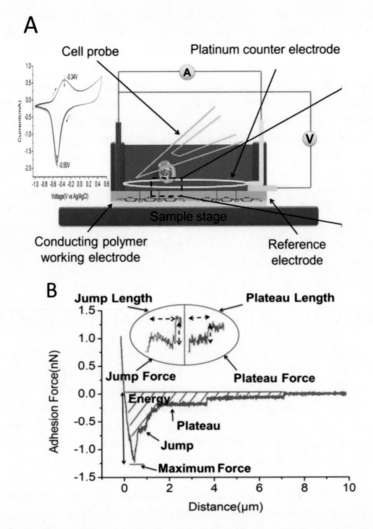

Figure 12. (A) Schematic of Electrochemical-SCFS combined with a 3-electrode electrochemical cell mounted on a Bio-AFM. The electrochemical cell is temperature controlled at 37 °C, consisting of the PPy/DBSA substrate (orange) as the working electrode, a platinum wire ring (silvery grey) as the counter electrode, small Ag/AgCl reference electrode (yellow), and controlled by an external potentiostat. Cyclic voltammograms performed in CO_2 independent medium indicate an oxidation peak at approximately -340mV and reduction peak at -500mV (left image). During the SCFS measurements, the adhesion forces are detected between a single cell attached to the tip and the polymer substrate whilst applying a constant voltage. (B) Force verus distance (retraction) curve for the adheisve interaction between a single L929 cell and native polymer (no applied voltage) of PPy/DBSA. The peak force is given as the maximum force values required to detach the cells. Plateau and jump interactions are evident following the bulk detachment of the cell. (Inset) Plateau interactions shows a constant force over a given distance, while jump interactions typically show a non-linear increase in force that is characteristic of an elastic response of a biological molecule or bond under strain. Adapted from [100].

Using the electrochemical-AFM setup in Figure 9A, the protein adhesion can also be quantified as a function of electrical stimulation or switching electrochemical redox properties. For the FN protein-PPy interaction, applying a positive bias to the PPy causes a strong electrostatic attraction between the majority of negatively charged protein domains and positively charged polymer, resulting in an order of magnitude higher adhesion force of ≈ 1-2 nN (Figure 11) [98]. This electrochemically induced adhesion is stronger and non-specific but can be reversibly switched to smaller piconewton adhesion forces by applying an opposite negative bias to the polymer. Rather than applying a constant potential, this work applied cyclic voltammetry where the pull-off adhesion is plotted as a function of the change in voltage and current (Figure 11) and becomes kinetically dependent on the scan-rate.

Single molecule AFM force sensitivity is helpful for deconvoluting complex interactions that operate on nanometer lengthscales across the polymer surface and are expected to play a critical role in fundamental advances in this area, including conducting biomaterials and electrochemical-based protein sensors. Practical uses can also be found, for example, by supporting high-throughput bacteriophage library screening methods with AFM as a next step in assessing binding kinetics and mechanistics of those ligands identified as having a high affinity. This was demonstrated by identifying a T59 phage peptide sequence with high affinity to PPy/Cl and applying force spectroscopy using peptide functionalized tips on the polymer to quantify binding constants [29].

BIO-AFM CELLULAR TECHNIQUES AND FUTURE DIRECTIONS

A major challenge in the study of dynamic, OCPs in this area, or any biomaterial for that matter, is the inability of conventional cell biology methods to quantitatively corroborate cell adhesion to developmental processes such as growth, proliferation and differentiation. Routine histological approaches typically define adhesion as the number of fixed and stained cells remaining behind on the substrate after washing. These measurements are non-quantitative (no actual adhesion forces are measured), generally not reproducible (e.g., inconsistency in washing step), convoluted by non-specific interactions and very difficult to relate to the involvement of cell adhesion proteins (e.g., focal adhesions) that are often fluorescently identified. Furthermore, the analysis of fixed and stained dead cells, makes it impossible to resolve dynamic adhesion processes. As such, there is somewhat of a vacuum in our knowledge of the forces involved in cell-material adhesion, especially at the dynamic, molecular level.

In the area of conducing polymers, we have recently taken steps toward addressing this by implementing Electrochemical-Single Cell Force Spectroscopy (EC-SCFS) to ultimately probe the entirety of the living cell-protein-OCP interface [100], which has been hitherto very difficult to achieve. The technique relies upon the attachment of a single living cell to an AFM probe, which is brought into contact with the electrode surface and adhesion forces measured as the probe is retracted during an applied voltage (Figure 12A). Significantly, this enables dynamic changes in binding forces of single integrin bonds, and formation of their complexes, to be detected with ultra-high sensitivity and resolution.

Without the use of adsorbed proteins or serum, the adhesion occurs primarily between the cell surface glycocalyx and conducting polymer. In this system, the EC-SCFS shows that the cell adhesion increases on DBSA doped PPy as they are electrically switched from oxidized to reduced states [100]. As the polymer is reduced, an increase in surface sulfonation and hydrophilicity correlates with an increase in single cell adhesion. At the molecular level, the glycocalyx interactions involve molecules that have either a weak/absent linkage to the intracellular cytoskeleton, resulting in interactions with membrane tethers, or those that have stronger cytoskeletal-linkages (Figure 12B). For the latter, binding forces of $\approx$ 30-40 pN with a narrow force distribution indicate interactions at the single molecule level. Electrical switching modifies the single molecule bond properties, including both the force and interaction length (i.e., bond stiffness) of the glycocalyx-polymer interactions. This may be due to switching of specific surface chemical groups, direct electrical effects at the electrode-electrolyte interface, or other redox-induced process such as ion/water uptake into the polymer. This work provides a platform to enable insight into the effect of electrical switching on molecular-level interactions between living cells and conducting polymers. More broadly, the approach can be applied to elucidating the bond properties and kinetics (e.g., dissociation constants and energy barriers) of specific cell adhesion molecules such as integrins in the presence of electrical fields and directly at electrode interfaces.

An alternative approach is to integrate the OCP electrode onto the AFM tip, which is then used to probe living cells [101]. In contrast to EC-SCFS, this approach effectively inverts the measurement by placing the OCP electrode up onto the tip, which now operates as a nanoscale electrode (Figure 13A-C). AFM tips modified with an integrated platinum electrode and subsequently coated with PPy/pTS via pulsed polymerization show adhesive interactions involving the stretching of PPy chains [101]. When a potential of +0.4V is applied to the PPy electrode tip its interaction with a glass surface shows repulsive forces and 'pull-off-adhesion of $\approx$ 0.5 nN but the latter is not present at -0.4V. Over-oxidizing the PPy tip with +1.0V induces an interaction that indicates the liberation and extension of PPy chains originating from the integrated PPy electrode tip. Lastly, the PPy-modified probes are suitable for probing force interactions at living cell surfaces (Figure 13F), while applying nanoscale electrical stimulation that cause electrical contractions of the cells, in this case muscle myotubes (Figure 12D-E).

CONCLUSION

Over the past 20 years it has been shown that electrical stimulation through OCP has a significant and beneficial effect on mammalian cells. While significant effects on the macroscopic scale have been observed, it is expected that molecular level insights into this fascinating phenomenon in a biologically relevant environment will provide major advancements. Newly developed Bio-AFM approaches, such as those described above and other emerging and exciting capabilities like High-Speed AFM [102], will enable real-time detection of any single protein, or single cell surface receptor, interaction with OCPs under electrical stimulation. This will have a significant impact on those concerned with the design of OCPs to amplify specific effects arising from electrical stimulation. We have no doubt AFM-based approaches will be embraced not just by those involved in the development of OCPs but any electromaterial or surface wherein electrical stimulation is used to control biomolecular and cellular interactions.

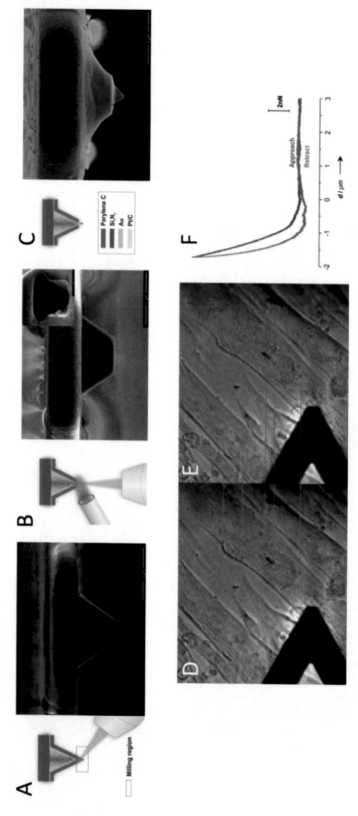

Figure 13. (A–C) Schematic representation and scanning electron microscopy (SEM) images of the individual fabrication steps of AFM-SECM probes with conical platinum/carbon (Pt/C) electrodes. (A) After gold layer deposition and insulation with Parylene C, (B) after Focused Ion Beam (FIB)-milling (inset shows the exposed gold electrode), and (C) after ion beam-induced Pt/C deposition (all scale bars are 3 μm). Schemes on the left side illustrate a cross-section through the tip at the corresponding fabrication step and direction of the ion beam. (D-E) Bright field images of C2C12 muscle cells (myotubes) before (D) and after (E) stimulation with a PPy-modified tip (A,B). Red arrows show positions before and after cell stimulation with a potential of - 1.5 V vs. Ag/AgCl was applied. (F) Force curve obtained at the cell before the stimulation is shown in (D). Note: The shown images are part of a recorded video, which shows cell movement. Adapted with permission from [101].

ACKNOWLEDGEMENTS

The authors are grateful for the support of the Australian Research Council under the Australian Research Council Centre of Excellence Scheme and Australian Research Fellowship of Assoc. Prof. Michael Higgins and Laureate Fellowship of Prof Gordon Wallace.

REFERENCES

[1]   Wallace, G.G., et al., *Organic Bionics.* 2012: John Wiley & Sons.

[2]   Weiland, J.D., D.J. Anderson, and M.S. Humayun, In vitro electrical properties for iridium oxide versus titanium nitride stimulating electrodes. *Biomedical Engineering, IEEE Transactions on,* 2002. *49*(12): p. 1574-1579.

[3]   Geddes, L. and R. Roeder, Criteria for the selection of materials for implanted electrodes. *Annals of biomedical engineering,* 2003. *31*(7): p. 879-890.

[4]   Franks, W., et al., Impedance characterization and modeling of electrodes for biomedical applications. *Biomedical Engineering, IEEE Transactions on,* 2005. *52*(7): p. 1295-1302.

[5]   Cogan, S.F., Neural stimulation and recording electrodes. *Annu. Rev. Biomed. Eng.,* 2008. *10*: p. 275-309.

[6]   Merrill, D.R., M. Bikson, and J.G. Jefferys, Electrical stimulation of excitable tissue: design of efficacious and safe protocols. *Journal of neuroscience methods,* 2005. *141*(2): p. 171-198.

[7]   Polikov, V.S., P.A. Tresco, and W.M. Reichert, Response of brain tissue to chronically implanted neural electrodes. *Journal of neuroscience methods,* 2005. *148*(1): p. 1-18.

[8]   Green, R.A., et al., Conducting polymers for neural interfaces: challenges in developing an effective long-term implant. *Biomaterials,* 2008. *29*(24): p. 3393-3399.

[9]   Wallace, G.G. and G.M. Spinks, Conducting polymers: A bridge across the bionic interface. *Australian Institute for Innovative Materials-Papers,* 2007.

[10]   Hassler, C., T. Boretius, and T. Stieglitz, Polymers for neural implants. *Journal of Polymer Science Part B: Polymer Physics,* 2011. *49*(1): p. 18-33.

[11]   Wallace, G.G., S.E. Moulton, and G.M. Clark, Electrode-Cellular Interface. *Science,* 2009. *324*(5924): p. 185-186.

[12]   Ludwig, K.A., et al., Chronic neural recordings using silicon microelectrode arrays electrochemically deposited with a poly (3, 4-ethylenedioxythiophene)(PEDOT) film. *Journal of neural engineering,* 2006. *3*(1): p. 59.

[13]   Khodagholy, D., et al., Highly conformable conducting polymer electrodes for in vivo recordings. *Advanced Materials,* 2011. *23*(36): p. H268-H272.

[14]   Guimard, N.K., N. Gomez, and C.E. Schmidt, Conducting polymers in biomedical engineering. *Progress in Polymer Science,* 2007. *32*(8): p. 876-921.

[15]   Cui, X., et al., Surface modification of neural recording electrodes with conducting polymer/biomolecule blends. *Journal of biomedical materials research,* 2001. *56*(2): p. 261-272.

[16] Gilmore, K.J., et al., Skeletal muscle cell proliferation and differentiation on polypyrrole substrates doped with extracellular matrix components. *Biomaterials,* 2009. *30*(29): p. 5292-5304.

[17] Hodgson, A., et al., Reactive supramolecular assemblies of mucopolysaccharide, polypyrrole and protein as controllable biocomposites for a new generation of 'intelligent biomaterials'. *Supramolecular Science,* 1994. *1*(2): p. 77-83.

[18] Ateh, D., H. Navsaria, and P. Vadgama, Polypyrrole-based conducting polymers and interactions with biological tissues. *Journal of the royal society interface,* 2006. *3*(11): p. 741-752.

[19] Janata, J. and M. Josowicz, Conducting polymers in electronic chemical sensors. *Nature materials,* 2003. *2*(1): p. 19-24.

[20] Liu, X., et al., Electrical stimulation promotes nerve cell differentiation on polypyrrole/poly (2-methoxy-5 aniline sulfonic acid) composites. *Journal of neural engineering,* 2009. *6*(6): p. 065002.

[21] Kotwal, A. and C.E. Schmidt, Electrical stimulation alters protein adsorption and nerve cell interactions with electrically conducting biomaterials. *Biomaterials,* 2001. *22*(10): p. 1055-1064.

[22] Zhang, Z., et al., Electrically Conductive Biodegradable Polymer Composite for Nerve Regeneration: Electricity-Stimulated Neurite Outgrowth and Axon Regeneration. *Artificial organs,* 2007. *31*(1): p. 13-22.

[23] Thompson, B.C., et al., Conducting polymers, dual neurotrophins and pulsed electrical stimulation—dramatic effects on neurite outgrowth. *Journal of Controlled Release,* 2010. *141*(2): p. 161-167.

[24] Schmidt, C.E., et al., Stimulation of neurite outgrowth using an electrically conducting polymer. *Proceedings of the National Academy of Sciences,* 1997. *94*(17): p. 8948-8953.

[25] Serra Moreno, J., et al., Polypyrrole-polysaccharide thin films characteristics: Electrosynthesis and biological properties. *Journal of Biomedical Materials Research Part A,* 2009. *88*(3): p. 832-840.

[26] Garner, B., et al., Polypyrrole–heparin composites as stimulus-responsive substrates for endothelial cell growth. *Journal of biomedical materials research,* 1999. *44*(2): p. 121-129.

[27] Svennersten, K., et al., Mechanical stimulation of epithelial cells using polypyrrole microactuators. Lab on a Chip, 2011. *11*(19): p. 3287-3293.

[28] Thompson, B.C., et al., Optimising the incorporation and release of a neurotrophic factor using conducting polypyrrole. *Journal of Controlled Release,* 2006. *116*(3): p. 285-294.

[29] Sanghvi, A.B., et al., Biomaterials functionalization using a novel peptide that selectively binds to a conducting polymer. *Nature materials,* 2005. *4*(6): p. 496-502.

[30] Stauffer, W.R. and X.T. Cui, Polypyrrole doped with 2 peptide sequences from laminin. *Biomaterials,* 2006. *27*(11): p. 2405-2413.

[31] Collier, J.H., et al., Synthesis and characterization of polypyrrole–hyaluronic acid composite biomaterials for tissue engineering applications. *Journal of biomedical materials research,* 2000. *50*(4): p. 574-584.

[32] Moreno, J.S., et al., Synthesis and characterization of new electroactive polypyrrole–chondroitin sulphate A substrates. *Bioelectrochemistry,* 2008. *72*(1): p. 3-9.

[33] Wong, J.Y., R. Langer, and D.E. Ingber, Electrically conducting polymers can noninvasively control the shape and growth of mammalian cells. *Proceedings of the National Academy of Sciences,* 1994. *91*(8): p. 3201-3204.

[34] Zhou, D., C. Too, and G. Wallace, Synthesis and characterisation of polypyrrole/heparin composites. *Reactive and Functional Polymers,* 1999. *39*(1): p. 19-26.

[35] Sugahara, K., et al., Recent advances in the structural biology of chondroitin sulfate and dermatan sulfate. *Current opinion in structural biology,* 2003. *13*(5): p. 612-620.

[36] Hook, M., et al., Cell-surface glycosaminoglycans. *Annual review of biochemistry,* 1984. *53*(1): p. 847-869.

[37] Johansson, S., et al., Fibronectin-integrin interactions. *Front Biosci,* 1997. *2*: p. d126-d146.

[38] Ruoslahti, E., Fibronectin and its receptors. *Annual review of biochemistry,* 1988. *57*(1): p. 375-413.

[39] Gomez, N. and C.E. Schmidt, Nerve growth factor-immobilized polypyrrole: Bioactive electrically conducting polymer for enhanced neurite extension. *Journal of Biomedical Materials Research Part A,* 2007. *81*(1): p. 135-149.

[40] Lundin, V., et al., Control of neural stem cell survival by electroactive polymer substrates. *PloS one,* 2011. *6*(4): p. e18624.

[41] Svennersten, K., et al., Electrochemical modulation of epithelia formation using conducting polymers. *Biomaterials,* 2009. *30*(31): p. 6257-6264.

[42] Wan, A.M., et al., Electrical control of cell density gradients on a conducting polymer surface. *Chemical Communications,* 2009(35): p. 5278-5280.

[43] Binnig, G., C.F. Quate, and C. Gerber, Atomic force microscope. *Physical review letters,* 1986. *56*(9): p. 930.

[44] Hansma, P.K., et al., Tapping mode atomic force microscopy in liquids. *Applied Physics Letters,* 1994. *64*(13): p. 1738-1740.

[45] Lehenkari, P., et al., Adapting atomic force microscopy for cell biology. *Ultramicroscopy,* 2000. *82*(1): p. 289-295.

[46] Heinz, W.F. and J.H. Hoh, Spatially resolved force spectroscopy of biological surfaces using the atomic force microscope. *Trends in biotechnology,* 1999. *17*(4): p. 143-150.

[47] Butt, H.-J., B. Cappella, and M. Kappl, Force measurements with the atomic force microscope: Technique, interpretation and applications. *Surface science reports,* 2005. *59*(1): p. 1-152.

[48] Higgins, M., et al., Electrochemical AFM. *Imaging & Microscopy,* 2009. *11*(2): p. 40-43.

[49] Avlyanov, J.K., J.Y. Josefowicz, and A.G. MacDiarmid, Atomic force microscopy surface morphology studies of 'in situ'deposited polyaniline thin films. *Synthetic metals,* 1995. *73*(3): p. 205-208.

[50] Silk, T., et al., AFM studies of polypyrrole film surface morphology II. Roughness characterization by the fractal dimension analysis. *Synthetic metals, 1998. 93*(1): p. 65-71.

[51] Silk, T., et al., AFM studies of polypyrrole film surface morphology I. The influence of film thickness and dopant nature. *Synthetic metals,* 1998. *93*(1): p. 59-64.

[52] Higgins, M.J., S.T. McGovern, and G.G. Wallace, Visualizing dynamic actuation of ultrathin polypyrrole films. *Langmuir,* 2009. *25*(6): p. 3627-3633.

[53]   Sharifi-viand, A., M.G. Mahjani, and M. Jafarian, Determination of fractal rough surface of polypyrrole film: AFM and electrochemical analysis. *Synthetic Metals*, 2014. *191*: p. 104-112.

[54]   Paramo-Garcia, U., N. Batina, and J. Ibanez, The effect of pH on the morphology of electrochemically-grown polypyrrole films: an AFM study. *Int. J. Electrochem. Sci*, 2012. *7*: p. 12316-12325.

[55]   Gelmi, A., M.J. Higgins, and G.G. Wallace, Physical surface and electromechanical properties of doped polypyrrole biomaterials. *Biomaterials*, 2010. *31*(8): p. 1974-1983.

[56]   Molino, P.J., et al., Influence of biopolymer loading on the physiochemical and electrochemical properties of inherently conducting polymer biomaterials. *Synthetic Metals*, 2015. *200*: p. 40-47.

[57]   Hwang, B.J., R. Santhanam, and Y.L. Lin, Nucleation and growth mechanism of electroformation of polypyrrole on a heat-treated gold/highly oriented pyrolytic graphite. *Electrochimica acta*, 2001. *46*(18): p. 2843-2853.

[58]   Longo, G., et al., Morphological characterization of innovative electroconductive polymers in early stages of growth. *Surface and Coatings Technology*, 2012. *207*: p. 286-292.

[59]   Innocenti, M., et al., In situ atomic force microscopy in the study of electrogeneration of polybithiophene on Pt electrode. *Electrochimica acta*, 2005. *50*(7): p. 1497-1503.

[60]   Marandi, M., et al., AFM study of the adsorption of pyrrole and formation of the polypyrrole film on gold surface. *Electrochemistry Communications*, 2010. *12*(6): p. 854-858.

[61]   Cohen, Y., M. Levi, and D. Aurbach, Micromorphological dynamics of polypyrrole films in propylene carbonate solutions studied by in situ AFM and EQCM. *Langmuir*, 2003. *19*(23): p. 9804-9811.

[62]   Passeri, D., et al., Indentation modulus and hardness of polyaniline thin films by atomic force microscopy. *Synthetic Metals*, 2011. *161*(1): p. 7-12.

[63]   Passeri, D., et al., Indentation modulus and hardness of viscoelastic thin films by atomic force microscopy: A case study. *Ultramicroscopy*, 2009. *109*(12): p. 1417-1427.

[64]   Chiarelli, P., A. Della Santa, and A. Mazzoldi. Actuation properties of electrochemically driven polypyrrole free-standing films. in ICIM'94- International Conference on Intelligent Materials, 2 nd, Colonial Williamsburg, VA. 1994.

[65]   Pelto, J.M., et al., Surface properties and interaction forces of biopolymer-doped conductive polypyrrole surfaces by atomic force microscopy. *Langmuir*, 2013. *29*(20): p. 6099-6108.

[66]   Hassarati, R.T., et al., Stiffness quantification of conductive polymers for bioelectrodes. *Journal of Polymer Science Part B: Polymer Physics*, 2014. *52*(9): p. 666-675.

[67]   Shanmugham, S., et al., Polymer nanowire elastic moduli measured with digital pulsed force mode AFM. *Langmuir*, 2005. *21*(22): p. 10214-10218.

[68]   Lee, S., et al., Fabrication and mechanical properties of suspended one-dimensional polymer nanostructures: polypyrrole nanotube and helical polyacetylene nanofibre. *Nanotechnology*, 2006. *17*(4): p. 992.

[69]   Cuenot, S., S. Demoustier-Champagne, and B. Nysten, Elastic modulus of polypyrrole nanotubes. *Physical Review Letters*, 2000. *85*(8): p. 1690.

[70] Cuenot, S., et al., Surface tension effect on the mechanical properties of nanomaterials measured by atomic force microscopy. *Physical Review* B, 2004. *69*(16): p. 165410.

[71] Suárez, M.F. and R.G. Compton, In situ atomic force microscopy study of polypyrrole synthesis and the volume changes induced by oxidation and reduction of the polymer. *Journal of Electroanalytical Chemistry,* 1999. *462*(2): p. 211-221.

[72] Smela, E. and N. Gadegaard, Surprising volume change in PPy (DBS): an atomic force microscopy study. *Advanced Materials,* 1999. *11*(11): p. 953-957.

[73] Smela, E. and N. Gadegaard, Volume change in polypyrrole studied by atomic force microscopy. *The Journal of Physical Chemistry B,* 2001. *105*(39): p. 9395-9405.

[74] Higgins, M.J., et al., Reversible shape memory of nanoscale deformations in inherently conducting polymers without reprogramming. *The Journal of Physical Chemistry B,* 2011. *115*(13): p. 3371-3378.

[75] Svirskis, D., et al., Evaluation of physical properties and performance over time of an actuating polypyrrole based drug delivery system. *Sensors and Actuators B: Chemical,* 2010. *151*(1): p. 97-102.

[76] Barisci, J.N., et al., Characterisation of the topography and surface potential of electrodeposited conducting polymer films using atomic force and electric force microscopies. *Electrochimica acta,* 2000. *46*(4): p. 519-531.

[77] Semenikhin, O.A., et al., Atomic force microscopy and Kelvin probe force microscopy evidence of local structural inhomogeneity and nonuniform dopant distribution in conducting polybithiophene. *The Journal of Physical Chemistry,* 1996. *100*(48): p. 18603-18606.

[78] O'Neil, K. and O. Semenikhin, Current-Sensing Atomic Force Microscopic Study of Doping Level Distribution in Doped Poly [2, 2'-Bithiophene*]. Electrochimica Acta,* 2014. *122*: p. 72-78.

[79] Nardes, A.M., et al., Microscopic understanding of the anisotropic conductivity of PEDOT: PSS thin films. *Advanced Materials,* 2007. *19*(9): p. 1196-1200.

[80] Hassenkam, T., D.R. Greve, and T. Bjornholm, Direct visualization of the nanoscale morphology of conducting polythiophene monolayers studied by electrostatic force microscopy. *Advanced Materials,* 2001. *13*(9): p. 631-634.

[81] Han, D.-H., H.J. Lee, and S.-M. Park, Electrochemistry of conductive polymers XXXV: Electrical and morphological characteristics of polypyrrole films prepared in aqueous media studied by current sensing atomic force microscopy. *Electrochimica acta,* 2005. *50*(15): p. 3085-3092.

[82] Park, J.G., et al., Electrical resistivity of polypyrrole nanotube measured by conductive scanning probe microscope: The role of contact force. *Applied Physics Letters,* 2002. *81*(24): p. 4625-4627.

[83] Samitsu, S., et al., Conductivity measurements of individual poly(3,4-ethylenedioxythiophene)/poly(styrenesulfonate) nanowires on nanoelectrodes using manipulation with an atomic force microscope. *Applied Physics Letters,* 2005. *86*(23): p. 233103.

[84] Hentschel, C., et al., *Conductance measurements of individual polypyrrole nanobelts.* Nanoscale, 2015.

[85] Saha, S., et al., Current–voltage characteristics of conducting polypyrrole nanotubes using atomic force microscopy. *Nanotechnology,* 2004. *15*(1): p. 66.

[86] Jiang, L., et al., Enhanced electrical conductivity of individual conducting polymer nanobelts. *Small*, 2011. *7*(14): p. 1949-1953.

[87] Cho, S.H. and S.-M. Park, Electrochemistry of conductive polymers 39. Contacts between conducting polymers and noble metal nanoparticles studied by current-sensing atomic force microscopy. *The Journal of Physical Chemistry B*, 2006. *110*(51): p. 25656-25664.

[88] Zlatanova, J., S.M. Lindsay, and S.H. Leuba, Single molecule force spectroscopy in biology using the atomic force microscope. *Progress in biophysics and molecular biology*, 2000. *74*(1): p. 37-61.

[89] Hinterdorfer, P. and Y.F. Dufrêne, Detection and localization of single molecular recognition events using atomic force microscopy. *Nature methods*, 2006. *3*(5): p. 347-355.

[90] Riener, C.K., et al., Simple test system for single molecule recognition force microscopy. *Analytica Chimica Acta*, 2003. *479*(1): p. 59-75.

[91] Schmitt, L., et al., A metal-chelating microscopy tip as a new toolbox for single-molecule experiments by atomic force microscopy. *Biophysical Journal*, 2000. *78*(6): p. 3275-3285.

[92] Higgins, M.J. and G.G. Wallace, Surface and biomolecular forces of conducting polymers. *Polymer Reviews*, 2013. *53*(3): p. 506-526.

[93] Wang, J. and A.J. Bard, On the absence of a diffuse double layer at electronically conductive polymer film electrodes. Direct evidence by atomic force microscopy of complete charge compensation. *Journal of the American Chemical Society*, 2001. *123*(3): p. 498-499.

[94] Campbell, S.D. and A.C. Hillier, Nanometer-scale probing of potential-dependent electrostatic forces, adhesion, and interfacial friction at the electrode/electrolyte interface. *Langmuir*, 1999. *15*(3): p. 891-899.

[95] Gelmi, A., M.J. Higgins, and G.G. Wallace, Quantifying fibronectin adhesion with nanoscale spatial resolution on glycosaminoglycan doped polypyrrole using Atomic Force Microscopy. *Biochimica et Biophysica Acta (BBA)-General Subjects*, 2013. *1830*(9): p. 4305-4313.

[96] Agnihotri, A. and C.A. Siedlecki, Adhesion mode atomic force microscopy study of dual component protein films. *Ultramicroscopy*, 2005. *102*(4): p. 257-268.

[97] Lee, J.-E., et al., Nanoscale Analysis of a Functionalized Polythiophene Surface by Adhesion Mapping. *Analytical chemistry*, 2014. *86*(14): p. 6865-6871.

[98] Gelmi, A., M. Higgins, and G.G. Wallace, Resolving Sub-Molecular Binding and Electrical Switching Mechanisms of Single Proteins at Electroactive Conducting Polymers. *Small*, 2013. *9*(3): p. 393-401.

[99] Gelmi, A., et al., Optical switching of protein interactions on photosensitive–electroactive polymers measured by atomic force microscopy. *Journal of Materials Chemistry B*, 2013. *1*(16): p. 2162-2168.

[100] Zhang, H., et al., Quantifying Molecular-Level Cell Adhesion on Electroactive Conducting Polymers using Electrochemical-Single Cell Force Spectroscopy. *Scientific Reports*, 2015 (accepted).

[101] Knittel, P., M. Higgins, and C. Kranz, Nanoscopic polypyrrole AFM–SECM probes enabling force measurements under potential control. *Nanoscale,* 2014. *6*(4): p. 2255-2260.

[102] Ando, T., et al., High-speed AFM and nano-visualization of biomolecular processes. *Pflügers Archiv-European Journal of Physiology,* 2008. *456*(1): p. 211-225.

In: Innovations in Nanomaterials   ISBN: 978-1-63483-548-0
Editors: Al-N. Chowdhury, J. Shapter, A. B. Imran   © 2015 Nova Science Publishers, Inc.

Chapter 4

NANOMATERIALS IN CELL IMAGING

Fulya Ekiz Kanik[1], Alexandra Sneider[2], Prakash Rai[1,2], Dilek Odaci Demirkol[3] and Suna Timur[3,]*

[1]University of Massachusetts Lowell, Biomedical Engineering and Biotechnology
Program, Lowell, MA, US
[2]University of Massachusetts Lowell, Department of Chemical Engineering,
Lowell, MA, US
[3]Ege University, Faculty of Science, Department of Biochemistry,
Bornova, Izmir, Turkey

ABSTRACT

The application of nanoparticles in biomedicine offers excellent prospects for the progress of novel non-invasive approaches for the diagnosis and treatment of various diseases. These materials encompass a wide range of substances, both organic (e.g., dendrimers, liposomes, polymers, carbon nanotubes, graphene) and inorganic materials (e.g., quantum dots, metal oxides). Biofunctionalization with the specific biomolecules as well as targeting ligands allows adaptation of nanoparticles to the various cell imaging and therapy applications (e.g., photodynamic therapy, neutron capture therapy and magneto therapy). In this chapter, we provide an overview of the different types of nanoparticles used in biomedical applications, including synthetic approaches as well as targeting strategies explored for their application in cell imaging, diagnosis and therapy.

1. INTRODUCTION

Recent advances in synthesis and functionalization of nanoparticles have brought a significant increase in their biomedical applications including cell imaging, gene and drug delivery, targeted therapy as well as sensing of biomarkers etc. In this chapter, synthesis, bioconjugation and applications of various nanoparticles in cell imaging will be reviewed.

---

[*] Corresponding Author address: Email: sunatimur@yahoo.com.

Targeting strategies and biofunctionalities of the nanoparticle surfaces will be also discussed in detail. Common nanoparticles in cell imaging include fluorescent particles such as gold nanoparticles and quantum dots for optical imaging and magnetic particles for MRI. One major advantage of nanoparticles is their potential use as non-invasive diagnostic tools. The other one is the capacity to combine multimodalities on a sensing material, allowing higher sensitivity and deeper insight gain into *in vivo* processes. Nanomaterials are also important as vehicles for drug-delivery enabling imaging guided therapy which is called as "see and treat" to "detect and prevent" [1]. Multi-functionality of these smart materials enables the precise, less-invasive diagnosis and theranostic methodologies. Figure 1 summarizes the aspects of multifunctional nanoparticles in cell imaging and medicine.

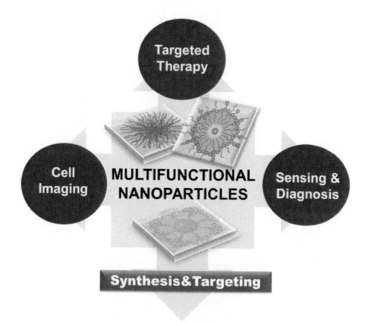

Figure 1. State of the art of multifunctional nanoparticles in medicine.

2. MAJOR CLASSES OF NANOMATERIALS USED AS CELL IMAGING PROBES

The types of commonly used nanomaterials are liposomes, dendrimers, carbon nanotubes (CNTs) and graphene, gold nanoparticles (AuNPs), quantum dots (QDs) and magnetic nanoparticles (Figure 2). Nanomaterials with their unique and adaptable properties and effectiveness overcome many problems such as toxicity issues making them ideal in cellular and molecular imaging. These novel and cutting-edge nanostructures have exhibited adequacy for use in applications in imaging and diagnosis. As a platform, nanomaterials are convenient for developing imaging agents due to their modifiable surfaces for targeting and stability purposes and physicochemical properties for tunable plasma circulation. A comprehensive examination of these nanostructures is given in the following sections.

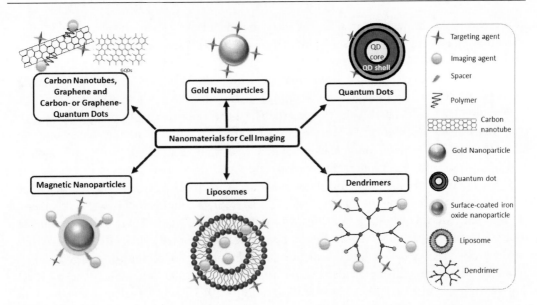

Figure 2. Structural representations of nanoparticle classes; possible main components of typical nanocarriers.

a. Liposomes

Liposomes are attractive materials as nanocarriers both for imaging and drug delivery purposes. Due to not only their biocompatibility and biodegradability, but also their ability to encapsulate many agents and easy modification, liposomes are strong candidates for delivery of imaging agents and drugs. Liposomes are self-closed, spherical, bilayer vesicles made up of phospholipids with an aqueous core. They have ability to encapsulate both hydrophilic and hydrophobic agents and drugs. These encapsulations can be formed covalently or non-covalently depending on the cargo. Moreover, these nanostructures are applicable for co-delivery of diagnostic and therapeutic agents with the help of their physicochemical properties. The clinical applications of liposome formulations are well-known and in development for more than 30 years and some of them are approved by regulatory agencies for delivery of range of agents [2-4].

Through tunneling of surface chemistry, kinds of different liposomes can be prepared depending on the purposes meanwhile surface components and physicochemical properties such as size, surface charge are manipulated. Liposomes can be used in the delivery of both hydrophobic and hydrophilic agents. Agents can be either loaded in the aqueous core or in the lipid membrane or both at the same time. Water-soluble agents are loaded in the aqueous compartment and lipid-soluble agents are integrated in the lipid bilayer [5]. The lipid bilayers contain hydrophilic glycerol-phosphate-alcohol head groups of phospholipids on both outer and inner shell of the liposomes. Compared other micelle type nanostructures, lipid bilayer allow charged molecules or hydrophilic drugs to be embedded in the interior core with the help of this special design of phospholipids. Hence, the pharmaceutical agents incorporated in the liposomes can be stored safely without any effect of exterior conditions and release when needed. Moreover, straightforward production of liposomes also makes them advantageous and capable to upscale. Production of liposomes using extrusion technique utilizing filters in certain pore sizes in nanometer size was also considered as a great advance in liposome production other than sonication or homogenization [6, 7]. Liposomes can be prepared as multilamellar vesicles with a size range from 500 to 5000 nm and formed by several bilayers, small unilamellar vesicles around 100 nm in size with a single bilayer and large unilamellar vesicles with a size range from 200 to 800 nm [2].

The problems associated with liposome delivery and drug release in preclinical *in vivo* trials were solved one by one. For instance, changing the fluidity of the liposomes with the introduction of cholesterol, drug leakage and release were controlled [8-10]. As another challenging problem, the primary liposomes were rapidly cleaned from circulation in the body by uptake into the cells of the mononuclear phagocyte system (MPS) such as macrophages, mainly in the liver and spleen [11-13]. Being a solution to this serious problem, the liposomes were covered or "stealth" with polyethylene glycol (PEG) polymer and the surface hydrophilicity was increased; hence, liposomes were not only sterically stabilized, but also long-circulated with increased half-life and reduced clearance into the MPS [14-17]. Soon after conquering all the challenges, the first PEGylated and doxorubicin encapsulated liposome (Doxil, the first FDA-approved nano-drug (1995)) was developed and used in clinical trials for the treatment of Kaposi's sarcoma [18]. Starting with Doxil, since 1995 there are other liposomal formulations for many drugs approved by FDA and in the markets. These show the applicability, reliability and success of the liposomal drug carrier systems. Moreover, with targeting technology, liposomes can selectively sent to related tissue and up taken by those cells. This increases the success of imaging performance, therapy and decreases the toxicity of the liposomes and release of agents to the other tissue in the body.

b. Dendrimers

Dendrimers are hyperbranched nanostructured materials. Their size and number of branching can be controlled by controlling the number of generations during polymerization. The branches of these polymeric macromolecules can be functionalized with imaging agents as well as targeting moieties and drugs. Since the end groups are well-oriented and open to reactions, there are many possibilities for functionalization.

Dendrimers can be designed using either a divergent or convergent way [19]. Through divergent approach, the synthesis of the macromolecular structures starts from a multivalent core unit. Consecutive layers which are the units constituting the branches in the structure are added to the main core. Each layer of units are added in a stepwise order till the exterior surface [20]. On the other hand, through the convergent approach, first the exterior part of the macromolecule is designed and the synthesis starts from the building units of the surface groups. After having different-sized branches, finally they are combined and whole structure is synthesized [20, 21]. This approach is advantageous when a different branched or nonsymmetrical dendrimers are needed. Moreover, since the number of synthesis steps and coupling reactions are less when compared to divergent approach, structural defects are minimized which provides more structural control.

Well-organized, monodispersed and compact molecular structures of dendrimers with cavities provide a range of possible loading of agents with great loading efficiency. Functional end groups or interior cavities during polymerization processes can be used for packing of imaging agents or therapeutics. They can be easily conjugated to imaging agents.

Tectodendrimers are class of dendrimers bearing different groups of agents on different units of the dendrimer complex. These units can contain imaging and targeting agents, agents for diagnosis of different stages of diseases and therapeutic molecules [22]. With the help of small deviations in the structures and surface properties, many derivatives of them can be achieved. Through amino acid substitution, polyethylene glycol addition and acetylation steps can also improve biocompatibility and lower the immune system response in the biological systems [23]. Moreover, their high water-solubility is a big advantage for cell imaging studies. In addition to rapid functionalization, they can be rapidly cleared from the systems due to their smaller size (< 5 nm) which makes them good candidate as an imaging agent carrier without any biodegradability issues [24].

c. Carbon Nanotubes, Graphene and Carbon-, Graphene-Quantum Dots

Carbon nanotubes (CNTs) are one-dimensional cylindrical tubes made of purely carbon in a hexagonal lattice arrangement with remarkable mechanical, electrical and optical properties. Consequently, they are one of the most studied and engineered nanostructures in biomedical applications. Nanotubes can be either formed as single-walled (SWCNT), rolled-up of one layer with an internal diameter of 1-2 nm or multi-walled (MWCNT), concentric multi layers of graphene with a diameter of 2-25 nm [22, 25]. Due to their greater strength and stability, they are considered as stable nanocarriers for imaging agents along with the drugs.

In the synthesis of both CNT types, electrical arc discharge, laser ablation and chemical vapor deposition (CVD) methods can be used. All these three methods enable large-scale production of CNTs with a low-cost efficiency [25]. Their stable structures can be tuned either with covalent or non-covalent modifications. Non-covalent modifications can be done by modifying sidewalls of nanotubes with additives such as polymers. In addition, inner core cavity of nanotubes can be also modified with certain molecules or structures. Depending on the purpose, body or ends of CNTs can be tailored while keeping the whole structure stable and functioning. High surface area and mechanical integrity, easy functionalization of sidewalls with organic or inorganic materials lead them to become an advantageous material to be engineered for potential biomedical applications. SWCNTs and MWCNTs can be functionalized with a wide range of biomolecules including fluorescent probes, drugs and antibiotics [26]. Functionalization is used for both targeting the cells and also for imaging purposes. Unmodified, pristine CNTs are water-insoluble forms and they are considered as more toxic *in vitro* compared to modified and water-dispersible functionalized ones [22]. It has been showed that functionalized CNTs are more likely to be uptaken intracellularly by wide range of cells with a nano-needle-like character passively penetrating cell membranes of the cells [27].

Graphene is recently merged as a captivating material with many fascinating properties such as electrical and optical properties and biocompatibility [28]. Graphene, unlikely CNTs, is two dimensional single or fewer layers of sp^2 hybridized carbon atoms [29]. Due to its fascinating properties, graphene has become an important tool in from designing electronic and optoelectronic devices, solar cells to medical devices, medical imaging and drug delivery. The strong carbon-carbon bonding in its structure makes graphene a tough material for many applications. On the other hand, relatively weak interlayer Van der Waals interactions turn it to be a soft material compared to other forms of carbon such as diamond. Moreover, these bonding make it different than tubular forms of carbon with powerful properties. Since graphene contains π-conjugated aromatic structure and free π electrons in addition to reactive sites, it is feasible for various kind of modifications as well as attachments, which makes graphene considerable for cell targeting, imaging and drug delivery applications.

Depending on number of layers and chemical modification, graphene nanostructures can be roughly classified as single layer, bilayer and multi-layer graphene and graphene oxide derivatives [30]. Single layer of carbons makes single layer graphene which can be synthesized via repeated mechanical exfoliation or controlled growth on substrates by chemical vapor deposition [29, 31]. Since it exhibits a highly reactive surface, it is hard to isolate and keep either in suspension or gas phase. Thus, multi-layer and graphene oxides forms of it are more common in many applications. On the other hand, because of attractive properties of graphene oxide (GO), it is more preferably used in biomedical applications. GO is perfectly biocompatible [32] and easily and stably dispersed in aqueous solutions [33]. Owing to its large surface area and π-conjugated structure, many drugs and imaging agents can be attached to the sides of it using the functional ends or stacked on the surface of it via Van der Waals interactions or π-π stacking using unreacted areas containing free π electrons. Furthermore, the presence of epoxide, hydroxyl and carboxyl functional groups on GO enables not only various modifications but also attachment of several agents and drugs via strong hydrogen binding. GO is achieved via chemical modifications of graphene by high oxidation it ending up with carboxylic acid, epoxide and hydroxyl groups on the single atom-thick graphene layer. While carboxylic acid groups afford colloidal stability, epoxide and hydroxyl groups contribute surface interactions such as hydrogen bonding with their polar characteristics [34]. Through rough oxidation of crystalline graphite, multi-layer GO can be produced. With repeated treatment and extreme conditions, single layer GO can be achieved.

Carbon-based quantum dots including graphene quantum dots (GQDs) and carbon quantum dots (CQDs, C-dots or CDs) are a novel class of carbon nanomaterials below 10 nm. They exhibit strong fluorescent performance, biocompatibility and low cytotoxicity due to their small size and nanometer dimensional properties [35]. Moreover, they can also be used as fluorescent inks which also makes them a good candidate for imaging and staining [36, 37]. They were first obtained during the purification of single-walled carbon nanotubes through preparative electrophoresis in 2004 [38]. Then, carbon dots were synthesized via laser ablation of graphite powder and cement in 2006 [39]. In time, several ways for synthesis of CDs were discovered such as electrochemical or acidic oxidation, microwave and ultrasonic ways. However, CDs generally exhibit lower fluorescence quantum-yield when they are compared with the conventional semiconductor QDs. In order to improve their fluorescence emission, doped CDs were used in different applications [40].

Carbon-based quantum dots with attractive properties have gradually become a rising star as a novel nanocarbon material because of their abundant and inexpensive nature. Wide attention has been focused on carbon-based quantum dots due to their good solubility and strong luminescence [41]. On the other hand, it is a fact that heavy metals are highly toxic even at moderately low levels, which causes problems in the clinical studies. This prompted the creation of CQDs to replace semiconductor quantum dots due to their low toxicity, biocompatibility as well as chemical inertness in addition to having similar fluorescence properties [42].

Moreover, graphene-quantum dots also started to take place in biomedical applications such as bioimaging [43], drug delivery [44], and biosensors [45]. Due to being sp^2 carbon-based material and easy generation of graphene oxides, they are attractive for covalent or physical modifications, which makes them applicable and interesting for cell imaging and drug delivery applications. However, their cytotoxic effects are not yet been dissolved clearly. Their biodistribution and potential toxicity is still under investigation [46].

d. Gold Nanoparticles

Colloidal nanosized gold nanoparticles (Au NPs) have been used since ancient times due to their distinct optical, therapeutic and thermal properties. Au NPs are great candidates for imaging, targeting, drug delivery and therapy applications by virtue of their unique physicochemical properties. First of all, their synthesis is well-established and easy to conduct. This brings many advantages along with time and cost efficiency, well-applicability and reliability. Secondly, they are inert, non-toxic and biocompatible in applicable sizes [47]. They can be prepared in a size range of 1.0 to 150 nm. Moreover, they are well-qualified for many kinds of functionalization depending on the applications. Especially, thiol chemistry is very useful in the case of Au NP functionalization.

As their usage, synthesis of Au NPs is also confirmed and straightforward with well-established protocols. They can be easily synthesized in aqueous solutions. Commonly, the route to achieve gold nanoparticles starts with reducing of gold salts such as $AuCl_3$ with certain reducing agents such as citric acid to have gold crystals via nucleation and gold ions turn to nanoparticles. Moreover, a stabilizing agent is to have colloidal stable, non-aggregating nanoparticle solution in long term [48]. Commonly, in aqueous media, reducing agent also serves as stabilizing agent. Depending on growing process of gold nanoparticles, different shapes and sizes can be achieved [49]. Moreover, easy characterization using UV-vis spectrophotometry, TEM or FT-IR techniques enables synthesis and characterization processes more accessible and facile.

Au NPs solutions contain stabilizing agents; hence, the nanoparticles are mostly surrounded by these stabilizing molecules. Nevertheless, modification of Au NPs is still considered as easier compared to other nanomaterials. Due to the no need for extra chemicals and harsh conditions, thiol chemistry is generally preferred and applied in functionalization of gold nanoparticles. Utilizing the high affinity of gold towards thiol containing materials, attachment or conjugation of thiol moiety containing groups is found as accessible, simple and efficient. As a result, gold nanoparticles can be applied in many fields including biomedical areas with the help of easy-modification. Biological ligands, targeting moieties, drugs, imaging agents, stabilizing agents or immunostains can be easily attached to nanoparticles.

In addition to their easy-functionalization, Au NPs can behave as different probes and materials in biomedical applications. Other than their labelling ability and behaving as contrast agent, they can also function as heat source and be used in manipulating tissues around them [50]. Since Au NPs have the ability to transfer thermal energy during controlled heating or photo-induced heating, along with imaging, they can also be utilized in photothermal therapy.

e. Quantum Dots

Quantum dots (QDs) are inorganic nanocrystals and fluorescent semiconductors. Their size changes in a range of 2-10 nm [51]. They can absorb the light at a broad range of wavelengths and emits at a narrow emission band generally in visible or near infrared region of the spectrum. They can be excited with a single wavelength of laser. As they are decreased n the size, they can absorb more light which increases their ability to absorb as means of the quantum confinement effect having the size of the crystal smaller than Bohr exciton radius bringing fascinating electronic and optical characteristics [51].

The structure of QDs includes of an inorganic core, an organic shell and an additional organic coating. The inner inorganic core is responsible for the color of emitted light. Organic coating can include various conjugated materials such as biomolecules or targeting agents. Due to their powerful and stable fluorescence property, they can easily be used as imaging probes. Best suitable QDs for biomedical applications consist of CdSe core and ZnS as inorganic shell. [41]. ZnS shell increases the stability and efficiency of overall structure by protecting the core from oxidation, preventing the interference of core material into the surrounding solution [52]. Colloidal QDs can be made of ZnS, CdS, ZnSe, CdTe and PbSe. Depending on the application, correct match of ingredients, thickness of the shells and synthesis steps should be considered and clarified. QDs can be achieved by high-temperature growth solvents/ ligands along with pyrolysis of organometallic precursors to have native CdSe QDs with high efficiency [53]. They can be further coveted with semiconducting materials such as ZnS. However, after having QDs via high-temperature route, the nanomaterials are water-insoluble and not applicable to biomedical applications. They require additional steps to transfer in aqueous media through specific conjugations.

QDs are useful for diagnostic and therapeutic purposes. They can behave as highly sensitive probes for diagnosis. However, due to the potential toxicity and elimination problems from the organisms, their use in clinical applications is limited. The modifications to overcome these problems results in raise in the size of nanoparticle whereas bring the clearance problems due to the bigger size resulting in accumulation and toxicity.

f. Magnetic Nanoparticles

Magnetic nanoparticles (MNPs) including metallic, bimetallic and super paramagnetic iron oxide nanoparticles have become significant due to their magnetic property enabling them to be tracked in radiology, magnetic resonance imaging (MRI) [54]. Among these magnetic nanoparticles, super paramagnetic iron oxide (SPIONs) nanoparticles are commonly preferred in biomedical applications due to their innocuous character and non-toxicity [55]. So far, many of super paramagnetic iron oxide nanoparticles are in clinical trials and some of them are already approved for clinical uses as imaging agents such as Lumiren® for bowel imaging [56] and Combidex® for lymph node metastasis imaging [57]. Controlling size, shape, coating and additional modifications for targeting and improvement of their physicochemical properties can make these nanoparticles more target-oriented and better imaging agents.

Magnetic nanoparticles can be made of pure iron and cobalt metals, alloys, iron oxides including magnetite [58]. Moreover, iron oxide nanoparticles can be doped with different cations such as Ni, Co to make the nanoparticles more sufficient as means of contrast. However, these additions make them possibly more toxic for clinical applications. They acquire enhanced coatings. Super paramagnetic iron oxide nanoparticles can be synthesized either by mechanical attrition or chemical synthesis. For better size and shape distribution, chemical synthesis pathways are more preferred [59]. Co-precipitation of Fe^{2+}/Fe^{3+} solutions is one of the mostly employed techniques having an efficient nucleation and growth mechanisms resulting in highly monodisperse nanoparticles.

SPIONs are useful in both diagnosis and therapy. They can be used in MRI as contrast agent and additionally, using thermal therapy they have ability to kill malignant cells or tissue when they are directed to a target tissue or disease site [59]. The surface of the nanoparticles can be tailored with functional units and targeting moieties to increase biocompatibility and biodegradability. Furthermore, since when the SPIONs are degraded in the organisms, they leak free iron ions, these ions easily get involved in body's iron pathways and degraded through these recycle mechanisms. This makes them attractive for all purposes.

3. TARGETING STRATEGIES

Tumors exist in a microenvironment [60, 61]. Cancerous cells develop the ability to undergo ligand independent signaling, limitless replication, and sustained angiogenesis [62-65]. These properties allow cancerous cells to continue to proliferate and metastasize inside the host. Unmediated delivery of chemotherapeutic drugs intravenously may result in toxicity to normal tissue due to poor solubility in water. The goal of nanoparticle mediated drug delivery is to maximize the effectiveness and minimize the toxicity/lethality of the drug [61, 66]. This section is a high level overview of targeting strategies. The term nanoparticles will refer to organic and inorganic nanomaterials, including liposomes, CNTs, QDs, graphene, gold nanoshells, dendrimers, etc. [61, 66].

Nanoparticles are administered to the host with the aim of completing multiple functions including drug/gene delivery (therapeutic potential) and imaging (diagnostic potential). Through variations in composition, size, and surface charge, nanoparticles assist in improving drug delivery to cancerous cells by increasing drug solubility, extending circulation time, reducing drug degradation during administration, and decreasing toxicity [67-70]. Surfactants, which can limit the toxicity and increase solubility of the drugs, are often used for drug delivery [68, 71-73]. Polyethylene glycol (PEG) is a commonly used surface modifier that improves solubility of the nanoparticle/drug complex in the host by reducing macrophage uptake [61, 68, 69, 74-76]. N-(2-hydroxypropyl) methacrylamide (HPMA) is also used to facilitate nanoparticle drug delivery [69].

a. Therapeutic Potential

Nanoparticles are functionally designed to target tissue passively or actively. The following is a discussion of both methods.

i. Passive Targeting

Passive targeting of nanoparticles utilizes the Enhanced Permeability and Retention (EPR) effect. EPR dependent nanoparticles rely on the extravasation of nanoparticles from blood circulation and the increased permeability of cancerous vasculature to accumulate and circulate the nanoparticles within the tumor [61, 66, 67, 73, 77, 78]. The success of utilizing the EPR effect relies on maximizing the circulation half-life of the nanoparticles while reducing side effects [79].

Successful application of the EPR effect may be observed in solid tumors [80]. Experimentally, breast cancer treatment options include passively targeted, PEGylated liposomal doxorubicin [81-83]; and Genexol-PM, polymeric micelle formulated paclitaxel [84, 85]. Human colorectal cancer cell line HT-29 xenograft has shown improved antitumor activity and reduced paclitaxel toxicity with the accumulation of NK105, a paclitaxel loaded micellar nanoparticle [86].

Despite the successful experiments above, the effectiveness of passively targeted nanoparticles is dependent on multiple factors including the drug, type of cancer, nanoparticle formation, delivery method, and drug assay [66, 73]. Experimental results show that on average less than 5% of the administered passively targeted nanoparticles reach the tumor [66, 73, 87-89]. Additionally, without ligand specific targeting, the nanoparticles indiscriminately accumulate in major organs and in the tumor tissue [67, 73].

ii. Active Targeting

Active targeting of nanoparticles requires the addition of a ligand specific to the cancerous tissue in an effort to reduce accumulation of the drug loaded nanoparticles in normal tissue. Ligands are selected, with varying degrees of specificity, based on cancerous overexpression of certain receptors. Once introduced into the body, the ligand binds to the target receptor, forming an endosome, and undergoes receptor-mediated endocytosis [68, 90]. The endosome then travels to target organelles, increases in acidy, and activates the lysozymes, which then release the drug into the cytoplasm [90].

Ligands include monoclonal antibodies, folic acid, transferrin, antibodies against transferrin receptors, oligonucleotides, peptides, glycoproteins; and can target Vascular Endothelial Growth Factor (VEGF), Epidermal Growth Factor (EGF), Folate receptors, etc. [67, 69, 78, 91-95]. Folic Acid is used to target folate receptors expressed in ovarian, breast, brain, kidney, myeloid cells, and lung cancer [95]. VEGF ligands, which bind to VEGF tyrosine kinase receptors, have been shown to inhibit pathological angiogenesis [64, 65]. In prostate cancer, Histone Deacetylase Inhibitors (HDACi) combined with nonsteroidal anti-androgen moieties, which bind to the androgen receptor, are able to decrease toxicity and improve cancerous tissue targeting in both hormone dependent and independent cancer cell lines [96]. Quinone warheads with estrogen receptor targeted ligands (ER+) were able to increase cytotoxicity and apoptosis in breast cancer cells [97]. Bergstrom et al. utilized T-43 monoclonal antibody as an inhibitor to target T-24 bladder tumor cancer cell lines *in vitro* using MGH-U1, causing photodynamic therapy (PDT) induced apoptosis of cancerous cells [98].

b. Comparison of Two Targeting Strategies

These targeting strategies simultaneously affect tumor tissue and the tissue microenvironment. The success of passively targeted nanoparticles depends on the circulation time of the particles in the blood stream to allow for accumulation. An advantage of passively targeted nanoparticles is the cost and ease of mass production [99]. Concerns with passive targeting include cytotoxicity as a result of undefined locations for accumulation, and optimizing the circulation time while providing an effective drug dose. Actively targeting cancer receptors does not eliminate normal tissue toxicity, but the PDT experiments demonstrate the increased antitumor activity as a result of targeting ligands to cancer expressed receptors. Concerns with active targeting include decreased blood circulation time and tumor penetration; and lysosomal degradation at administration [100]. Issues with both passive and actively targeted nanoparticles include maximizing accumulation in the capillaries of tumorous tissue over the liver and spleen; reducing degradation upon administration; and increasing circulation while reducing toxicity [101].

Specific Photodynamic therapy (PDT) experiments illuminate direct differences in passive verse active targeting of the same cancerous tissue. PDT, a treatment method utilizing light irradiation between 650-850 nm to noninvasively target cancerous tissue, relies on nanoparticles to reduce cytotoxicity and third generation photosensitive agents to photochemically induce cell apoptosis in tumors [67, 102-107]. The following experiments illustrate the dramatic increase in PDT induced antitumor activity using actively targeted nanoparticles. In a study targeting osteosarcoma, while both targeted and non-targeted Curcumin-C6 liposomes generated a cytotoxic effect 1.5 times more than Curcumin liposomes, the folate PEGylated targeted Curcumin-C6 liposomes also demonstrated a significant decrease in the tumor size [76]. The Bhatti et al. study revealed that targeting of HER-2 breast cancer cell lines was improved eight fold when using targeted versus non-targeted verteporfin in conjunction with single-chain Fvs with pyropheophorbide-α [108]. Gijsens et al. experimentally determined that transferrin receptor targeted aluminum phthalocyanine tetrasulfonate (AlPcS4) was ten times more cytotoxic than non-targeted AlPcS4 in HeLa cells [109].

4. FUNCTIONALIZATION, BIOCONJUGATION AND BIOMOLECULES

Nanomaterials in their native forms are generally hard to dissolve or disperse homogeneously in aqueous media or they attract immune response in the body and cause toxicity. In order to make them applicable and useful for biomedical applications or to use them as markers, modification and conjugation is a huge step towards these goals. For these purposes, while designing nanomaterials, they are functionalized with various functional groups reactive towards the other molecules to be able to modify the nanomaterials with diverse biomolecules, contrast agents ligands, antibodies, peptides, drugs etc. Moreover, with the help of bioconjugation, multifunctional or hybrid materials can be achieved. Hence, dual properties or all-in-one type nanomaterials can be accomplished.

Nanomaterials can be modified depending on their surface characteristics, shape, structure and physicochemical properties. Covalent and non-covalent conjugation approaches can be utilized. In the presence of different functional groups such as carboxylic acid ends or thiol groups, different conjugation chemistries can be preferred through covalent binding and conjugation to the nanoparticles is performed easily using functional groups of the biomolecules. On the other hand, if the materials consist aromatic units available for stacking, then, units can be conveniently conjugated or adsorbed each other through π-π stacking interactions as non-covalent binding. Most commonly used covalent conjugation routes are carbodiimide-succinimide, EDC-NHS (N-(3-dimethylaminopropyl)-N-ethylcarbodiimide hydrochloride - N-hydroxysuccinimide) and thiol chemistries. EDC-NHS chemistry is used for conjugation of carboxyl and amine groups. The conjugation starts with the activation of carboxyl ends with the addition of EDC and the reactive intermediate is stabilized with the help of NHS. It is finalized with amide formation which is a standard and common conjugation [110]. This chemistry is useful for bioconjugations due to the presence of high amount of amines in the biomolecule structures such as peptides. In addition to EDC-NHS chemistry, thiol-Au and thiol-maleimide conjugations is other most common and used chemistries in bioconjugations. SH- moiety containing biological units can be easily attached to gold containing structures due to the high affinity in Au-S conjugation. Furthermore, maleimide group-functionalized materials can be coupled with sulfhdryls easily to achieve a conjugation since this is a one direction, single-step reaction [110].

As examples of liposome modifications, dual conjugations can be performed to enhance targeting efficiency. Tang et al. designed liposomes for better targeting and cell uptake in which they decorated the same liposomes with conjugation of both specific ligand transferrin and cell-penetrating peptide TAT [111]. They used covalent conjugation and synthesized transferrin and TAT conjugated cholesterol to be used in the body of the liposomes. Dendrimers are considerably available structures for many kinds of reactions and conjugations. During the synthesis of dendrimers, many additional conjugations can be performed simultaneously. For example, a fluorescent dye (FITC) and targeting moiety, folic acid, can be attached covalently using simple EDC chemistry to dendrimer unit having two important component on the nanostructure [112]. Then, through attachment of complementary DNA oligonucleotides to these structures, it is possible to link them together to produce clustered molecules for better targeting and uptake by the cells [112]. Carbon nanotubes can be modified and functionalized covalently such as via oxidation producing carboxyl groups on the ends and non-covalently such as via π-π stacking or hydrophobic interactions. Their large surface area enables attachment of multi cargos at a time using either covalent or non-covalent interactions. Conjugation of several polymers like polyethylene glycol (PEG) for long circulation in blood stream is another possible conjugation for these unique structures [113]. Moreover, quantum dots are also nanoparticles available for various bioconjugations where a broad range of conjugations can be employed such as carbodiimide or thiol-maleimide coupling, avidin-biotin chemistry [114]. For instance, specific dyes such as amine-terminated rhodamine dyes can be attached to carboxyl-terminated QDs via EDC coupling and FRET-based sensing studies can be performed [115]. Another study reported that magnetic nanoparticles are also convenient to be functionalized to enhance their superiority with organic coatings such as PEG, chitosan or conjugation through reactive ligands such as epoxide, maleimide, hydrazide to attach dyes, peptides or antibodies via stable or cleavable linkages [58].

5. NANOTOXICOLOGY

Studies of the potential threats of nanostructured materials have been extensively carried out using cell models and a range of *in vitro* methodologies. The *in vitro* cell-based approaches are more attractive for nanomaterial, especially due to ethical aspects and the expense of animal testing. The requirement for more toxicological information concerning the risks of exposure to nano-based structures is reflected in the increasing number of studies oriented toward the evaluation of nanotoxicity and clarification of the mechanisms underlying these harmful effects to health. To evaluate cytotoxic activity of nanomaterials various *in vitro* methods are available in the literature. As for the cytotoxicity activity, testing approaches used are mainly based on the 3-(4,5-dimethylthiazol-2-yl)-2,5-diphenyl-tetrazolium bromide (MTT), 3-(4,5-dimethylthiazol-2-yl)-5-(3-carboxylmethoxyphenyl)-2-(4-sulfonyl)-2H-tetrazolium) (MTS), 2,3-bis-(2-methoxy-4-nitro-5-sulfophenyl)-2H-tetrazolium-5-carboxanilide) (XTT), and the water-soluble tetrazolium salts (WST-8 and WST-1) assays. Furthermore, the neutral red uptake assay, trypan blue staining, observation of cell membrane integrity via lactate dehydrogenase activity, luminescent cell viability assay, clonogenic assay that were applied to count the colony numbers were also widely used. Staining either the whole cell or specific cellular components with fluorescent dyes could be considered another method which allows quick detection via flow cytometry and can be combined with microscopy-based analysis. It is also very important that nanomaterials interfere with several assay systems, leading some researchers to dedicate efforts to understanding the specific behavior of different nanomaterials with cell viability endpoints. Three-dimensional (3D) cell culture systems are considered as clinically relevant *in vitro* models for mimicking the features of tissue *in vivo*. In addition, nanomaterials could affect the enzymatic activity as well as cause alterations on the normal cell cycle. On the other hand, induction of apoptosis and necrosis, oxidative stress, injury in specific cell organelles such as mitochondria and lysosome, as well as DNA damage and genotoxicity, inflammatory responses, could also occurr during the exposure of nanomaterials. Thus, all these effects should be well considered and examined for the detailed evaluation of nanotoxicity [116].

6. CELL IMAGING

In recent years, imaging of cells using nanoparticles has become widely accessible to develop new sensors, to understand biological structure and function, and visualize/analyze cancer and healthy cells. Addition of biological molecules to backbone of nanoparticles gives chance to target nanocarrier. Also, loaded drugs can be delivered into target region in higher local concentrations. This strengthens therapeutic effects of the drug.

Combining of nanomaterials to biomolecules for cell imaging provides both advantages of two parts: unique optical, electronic, or mechanical properties of nanomaterials and selectivity/specificity of biological molecules. Various types of organic and inorganic nanomaterials such as gold nanoparticles, semi-conducting nanoparticles, dendrimers, carbon nanotubes and vesicular structures has been decorated with biomolecules such as antibody, aptamer, folic acid and peptide. The designed bio-conjugates have some implementations for different aims: cancer diagnosis, molecular imaging, and targeted therapy, wound healing etc.

Some features of biological molecule covered nanomaterials such as passing biological barriers (bio-barriers), accumulation in tumors and specifically recognizing single cancer cells increases the applicability of nanomaterials in cancer targeting and medical imaging. Table 1 summarizes the used targeting ligands to cover nanomaterials and loaded drugs for imaging of cells and therapeutic effects of drug incorporated bioconjugates.

Table 1. Various nanoparticles used in cell sensing and drug delivery in the literature

Nanomaterial	Ligand	Drug	*In vitro* Application	Effects	Ref.
AuNP	Transferrin	-	Hs578T; 3T3	Imaging; Photothermal therapy	117
AuNP	Anti-MTDH	-	MCF-7	Imaging	118
AuNP	PSMA	-	LNCaP	Targeting	119
AuNP	-	Methotrexate	JAR	Therapeutic	120
AuNP	Anti-EGFR	-	A431	Photothermal therapy	121
Fe$_3$O$_4$@Au NPs	-	DOX	MCF-7	Therapeutic	122
AuNP	FA-Glu	-	KB; A549	Dual-ligand targeting	123
AuNP	Rhodamine B linked beta-cyclodextrin; Biotin	Paclitaxel	HeLa, A549, MG63; NIH3T3	Theranostic	124
AuNP	Cell-penetrating peptide (VG-21)	-	HEp-2, HeLa, Cos-7; Vero	Imaging	125
AuNP	Type 1 ribotoxin-curcin; folic acid	-	Glioma (human brain glioblastoma); HCN-1A	Multimodal therapeutic	126
QD/PAMAM	Ab	-		Imaging	127
QD	Ab	-		Imaging	128
QD	Lectin	-	Caco-2;MCF-7;A-549	Imaging	129
QD	Folic acid decorated bacteria	-	Whole animal	Bio-imaging; Diagnosis	130
QD/PAMAM	DNA Aptamer	-	glioblastoma	Imaging	131
QD	Folate–PEG–PAMAM	-	HeLa	Imaging	132
QD	HER2; RNase A	-	MGC-803	Imaging; Therapy	133
CNT/Polymer	HER2	-	MCF-7; PC-3; HMEC	Imaging	134
CNT	Folate/Iron	DOX	HeLa	Dual targeted nanocarrier	135
CNT	PEG; angiopep-2	DOX	Glioma	Targeting	136
CNT	EGFR	7-Ethyl-10-hydroxy-camptothecin	HCT116, HT29, and SW620	Targeting	137
CNT	Hyaluronic acid	Salinomycin	CD44++	Therapy	138
CNT/Polymer	Folic acid	-	HeLa	Imaging	139
PAMAM	Alexa Fluor 555	-	Capan-1	Imaging	140
PAMAM	Transferrin; Tamoxifen	DOX	C6 glioma	Dual-targeting nanocarrier	141
PAMAM	N-Acetylgalacto-samine	-	HepG2 and MCF-7	Imaging	142
Phytosom/AuNP	-	-	Vero	Imaging; Wound healing	143
Liposome	-	Curcumin	MCF7	Treatment	144
Liposome	EGFR	-	H1299	Imaging; Targeting	145
GQD-HA	HA	DOX	A549	Imaging; Targeting; Treatment	43

AuNP: Gold nanoparticle; Anti-MTDH: anti-metadherin; PSMA: Prostate-specific membrane antigen; DOX: Doxorubicin; FA: Folic acid; Glu: Glucose; QD: Quantum dot; PAMAM: Polimidoamine dendrimer; CNT: Carbon nanotube; EGFR: Epidermal growth factor receptor, GQD-HA: hyaluronic acid attached graphene-quantum dots.

Imaging for diagnostic purposes requires certain high contrast of the interested area compared to surrounding environment. To acquire contrast in a particular area, best way is the use of nanoparticles directed to this specific area. Liposomes are one of the best choice of nanomaterials among the similar structures with their high encapsulation efficiency and ability to encapsulate both hydrophilic and hydrophobic agents. Label can be added during the preparation of liposomes; hence, the particular cells can be labeled using these liposomes. Moreover, imaging agents can be also covalently attached to liposomes. To prepare labeling markers using liposomes, QDs Mn and Gd [146], radionuclides Ga-67 [147] have been incorporated with the liposomes either in aqueous core or lipid bilayer. As a result, structural and physicochemical superiorities of liposomes were combined with the imaging power of other unique materials for every kind of purposes.

Diagnostic functions of nanoparticles include acting as imaging agents, for example with Magnetic Resonance Imaging (MRI) and Ultrasound machines, as a means of cancer detection [67, 148, 149]. Superparamagnetic iron oxide nanoparticles are used as an MRI contrast agent to delineate neoplastic and normal brain tissue [150-152]. Contrast enhanced functional ultrasound imaging utilizes near-infrared-fluorophore-labeled polymeric drug carrier (pHPMA-Dy750) as an effective indicator of EPR tumor particle accumulation in CT26 tumor bearing mice [153]. El-Sayed et al. released a study showing the effectiveness of gold nanoparticles targeted to monoclonal anti-epidermal growth factor receptor, as an indicator for Surface Plasmon Resonance [147].

Furthermore, quantum dots are mostly used for cell imaging in biological applications. It is possible to targeted organelles with QDs using peptides; for instance, nucleus localization signal (NLS) and mitochondrial localization signal (MLS) help to label and visualize intracellular compartments, nucleus and mitochondria [113]. The advantage of use of QDs for cell imaging, labeling of the cells is easy and long-life. Even under continuous illumination, multicolor and extended visualization as well as live-dead cell differentiation can be achieved [52]. Multicolor imaging can be performed labeling cell-surface proteins selectively using antibody-conjugated QDs [148, 154]. Additionally, mentioned specific labeling empowers monitoring of cellular development over a certain period of time. By this way, cellular movement, differentiation and behavior also can be followed eventually [148, 154].

Dendrimers have also been widely investigated as MRI agents. Dendritic structures incorporating many paramagnetic centers are well adapted to medical use because of their proper pharmacokinetic features in combination with shortening of proton relaxation times and signal intensification. Various Gd^{3+} complexes were recently used to functionalized PAMAM dendrimers and applied to tissue imaging as well as the monitoring of the state of angiogenesis. Additionally, dendrimeric structures can be evaluated an ideal platform for multimodal imaging studies besides the targeted therapy applications [155, 1].

AuNPs have been decorated with various targeting ligands such as folic acid, human transferrin to increase the cell uptake by tumors. Different drugs were also loaded these targeted nanoparticles to develop specific delivery and release [147]. On the other hand, Au nanocubes are also reported as useful cancer imaging due to their highest photoluminescence quantum yield among AuNPs [1].

Among various nanostructured materials, CNTs are very promising materials for diagnostic, and drug delivery applications due to their unique structural and mechanical properties. Due to their high specific surface areas, CNTs can easily be modified with desired biomolecules via adsorption, encapsulation and chemical attachment processes [134]. SWCNT-based bioconjugates have the ability to deliver bioactive molecules across cell membranes and also into the cell nuclei. It is reported that nanotubes could release drugs into the cells without giving any damages to the healthy cells of the tissue. Therefore, SWCNTs appear to be promising candidates as imaging-guided drug delivery systems [139]. Recently, graphene oxide which exhibits excellent water solubility and biocompatibility in addition to the characteristic G band in Raman spectra, has been reported as a flexible Raman probe to image cells or tissues through Raman mapping after decorated with Au NPs. The cell internalization mechanism of the hybrid was investigated using Hela 229 cells [156].

Carbon- and graphene-quantum dots have also become emerging materials in cellular applications. Graphene-quantum dots were decorated with hyaluronic acid (HA) (GQD-HA) for *in vitro* and *in vivo* targeting of CD44-receptor overexpressed A549 cells and tumor-bearing balb/c female mice [43]. It has been shown that *in vitro* cancer cells showed a strong fluorescence during confocal laser scanning microscopy studies owing to targeted delivery of GQD-HA. Moreover, *in vivo* biodistribution of the nanomaterials can be easily tracked with the help of bright fluorescence arising from the GQD-HA agents.

In addition to molecular imaging, nanoparticles may also be used to label cells or DNA, tumor vessel imaging, and immunohistochemistry [157]. A major concern with the development of biodistribution imaging agents is toxicity. Silica and PLGA nanoparticles are favored for long term circulation due to their low nanotoxicity [157]. Silica nanoparticles encapsulating Fluorophores (FITC) and Rubpy have proven to be effective biologic labels [158, 159]. Scientists/Engineers are also extensively exploring Quantum Dots as an effective imaging method for tracking cancer cells [160-162].

Some of cell images in the presence of various nanomaterials in the literature have been shown in Figure 3.

Quantum dots are extensively used as fluorescent biomarkers and probes. Rakovish et al. (2014) generated ultra-small nanoprobes based on QDs by modifying anti-HER2 (human epidermal growth factor receptor 2) antibodies (sdAbs) and labeled lung cancer cells *in vitro* [163]. It was shown that sdAbs-QD conjugates exhibited superior staining in lung cancer cells with differential HER2 expression. In addition, QDs can be modified with other ligands such as radiolabels for different purposes. In a study, Tu et al. (2011) synthesized and used macrocyclic ligand-$^{64}Cu^{2+}$ complex modified dextran-coated silicon quantum dots for *in vivo* positron emission tomography (PET) [164]. By this way, they were able to track the nanomaterials easily and determine the nanomaterial disposition and fate in the body as seen in Figure 3-B.

Dendrimers are synthetic nanoprobes which can be tailored according to application. Drugs or therapeutic pieces of nucleic acid can be carried to the disease side. Also, since modification is common in dendrimer synthesis, with fluorescent labeling, they can be easily tracked. For example, siRNA transfer was achieved successfully with a target specificity to folate receptor (FR)-overexpressing KB cancer cells *in vitro* (Figure 3-C) [165].

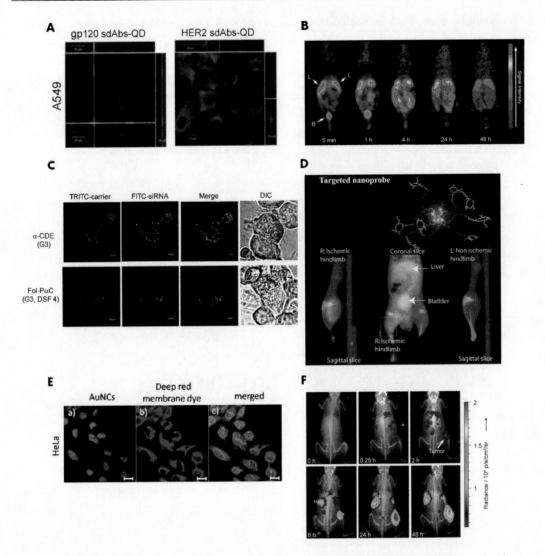

Figure 3. Various nanoparticles used in cell imaging in the literature. **A)** HER2 protein expression in positive lung cancer cells (A549) labeled with HER2-specific sdAbs-QD and gp120-specific sdAbs-QD conjugates analyzed with confocal microscopy. Reproduced with permission from [163]. Copyright © 2014, American Chemical Society. **B)** *In vivo* PET images of mice at 5 min, 1 h, 4 h, 24 h and 48 h after injection of ^{64}Cu-DO3A conjugated dextran SiMn QDs L, liver; B, bladder. Reproduced with permission from [164]. Copyright © 2011, American Chemical Society. **C)** Intracellular distribution of FITC-siRNA complexes with TRITC-α-CDE (G3) and TRITC-Fol-PαC (G3, DSF 4) in KB cells. Reproduced with permission from [165]. Copyright © 2012, American Chemical Society. **D)** Noninvasive PET/CT images of angiogenesis induced by hind limb ischemia in a murine model, uptake of αvβ3-targeted dendritic nanoprobes. Reproduced with permission from [166]. **E)** Confocal laser scanning microscope images of HeLa cells after treatment with green emitting AuNCs, deep red plasma membrane dye exhibits red emission, used as for cell membrane staining. Reproduced with permission from [167]. Copyright © 2014, American Chemical Society. **F)** Bioluminescence imaging of a mouse bearing EMT-6 tumors after tail vain injection of the PEGylated ^{198}Au-doped AuNCs. Reproduced with permission from [168]. Copyright © 2013, American Chemical Society.

Polyamidoamine (PAMAM) dendrimer was conjugated with polyethylene glycol (PEG)-appended α-cyclodextrins (CyDs) which is further modified with folic acid (Fol-PαC). In the study, it was shown that siRNA was transferred into the cells with a high transfer activity and fluorescent-labeled siRNA and TRITC-labeled Fol-PαC were actually accumulated in tumor cells. Moreover, dendrimers are used *in vivo* applications as nanoprobes. Almutairia et al. (2008) developed a biodegradable αvβ3 integrin-targeted positron-emitting dendritic nanoprobe for noninvasive imaging of angiogenesis (Figure 3-D) [166]. It has been shown that targeted-dendritic nanoprobes was higher accumulated in ischemic hind limb as compared with control hind limb.

Gold nanoparticles are widely used in various applications as imaging agents. Venkatesh and coworkers synthesized purine-capped water-soluble green fluorescent gold nanoclusters (AuNCs) and used in nuclei staining in vitro cell culture applications [167]. Highly photostable and biocompatible AuNCs showed high specificity towards cell nuclei in staining (Figure 3-E). In another study, radioluminescent Au nanocages were prepared and used *in vivo* imaging [168]. Use of ^{198}Au in preparation of ^{198}Au -doped gold nanocages (AuNCs) enabled accurate real-time imaging *in vivo* as seen in Figure 3-F. Time-dependent imaging showed that successful multimodality imaging and the nanocages were applicable for both small animal and clinical imaging.

CONCLUSION AND FUTURE DIRECTION

The future of nanoparticle imaging and drug delivery exists in "Theranostics," an effective combination of both the therapeutic and diagnostic benefits. The goal of theranostic nanoparticles is to identify tumor tissue, deliver targeted drug or gene therapy, and observe the tissue post-delivery [169]. Improvements in targeted nanomedicine may even allow for personalized treatment on a cellular level [170]. Continued importance will be placed on reducing toxicity, improving circulation, overcoming drug resistance, rapid response, minimizing cost, and supporting mass production.

REFERENCES

[1]   Barreto JA, O'Malley W, Kubeil M, Graham B, Stephan H, Spiccia L. Nanomaterials: Applications in Cancer Imaging and Therapy. *Adv. Mater.* 2011;23:18–40.
[2]   Torchilin VP. Recent advances with liposomes as pharmaceutical carriers. *Nat. Rev. Drug Discov.* 2005;4:145-160.
[3]   Gabizon AA. Pegylated Liposomal Doxorubicin: Metamorphosis of an Old Drug into a New Form of Chemotherapy. *Cancer Invest.* 2001;19:424-436.
[4]   Safra T, Muggia F, Jeffers S, Tsao-Wei DD, Groshen S, Lyass O, Henderson R, Berry G, Gabizon A. Pegylated liposomal doxorubicin (doxil): Reduced clinical cardiotoxicity inpatients reaching or exceeding cumulative doses of 500 mg/m2. *Ann. Oncol.* 2000;11(8):1029-1033.
[5]   Allen TM, Cullis PR. Liposomal drug delivery systems: from concept to clinical applications. *Adv. Drug Deliv. Rev.* 2013;65:36–48.

[6]   Mayer LD, Hope MJ, Cullis PR. Vesicles of variable sizes produced by a rapid extrusion procedure. *Biochim. Biophys. Acta* 1986;858:161–168.

[7]   Huang CH. Phosphatidylcholine vesicles. Formation and physical characteristics. *Biochem.* 1969;8:344–352.

[8]   Allen TM, Cleland LG. Serum induced leakage of liposome contents. *Biochim. Biophys. Acta* 1980;597:418–426.

[9]   Cullis PR. Lateral diffusion rates of phosphatidylcholine in vesicle membranes: effects of cholesterol and hydrocarbon phase transitions. *FEBS Lett* 1970;70:223–228.

[10]   McIntosh TJ. The effect of cholesterol on the structure of phosphatidylcholine bilayers. *Biochim. Biophys. Acta* 1978;513:43–58.

[11]   Kimelberg HK, Tracy TF, Biddlecome SM, Bourke RS. The Effect of Entrapment in Liposomes on the in Vivo Distribution of [3H]Methotrexate in a Primate. *Cancer Res.* 1976;36:2949–2957.

[12]   Gregoriadis G, Neerunjun D. Control of the Rate of Hepatic Uptake and Catabolism of Liposome-Entrapped Proteins Injected into Rats. Possible Therapeutic Applications. *Eur. J. Biochem.* 1974;47:179–185.

[13]   Juliano RL, Stamp D. The effect of particle size and charge on the clearance rates of liposomes and liposome encapsulated drugs. Biochem. Biophys. Res. Commun. 1975, 63, 651–658.

[14]   Allen TM. Liposomes in the Therapy of Infectious Disease and Cancer. *UCLA Symposium in Molecular and Cellular Biology* CRC Press: California, 1989; pp 405–415.

[15]   Abuchowski A, McCoy JR, Palczuk NC, Van Es T, Davis FF. Effect of covalent attachment of polyethylene glycol on immunogenicity and circulating life of bovine liver catalase. *J. Biol. Chem.* 1977;252:3582–3586.

[16]   Klibanov AL, Maruyama K, Torchilin VP, Huang L. Amphipathic polyethyleneglycols effectively prolong the circulation time of liposomes. *FEBS Lett.* 1990;*268*:235–237.

[17]   Blume G, Cevc G. Liposomes for sustained drug release in vivo. *Biochim. Biophys. Acta* 1990;1029:91–97.

[18]   James ND, Coker RJ, Tomlinson D, Harris JR, Gompels M, Pinching AJ, Stewart JS. Liposomal doxorubicin (Doxil): an effective new treatment for Kaposi's sarcoma in AIDS. *Clin. Oncol.* 1994;6(5):294–296.

[19]   Boas U, Heegaard PMH. Dendrimers in drug research. *Chem. Soc. Rev.* 2004; 33:43-63.

[20]   Grayson SM, Frechet JMJ. Convergent Dendrons and Dendrimers: from Synthesis to Applications. *Chem. Rev.* 2001;101:3819-3868.

[21]   Onitsuka K, Fujimoto M, Kitajima H, Ohsiro N, Takei F, Takahashi S. Convergent Synthesis of Platinum–Acetylide Dendrimers. *Chem. Eur. J.* 2004;10:6433-6446.

[22]   Surendiran A, Sandhiya S, Pradhan SC, Adithan C. Novel applications of nanotechnology in medicine. *Indian J. Med. Res.* 2009;130:689–701.

[23]   Clementi C, Miller K, Mero A, Satchi-Fainaro R, Pasut G. Dendritic poly(ethylene glycol) bearing paclitaxel and alendronate for targeting bone neoplasms. *Mol. Pharm.* 2011; 8:1063-1072.

[24]   Peer D, Karp JM, Hong S, Farokhzad OC, Margalit R, Langer R. Nanocarriers as an emerging platform for cancer therapy. *Nat. Nanotechnol.* 2007;2:751-760.

[25] Kunzmann A, Andersson B, Thurnherr T, Krug H, Scheynius A, Fadeel B. Toxicology of engineered nanomaterials: focus on biocompatibility, biodistribution and biodegradation. *Biochim. Biophys. Acta* 2011;1810:361–373.

[26] Kostarelos K, Lacerda L, Pastorin G, Wu W, Wieckowski S, Luangsivilay J, Godefroy S, Pantarotto D, Briand JP, Muller S, Prato M, Bianco A. Cellular uptake of functionalized carbon nanotubes is independent of functional group and cell type. *Nat. Nanotechnol.* 2007;2:108 – 113.

[27] Lopez CF, Nielsen SO, Moore PB, Klein ML. Understanding nature's design for a nanosyringe. *PNAS* 2004;101:4431 – 4434.

[28] Allen MJ, Tung VC, Kaner RB. Honeycomb Carbon: A Review of Graphene. *Chem. Rev.* 2010;110:132-145.

[29] Novoselov KS, Geim AK, Morozov SV, Jiang D, Zhang Y, Dubonos SV, Grigorieva IV, Firsov AA. Electric Field Effect in Atomically Thin Carbon Films *Science* 2004;306,:666–669.

[30] Goenka S, Sant V, Sant S. Graphene-based nanomaterials for drug delivery and tissue engineering. *J. Control Release* 2014;173:75–88.

[31] Sprinkle M, Ruan M, Hu Y, Hankinson J, Rubio-Roy M, Zhang B, Wu X, Berger C, de Heer WA. Scalable templated growth of graphene nanoribbons on SiC. *Nat. Nanotechnol.* 2010;5:727–731.

[32] Kuila T, Bose S, Khanra P, Mishra AK, Kim NH, Lee JH. Recent advances in graphene-based biosensors. *Biosens. Bioelectron.* 2011;26:4637 – 4648.

[33] He S, Song B, Li D, Zhu C, Qi W, Wen Y, Wang L, Song S, Fang H, Fan CA. Graphene Nanoprobe for Rapid, Sensitive, and Multicolor Fluorescent DNA Analysis. *Adv. Funct. Mater.* 2010;20:453 – 459.

[34] Kim J, Cote LJ, Kim F, Yuan W, Shull KR, Huang J. Graphene Oxide Sheets at Interfaces. *J. Am. Chem. Soc.* 2010;132:8180–8186.

[35] Liu H, Ye T, Mao C. Fluorescent Carbon Nanoparticles Derived from Candle Soot. *Angew. Chem.* 2007;46:6473–6475.

[36] Jia XF, Li J, Wang EK. One-pot green synthesis of optically pH-sensitive carbon dots with upconversion luminescence. *Nanoscale* 2012;4:5572–5575.

[37] Zhuo Y, Miao H, Zhong D, Zhu S, YangX. One-step synthesis of high quantum-yield and excitation-independent emission carbon dots for cell imaging. *Mater. Lett.* 2015;139:197–200.

[38] Xu XY, Ray R, Gu YL, Ploehn HJ, Gearheart L, Raker K, Scrivens WA. Electrophoretic Analysis and Purification of Fluorescent Single-Walled Carbon Nanotube Fragments. *J. Am. Chem. Soc.* 2004;126:12736–12737.

[39] Hu S-L, Niu K-Y, Sun J, Yang J, Zhao N-Q, Du X-W. One-step synthesis of fluorescent carbon nanoparticles by laser irradiation. *J. Mater. Chem.* 2009;19:484 – 488.

[40] Wang X, Cao L, Bunker CE, Meziani MJ, Lu FS, Guliants EA, Sun Y-P. Fluorescence Decoration of Defects in Carbon Nanotubes. *J. Phys. Chem.* C 2010;114:20941–20946.

[41] Wang Y, Hu A. Carbon quantum dots: synthesis, properties and applications. *J. Mater. Chem.* C 2014;2:6921 – 6939.

[42] Lim SY, Shen W, Gao Z. Carbon quantum dots and their applications. *Chem. Soc. Rev.* 2015;44:362 – 381.

[43]   Abdullah Al N, Lee JE, In I, Lee H, Lee KD, Jeong JH, Park SY. Target delivery and
       cell imaging using hyaluronic acid-functionalized graphene quantum dots. *Mol. Pharm.*
       2013;10:3736–3744.

[44]   Wang Z, Xia J, Zhou C, Via B, Xia Y, Zhang F, Li Y, Xia L, Tang J. Synthesis of
       strongly green-photoluminescent graphene quantum dots for drug carrier. *Colloids
       Surf.* B 2013b;112:192–196.

[45]   Qian ZS, Shan XY, Chai LJ, Ma JJ, Chen JR, Feng H. DNA nanosensor based on
       biocompatible graphene quantum dots and carbon nanotubes. *Biosens. Bioelectron.*
       2014;60C:64–70.

[46]   Nurunnabi M, Khatun Z, Huh KM, Park SY, Lee DY, Cho KJ, Lee Y. In vivo
       biodistribution and toxicology of carboxylated graphene quantum dots. *ACS Nano*
       2013;7:6858–6867.

[47]   Connor EE, Mwamuka J, Gole A, Murphy CJ, Wyatt MD. Gold nanoparticles are taken
       up by human cells but do not cause acute cytotoxicity. *Small* 2005;1:325–327.

[48]   Sperling RA, Gil PR, Zhang F, Zanella M, Parak WJ. Biological applications of gold
       nanoparticles. *Chem. Soc. Rev.* 2008;37:1896–1908.

[49]   Sonnichsen C, Alivisatos AP. Gold nanorods as novel nonbleaching plasmon-based
       orientation sensors for polarized single-particle microscopy. *Nano Lett.* 2005;5:301–
       304.

[50]   Pissuwan D, Valenzuela SM, Cortie MB. Functionalised gold nanoparticles for
       controlling pathogenic bacteria. *Trends Biotechnol.* 2006;24:62–67.

[51]   Wang Y, Chen L. Quantum dots, lighting up the research and development of
       nanomedicine. *Nanomedicine: NBM* 2011;7:385-402.

[52]   Medintz IL, Uyeda HT, Goldman ER, Mattoussi H. Quantum dot bioconjugates for
       imaging, labelling and sensing. *Nat. Mater.* 2005;4:435-446.

[53]   Murray CB, Norris DJ, Bawendi MG. Synthesis and characterization of nearly
       monodisperse CdE (E = sulfur, selenium, tellurium) semiconductor nanocrystallites. *J.
       Am. Chem. Soc.* 1993;115:8706–8715.

[54]   Sun C, Lee JSH, Zhang MQ. Magnetic nanoparticles in MR imaging and drug delivery.
       Adv. Drug Deliv. Rev. 2008; 60:1252–1265.

[55]   Lewinski N, Colvin V, Drezek R. Cytotoxicity of nanoparticles. *Small* 2008;4:26–49.

[56]   Wang YXJ, Hussain SM, Krestin GP. Superparamagnetic iron oxide contrast agents:
       physicochemical characteristics and applications in MR imaging. *Eur. Radiol.* 2001;11:
       2319–2331.

[57]   Harisinghani MG, Barentsz J, Hahn PF, Deserno WM, Tabatabaei S, van de Kaa CH,
       de la Rosette J, Weissleder R. Noninvasive detection of clinically occult lymph-node
       metastases in prostate cancer. *N. Eng. J. Med.* 2003;348:2491-2495.

[58]   Veiseh O, Gunn JW, Zhang M. Design and fabrication of magnetic nanoparticles for
       targeted drug delivery and imaging. *Adv. Drug Deliv.* Rev. 2010;62:284–304.

[59]   Huang HC, Barua S, Sharma G, Dey SK, Rege K. Inorganic nanoparticles for cancer
       imaging and therapy. *J. Control. Release* 2011;1553:44-57.

[60]   Hanahan D, Weinberg R. Hallmarks of Cancer: The Next Generation. *Cell* 2011;144:
       646-674.

[61]   Milane L, Ganesh S, Shah S, Duan Z, Amiji M. Multi-modal strategies for overcoming
       tumor drug resistance: hypoxia, the Warburg effect, stem cells, and multifunctional
       nanotechnology. *J. Cont. Rel.* 2011;155:237-247.

[62] Hanahan D, Weinberg R. The hallmarks of cancer. *Cell* 2000;100:57-70.

[63] Kohandel M, Haselwandter CA, Kardar M, Sengupta S, Sivaloganathan S. Quantitative model for efficient temporal targeting of tumor cells and neovasculature. Comput. Math. Model Med. 2011;2011: doi:10.1155/2011/790721.

[64] Donovan EA, Kummar S. Targeting VEGF in cancer therapy. *Curr. Prob. Cancer* 2006;30:7-32.

[65] Duda DG, Batchelor T, Willett CG, Jain RK. VEGF-targeted cancer therapy strategies: current progress, hurdles and future prospects. *Trend Molecul. Med.* 2007;13(6):223-230.

[66] Kwon IK, Lee SC, Han B, Park K. Analysis on the current status of targeted drug delivery to tumors. *J. Control. Release* 2012;164:108-114.

[67] Paszko E, Ehrhardt C, Senge MO, Kelleher DP, Reynolds JV. Nanodrug applications in photodynamic therapy. *Photodiagn. Photodyn.* 2011;8(1):14-29.

[68] Xie S, Tao Y, Pan Y, Qu W, Cheng G, Huang L, Chen D, Wang X, Liu Z, Yuan Z. Biodegradable nanoparticles for intracellular delivery of antimicrobial agents. *J. Control. Release* 2014;187:101-117.

[69] Zhang L, Gu FX, Chan JM, Wang AZ, Langer RS, Farokhzad OC. Nanoparticles in Medicine: Therapeutic Applications and Developments. *Clin. Pharmacol. Ther.* 2008;83: 761-769.

[70] Torchilin VP. Multifunctional nanocarriers. *Adv. Drug Deliver. Rev.* 2006;58:1532-1555.

[71] Maeda H, Wu J, Sawa T, Matsumura Y, Hori K. Tumor vascular permeability and the EPR effect in macromolecular therapeutics: a review. *J. Control. Release* 2000;65:271-284.

[72] Torchilin VP. Structure and design of polymeric surfactant-based drug delivery systems *J. Control. Release* 2001;73:137-172.

[73] Hollis CP, Weiss HL, Leggas M, Evers BM, Gemeinhart RA, Li T. Biodistribution and bioimaging studies of hybrid paclitaxel nanocrystals: lessons learned of the EPR effect and image-guided drug delivery. *J. Control. Release* 2013;172:12-21.

[74] Shen M, Huang Y, Han L, Qin J, Fang X, Wang J, Yang VC. Multifunctional drug delivery system for targeting tumor and its acidic microenvironment. *J. Control. Release* 2012;161:884-892.

[75] Fontana G, Licciardi M, Mansueto S, Schillaci D, Giammona G. Amoxicillin-loaded polyethylcyanoacrylate nanoparticles: influence of PEG coating on the particle size, drug release rate and phagocytic uptake. *Biomaterials* 2001;22:2857–2865.

[76] Dhule SS, Penfornis P, He J, Harris MR, Terry T, John V, Pochampally R. The combined effect of encapsulating curcumin and C6 ceramide in liposomal nanoparticles against osteosarcoma. *Mol. Pharm.* 2014;11:417-427.

[77] Yin H, Liao L, Fang J. Enhanced Permeability and Retention (EPR) Effect Based Tumor Targeting: The Concept, Application and Prospect. *JSM Clin. Oncol. Res.* 2014;2:1010 – 1015.

[78] Torchilin VP. Targeted pharmaceutical nanocarriers for cancer therapy and imaging. *AAPS J.* 2007;9 (2):128-147.

[79] Torchilin VP. Tumor delivery of macromolecular drugs based on the EPR effect. *Adv. Drug Deliv. Rev.* 2011;63(3):131-135.

[80] Maeda H, Bharate GY, Daruwalla J. Polymeric drugs for efficient tumor-targeted drug delivery based on EPR-effect. *Eur. J. Pharm. Biopharm.* 2009;71:409-419.

[81] Gabizon AA. Pegylated liposomal doxorubicin: metamorphosis of an old drug into a new form of chemotherapy. *Cancer Invest.* 2001;19:424-436.

[82] O'Shaughnessy JA. Pegylated liposomal doxorubicin in the treatment of breast cancer. *Clin. Breast Cancer* 2003;4:318-328.

[83] Perez AT, Domenech GH, Frankel C, Vogel CL. Pegylated liposomal doxorubicin (Doxil) for metastatic breast cancer: the Cancer Research Network, Inc., experience. *Cancer Invest.* 2002;20:22-29.

[84] Lee KS, Chung HC, Im SA, Park YH, Kim SB, Rha SY, Lee MY, Ro J. Multicenter phase II trial of Genexol-PM, a Cremophor-free, polymeric micelle formulation of paclitaxel, in patients with metastatic breast cancer. *Breast Cancer Res.* 2008;108(2):241-250.

[85] Kim TY, Kim DW, Chung JY, Shin SG, Kim SC, Heo DS, Kim NK, Bang YJ. Phase I and pharmacokinetic study of Genexol-PM, a cremophor-free, polymeric micelle-formulated paclitaxel, in patients with advanced malignancies. *Clin. Cancer Res.* 2004;10:3708-3716.

[86] Hamaguchi T, Matsumura Y, Suzuki M, Shimizu K, Goda R, Nakamura I, Nakatomi I, Yokoyama M, Kataoka K, Kakizoe T. Brit. NK105, a paclitaxel-incorporating micellar nanoparticle formulation, can extend in vivo antitumour activity and reduce the neurotoxicity of paclitaxel. *J. Cancer* 2005;92:1240-1246.

[87] Bae YH, Park K. Targeted drug delivery to tumors: myths, reality and possibility. *J. Control. Release* 2011;153:198-205.

[88] von Maltzahn G, Park JH, Lin KY, Singh N, Schwöppe C, Mesters R, Berdel WE, Ruoslahti E, Sailor MJ, Bhatia SN. Nanoparticles that communicate in vivo to amplify tumour targeting. *Nature Mater.* 2011;10:545-552.

[89] Kirpotin DB, Drummond DC, Shao Y, Shalaby MR, Hong K, Nielsen UB, Marks JD, Benz CC, Park JW. Antibody targeting of long-circulating lipidic nanoparticles does not increase tumor localization but does increase internalization in animal models. *Cancer Res.* 2006;66:6732-6740.

[90] Cho K, Wang X, Nie S. Therapeutic nanoparticles for drug delivery in cancer. *Clinical Cancer Res.* 2008;14:1310-1316.

[91] Suzuki H, Sato M, Umezwa Y. Accurate targeting of activated macrophages based on synergistic activation of functional molecules uptake by scavenger receptor and matrix metalloproteinase. *ACS Chem.* Biol. 2008;3(8):471-479.

[92] Thomas SM, Sahu B, Rapireddy S, Bahal R, Wheeler SE, Procopio EM, Kim J, Joyce SC, Contrucci S, Wang Y, Chiosea SI, Lathrop KL, Watkins S, Grandis JR, Armitage BA, Ly DH. Antitumor effects of EGFR antisense guanidine-based peptide nucleic acids in cancer models. A*CS Chem. Biol.* 2013;8(2):345-352.

[93] Govan JM, Uprety R, Thomas M, Lusic H, Lively, MO, Deiters A. *ACS Chem. Biol.* 2013;8 (10):2272-2282.

[94] Morgan J, Gray AG, Huehns ER. Brit. Specific targeting and toxicity of sulphonated aluminium phthalocyanine photosensitised liposomes directed to cells by monoclonal antibody in vitro. *J. Cancer* 1989;59:366-370.

[95] Lu Y, Low PS. Folate-mediated delivery of macromolecular anticancer therapeutic agents. *Adv. Drug Deliv. Rev.* 2002;54:675-693.

[96] Gryder BE, Akbashev MJ, Rood MK, Raftery ED, Meyers WM, Dillard P, Khan S, Oyelere AK. Selectively targeting prostate cancer with antiandrogen equipped histone deacetylase inhibitors. *ACS Chem. Biol.* 2013;8(11):2550-2560.

[97] Peng KW, Wang H, Qin Z, Wijewickrama GT, Lu M, Wang Z, Bolton JL, Thatcher GR. Selective estrogen receptor modulator delivery of quinone warheads to DNA triggering apoptosis in breast cancer cells. *ACS Chem. Biol.* 2009;4(12):1039-1049.

[98] Bergstrom LC, Vucenik I, Hagen IK, Chernomorsky SA, Poretz RD. In-vitro photocytotoxicity of lysosomotropic immunoliposomes containing pheophorbide a with human bladder carcinoma cells. *J. Photochem. Photobiol. B.* 1994;24:17-23.

[99] Allen T. Ligand-targeted therapeutics in anticancer therapy. *Nat. Rev. Cancer* 2002;2:750-763.

[100] Chen WC, Zhang AX, Li SD. Limitations and niches of the active targeting approach for nanoparticle drug delivery. *Europ. J. Nanomed.* 2012;4(2-4):89-93.

[101] Nguyen KT. Targeted Nanoparticles for Cancer Therapy: Promises and Challenges. *Nanomed. Nanotechnol.* 2011;2(5).

[102] Randles EG, Bergethon PR. A photo-dependent switch of liposome stability and permeability. *Langmuir* 2013;29(5):1490-1497.

[103] Akens MK, Wise-Milestone L, Won E, Schwock J, Yee AJM, Wilson BC, Whyne CM. In vitro and in vivo effects of photodynamic therapy on metastatic breast cancer cells pre-treated with zoledronic acid. *Photodiagn. Photodyn. Ther.* 2014;11(3):426-433.

[104] O'Connor AE, Gallagher WM, Byrne AT. Porphyrin and nonporphyrin photosensitizers in oncology: preclinical and clinical advances in photodynamic therapy. *Photochem. Photobiol.* 2009;85:1053-1074.

[105] Josefsen LB, Boyle RW. Photodynamic therapy: novel third-generation photosensitizers one step closer? *Brit. J. Pharmacol.* 2008;154:1-3.

[106] Huntosova V, Alvarez L, Bryndzova L. Nadova Z, Jancura D, Buriankova L, Bonneau S, Brault D, Miskovsky P, Sureau F. Interaction dynamics of hypericin with low-density lipoproteins and U87-MG cells. *Int. J. Pharm.* 2010;389:32-40.

[107] Leonard KA, Nelen MI, Simard TP. Davies SR, Gollnick SO, Oseroff AR, Gibson SL, Hilf R, Chen LB, Detty MR. Synthesis and evaluation of chalcogenopyrylium dyes as potential sensitizers for the photodynamic therapy of cancer. *J. Med. Chem.* 1999;42:3953-3964.

[108] Bhatti M, Yahioglu G, Milgrom LR, Garcia-Maya M, Chester KA, Deonarain MP. Targeted photodynamic therapy with multiply-loaded recombinant antibody fragments. *Int. J. Cancer* 2008;122:1155-1163.

[109] Gijsens A, Derycke A, Missiaen L. De Vos D, Huwyler J, Eberle A, de Witte P. Targeting of the photocytotoxic compound AlPcS4 to Hela cells by transferrin conjugated PEG-liposomes. *Int. J. Cancer* 2002;101:78-85.

[110] Hermanson GT. *Bioconjugate Techniques*, 2nd ed., Academic Press: San Diego, 2008; pp. 969-1002.

[111] Tang J, Zhang L, Liu Y, Zhang Q, Qin Y, Yin Y, Yuan W, Yang Y, Xie Y, Zhang Z, He, Q. Synergistic targeted delivery of payload into tumor cells by dual-ligand liposomes co-modified with cholesterol anchored transferrin and TAT. *Int. J. Pharm.* 2013;454:31–40.

[112] Choi Y, Thomas T, Kotlyar A, Islam MT, Baker Jr JR. Synthesis and functional evaluation of DNA-assembled polyamidoamine dendrimer clusters for cancer cell-specific targeting. *Chem. Biol.* 2005;12:35–43.

[113] Biju V. Chemical modifications and bioconjugate reactions of nanomaterials for sensing, imaging, drug delivery and therapy. *Chem. Soc. Rev.* 2014;43:744-764.

[114] Blanco-Canosa, JB, Wu M, Susumu K, Petryayeva E, Jennings TL, Dawson PE, Algar WR, Medintz IL. Recent Progress in The Bioconjugation Of Quantum Dots. *Coordin. Chem. Rev.* 2014;263:101–137.

[115] Susumu K, Uyeda HT, Medintz IL, Pons T, Delehanty JB, Mattoussi H. Enhancing the stability and biological functionalities of quantum dots via compact multifunctional ligands. *J. Am. Chem. Soc.* 2007;129:13987–13996.

[116] Nogueira DR, Mitjans, M, Rolim CMB and Vinardell, MP. Mechanisms Underlying Cytotoxicity Induced by Engineered, Nanomaterials: A Review of In Vitro Studies, *Nanomaterials* 2014;4:454-484.

[117] Li JL, Wang L, Liu XY, Zhang ZP, Guo HC, Liu WM, Tang SH. In vitro cancer cell imaging and therapy using transferrin-conjugated gold nanoparticles. *Cancer Lett.* 2009;274:319–326.

[118] Unak G, Ozkaya F, Medine EI, Kozgus O, Sakarya S, Bekis R, Unak P, Timur S. Gold nanoparticle probes: Design and in vitro applications in cancer cell culture. *Colloid. Surface.* B 2012;90:217–226.

[119] Kasten BB, Liu T, Nedrow-Byers JR, Benny PD, Berkman CE. Targeting prostate cancer cells with PSMA inhibitor-guided gold nanoparticles. *Bioorgan. Med. Chem. Lett.* 2013;23:565–568.

[120] Tran NTT, Wang TH, Lin CY, Tai Y. Synthesis of methotrexate-conjugated gold nanoparticles with enhanced cancer therapeutic effect. *Biochem. Eng. J.* 2013;78:175–180.

[121] Raji V, Kumar J, Rejiya CS, Vibin M, Shenoi VN, Abraham A. Selective photothermal efficiency of citrate capped gold nanoparticles for destruction of cancer cells. *Experim. Cell Res.* 2011;317:2052–2058.

[122] Elbialy NS, Fathy MM, Khalil WM, Preparation and characterization of magnetic gold nanoparticles to be used as doxorubicin nanocarriers. *Phys. Medica* 2014;30(7):843-848.

[123] Li X, Zhou H, Yang L, Du G, Pai-Panandiker AS, Huang X, Yan B. Enhancement of cell recognition in vitro by dual-ligand cancer targeting gold nanoparticles. *Biomaterials* 2011;32:2540-2545.

[124] Heo DN, Yang DH, Moon HJ, Lee JB, Bae MS, Lee SC, Lee WJ, Sun IC, Kwon IK. Gold nanoparticles surface-functionalized with paclitaxel drug and biotin receptor as theranostic agents for cancer therapy. *Biomaterials* 2012;33:856-866.

[125] Tiwari PM, Eroglu E, Bawage SS, Vig K, Miller, ME, Pillai S, Dennis VA, Singh SR. Enhanced intracellular translocation and biodistribution of gold nanoparticles functionalized with a cell-penetrating peptide (VG-21) from vesicular stomatitis virus. *Biomaterials* 2014;35(35):9484-9494.

[126] Mohamed MS, Veeranarayanan S, Poulose AC, Nagaoka Y, Minegishi H, Yoshida Y, Maekawa T, Kumar DS. Type 1 ribotoxin-curcin conjugated biogenic gold nanoparticles for a multimodal therapeutic approach towards brain cancer. *Biochim. Biophys. Acta* 2014;1840:1657–1669.

[127] Akin, M, Bongartz, R, Walter, JG, Demirkol, DO, Stahl, F, Timur S, and Scheper T. PAMAM-functionalized water soluble quantum dots for cancer cell targeting. *J. Mater. Chem.* 2012;22(23):11529-11536.

[128] Ag D, Dogan LE, Bongartz R, Seleci M, Walter, JG, Demirkol, DO, Stahl F, Ozcelik S, Timur S and Scheper T. Biofunctional Quantum dots as Fluorescence Probe for Cell Specific Targeting, *Colloid Surf.* B. 2014;114(1):96-103.

[129] Akca O, Unak P, Medine EI, Sakarya S, Yurt Kılcar A, Ichedef C, Bekıs R, Tımur S. Radioiodine Labeled Cdse/Cds Quantum Dots:Lectın Targeted Dual Probes, Radiochim. *Acta* 2014;102(9):849-859.

[130] Liu Y, Zhou M, Luo D, Wang L, Hong Y, Yang Y, Sha Y, Bacteria-mediated in vivo delivery of quantum dots into solid tumor. *Biochem Biophys Res Commun.* 2012;425(4):769-774.

[131] Li Z, Huang P, He R, Lin J, Yang S, Zhang X, Ren Q, Cui D. Aptamer-conjugated dendrimer-modified quantum dots for cancer cell targeting and imaging. *Materials Letters* 2010;64:375–378.

[132] Zhao Y, Liu S, Li Y, Jiang W, Chang Y, Pan S, Fang X, Wang YA, Wang J. Synthesis and grafting of folate–PEG–PAMAM conjugates onto quantum dots for selective targeting of folate-receptor-positive tumor cells. *J. Colloid Int. Science* 2010;350:44–50.

[133] Ruan J, Song H, Qian Q, Li C, Wang K, Bao C, Cui D. Biomaterials HER2 monoclonal antibody conjugated RNase-A-associated CdTe quantum dots for targeted imaging and therapy of gastric cancer. 2012;9:7093-7102.

[134] Yuksel M, Colak DG, Akin M, Cianga I, Kukut M, Medine EI, Can M, Sakarya S, Unak P, Timur S, Yagci Y. Nonionic, water self-dispersible "hairy-rod" poly(p-phenylene)-g-poly(ethylene glycol) copolymer/carbon nanotube conjugates for targeted cell imaging. *Biomacromol.* 2012;13:2680–2691.

[135] Li R, Wu R, Zhao L, Hu Z, Guo S, Pan X, Zou H. Folate and iron difunctionalized multiwall carbon nanotubes as dual-targeted drug nanocarrier to cancer cells. *Carbon* 2011;49:1797–1805.

[136] Ren J, Shen S, Wang D, Xi Z, Guo L, Pang Z, Qian Y, Sun X, Jiang X: The targeted delivery of anticancer drugs to brain glioma by PEGylated oxidized multi-walled carbon nanotubes modified with angiopep-2. *Biomaterials* 2012;33:3324-3333.

[137] Lee PC, Chiou YC, Wong JM, Peng CL, Shieh MJ. Targeting colorectal cancer cells with single-walled carbon nanotubes conjugated to anticancer agent SN-38 and EGFR antibody. *Biomaterials* 2013;34:8756-8765.

[138] Yao H, Zhang Y, Sun L, Liu Y. The effect of hyaluronic acid functionalized carbon nanotubes loaded with salinomycin on gastric cancer stem cells. *Biomaterials* 2014;35:9208-9223.

[139] Ag D, Seleci M, Bongartz R, Can M, Yurteri S, Cianga I, Stahl F, Timur S, Scheper T, Yagci Y. From invisible structures of SWCNTs toward fluorescent and targeting architectures for cell imaging. *Biomacromol.* 2013;14:3532–3541.

[140] Opitz AW1, Czymmek KJ, Wickstrom E, Wagner NJ. Uptake, efflux, and mass transfer coefficient of fluorescent PAMAM dendrimers into pancreatic cancer cells. *Biochim. Biophys. Acta.* 2013;1828:294–301.

[141] Li Y, He H, Jia X, Lu WL, Lou J, Wei Y: A dual-targeting nanocarrier based on poly(amidoamine) dendrimers conjugated with transferrin and tamoxifen for treating brain gliomas. *Biomaterials* 2012;33:3899-3908.

[142] Medina SH, Tekumalla V, Chevliakov MV, Shewach DS, Ensminger WD, El-Sayed MEH. N-acetylgalactosamine-functionalized dendrimers as hepatic cancer cell-targeted carriers. *Biomaterials* 2011;32:4118-4129.

[143] Demir B, Barlas FB, Guler E, Gumus PZ, Can M, Yavuz M, Coskunol H, Timur S. *RSC Adv.* 2014;4:34687-34695.

[144] Hasan M, Belhaj N, Benachour H, Barberi-Heyob M, Kahn CJ, Jabbari E, Linder M, Arab-Tehrany E. Liposome encapsulation of curcumin: physico-chemical characterizations and effects on MCF7 cancer cell proliferation. *Int. J. Pharm.* 2014;461:519–528.

[145] Song S, Liu D, Peng J, Sun Y, Li Z, Gu JR, Xu Y. Peptide ligand-mediated liposome distribution and targeting to EGFR expressing tumor in vivo. *Int. J. Pharm.* 2008;363:155–161.

[146] Theek B, Gremse F, Kunjachan S, Fokong S, Pola R, Pechar M, Deckers R, Storm G, Ehling J, Kiessling F, and Lammers T. Characterizing EPR-mediated passive drug targeting using contrast-enhanced functional ultrasound imaging. *J Control Release.* 2014;182:83‑89.

[147] El-Sayed, I. H.; Huang, X.; El-Sayed, M. A. Surface plasmon resonance scattering and absorption of anti-EGFR antibody conjugated gold nanoparticles in cancer diagnostics: applications in oral cancer. *Nano Lett.* 2005;5:829-834.

[148] Kateb B, Chiu K, Black KL, Yamamoto V, Khalsa B, Ljubimova JY, Ding H, Patil R, Portilla-Arias JA, Modo M. Nanoplatforms for constructing new approaches to cancer treatment, imaging, and drug delivery: What should be the policy? *Neuroimage* 2011;54:106-124.

[149] Meetoo D. The revolution of the big future with tiny medicine. *Brit. J. Nursing.* 2009;18:1201-1206.

[150] Sun C, Veiseh O, Gunn J. In Vivo MRI Detection of Gliomas by Chlorotoxin-Conjugated Superparamagnetic Nanoprobes. *Small* 2008;4:372-379.

[151] Yu MK, Park J, Jon S. Targeting Strategies for Multifunctional Nanoparticles in Cancer Imaging and Therapy. *Theranostics* 2012;2:3-44.

[152] Weinstein JS, Varallyay CG, Dosa E, Gahramanov S, Hamilton B. High-resolution steady-state cerebral blood volume maps in patients with central nervous system neoplasms using ferumoxytol, a superparamagnetic iron oxide nanoparticle. *J. Cerebr. Blood F. Met.* 2010;30:15-35.

[153] Dubertret B, Skourides P, Norris DJ, Noireaux V, Brivanlou AH, Libchaber A. In Vivo Imaging of Quantum Dots Encapsulated in Phospholipid Micelles. *Science* 2002;298:1759–1762.

[154] Jaiswal JK, Mattoussi H, Mauro JM, Simon SM. Long-term multiple color imaging of live cells using quantum dot bioconjugates. *Nature Biotechnol.* 2002;21:47–51.

[155] Villaraza A J L, Bumb A, Brechbie MW. Organic Radical Contrast Agents for Magnetic Resonance Imaging. *Chem. Rev.* 2010;110:2921–2959.

[156] Liu Q, Wei L, Wang J, Peng F, Luo D, Cui R, Niu Y, Qin X, Liu Y, Sun H, Yang J, Li Y. Cell imaging by graphene oxide based on surface enhanced Raman scattering. *Nanoscale* 2012;4:7084–7089.

[157] Liu Y, Miyoshi H, Nakamura M. Nanomedicine for drug delivery and imaging: A promising avenue for cancer therapy and diagnosis using targeted functional nanoparticles. *Int. J. Cancer.* 2007;120:2527-2537.

[158] Makarova OV, Ostafin AE, Miyoshi H, Norris JR, Meisel D. Adsorption and Encapsulation of Fluorescent Probes in Nanoparticles. *J. Phys. Chem. B.* 1999;103:9080-9084.

[159] Zhao X, Bagwe RW, Tan W. Development of Organic-Dye-Doped Silica Nanoparticles in a Reverse Microemulsion. *Adv. Mater.* 2004;16:173-176.

[160] Medintz IL1, Uyeda HT, Goldman ER, Mattoussi H. Quantum dot bioconjugates for imaging, labelling and sensing. *Nature Mater.* 2005;4:435-446.

[161] Jaiswal JK, Mattoussi H, Mauro JM, Simon SM. Long-term multiple color imaging of live cells using quantum dot bioconjugates. *Nature Biotechnol.* 2002;21:47-51.

[162] Zrazhevskiy P, Gao X. Quantum dot imaging platform for single-cell molecular profiling. *Nature Commun.* 2013;4:1619.

[163] Rakovich TY, Mahfoud OK, Mohamed BM, Prina-Mello A, Crosbie-Staunton K, Broeck TVD, Kimpe LD, Sukhanova A, Baty D, Rakovich A, Maier SA, Alves F, Nauwelaers F, Nabiev I, Chames P, Volkov Y. Highly Sensitive Single Domain Antibody Quantum Dot Conjugates for Detection of HER2 Biomarker in Lung and Breast Cancer Cells. *ACS Nano.* 2014;8:5682–5695.

[164] Tu C, Ma X, House A, Kauzlarich SM, Louie AY. PET Imaging and Biodistribution of Silicon Quantum Dots in Mice, *ACS Med. Chem. Lett.* 2011;2:285–288.

[165] Arima H, Yoshimatsu A, Ikeda H, Ohyama A, Motoyama K, Higashi T, Tsuchiya A, Niidome T, KatayamaY, Hattori K, Takeuchi T. Folate-PEG-Appended Dendrimer Conjugate with α-Cyclodextrin as a Novel Cancer Cell-Selective siRNA Delivery Carrier. Mol. *Pharmaceutics.* 2012: 9;2591−2604.

[166] Almutairi A, Rossin R, Shokeen M, Hagooly A, Ananth A, Capoccia B, Guillaudeu S, Abendschein D, Anderson CJ, Welch MJ, Frechet JMJ. Biodegradable dendritic positron-emitting nanoprobes for the noninvasive imaging of angiogenesis. *PNAS.* 2009;106:685–690.

[167] Venkatesh V, Shukla A, Sivakumar S, Verma S. Purine-Stabilized Green Fluorescent Gold Nanoclusters for Cell Nuclei Imaging Applications. *ACS Appl. Mater. Interfaces* 2014;6:2185−2191.

[168] Wang Y, Liu Y, Luehmann H, Xia X, Wan D, Cutler C, Xia Y. Radioluminescent Gold Nanocages with Controlled Radioactivity for Real-Time in Vivo Imaging. *Nano Lett.* 2013;13:581−585.

[169] Sumer B, Gao J. Theranostic nanomedicine for cancer. *Nanomedicine* 2008;3:137-140.

[170] Muthu MS, Leong DT, Mei L, Feng SS. Nanotheranostics - Application and Further Development of Nanomedicine Strategies for Advanced Theranostics. *Theranostics* 2014;4:660-677.

In: Innovations in Nanomaterials
Editors: Al-N. Chowdhury, J. Shapter, A. B. Imran

ISBN: 978-1-63483-548-0
© 2015 Nova Science Publishers, Inc.

Chapter 5

NANOMATERIALS AND THEIR APPLICABILITY AS MEMBRANES' FILLERS

Evangelos P. Favvas[*]

Institute of Nanoscience and Nanotechnology, NCSR "Demokritos", Terma Patriarchou Grigoriou and Neapoleos, Aghia Paraskevi, Attica, Greece

ABSTRACT

Development of both polymeric and inorganic gas separation membranes is one of the fastest growing sections of membrane technology. However, until recently, polymeric materials have been somewhat deficient in terms of the properties required for the current membrane technology. Similar behavior is also presented by the inorganic materials which, despite their advances in mechanical, thermal and chemical stability, they haven't been established yet in many industrial separation processes. To this end a plurality of new, even "smart", materials and production protocols have been suggested for the preparation of more efficient membranes in both polymeric and inorganic structures. Nonetheless the solution, if it exists, lies in specific in the use of the nanomaterials, as the second phase of the existent, or new, polymeric and inorganic membrane materials. By this chapter, and based on this idea, we present the advantages of the nanomaterials, especially where they are used as membrane's filler materials. Furthermore an overall presentation of the critical "stations" of the field is presented and discussed. A special focus is given on the recent advances in the area of gas separation processes and especially in the case of mixed matrix membranes. Finally, our new experimental data of MWCNTs based mixed matrix hollow fiber membranes are also reported and discussed.

1. INTRODUCTION

Nanomaterials are in general the materials with at least one external dimension in the size range from approximately 1nm to 100nm [1-3]. Furthermore nanoparticles are these objects which have all three external dimensions at this scale [4]. Figure 1 shows the scale of things,

---

[*] E-mails: e.favvas@inn.demokritos.gr & e.favvas@gmail.com.

natural and manmade. As we can see (Figure 1) the threshold from the "microword" to the "nanoword" is noted at the 0.1µm. There are several definitions of nanotechnology and of the products of nanotechnology, often these have been generated for specific purposes. According to the Scientific Committee on Emerging and Newly Identified Health Risks (SCENIHR) [5], the underlying scientific concepts of nanotechnology have been considered more important than the semantics of a definition, so these are considered first. The Committee considers that the scope of nanoscience and nanotechnology used by the UK Royal Society and Royal Academy of Engineering in their 2004 report (Royal Society and Royal Academy of Engineering 2004) adequately expresses these concepts. This suggests that the range of the nanoscale is from the atomic level, from ~ 0.2nm up to ~ 100nm.

It is within this range that materials can have substantially different properties compared to the same substances at larger sizes, both because of the substantially increased ratio of surface area to mass, and also because quantum effects begin to play a role at these dimensions, leading to significant changes in several types of physical property. The present opinion uses the various terms of nanotechnology in a manner consistent with the recently published Publicly Available Specification on the Vocabulary for Nanoparticles of the British Standards Institution (BSI 2005), in which the following definitions for the major general terms are proposed [5]:

- Nanoscale: having one or more dimensions of the order of 100nm or less.
- Nanoscience: the study of phenomena and manipulation of materials at atomic, molecular and macromolecular scales, where properties differ significantly from those at a larger scale.
- Nanotechnology: the design, characterization, production and application of structures, devices and systems by controlling shape and size at the nanoscale.
- Nanomaterials: materials with one or more external dimensions, or an internal structure, which could exhibit novel characteristics compared to the same material without nanoscale features.
- Nanoparticle: particle with one or more dimensions at the nanoscale. (Note: In the present report, nanoparticles are considered to have two or more dimensions at the nanoscale).
- Nanocomposite: composite in which at least one of the phases has at least one dimension at the nanoscale.
- Nanostructured: having a structure at the nanoscale.

Overall the particle size of a material plays an important role in its properties, such as porosity, chemi-resistance, electrical resistance, photocatalytic properties, transparency, solubility (mainly for salts), strength etc. Below this size, even of the same nature of its material, the properties changed dramatically [8, 9]. A characteristic case reported by Emil Roduner [10] which is the example of the gold. As Roduner reports: "*Gold is a shiny, yellow noble metal that does not tarnish, has a face centred cubic structure, is non-magnetic and melts at 1336 K. However, a small sample of the same gold is quite different, providing it is tiny enough: 10nm particles absorb green light and thus appear red. The melting temperature decreases dramatically as the size goes down. Moreover, gold ceases to be noble, and 2-3nm nanoparticles are excellent catalysts which also exhibit considerable magnetism. At this size they are still metallic, but smaller ones turn into insulators. Their equilibrium structure changes to icosahedral symmetry, or they are even hollow or planar, depending on size etc."* [10]. Therefore, the properties of the nanomaterials (especially the nanoparticles) must be studied at various size ranges because their characteristics are strong depended on their dimensions.

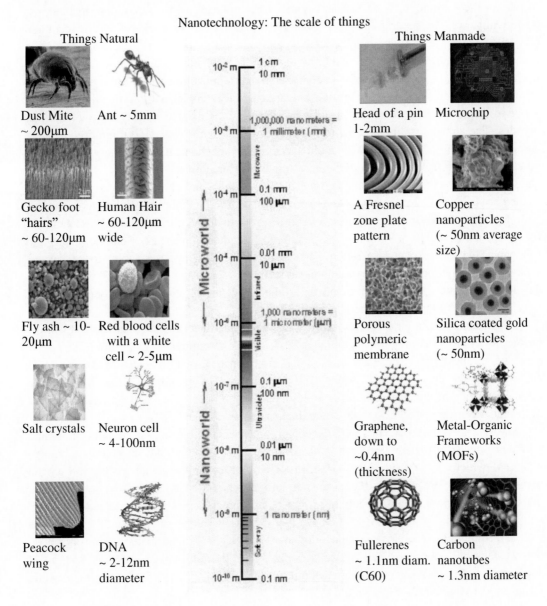

Figure 1. Nanotechnology, the scale of things [6, 7].

Therefore, and thanks to these interesting and unexpected properties, the nanomaterials have been instituted one of the most promising research areas in the field of the preparation of the mixed matrix membranes.

2. NANOMATERIALS AS MEMBRANE FILLERS

Membranes, both polymeric and inorganic, have been used in various industrial applications such as reverse osmosis (RO), gas separation, microfiltration, desalination, water purification, heavy metal removal, hemodialysis etc. Additionally, polymeric and inorganic membranes constitute a wide research field for applications in separation technology. Specifically, for gas separation, the permeability and selectivity coefficients are the main properties which determine their efficiency and suitability for a specific process, while porosity, temperature resistance, swelling phenomena and electrical conductivity play an important role and depending on the intended application must be optimal [11]. Due to many advantages of membrane technology in comparison, for example, with cryogenic process, a lot of research and demonstration projects have been implemented worldwide over the last decades. The advantages include their high stability, efficiency, low energy requirement and ease of operation. Membranes with good thermal and mechanical stability combined with good permeability/selectivity properties are important for industrial processes [12]. To this end the improvement of the thermal and mechanical stability in conjunction with the increase of the permeability factors is one of the major targets for the membrane research community. Note that the gas separation occupies a central position in the chemical and petrochemical feedstock industry [13]. Some typical examples of membrane gas separations are the following [14]: *i*) hydrogen separation, *ii*) air separation, *iii*) acid gases separation, *iv*) water separations, *v*) hydrocarbons separation and *vi*) helium separation. Over the last decades, the development of new type of membranes is on the top of the research actions because of the large economical interest of all these separations.

In particular, a plurality of new promising membrane materials such as block, even di-block and tri-block, copolymer membranes [14], zeolite membranes [15], carbon molecular sieve membranes [16], AlPO membranes [17], mixed matrix membranes (MMMs) [18] etc., are reported in the literature. Among all these types of membranes the MMMs are traditionally one of the most promising candidates for gas separations, thanks to their ability in the enhancement of the properties of the final structure.

MMMs are composed of homogeneously interpenetrating polymeric and inorganic particles. Specifically, the incorporation of many types of nanomaterials into the membrane's matrices, which is a very complicated process, can provide special properties to the membrane's structure and the fluid transport properties [19]. Generally, the most popular dispersed inorganic particles in polymer matrices are metal oxides nanoparticles (e.g., MgO and TiO_2) [20-27], ZnO [28], Al_2O_3 [29, 30], Au [31], zero-valent iron (ZVI) [32], Pd [33] as well as zeolites [34, 35], carbon molecular sieves [36-43], silicalate materials [44-49], non-porous silica [50], porous silica [51], zeolites [52-54], both multi-walled (MWCNTs) [55-61] and single-walled (SWCNTs) carbon nanotubes [55, 62-64], clays [65-71], metal organic frameworks (MOFs) [72-78] and others.

3. THE USED NANOMATERIALS: PREPARATION AND PROPERTIES

Numerous of nanomaterials have been reported as candidates as membrane fillers. In this chapter the most interesting inorganic nanoparticles will be presented followed by their preparation route and their basic properties. Note that some are natural products and some others are synthesized in the laboratory. In the first case a purification process of the material is necessary before its use. The most reported conventional and alternative fillers are:

Conventional

1. *Metal Oxides*: Metal oxide nanoparticles such as magnesium oxide (MgO) and titanium oxide (TiO_2) are emerging materials due to their potential applications for membrane-based separation properties [19]. As Aramendia et al. reports, MgO can bee obtained mainly by the thermal decomposition of magnesium hydroxide [79]. Specifically, the following recipe reported for the synthesis of MgO: Two different precursors, namely: (a) $Mg_5(OH)_2(CO_3)_4 \cdot 4H_2O$ (Merck Art. 5827), which yielded the catalysts MgO(I)AIR and MgO(I)VAC, and (b) $Mg(OH)_2$ (Merck Art. 5870), which provided the solid MgO(II)AIR were used for the metal oxide production. All the solids tested were obtained by calcination in a ceramic crucible, either in the air or in vacuo (hence AIR or VAC labels), by heating from room temperature to 600°C at a rate of 4°C/min and then maintaining the final temperature for 2 h, after which the solids were allowed to cool to room temperature. Catalyst BM50 was prepared by suspending 29.0g of $Mg(OH)_2$ and 0.35 g of B_2O_3 in 200 ml of distilled water and sonicating the mixture for 1 h. The resultant solid was dried in a stove at 120°C for 2 h and calcined by using the above described temperature programme. The final Mg:B ratio thus obtained was 50:1 or carbonate [80] as well as, recently, by the sol-gel process [81]. Furthermore, metal oxides nanoparticles can be synthesized by microwave radiation [82] and consequently Figure 2 shows these MgO nanoparticles.

One of the major targets for the metal oxides nanoparticles uses is the bactericidal applications. The unusual crystal morphologies that possess numerous edge/corner and other reactive surface sites in conjunction with the extremely high surface areas are some of their main characteristics which enable the metal oxides as promising materials not only for their interesting bactericidal properties [83] but also for their applicability to use it as fillers in nanocomposites materials. In this approach, numerous nanoparticles include silver, iron, zirconium, silica, aluminium, titanium, and magnesium have been suggested as filler materials for the preparation of composite membranes. These nanoparticles can affect the permeability, the selectivity, the hydrophilicity, the conductivity, the mechanical strength, the thermal stability and the antibacterial properties of a series of polymeric membranes [84].

2. *Silica based materials*: Silica based materials have been some of the most reported materials especially in the field of inorganic chemistry and nanotechnology. The most known silica-based mesoporous nanomaterials are: MCM-, SBA-, MCF-, MSU-, KSW-, FSM-, HMM- which have candidate application in catalysis and drug delivery. From all these materials the MCM-41 and SBA-15 are the most studied and reported structures [85, 86]. Two characteristic SEM pictures of MCM-41 and MCF-LA are posted below (Figure 3) where the differences in the particle size as well as in their shape are clearly observed. As a conclusion the synthesis route plays an

important role to both size and crystallinity properties of prepared metal nanoparticles.

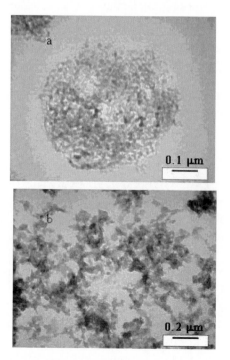

Figure 2. TEM images of the a) as-prepared MgO nanoparticles and b) MgO nanoparticles after crystallization (reproduced with permission from [82]).

The typical synthesis recipe of M41S materials (silicate/alumininosilicate mesoporous materials) involves a source of silica or other metal oxide precursors, the structure-directing surfactants, a solvent and a catalyst (an acid or a base) [88]. They are basically prepared through silica formation around template micelle assemblies followed by template removal by appropriate methods as calcination and solvent evaporation [89].

Table 1 shows the abbreviations and the full names of eight different mesoporous materials.

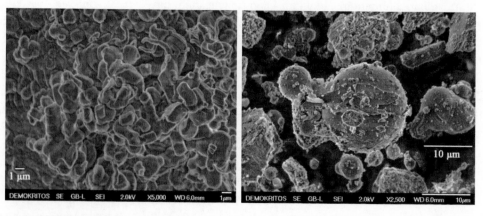

Figure 3. SEM images of the MCM-41 (left) and MCF-LA (right) silica based mesoporous amorphous nanomaterials (From [87]).

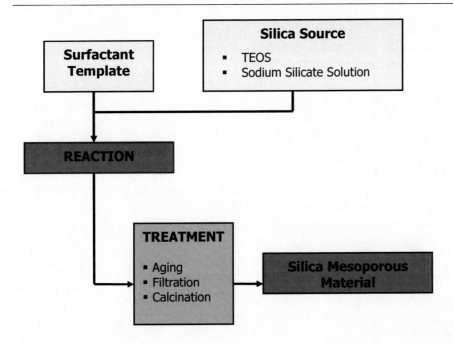

Scheme. Basic reaction scheme during silica mesoporous materials synthesis (From Reference [87]).

Table 1. Different Mesoporous Materials (From Reference [89])

Mesoporous Material Abbreviation	Full Name
MSU	Michigan State University
SBA	Santa Barbara Amorphous
MCM	Mobil Crystalline Matter/Mobil Composite Matter
HMS	Hollow Mesoporous Silica
OMS	Ordered Mesoporous Silica
TUD	Technische Universiteit Delft
MCF	Meso Cellular Form
FSM	Folded Sheet Mesoporous

The main objective of the use of the mesoporous silica based materials as membrane fillers is to achieve the increase of the permeability and at the same time the maintenance of the selectivity coefficients. The advantages of silica based micro-mesoporous materials for their use in the membrane technology is that their plurality in both porous and particle size in conjunction with their ability to be both organo- and hydrophilic are the basic tools for a good dispersion in the polymeric membrane's matrices without the creation of voids and cracks in the final membrane structure [90].

3. *Carbon molecular sieves (CMS)*: Porous carbons with high surface areas, large pore volumes, and chemical inertness are useful in many material application areas, including water and air purification, adsorption, catalysis, and energy storage [91]. A popular route for the preparation of carbon molecular sieves (CMS) is by control carbonization of polymer precursors [92, 93] or by hydrocarbons pyrolysis over

stable substrates [94]. Figure 4 illustrates a representation of a carbon molecular sieve with slit like pores as well as a SEM picture of an ultra-microporous carbon molecular sieve prepared by pyrolysis from BTDA-TDI/MDI co-polyimide.

The carbon molecular sieves can be used as membrane fillers due to their good gas selectivity properties. In specific, and as Vu et al. reported [92], the carbon molecular sieves appear to have numerous advantages relative to other fillers. For example from zeolites, because of the better affinity to glassy polymers (for example Matrimid® 5218 and Ultem®), achieving good adhesion and polymer–sieve contact with minimal preparation and casting modifications. In contrast, mixed matrix work with zeolites has identified poor polymer–sieve as a significant hurdle for successful film formation. Furthermore, modelling studies for the prediction of both pure and mixed gas permeation behaviour of the mixed matrix membranes are reported in the literature [96].

Figure 4. Left: 3D representation for which full knowledge of its structure and adsorption phenomena (reproduced with permission from [95]) and right: CMS derived from P84 polyimide precursor (original unpublished picture).

In the last decade a new class of carbon molecular sieves has been demonstrated in the literature. These new CMS have more organized nanostructures and present improved properties. They can be prepared through the use of mesostructured silica as a host to template the carbon structure. The subsequent removal of the silica host by dissolution methods allowed the carbon replica to be recovered intact [97, 98]. By this technique a series of new interesting and promising carbon microporous materials, for example CMK-3, have been prepared and have been tested at various interesting applications such as hydrogen storage [99], fuel cells [100], CH_4, CO_2 adsorption [101] etc.

4. Zeolites: Zeolites are microporous, aluminosilicate minerals commonly used as commercial adsorbents and catalysts [102]. The term zeolite was originally coined in 1756 by Swedish mineralogist and chemist Axel Fredrik Cronstedt, who observed that upon rapidly heating the material stilbite, it produced large amounts of steam from water that had been adsorbed by the material. Based on this, he called the material zeolite, from the Greek ζέω (zéō), meaning "to boil" and λίθος (líthos), meaning "stone" [103, 104].

Figure 5 illustrates a photo of a clinoptilolite-Na natural zeolites as well as a model of the microporous structure of a zeolites ZSM-5.

Zeolites are some of the most studied materials mainly because of their good adsorption properties and their accurate nanostructure. The main applications of zeolites materials can be found in catalysis, separation, purification and storage for both gas and liquid phase processes. In addition, zeolites have been proposed as additive materials in membrane technology. For example, Karatay et al. [107, 108] studied the effect of low molecular weight additive (LMWA) loading on the gas permeation properties of pure PES and PES/SAPO-34 membranes. The researchers conclude that the incorporation of zeolite SAPO-34 to neat PES membranes resulted in higher permeabilities for all gases with slight decrease in CO_2/CH_4 and H_2/CH_4 ideal selectivities compared to pure PES membranes. Furthermore, PES/SAPO-34/HMA membranes had significantly higher ideal selectivities than pure PES, PES/SAPO-34, and PES/HMA membranes, for example the H_2/CH_4 selectivity of HMA/SAPO-34/PES membrane was 175.8. Additionally, Hudiono et al. [109], as well as many other researchers, have studied the effect of zeolites nanoparticles on the separation properties of mixed matrix membranes.

Alternatives:

5. *Layered silicate*: These materials commonly used in nanocomposites belong to the structural family known as the 2:1 phyllosilicates (see also talc and mica minerals). The layer thickness is around 1nm and the lateral dimensions of these layers may fluctuate from 300 Å to several microns and even larger depending on the particular silicate. These layers organize themselves to form stacks with a regular van der Walls gap in between them called the interlayer or the gallery [110]. Layered silicates materials can be both natural (mineral) and synthetic. In the case of mica-type layered silicates it has been recently demonstrated that nanocomposites (both intercalated and delaminated) can be synthesized by direct melt intercalation even with high molecular weight polymers [111-113]. Figure 6 shows the 3D structure of layered silicate as well as an SEM picture of a layered silicate HUS-7 [114].

Due to the ability to exploit the high aspect ratio of exfoliated selective flakes/layers the composite layered silicate/polymer membranes have won the high interest of the scientific community [116]. Even if no direct and objective evidence of the "nano" structure is given, a large number of the nanocomposites membranes' properties (gas permeability, enhanced mechanical properties, thermal stability) tends to demonstrate that the behavior of these layered silicate-based composites is in the range of what is usually observed for nanocomposites [110].

6. *Carbon nanotubes (CNTs)*: Since their first observation nearly two decades ago by Sumio Iijima [117] carbon nanotubes have been the focus of considerable research. Carbon nanotubes (CNTs) are very interesting and novel graphite based materials which are made up by the rolling of the graphite sheets into a tube. As a result, a hollow cylinder is produced which is usually capped at least at one's end. In general, the carbon nanotubes (CNTs) can be synthesized as singular tubes, which are

referred as single walled carbon nanotubes (SWCNTs) or as a series of concentric cylinders of different diameters spaced around a common axis called multiwalled carbon nanotubes (MWCNTs) [118]. In Figure 7 a common 3D representation as well as TEM images of SWCNTs and MWCNTs are mentioned.

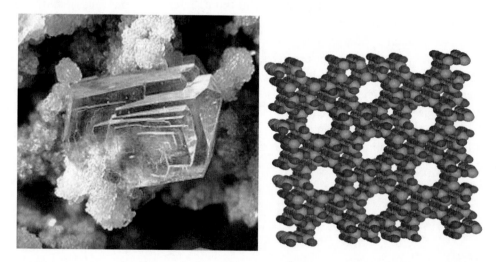

Figure 5. Left: Clinoptilolite-Na picture (Reproduced with permission from [105]) and right: the microporous molecular structure of a zeolite, ZSM-5 (Reproduced with permission from [106]).

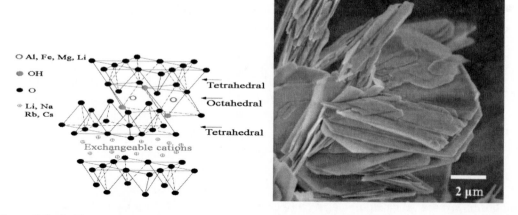

Figure 6. Left: The structure of 2:1 layered silicates (Reproduced with permission from [115]). Right: SEM picture of layered silicate HUS-7 (Reproduced with permission from [114]).

Due to their advanced properties the CNTs are suggested for a plurality of technological applications including the medical industry, the material science, the surface chemistry, the mechanical and chemical engineering. These materials are thoroughly discussed in section 4 below.

7. *Metal organic frameworks (MOFs)*: Metal organic frameworks (MOFs) are new polymeric porous materials, consisting of metal ions linked together by organic bridging ligands [120, 121]. MOFs have attracted much research interest in recent years, and are emerging as a new family of molecular sieves [122, 123]. MOFs are

novel porous crystalline materials consisting of metal ions or clusters interconnected by a variety of organic linkers. Furthermore, to promising applications in adsorptive gas separation and storage or in catalysis, their unique properties, such as their highly diversified structures, large range in pore sizes, very high surface areas, and specific adsorption affinities, make MOFs excellent candidates for use in the construction of molecular sieve membranes with superior performance [124-126]. The preparation of MOFs' based MMMs is similar to the one applied for the synthesis of other MMMs. Following, Figure 8 represents the outline of chemical-3D structure as well as the cubic crystals of MOF-5 (from SEM picture).

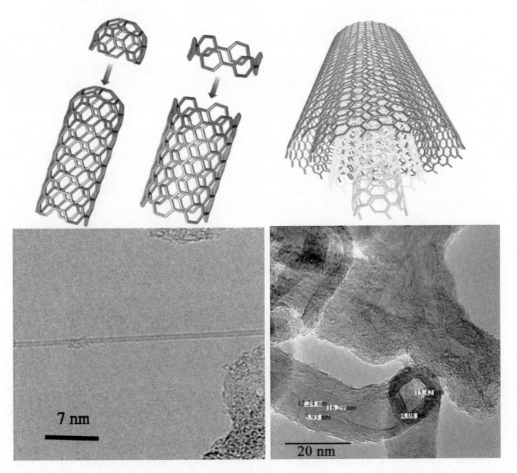

Figure 7. Up: Representations, and down: TEM images of SWCNTs (left: reproduced with permission from [119]) and MWCNTs (right: original unpublished picture).

Work where varieties of MOFs have been used as filler materials in membrane preparation technology are now very common. Note that the research trend in mixed matrix membranes (MMMs) has shifted towards the utilization of MOFs [129]. The main mechanism of the gas permeation transport in MOFs' MMMs is governed by the solution-diffusion mechanism because of their higher fraction of continuous polymer matrix. It is important to mention that the use of MOFs materials in the membrane technology is still an open discussion in the community mainly because of the pure stability of these materials. The question: *"MOF membranes: Bright industrial future or a laboratory curiosity?"* [130] is still unanswered.

8. *Graphene:* The 2010 Nobel Prize in Physics awarded jointly to Andre Geim and Konstantin Novoselov "for groundbreaking experiments regarding the two dimensional material graphene" recognized the potential impact of fullerenes, carbon nanotubes, graphene, and other carbon nanostructures in future nanotechnology-based discoveries [131]. Graphene, a monolayer of sp2-bonded carbon atoms, is a quasi-two-dimensional (2D) material and has been attracting great interest because of its distinctive band structure and physical properties [132]. Figure 9 shows a 3D approach of the pillared graphene, a novel 3-D network nanostructure proposed for enhanced hydrogen storage, as well as a TEM image of exfoliated graphenes' sheets.

Graphene as well as graphene oxide (GO) materials have been used as membrane materials and fillers for the preparation of proton exchange membranes [134], gas separation membranes [135], nanofiltration membranes [136], ultra-fast molecular sieve membranes [137] etc. Between these two materials, Graphene and GO, the graphene-derived sheets, graphene oxide sheets, are heavily oxygenated, bearing hydroxyl and epoxide functional groups on their basal planes, in addition to carbonyl and carboxyl groups located at the sheet edges [138]. The presence of these functional groups makes GO sheets strongly hydrophilic, which allows graphite oxide to readily swell and disperse in water [139, 140]. Therefore GO are better fillers in the case of water solutes based polymeric membranes' matrices.

Overall must be noted that the properties of the graphene/graphene oxide nanocomposites membranes are reported improved and it is sure that these materials will be on the top of the research interest for the next decades.

4. CARBON NANOTUBES AS MEMBRANES' FILLERS

Furthermore nanotubes, especially carbon nanotubes, are one of the high-interest reported nanomaterials because they are used as filler for composite membrane preparation [141-143]. Carbon nanotubes (CNTs) are a fascinating class of materials thanks to their special properties that can be exploited on numerous applications. Specifically, their extremely high aspect ratio, which makes them to molecular-level needles, their electrical conductivity, their mechanical strength and their good thermal properties as well as their wave adsorption characteristics establish them promising materials for many applications [144-149]. The interest of using nanoparticles as fillers in membrane structures focuses mainly on their beneficial effect on fluxes and fouling resistance [150]. Significant improvement in separation properties compared to neat polymers is expected for the resultant MMMs [151]. Inorganic particles added to a polymer matrix can have the following three effects on the permeability: *i)* they can act as molecular sieves altering the permeability, *ii)* they can disrupt the polymeric structure increasing the permeability too and *iii)* they can act as barriers reducing the permeability [14, 152]. Hind et al. [153], in one of the pioneering works in the field, reported in 2004 that nitrogen permeability, when aligned carbon nanotubes incorporate into the polymer matrix, provides a clear indication of the potential of the nanotubes inner cores to act as a feasible channel for their transport. Moreover a plurality of composite membranes obtained mainly with randomly dispersed functionalized/modified carbon nanotubes were reported in the literature in the last ten years. Specifically, in 2007 Cong et al. [55] prepared composite poly(2,6-diphenyl-1,4-phenylene oxide (BPPO$_{dp}$) membranes using both pristine single-wall CNTs (SWNTs) and multi-wall CNTs (MWNTs). They concluded that the composite membranes had an increase in CO_2 permeability but a similar CO_2/N_2 selectivity compared to the corresponding pure polymer membrane. In 2009, Weng et al. [154] developed a series of MWCNT/PBNPI (PBNPI: poly(bisphenol A-co-4-nitrophthalic anhydride-co-1,3-phenylene diamine) nanocomposite membranes with a nominal MWCNT content between 1 and 15 wt% in order to separate H_2 from CH_4. The researchers found that this new class of membrane had increased permeability and enhanced selectivity, and a useful ability to filter gases and organic vapors at the molecular level. In 2010, Goh et al. [155] prepared matrix polyetherimide (PEI) based membranes by embedding dispersed MWCNTs using surfactants of different charges. The prepared MMMs exhibited improved thermal stability and mechanical strength, likewise the resulting membrane exhibited permeance improvement of O_2 and N_2 as much as 87.7% and 120% respectively compared to that of neat PEI membrane. Aroon et al. [59], in 2010, fabricated PI/Raw-MWCNTs and PI/chitosan-functionalized MWCNTs (PI/C-f-MWCNTs) mixed matrix membranes by phase inversion method. Gas permeation results showed that addition of 1 wt% (solid base) C-f-WCNTs into casting dope can increase the CO_2 and CH_4 permeabilities by 20.48 and 0.71 Barrer, respectively. The increase in permeabilities of CO_2, CH_4 and He was attributed to the presence of high diffusivity tunnels in the MWCNTs within the polyimide matrix. It was interesting that the CO_2/CH_4 selectivity was also increased by 51.4% (from 10.9 to 16.5) by addition of C-f-MWCNTs. Recently, (2011), Ge et al. [156] used MWCNTs, after treatment with sonication in a mixture of H_2SO_4/HNO_3 for 3 h at 60∘C at water bath, as filler materials in polyethersulfone nanocomposite membrane preparation. By this addition the gas permeation fluxes of the derived membranes increased by ~ 67% without sacrificing selectivity when 5 wt% of MWCNTs was introduced. During the same year, Ge et al. [157] used metal and carboxyl doped MWCNTs as filler materials in PES polymer matrix membranes and they found that the selectivity increases when the added CNTs are modified by Ru, but it remains similar or even decreases when carboxyl and Fe are employed to functionalize CNTs. On the other hand, by controlling the Ru modification site, higher gas selectivity was found when Ru particles were deposited mainly on the external surface of CNTs. Furthermore, Sieffert and Staudt [158] reported that commercial, hydroxyl functionalized-MWCNTs were obtained in a two step reaction and characterized. Different procedures to prepare homogenous hybrid films by ultrasound dispersion were investigated and appropriate instructions were found.

However, the investigations on MMMs show that these hybrid materials are not simple systems where the effect of the fillers can easily be understood. To this end, it is necessary an extended study of the optimum way where the inorganic particles must be added into the polymer matrices. The physicochemical properties of the inorganic fillers as well as the nanoparticle/polymer interface morphology [159] are individual factors which have to be investigated in order to facilitate the gas permeation properties of these news materials and, finally, to predict the changes of the MMM performance from the age of their design.

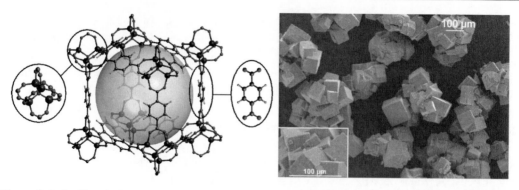

Figure 8. Left: Chemical structure of MOF-5 (reproduced with permission from [127]) and right: SEM image of MOF-5 cubic crystals (reproduced with permission from [128]), the inset shows an enlarged view.

Figure 9. Left: representation of pillared graphenes' sheets (reproduced with permission from [133]), center: SEM image of graphene sheets of about 8µm in length &, right: TEM image of graphenes' sheets (*SEM & TEM are original unpublished pictures*).

5. MIXED MATRIX HOLLOW FIBER MEMBRANES (MWCNTS/PI): EVALUATION STUDY IN GAS SEPARATION PERFORMANCE

In previous work [160] the gas permeance/selectivity performance of single co-polyimide hollow fiber membranes was reported. Furthermore, in our recent work [61, 161] MWCNTs were prepared, modified and dispersed into co-polyimide based hollow fiber membranes. Specifically, advanced experimental techniques are utilized for the characterization of the produced nanocomposite membranes together with the polymeric BTDA-TDI/MDI co-polyimide. TEM was employed to observe the morphological characteristics of the fillers. Nanoscale structural characteristics of the composite membranes were revealed by SANS. Furthermore, the thermal properties of the membranes were tested by DSC measurements, whereas FTIR was used for investigating the effect of the filler addition in the character or quantity of the particular bonds. Finally the effect of the filler concentration into the membrane matrix was carried out by single phase permeance experiments of He, H_2, CH_4, N_2, O_2, C_2H_6, CO_2 and C_3H_8 at 25, 60 and 100oC respectively.

5.1. MWCNTs/Polyimide Hollow Fiber Membranes: Preparation Procedures and Characterization Techniques

One polymeric (S1) and three nanocomposite membranes (C1, C2 and C3) with MWCNT concentrations 1, 2 and 4 wt% respectively were prepared via the wet spinning technique, using a spinning set up, in a way described in our previous work [16, 162]. Previously, MWCNTs were synthesized using a novel catalytic CVD (*Chemical Vapor Deposition*) technique and afterwards by the functionalization process certain functional groups, namely carboxyl and phenol groups were introduced on the external surface of the MWCNTs [163]. The TEM images were produced by a JEOL-JEM-2100F Transmission Electron Microscope.

The thermal properties of the samples were tested by differential scanning calorimetry (DSC) in a TA Instruments, Model MDSC 2920. The runs were undertaken using a heating ramp of 2°C/min, a cooling rate of 5°C/min with a temperature modulation of $\pm 0.32^{\circ}$C every 60 s.

Fourier transform infrared spectroscopy (FTIR) spectra were recorded using a horizontal ATR Trough plate crystal cell (Thermo Electron 6700 ATR diamond) equipped with a Nicolet 6700 FTIR (Thermo Electron Corporation) operating at room temperature. For spectra analysis, the samples were placed on the crystal cell, which was in turn, mounted on the spectrometer. The spectra in the range of 400-4000cm^{-1} were scanned and the automatic signals were collected. The background spectrum was recorded at room temperature with an empty cell. Before FTIR analysis, all samples were dried at 100°C in order to remove adsorbed water from the surface.

The small-angle neutron scattering (SANS) measurements were carried out at the PACE spectrometer (LLB, Saclay, France) in a Q-range varying from 0.03 to 2.2nm^{-1}, corresponding to two sample-detector distances (4.57 m and 0.67 m respectively). All measurements were performed under atmospheric pressure and at room temperature. The samples were in powdered form and were mounted in a 1mm thick sealed quartz container (provided by Hellma). Absolute values for the intensity, $I(Q)$, were obtained by correcting the raw data for sample transmission, scattering of the empty cell and instrumental background, as a final step, the data were calibrated against a water standard [164]. All corrections were treated by using the software PASINET (for a detailed procedure of data treatment, see ref. 164).

Finally, single gas permeability measurements of He, H_2, CH_4, N_2, O_2, C_2H_6, CO_2 and C_3H_8 were carried out in a high accuracy metallic apparatus [162] at 25, 60 and 100°C, the feed pressure was ~ 1 bar and the vacuum permeate pressure was 10^{-7} mbar. The experiments were performed using the constant pressure/variable volume method (see ref. 11 for details).

5.2. MWCNTs Preparation and Modification

Multi-walled carbon nanotubes were produced in a fluidized bed chemical vapor deposition vertical reactor (FBCVD) which has been tailored for the synthesis of high-purity MWCNTs using proprietary catalysts [165]. Temperature was controlled by a controller with three Pt/Pt–Rh thermocouples. The experimental device was completed by mass flow controllers and flow read-out units. After stabilization of the system at the operating temperature, which varied between 650°C and 800°C, the gaseous feed stream was supplied to the reactor. Diluted, in helium, ethylene or acetylene was employed as carbon precursor. The CNTs synthesis techniques involved novel reactor design and specifically sized catalysts that can achieve narrower CNTs diameter distribution and purities of up to 99%. A variety of MWCNT diameters could be produced in the FBCVD system that lie in the range from under 10nm up to 50nm. The MWCNTs used in this chapter ranged from 15-40nm.

Covalent modification of pristine MWCNTs was applied in order to optimize both the nanoscale morphology and their dispersion in organic solvents [166, 167]. The functionalization process introduced certain functional groups, namely phenol groups, on the external surface of the nanotubes [163, 168]. The functionalization procedure towards the – phenol moieties attachment, was converted through "wet chemistry" by the addition of p-aminophenol reactants to form sigma bonds with the p_z orbitals of the carbon atoms on the outer surface of the nanotube.

5.3. Physicochemical Characterization of MWCNTs

5.3.1. Morphological Properties of MWCNTs (TEM & SEM)

Figure 10 shows representative TEM micrographs of the phenol-functionalized MWCNTs selected as filler material for the preparation of nanocomposite hollow fiber membranes. TEM images show clearly that the used material is pure MWCNTs with a narrow size distribution and uniform stipulation of the building walls. Figure 10c shows a high magnification view of a MWCNT with a diameter of ~ 38nm. Another interesting finding is that the open channel (pore) is only ~8nm compared to the wall thickness (about 30nm). More precisely, TEM measurements also reveal that the outer diameter of the MWCNTs lies in the 25-50nm range, while the diameter of the pore varies in the range 7-10nm. The functionalization procedure towards the attachment of the phenol moieties was achieved through "wet chemistry" by the addition of p-aminophenol reactants in order to form sigma bonds with the p_z orbitals of the carbon atoms on the outer surface of the nanotube.

Figure 11 shows representative SEM micrographs of pristine MWCNTs produced from ethylene at 700°C (upper images) and subsequently functionalized with phenol groups (lower images). In all images, only bundles of multi-wall carbon nanotubes are observed (Figures 11(a) and 11(d)) and there is no sign of any impurities e.g., amorphous carbon, which would be present in granular form. Especially, as we can see in Figures 11(e) and 11(f) only the existing carbon nanotubes are illustrated as a typical form of ribbon complexes. The diameter of MWNTs varies from 15 to ~ 40nm.

5.3.2. Thermogravimetric Analysis (TGA)

Using Thermogravimetric Analysis (TGA) in air, the carbon content of pristine MWCNTs was determined to be around 98.5% (i.e., metal particle content 1.5%) with the amount of amorphous carbon in the carbon content being minimal (Figure 12).

The combustion of multi-wall carbon nanotubes takes place at temperatures above 550°C, whereas that of amorphous carbon at 300-350°C. TGA analysis of functionalized MWCNTs in inert atmosphere revealed a ~ 7.5% wt% percentage of phenol groups in the CNTs sample (Figure 12). By the TGA curve can also determined that in case of the pristine carbon nanotubes the effect of the temperature on the sample mass is equal to zero at temperatures up to 653°C and this is due to their high purity of the carbon nanotubes structure. If amorphous carbon or any other type of impurities existed the mass of the sample will be decrease as the temperature increase.

5.3.3. X-Ray Diffraction (XRD) Study

X-ray spectroscopy is a good tool for the structural characterization of carbon nanotubes. Figure 13 shows the powder XRD pattern (CuKα, λ = 1.5406 Å) of pristine and functionalized multi-wall carbon nanotubes samples. The peaks are indexed to the (002) and (101) reflections of hexagonal graphite. The 'd' value corresponding to the (002) peak ($2\vartheta = 26.3°$) is evaluated to be 3.38 Å, which is very close to that of graphite (3.35 Å) [169]. The presence of (002) peak in the XRD patterns suggests the multi-walled nature of carbon nanotubes. The pattern shows the main intense peak at $2\theta = 26.3°$ and compared to the normal graphite, $2\theta = 26.5°$, this peak shows a downward shift, which is attributed to an increase in the sp^2, C = C layers spacing [170]. The second diffraction peaks is at the angle 2θ of ~ 43.5° and indexed to the (101) reflection.

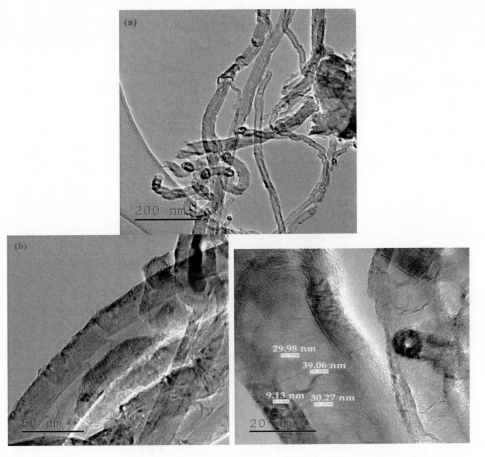

Figure 10. Low (a) and high (b & c) magnification TEM images of phenol-functionalized MWCNTs.

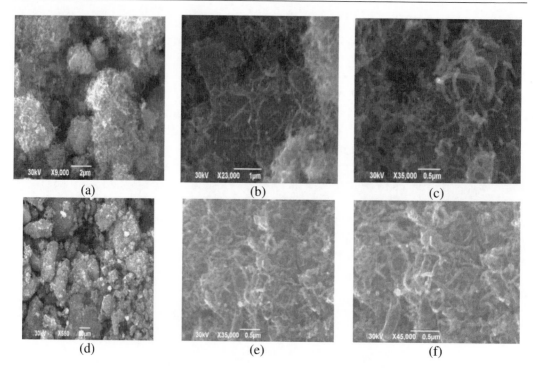

Figure 11. SEM pictures of the pristine (up) and functionalized (down) MWCNTs.

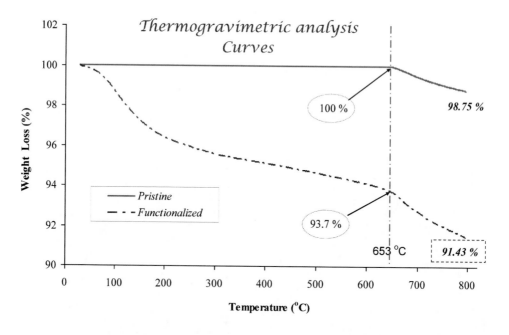

Figure 12. TGA curves for pristine and functionalized MWCNTs in air condition.

5.3.4. Raman Spectroscopy (RAMAN)

The nature of MWCNTs was also verified by Raman spectroscopy (Figure 14). Micro-Raman spectra were measured in backscattering configuration on a Renishaw in Via Reflex microscope using an Ar^+ ion laser ($\lambda = 514.5nm$) as excitation source.

Pristine MWCNTs exhibit narrower Raman bands, with the defect-activated D band being much less intense than the first order tangential G and this is indicative of CNTs with enhanced crystallinity and less amorphous carbon. The overtone at 2D (G' band) is activated independently of defects through a two-phonon double resonance process, and it is very sensitive to the stacking order of the graphitic walls and the crystallinity of the graphitic planes [171]. Two additional defect-induced modes of lower intensity were also observed: the D' band appearing as a shoulder to the G mode at ~ $1622cm^{-1}$ arising from a double resonance process similar to the D-band, and the relatively broad D + D' combination mode (~$2950cm^{-1}$ at 514.5nm) [172].

5.4. MWCNTs Dispersion Study in NMP Solvent

The precipitation growth of pristine, carboxyl and phenol-functionalized nanotubes was investigated optically for the MWCNTs/NMP solutions (0.85% w/v) (Figure 15).

Pristine nanotubes lead to precipitation of the system, mainly due to the strong van der Waals forces. In the case of carboxyl-functionalized nanotubes the system is more homogeneous however, traces of aggregates are visible in the area of air-liquid interface. Clearly, the best dispersion properties are observed for the phenol-functionalized MWCNTs. Further, thin films were prepared from the controlled evaporation of the above solutions (Figure 16). The film formed from the solution of pristine MWCNTs shows an aggregated morphology.

The sample with the carboxyl-functionalized MWCNTs gives a more uniform picture overall the sample area while the functionalized MWCNTs, with the phenol moieties on their external surface, show enhanced nanotube dispersion. Based on these results, which have been replicated at higher concentrations and for time periods from 24 up to 96 hours, phenol-functionalized MWCNTs were chosen as the best candidates for the preparation of nanocomposites hollow fiber membranes.

5.5. Preparation of Mixed Matrix Hollow Fiber Membranes

The produced functionalized multi-wall carbon nanotubes (MWCNTs) were used as filler materials in order to produce composite MWCNTs/polyimide hollow fiber membranes. In particular 1, 2 and 4 wt% concentrations of functionalized MWCNTs were dispersed in NMP solvent using a sonicator instrument at 50°C. Afterwards 28.5 wt% of commercial BTDA-TDI/MDI (P84) co-polyimide were added in the solution and mixed mechanically for eight hours isothermally (50°C). The solutions were filtered using a metal filter of 450 mesh and leave in overnight for outgassing. The extrusion was done using a special spinneret with the follow dimensions: needle ID = 0.5mm, needle OD = 0.7mm and orifice ID = 1.2mm. The spinning conditions were constant for all the prepared membranes and described follow: i) dope solution composition: 28.5% (P84/NMP), ii) bore fluid composition: 70/30 (NMP:H$_2$O), iii) dope flow rate: 5 ml/min, iv) bore fluid flow rate: 3.7 ml/min, v) air gap: 0cm, vi) room temperature: 20°C, vii) relative humidity: 36%, viii) take up velocity: 7.6 m/min, ix) dope solution temperature 50°C and x) the temperature of the coagulations baths was at 25°C. The MWCNTs concentration was 0% for the S1 membrane, 1% for the C1, 2% for the C2 and 4% for the C4.

The involved spinning set up is already described in our previous works [11, 16].

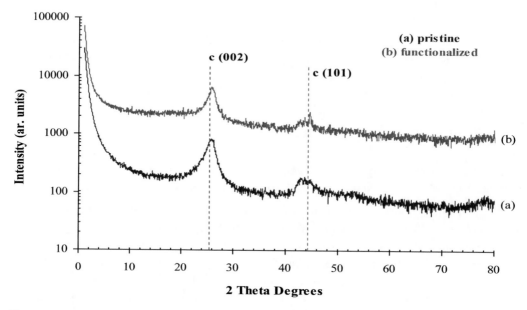

Figure 13. XRD patterns of pristine and functionalized MWCNTs.

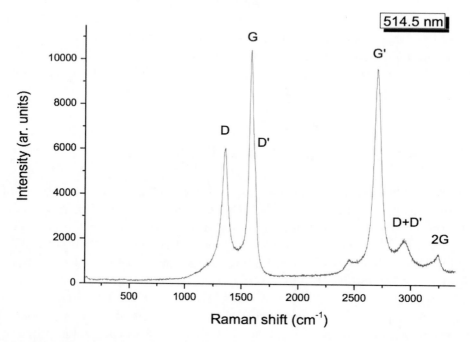

Figure 14. RAMAN spectra of pristine MWCNTs.

5.6. Characterization of Mixed Matrix Hollow Fiber Membranes

5.6.1. Morphological and Structural Characterization

The dimensions and asymmetric hollow fiber membranes were investigated by SEM. For the cross-sectional characterization, polymeric fiber samples were fractured after their immersion in liquid nitrogen.

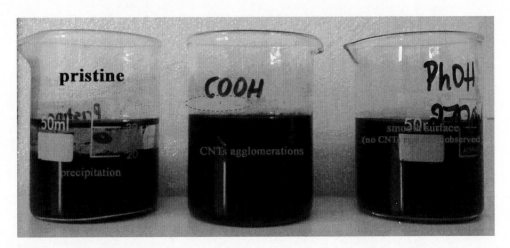

Figure 15. Precipitation profile of 0.85% w/v solutions of pristine, –COOH and –PhOH functionalized MWCNTs in NMP solvent after rest time of 96 h.

The produced specimens were mounted on the stub using a double-side conductive carbon adhesive tape. Fiber samples were sputter coated using an ion-sputtering device. Figures 17(a), 18(a), 19(a) and 20(a) illustrate the cross-section morphology of S1, C1, C2 and C3 hollow fiber membranes respectively which spun through a wet spinning process. It is worth mentioning that the inner "circuit" of all the hollow fibers is not uniform circumferentially and the shape seems like a hexagonal star. This materialization can happen due to the instability mechanisms during the spinning process. The solvent amount in the bore fluid plays an important role for this result and it is rendered as a different mass transfer driving force. Here it must be pointed out that the air–gap distance, the external coagulant, the take–up speed and the dope concentration contribute also significantly to the final membrane cross–section result. This phenomenon was studied in depth by other researchers previously [173-175]. Another common characteristic of hollow fiber membranes is the existence of a separating layer on the external surface of the samples. This layer, the outer layer, which is responsible for the separation characteristics, spreads perimetrically and has a length of about 2μm (see Figure 20(c)).

The average outer and inner diameters (OD and ID) of all the membranes are about 1200 and 800μm respectively (Figures 17(a), 18(a), 19(a) and 20(a)). The asymmetric structure of the membranes exists both in single and composites membranes. An interesting morphological feature is the existence of a quite dense interface layer at a distance of about 60μm from the inner surface of the fiber (Figures 17(b), 18(b) and 19(b)). This interface "separates" the fiber into two regions: the inner region with finger like voids (cavities) of 50μm length and 3-5μm width and the outer region with similar cavities of ~130μm length and 5-10μm width.

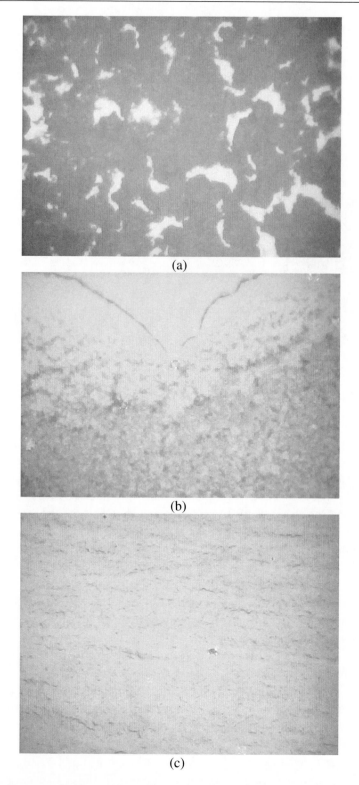

(a)

(b)

(c)

Figure 16. Optical images with a magnification scale of X80 for: (a) Pristine, (b) carboxyl and (c) phenol-functionalized multi-wall carbon nanotubes thin films (0.85% w/v in NMP).

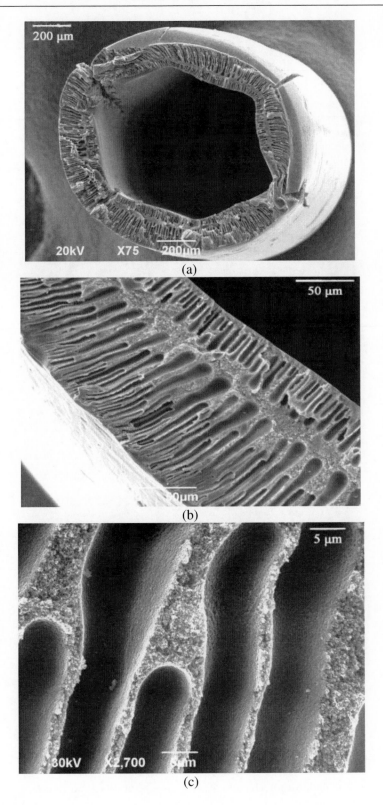

Figure17. (Continued)

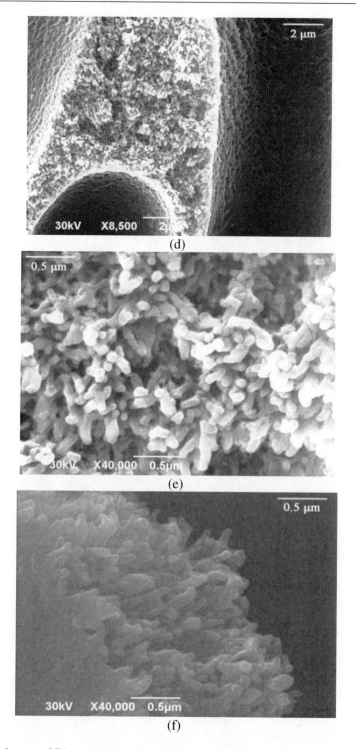

Figure 17. SEM pictures of S1 composite hollow fiber membrane.

The occurrence of such an interface layer has also been observed in other recent studies [176, 177] and reflects the boundary region between the opposite streams of the non solvent (water), diffusing simultaneously from the inner (bore liquid) and outer (coagulation bath) surface of the hollow fiber membrane.

The effect of the carbon nanotube fillers on the morphological characteristics of the hollow fiber membranes is unobservable. The filling MWCNTs doesn't change the "macroscopic" structure of the prepared membranes in the scale where the SEM technique can be determinated. It is clear that the structural characteristics of all the studied membranes are identical and the tracing of the filled MWCNTs is possible only in the case of C3 membrane, the membrane with the higher filler concentration of 4% wt% (see Figures 20(f) and 20(g)). These MWCNTs are shown in the Figures 20(f) and 20(g) where their diameter of 40nm is noticeable. In particular, the MWCNTs are well dispersed into the membrane matrices and it is not observed any agglomeration of them. The perfect dispersion of the fillers is established indirectly by gas permeability measurements where the effect of the MWCNTs concentration follows a linear relation with the permeance coefficients for the studied pure gases.

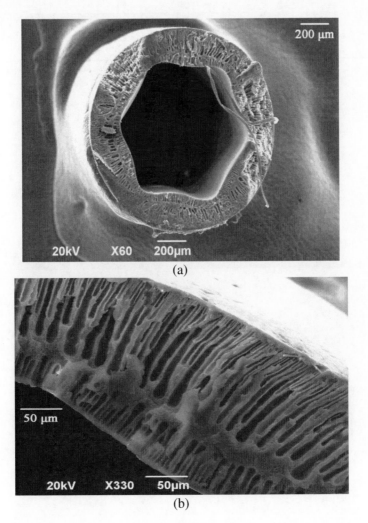

(a)

(b)

Figure 18. (Continued)

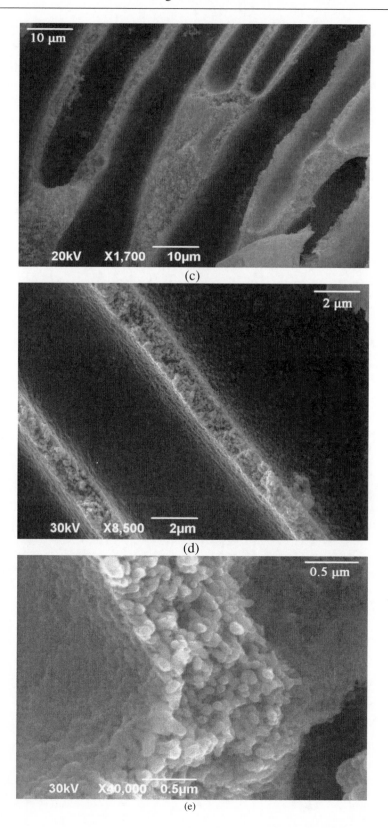

(c)

(d)

(e)

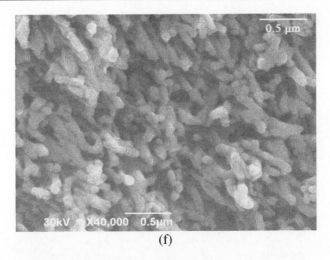

(f)

Figure 18. SEM pictures of C1 composite hollow fiber membrane.

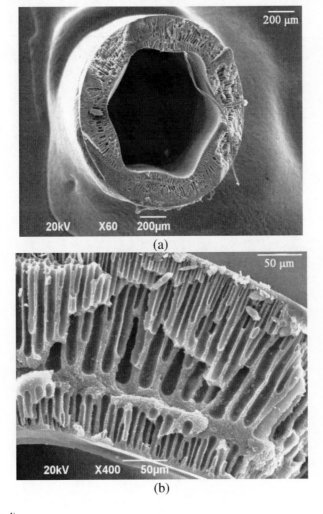

(a)

(b)

Figure 19. (Continued)

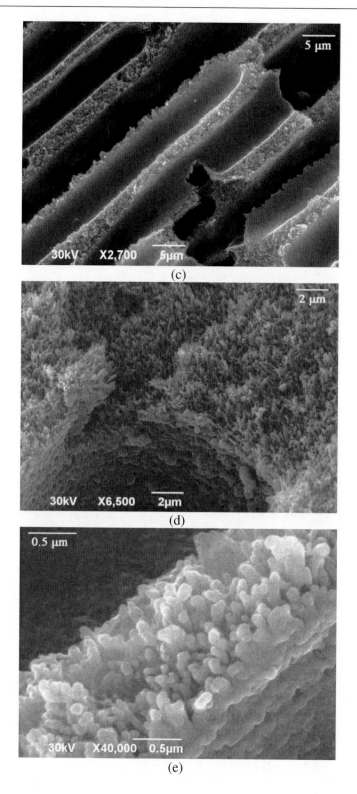

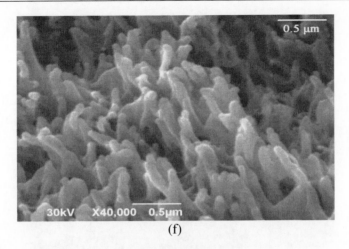

(f)

Figure 19. SEM pictures of C2 composite hollow fiber membrane.

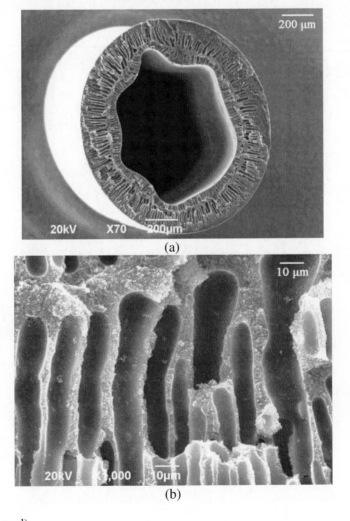

(a)

(b)

Figure 20. (Continued)

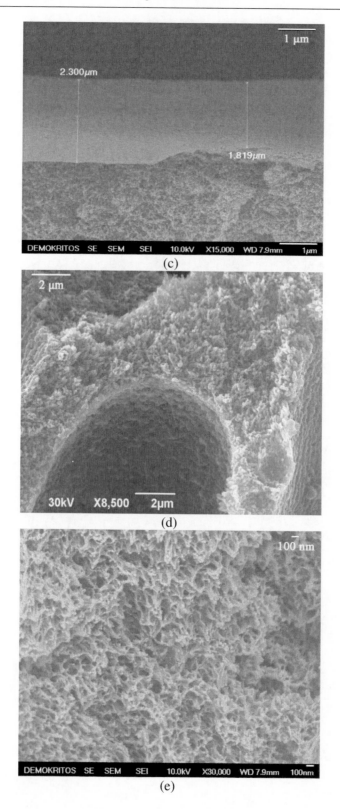

(c)

(d)

(e)

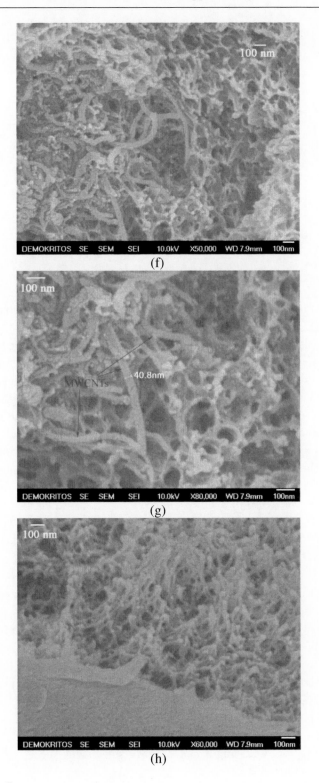

Figure 20. (Continued)

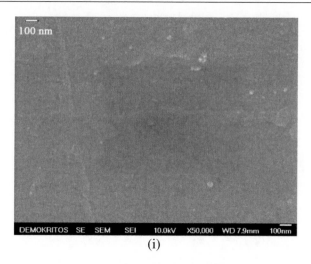

(i)

Figure 20. SEM pictures of C3 composite hollow fiber membrane.

Even in the case of C3 membrane (4 wt% MWCNTs) the external surface is clear of any MWCNTs "wing" (Figure 20(i)) due to the higher affinity of polymeric to polymeric chains during the phase inversion process, especially in the external region of the membrane where the extruded solution meets the water coagulant.

It should be noted that the separating layer (Figure 20(c)) is responsible for the gas separation properties of the membranes. In this region we can't see any differences by SEM, even at the higher resolution, because of the high density of the polymeric chains which are formed as a continuous "solid" phase. However the presented SEM pictures give information of the cross–section as well as of all the areas of the supporting structure of the membranes (separating layer). This information is important for the design of the hollow fiber membrane preparation route which is responsible for the final structural, mechanical and permeation/separation properties of the membranes.

5.6.2. Differential Scanning Calorimetry (DSC) & Fourier Transform Infrared Spectroscopy (FTIR) Analysis

Differential scanning calorimetry (DSC) measurements show that the presence of carbon nanotubes into the membrane matrix provides better thermal properties. In specific, the Tg of the pure polyimide membrane sample (S1) is 325°C and increases 1-4°C depending on the filler concentration. When 1 wt% MWCNTs was introduced into the polyimide matrix (C1 membrane), a small increase in Tg was observed (326°C). The Tg increase was more pronounced for the two samples with higher filler composition (328°C for the C2 sample and 330°C for the C3 sample) suggesting that the inorganic fillers restrict the movements of polymer chains [178]. This behavior could be explained in terms of a fair interfacial interaction between polymer and carbon nanotubes. Similar results have also been observed in other systems such as polymer-zeolites blends [179, 180]. Overall the DSC study implies good thermal stability for both single and MMMs suggesting successful ligature of the inorganic and organic moieties.

Figure 21 shows the FTIR spectra of all studied membranes. For comparison reasons, Figure 21(a) presents only the spectra of pristine (S1) and nanocomposite membranes (C1). As it can be seen (Figure 21(a) & (b)), the IR spectra of the four studied samples are almost identical for a broad range of frequencies except the region ~ 2600-3400cm^{-1}. In the region of the "identical area" of Figure 21 characteristic peaks of membrane (S1) and composites derivatives, C1, C2 and C3 are recorded. One common feature is that all samples exhibit the imide characterization peaks at 1780cm^{-1} (C = O asymmetric stretch), 1720cm^{-1} (C = O symmetric stretch), and 1360cm^{-1} (C-N stretch) [181, 182]. Another interesting finding is that the absorption peak at 1100cm^{-1} appears not only in pristine P84 membrane but also in all the composite derivatives this peak is due to the vibrations of the benzene rings.

In the frequency region 2600-3400cm^{-1}, the absorption peak at 3100cm^{-1} can be attributed to the asymmetric stretching vibrations of C-H groups in benzene rings (Figure 21(a)). In addition, the symmetric and asymmetric stretching vibrations of CH_2 groups give rise to the absorption peaks at 2930 and 2860cm^{-1} respectively. This behavior in next nearest neighbor wave numbers has also been reported for P84 polyimide film membranes [183]. Figure 21(b) shows that these peaks become weaker with slight frequency variations with the addition of phenol-functionalized MWCNTs into the matrix. The most probable explanation is that the vibrations of the C–H and CH_2 groups, with small energy potential, are cloaked from the existence of the MWCNTs which are "connected" with the polymer by weak van der Waals' forces.

5.6.3. Small-angle Neutron Scattering (SANS) Results

SANS is a powerful technique for investigating structural changes in materials (corresponding to length scales between 1- ~ 200nm) during various processes such as the formation of polymeric hollow fiber membranes [184], the effect of doping on the porosity of graphites [185], etc. In the present study, SANS measurements were carried out on pristine and functionalized MWCNTs, as well as on pristine and composite polymeric hollow fiber membranes. The SANS profiles of all samples decay following a power law ($I(Q) \propto Q^{-a}$). According to the theory, when $3 < a < 4$ the data are interpreted as scattering from surfaces Bale and Schmidt [186] derived $a = 6-D_s$ for a rough interface with a surface fractal dimension $2 < D_s < 3$. Apparently for a smooth surface $a = 4$ and $D_s = 2$ (the Porod's law [187] has been predicted. The scattering curves of pristine and (PH-OH-modified) functionalized MWCNTs are very similar and the calculated values of D_s are 2.55 and 2.44 respectively (Figure 22).

This result is reasonable as neither for pristine nor for functionalized MWCNTs a smooth surface is expected because MWCNTs have numerous carbonaceous impurities like carbon black, defects on their carbon networks and twists which lead to a rough surface [188, 189]. In addition, the presence of phenols in functionalized MWCNTs seems to contribute to a slight tendency for surface smoothening, possibly due to their attachment to the carbonaceous impurities and defects. The SANS curves of pristine and composite polymeric hollow fiber membranes are also presented in Figure 22. In the case of composite samples an intensity increase is observed as MWCNT concentration increases. This can be attributed to the contrast between the polymeric matrix and the nanotubes. The scattering increase is more pronounced for C1 and C3 membranes.

Except the intensity variation, the SANS profiles of all membranes are rather similar in a significant part of the Q-region ($0.03 < Q < 1.7nm^{-1}$), exhibiting a non-smooth surface texture with $D_s \sim 2.3$. The d-spacings ($d = Q/2\pi$), ~ 3.5 < d < ~ 210nm, give a rough estimation of the correspondent real space where the addition of nanotubes does not seem to cause substantial structural changes within the polymer matrix. The result suggests a uniform dispersion of nanotubes, if this was not true one might expect variations in the SANS signal, especially in the low-Q region, resulting from possible aggregation of nanotubes. In the vicinity of high-Q regime, however, the scattering curve from polymeric membrane S1 reaches a plateau while the SANS profiles from the three composite membranes exhibit only a small flat region followed by a shoulder ($Q \geq 1.7 \mathrm{nm}^{-1}$), indicative for the formation of small structures with approximate sizes $d \leq Q/2\pi \sim 3.5 \mathrm{nm}$ [190, 191]. Therefore these structures could be possibly attributed to the presence of pores created by the filler addition. From the curvature of the shoulder one may calculate the Guinier radius of gyration (R_G) of the created scatterers [187]. The values of R_G of all nanocomposite membranes are approximately 1nm. However, caution must be taken about the accuracy of the calculated radii of gyration, as Guinier approximation is generally valid for dilute and monodisperse systems.

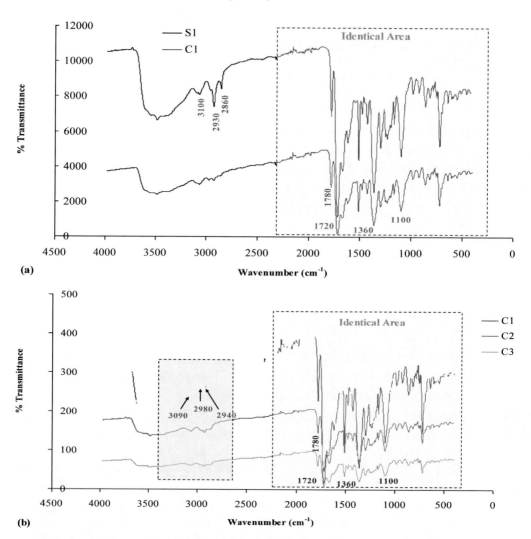

Figure 21. FTIR spectra of the single PI hollow fiber membrane and MWCNTs/PI composite membranes C1, C2 and C3. ((a): offset 50, 100 for each following sample from the bottom and (b): offset 0, 2, 4 for each following sample from the bottom).

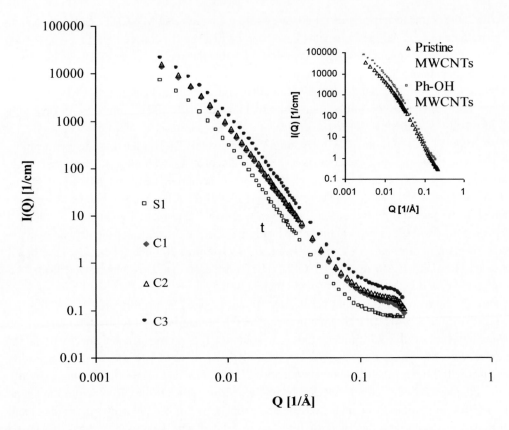

Figure 22. SANS profiles of pristine and composite polymeric membranes. Inset: SANS spectra from pristine and phenol-modified MWCNTs.

5.6.4. Gas Permeance Performance

Single gas permeation experiments for He, H_2, CH_4, N_2, O_2, C_2H_6, CO_2 and C_3H_8 were performed at 25, 60 and 100°C. The permeance values of the studied gases for the four prepared membranes are presented in Table 2. All permeance measurements took after degassing for 24 hours at 100°C. Before outgassing, the membranes were heated in a vacuum oven at 280°C for 5 hours in order to stabilize the structure after the re-conformation of the polymeric chains and the packing of the additive fillers [192]. Thermal treatment is a simple and effective post-treatment method to improve the stability as well as the selectivity of the membranes. This is due to the fact that annealing at high temperatures accelerates the thermal motion of polymer chains, which facilitate chain relaxation and rearrangement towards a dense and closer packing [193]. As a result, the minor defects in the pristine membrane could be repaired and the annealed membrane may have a smaller *d*-space and a higher transport resistance [194, 195]. This is the reason for the decrease of the recorded values in the cases of He, O_2 and N_2 relevant with those presented in our previous work [61] where the membranes were treated only with heating under helium atmosphere but without applying the simultaneous annealing process. Table 2 and Figure 23 illustrate the gas permeation measurements for the composite membranes having separating layers with thickness approximately 2-30μm. The permeance of each gas through each cast of membrane was measured at least for 3 runs and the presented values are the average of these runs. As we can see in Table 2, the permeance increases for all gases with the increase of MWCNT concentration into the membrane matrix. All measurements have been carried out for constant gas pressure (~1000 mbar) at all studied temperatures. One may also notice that the permeance of every gas in pristine polyimide membrane (S1) is lower compared with the three nanocomposites derivatives membranes and starts increasing linearly with filler concentration. This behavior has also been reported in the literature for similar composite membranes [196, 197]. In the case of helium, a non-adsorbed gas with small kinetic diameter, at 25°C the permeance values are 9.0, 16.4, 21.7 and 39.2 for S1, C1, C2 and C3 membranes respectively. The reason for the permeability enhancement on composite membranes is that fillers provide well-aligned easy channel to permeate the gases. In addition, permeance values of S1 membrane lie from 0.64 GPU (CO_2) up to 9 GPU (He) at 25°C (Table 2). These values are classified at the average values of what are reported in the literature in the case of single or polymer blend matrices (with no addition of inorganic particles) [198, 199]. On the other hand, the permeation values of the nanocomposites hollow fiber membranes at 25°C spreads from 3.6 (CO_2) up to 20 GPU (H_2) in the case of C1 membrane, from 6.4 (CO_2) up to 31.1 GPU (H_2) in the case of C2 membrane and from 11.2 (CO_2) up to 48.4 GPU (H_2) in the case of C3 membrane.

Further, the increase in helium permeance for composite C3 membrane is up to 4.4 times compared with that of pristine S1 membrane. This relationship almost follows the size of the kinetic diameter of the gas under study. In specific, the permeance enhancement factor increases with the molecular size for He, H2, CO2, O2 and N2 (Figure 24). Concerning molecules with bigger kinetic diameters such as CH4, C2H6, and C3H8 this factor is slightly smaller compared to N2. This result can be explained in terms of the difference in the shape of molecules. In particular, methane, ethane and propane are molecules with bigger diameters than nitrogen and oxygen but their shape is more related with sphere and, thus, this is the most probable reason for this behavior on the permeance enhancement factor. Similar results have also been reported for zeolite membranes where bigger molecules with spherical shape presented higher permeability values compared to other smaller molecules [200]. On the other hand, the temperature increase causes a decrease in permeances for CH4, O2, N2, C2H6, CO2 in S1, C1 and C3 membranes. Permeance decrease is also observed in C3H8 and O2 only in the case of C3 and C2 membranes respectively. This is not the case, however, for the smallest and highly-mobile He and H2 molecules where the permeances in C3 membrane do not follow the temperature variation. Finally the only cases where the dependence of temperature was positive on the permeances were for helium in S1, C1 and C2 and for C3H8 in S1 and C1 membranes respectively. These complex behaviors, as will be discussed later, can be attributed to the competitive mechanisms of solution and diffusion.

An interesting result is that the permeance values for all studied pure gases at 25°C increase significantly as the carbon nanotube content increases according to linear relations (Figure 25). This enhancement in He permeance implies that the functionalized surface of the added MWCNTs increases the free volume of the polymer matrix, especially within the separating, skin, layer of the membranes through which the gas molecules pass [37, 159]. At this low temperature, the stronger factor for the permeation mechanism is the existence of open path channels, such as pores, which are strongly dependent on the filler concentration and its dispersion into the matrix. This linear correlation between helium permeation and filler concentration is a strong evidence for the excellent dispersion of the MWCNTs, in agreement with SANS results.

Table 2. Permeance values of He, H₂, CH₄, N₂, O₂, C₂H₆, CO₂ and C₃H₈ for S1, C1, C2 and C3 single and nanocomposite hollow fiber membranes

Gas	Permeance (GPU)											
	S1			C1			C2			C3		
	25°C	60°C	100°C	25°C	60°C	100°C	25°C	60°C	100°C	25°C	60°C	100°C
He	9.0	17.2	31.6	16.4	23.6	30.9	21.7	31.6	46.6	39.2	37.2	40.5
H₂	7.6	15.2	25.2	20.0	25.2	30.6	31.1	46.4	49.4	48.4	40.4	50.5
CH₄	0.97	0.85	0.82	5.5	4.7	4.3	10.2	8.3	10.6	17.3	15.7	13.9
N₂	0.74	0.68	0.72	4.2	3.6	3.4	7.5	4.1	8.2	14.0	11.6	10.6
O₂	0.76	0.74	0.71	4.2	3.7	3.5	7.9	6.7	5.4	13.2	11.8	10.5
C₂H₆	0.74	0.62	0.56	4.1	3.4	3.2	7.5	5.7	7.8	12.8	11.9	10.0
CO₂	0.64	0.53	0.44	3.6	2.9	2.7	6.4	4.4	6.6	11.2	10.0	8.4
C₃H₈	1.3	2.0	2.9	4.0	4.2	4.7	6.7	6.3	8.9	11.4	10.0	9.2

Table 3. Comparison of Knudsen with the ideal selectivities of the studied membranes

Gases	Knudsen Selectivity	Permselectivity											
		S1			C1			C2			C3		
		25°C	60°C	100°C	25°C	60°C	100°C	25°C	60°C	100°C	25°C	60°C	100°C
He/N₂	2.64	12.16	25.29	43.89	3.90	6.56	9.09	2.89	7.71	5.68	2.80	3.21	3.82
He/CO₂	4.7	14.06	32.45	71.82	4.56	8.14	11.44	3.39	7.18	7.06	3.50	3.72	4.82
H₂/CH₄	2.83	7.84	17.88	30.73	3.64	5.36	7.12	3.05	5.59	4.66	2.80	2.57	3.63
CH₄/CO₂	1.66	1.52	1.60	1.86	1.53	1.62	1.59	1.59	1.89	1.61	1.54	1.57	1.65
O₂/N₂	0.94	1.03	1.09	0.99	1.00	1.03	1.03	1.05	1.63	0.66	0.94	1.02	0.99

Table 4. Activation energies of permeation (kJ/mol) in case of He, O₂, N₂, CO₂, CH₄, C₂H₆ and C₃H₈ gases

Activation Energy (kJ/mol)				
Gas	S1	C1	C2	C3
He	15.5	7.8	9.4	
O₂	-0.88	-2.5	-4.6	-2.8
N₂		-2.7		-2.8
CO₂	9.8	2.0		-2.7
CH₄	-2.1	-3.0		-2.7
C₂H₆	-3.5	-3.2		-3.1
C₃H₈	-4.6	-3.8		-3.6

As we can see in Table 3 the studied membranes and especially the S1 membrane exhibit significant separation factors for helium over nitrogen and carbon dioxide as well as for hydrogen over methane (Table 3). In specific, the ideal separation factors fluctuate from almost 44 to ~3 for He/N₂ mixtures for S1 at 100°C and C3 at 25°C respectively. Furthermore higher ideal separation coefficients were achieved in the case of He/CO₂ mixture, where the ideal selectivity recorded of about 72 for S1 membrane at 100°C while the smaller value was 3.5 for C3 membrane at 25°C. Note that while these ideal selectivities values were decreased from S1 to C1, to C2 and to C3 the gas permeation values were increased at the same time (for example the H_2 permeance at 25°C is 7.6, 20, 31.1 and 48.4 for S1, C1, C2 and C3 membrane respectively). This comparison between the increase of the gas permeance values and the decrease of the ideal separation coefficients can be easily observed in Tables 2 and 3.

It is worth mentioning that the suggested gas separations and especially the helium's gas couples (He/N_2 and He/CO_2) are very interesting and expensive. It is noteworthy that helium molecule is highly mobile and is the only gas which escapes from the earth atmosphere to the space [201, 202].

Further, the calculation of the activation energy of gas permeation through a membrane is a useful tool for understanding the "grade of the spontaneity" of the permeation mechanism. The activation energy for permeation is the sum of the activation energy for diffusion E_d and the heat of sorption ΔH_s (generally exothermic, i.e., ΔH_s takes negative values). As temperature rises, diffusivity of a penetrant increases in a membrane while its sorption coefficient decreases, causing variation in the permeability according to the follow equation [203]:

$$P_i = D_i \cdot S_i = \frac{(Flux)_i}{\Delta P_i / l}$$

where D is the diffusivity and S is the sorption coefficient of the gas penetrant "i" in the membrane, ΔP_i is the partial pressure difference and l is the membrane thickness. All activation energies, for He, O_2, N_2, CO_2, CH_4, C_2H_6 and C_3H_8, determined from the slope of each Arrhenius plot, are presented in Table 4. Based on this approximation, the activation energy values obtained from the experimental data of temperature dependence on the permeation fluxes generally exhibit a linear relationship between the log of flux versus reciprocal temperature, i.e.,

$$J = J_0 \cdot \exp(E_j / RT)$$

where E_j has been considered to be the activation energy for permeation [204]. As can be seen in our case, the ordering of activation energy values for He is: 15.5 (S1), 9.4 (C2) and 7.8 (C1), no activation energy value could be calculated for C3 membrane because there was no linearity in Arrhenius plot. Similar results were reported in our previous work [61, 161] as well as from many other groups, Dixon-Garrett et al. [205], Lin et al. [206], Bao et al. [207] and Budd et al. [208]. In addition, negative activation energies values were observed for oxygen, nitrogen, carbon dioxide, methane, ethane and propane. The most pronounced negative activation energy -4.6 kJ/mol occurred for C_3H_8 in the S1 membrane.

Such negative values have been reported for various polymeric and inorganic membranes in the literature [205, 207, 208] and have been explained in terms of the negative influence of temperature on solubility and, thus, on permeation [209]. Another interesting result is that the activation energies for C3 membrane take negative values for all gases except of helium and CO_2 for S1 and C1 membranes. This effect could be caused by swelling of the membrane matrix due to sorption of all these gases in the polymer, which leads to increased chain mobility, and, hence, lower gas E_d. The high concentration of the MWCNTs fillers provides a larger free volume within the polymer matrix resulting in negative activation energy of permeation because of the reduction of the flux resistance.

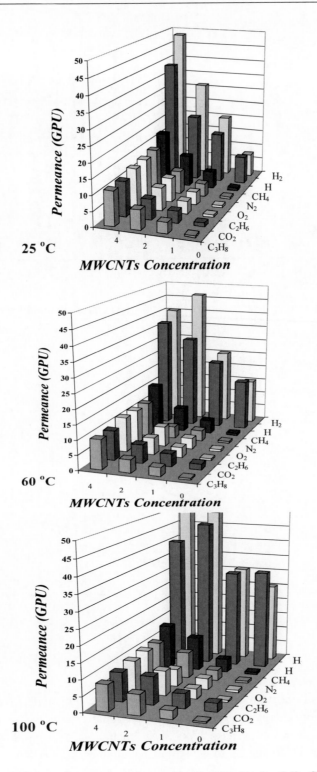

Figure 23. Effect of carbon nanotubes concentration on He, H_2, CH_4, N_2, O_2, C_2H_6, CO_2 and C_3H_8 permeances at 25, 60 and 100°C.

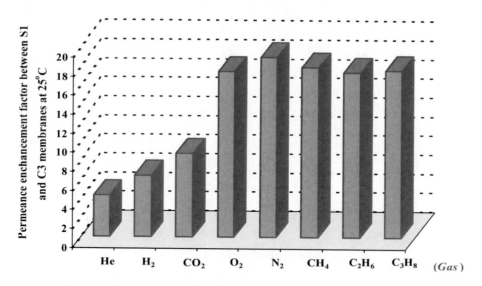

Figure 24. Permeance enhancement factor between pristine S1 and composite C3 membranes for He, H_2, CO_2, O_2, N_2, CH_4, C_2H_6, and C_3H_8 gases at 25°C.

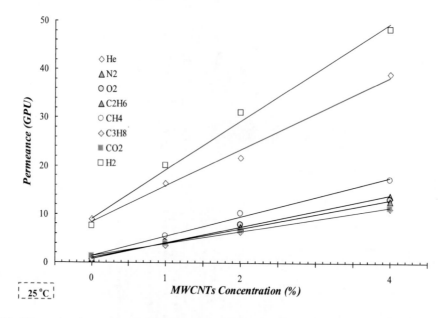

Figure 25. Effect of carbon nanotube concentration on He, H_2, CH_4, N_2, O_2, C_2H_6, CO_2 and C_3H_8 permeance at 25°C ($0.992 < R^2 < 0.999$).

Although the gas permeances in most cases increase with temperature, the apparent permeation activation energies increase non-proportionally with the increase in the gas molecular sizes. This happens because in polymeric membranes, the gas diffusion mechanism is completely different compared to inorganic membranes, especially in molecular sieve membranes. Here the activation energy for permeation is strongly dependent on the chemical affinity and the interactions between the gas molecules and the polymer chains as well as between gas molecules and the surface/interface of the fillers. Gases such as O_2, N_2, CH_4, C_2H_6, C_3H_8 etc., with their permeance values are not positively dependent on the temperature, present low and even negative permeation activation energies. The negative values of activation energies for these gases in our membranes result from the strange behavior of the temperature on the permeance values (Table 4). However, for CO_2 the situation is complicated because the effects of diffusion and solubility are opposing each other. Thus, the transport of CO_2 through polyimide can be diffusion-controlled when temperature is high and solubility-controlled when temperature is low [192]. Finally, it is also noteworthy that the results presented at Table 3 verify that the more permeable polymeric membranes are generally less selective and vice versa [160, 210, 211].

CONCLUSION

Nanocomposites are a new class of composites which are particle-filled polymers of which at least one dimension of the dispersed particles is in the nanometer range. Even though nanomaterials have been long known and extensively used as membrane fillers, there are still applications that would benefit from further research on this versatile topic. Apart from the several polymeric matrices in which nanomaterials have already been established because of their good dispersion and compatibility, there are also many membrane materials where nanomaterials have shown great potential as membrane additives. Especially in the field of polymeric membrane technology, many nanomaterials have proved their value for a multitude of applications including gas separations and water purification. Undoubtedly, it would be important for further research to keep exploring their potential in this field not only to improve their applicability as membranes' fillers, for the preparation of membranes, but also to take advantage of their catalytic properties that could lead to novel membrane reactors and new more efficient separation/catalytic processes. In this direction, nanomaterials seem to be competitive as additive materials not only at polymeric but also at carbon, glass, ceramic, metal etc., membrane matrices.

ACKNOWLEDGEMENT

The present work is a result in the framework of NSRF. The NANOSKAÏ Project (Archimedes Framework) of the Eastern Macedonia and Thrace Institute of Technology is co-financed by Greece and the European Union in the frame of operational program "Education and lifelong learning investing in knowledge society", Ministry of Education and Religious Affairs, Culture and Sports, NSRF 2007-2013.

REFERENCES

[1]   Kreyling, W. G., Semmler-Behnke, M., Chaudhry, Q. *Nano Today* 2010, 5, 165-168.
[2]   *The (US) National Nanotechnology Initiative Strategic Plan* December 2007, www.nano.gov/NNI Strategic Plan 2007.pdf.

[3] Borm, P. J. A., Robbins, D., Haubold, S., Kuhlbusch, T., Fissan, H., Donaldson, K., Schins, R., Stone, V., Kreyling, W., Lademann, J., Krutmann, J., Warheit, D., Oberdorster, E. *Particle and Fibre Toxicology* 2006, 3, 11-46.

[4] Buzea, C., Pacheco, I., Robbie, K. *Biointerphases* 2007, 2(4), MR17-MR71.

[5] Sci. Comm. Emerg. New. Identified Health Risks (SCENIHR). *The appropriateness of existing methodologies to assess the potential risks associated with engineered and adventitious products of nanotechnologies*, http://ec.europa.eu/health/ph_risk/ committees/04_ scenihr/docs/scenihr_o_003b.pdf, Eur. Comm., Brussels, 2006.

[6] http://www.circuitstoday.com/nanotechnology.

[7] www.wikipedia.com.

[8] Daniel, M. C., Astruc, D. *Chem. Rev.* 2004, 104, 293-346.

[9] Chen, X., Mao, S. S. *Chem. Rev.* 2007, 107, 2891-2959.

[10] Roduner, E. *Chem. Soc. Rev.* 2006, 35, 583-592.

[11] Favvas, E. P., Papageorgiou, S. K. Chapter Polyaniline: *A Promising Precursor Material for Membrane Technology Applications* In Trends in Polyaniline Research, Eds: Ohsaka, T., Chowdhury, A. L., Rahman, A., Islam, M., ISBN: 978-1-62808-424-5, Nova Science Publishers: New York, USA, 2013, pp. 261-286.

[12] Majeed, S., Fierro, D., Buhr, K., Wind, J., Du, B., Boschetti-de-Fierro, A., Abetz, V. *J. Membr. Sci.* 2012, 403, 101-109.

[13] Yampolskii Y., Freeman, B. (Eds.), *Membrane gas separation*, Wiley, ISBN 978-0-470-74621-9, Chichester, 2010.

[14] Buonomenna, M. G., Yave, W., Golemme, G. *RSC Advances* 2012, 2, 10745-10773.

[15] Caro, J., Noack, M., Kölsch, P., Schäfer, R. *Microp. Mesop. Mater.* 2000, 38, 3-24.

[16] Favvas, E. P., Kouvelos, E. P., Romanos, G. E., Pilatos, G. I., Mitropoulos, A. Ch., Kanellopoulos, N. K. *J. Porous Materials* 2008, 15, 625-633.

[17] Stoeger, J. A., Palomino, M., Agrawal, K. V., Zhang, X., Karanikolos, G. N., Valencia, S., Corma, A., Tsapatsis, M. *Angewandte Chemie: Inter. Ed.* 2012, 51, 2470-2473.

[18] Aroona, M. A., Ismail, A. F., Matsuura, T., Montazer-Rahmati, M. M. *Separ. Purif. Techn.* 2010, 75, 229-242.

[19] Goh, P. S., Ismail, A. F., Sanip, S. M., Ng, B. C., Aziz. M. *Separ. Purif. Technol.* 2011, 81, 243-264.

[20] Hosseini, S. S., Li, Y., Chung, T.-S., Liu, Y. *J. Membr. Sci.* 2007, 302, 207-217.

[21] Matteucci, S., Kusuma, V. A., Kelman, S. D., Freeman, B. D. *Polymer* 2008, 49, 1659-1675.

[22] Matteucci, S., Raharjo, R. D., Kusuma, V. A., Swinnea, S., Freeman, B. D. *Macromolecules* 2008, 41, 2144-2156.

[23] Matteucci, S., Kusuma, V. A., Sanders, D., Swinnea, S., Freeman, B. D. *J. Membr. Sci.* 2008, 307, 196-217.

[24] Matterucci, S., Kusuma, V. A., Swinnea, S., Freeman, B. D. *Polymer* 2008, 49, 757-773.

[25] Moghada, F., Omidkhah, M. R., Vasheghani-Farahani, E., Pedram, M. Z., Dorosti, F. *Separ. Purif. Technol.* 2011, 77, 128-136.

[26] Soroko, I., Livingston, A. *J. Membr. Sci.* 2009, 343, 189-198.

[27] Sotto, A., Boromand, A., Balta, S., Kim, J., Van Der Bruggen, B. *J. Mater. Chem.* 2011, 21, 10311-10320.

[28] Balta, S., Sotto, A., Kim, J., Luis, P., Benea, L., Van der Bruggen, B. *J. Membr. Sci.* 2012, 389, 155-161.

[29] Yan, L., Hong, S., Li, M.L., Li, Y.S. *Separ. Purif. Technol.* 2009, 66, 347-352.

[30] Yu, S. L., Shi, W. X., Lu, Y., Yang, J. X. *Water Sci. Technol.* 2011, 64, 1892-1897.

[31] Vanherck, K., Vankelecom, I., Verbiest, T. *J. Membr. Sci.* 2011, 373, 5-13.

[32] Xu, J., Bhattacharyya, D. *Environ. Prog.* 2005, 24, 358-366.

[33] Tanaka, D. A. P., Tanco, M. A. L., Nagase, T., Okazaki, J., Wakui, Y., Mizukami, F., Suzuki, T. M. *Adv. Mater.* 2006, 18, 630-632.

[34] Wijenayake, S. N., Panapitiya, N. P., Versteeg, S. H., Nguyen, C. N., Goel, S., Balkus Jr., K. J., Musselman, I. H. Ferraris, J. P. *Ind. Engin. Chem. Resear.* 2013, 52, 6991-7001.

[35] Varoon, K., Zhang, X., Elyassi, B., Brewer, D. D., Gettel, M., Kumar, S., Lee, J. A., Maheshwari, S., Mittal, A., Sung, C-Y., Cococcioni, M., Francis, L. F., McCormick, A. V., Mkhoyan, K. A., Tsapatsis, M. *Science* 2011, 334, 72-75.

[36] Vua, D. Q., Koros, W. J., Miller, S. J. *J. Membr. Sci.* 2003, 211, 311-334.

[37] Rafizah, W. A. W., Ismail, A. F. *J. Membr. Sci.* 2008, 307, 53-61.

[38] Zimmerman, C. M., Singh, A., Koros, W. J. *J. Membr. Sci.* 1997, 137, 145-154.

[39] Vu, D. Q., Koros, W. J., Miller, S. J. *J. Membr. Sci.* 2003, 211, 311-334.

[40] Vu, D. Q., Koros, W. J., Miller, S. J. *J. Membr. Sci.* 2003, 211, 335-348.

[41] Shimekit, B., Muktar, H., Maitra, S. *J. Appl. Sci.* 2010, 10, 1204-1211.

[42] Liu, J., Wang, H., Zhang, L. *Chem. Mater.* 2004, 16, 4205-4207.

[43] Rafizah, W. W. W., Ismail, A. F. *J. Membr. Sci.* 2008, 307, 53-61.

[44] Moadded, M., Koros, W. J. *J. Membr. Sci.* 1996, 111, 283-290.

[45] Kim, J. H., Lee, Y. M. *J. Membr. Sci.* 2001, 193, 209-225.

[46] Ferrari, M. C., Galizia, M., De Angelis, M. G., Sarti, G. C. *Ind. Eng. Chem. Res.* 2010, 49, 11920-22935.

[47] Gomes, D., Nunes, S. P., Peinemann, K.-V. *J. Membr. Sci.* 2005, 246, 13-25.

[48] Zornoza, B., Tellex, C., Coronas, J. *J. Membr. Sci.* 2011, 368, 100-109.

[49] Merkel, T. C., He, Z., Pinnau, I. *Macromolecules* 2003, 36, 6844-6855.

[50] Sanchez, C., Julián, B., Belleville, P., Popall, M. *J. Mater. Chem.* 2005, 15, 3559-3592.

[51] Zhang, Y., Balkus Jr., K. J., Musselman, I. H., Ferraris, J. P. *J. Membr. Sci.* 2008, 325, 28-39.

[52] Hillock, A. M. W., Miller, S. J., Koros, W. J. *J. Membr. Sci.* 2008, 314, 193-199.

[53] Li, Y., Liang, F., Bux, H., Yang, W., Caro, J. *J. Membr. Sci.* 2010, 354, 48-54.

[54] Caro, J., Noack, M. *Microp. Mesopor. Mater.* 2008, 115, 215-233.

[55] Cong, H., Zhang, J., Radosx, M., Shen, Y. *J. Membr. Sci.* 2007, 294, 178-185.

[56] Ge, L., Zhu, Z., Rudolph, V. *Sep. Purif. Technol.* 2011, 78, 76-82.

[57] Ge, L., Zhu, Z., Li, F., Liu, S., Wang, L., Tang, X., Rudolph, V. *J. Phys. Chem.* C. 2011, 115, 6661-6670.

[58] Goh, P. S., Ng, B. C., Ismail, A. F., Sanip, S. M., Aziz, M., Kassim, M. A. *Sep. Sci. Technol.* 2011, 46, 1250-1261.

[59] Aroon, M. A., Ismail, A. F., Montazer-Rahmati, M. M., Matsuura, T. *J. Membr. Sci.* 2010, 364, 309-317.

[60] Yin, J., Zhu, G., Deng, B. *J. Membr. Sci.* 2013, 437, 237-248.

[61] Favvas, E. P., Nitodas, S. F., Stefopoulos, A., Stefanopoulos, K. L., Papageorgiou, S. K., Mitropoulos, A. C. *Separ. Purif. Technol.* 2014, 122, 262-269.

[62] Kim, S., Chen, L., Johnson, J. K., Marand, E. *J. Membr. Sci.* 2007, 294, 147-158.

[63] Kim, S., Pechar, T. W., Marand, E. *Desalination* 2006, 192, 330-339.

[64] Das, R., Ali, M. E., Hamid, S. B. A., Ramakrishna, S., Chowdhury, Z. Z. *Desalination* 2014, 336, 97-109.

[65] Zulhairun, A. K., Ismail, A. F., Matsuura, T., Abdullah, M. S., Mustafa, A. *Chem. Engin. J.* 2014, 241, 495-503.

[66] Daraei, P., Madaeni, S. S., Salehi, E., Ghaemi, N., Ghari, H. S., Khadivi, M. A., Rostami, E. *J. Membr. Sci.* 436 (2013) 97-108

[67] Hashemifard, S. A., Ismail, A. F., Matsuura, T. *Chem. Engin. J.* 2011, 170, 316-325

[68] Sapalidis, A. A., Katsaros, F. K., Romanos, G. E., Kakizis, N. K., Kanellopoulos, N. K. *Composites:* Part B 2007, 38, 398-404.

[69] Zhao, F., Bao, X., McLauchlin, A. R., Gu, J., Wan, C., Kandasubramanian, B. *Appl. Clay Sci.* 2010, 47, 249-256.

[70] Defontaine, G., Barichard, A., Letaief, S., Feng, C., Matsuura, T., Detellier, C. *J. Colloid. Interface Sci.* 2010, 343, 622-627.

[71] Sapalidis, A. A., Katsaros, F. K., Steriotis, Th. A., Kanellopoulos, N. K. *J. Appl. Polym. Sci.* 2012, 123, 1812-1821.

[72] Bae, T.-H., Lee, J. S., Qiu, W., Koros, W. J., Jones, C. W., Nair, S. *Angew. Chem. Int. Ed.* 2010, 49, 9863-9866.

[73] Perez, E. V., Balkus Jr., K. J., Ferraris, J. P., Musselman, I. H. *J. Membr. Sci.* 2009, 328, 165-173

[74] Burmann, P., Zornoza, B., Téllez, C., Coronas, J. *Chem. Engin. Sci.* 2014, 107, 66-75.

[75] Ploegmakers, J., Japip, S., Nijmeijer, K. *J. Membr. Sci.* 2013, 428, 445-453.

[76] Zornoza, B., Tellez, C., Coronas, J., Gascon, J., Kapteijn, F. *Microp. Mesop. Mater.* 2013, 166, 67-78.

[77] Zornoza1, B., Seoane, B., Zamaro, J. M., Téllez1, C., Coronas, J. *Chem Phys Chem* 2011, 12, 2781-2785.

[78] Zornoza, B., Seoane, B., Zamaro, J. M., TŽllez, C., Coronas, J. *Procedia Engineering* 2012, 44, 2118-2120.

[79] Aramendia, M. A., Borau, V., Jimenez, C., Marinas, J. M., Porras, A., Urbano, F.J. *J. Mater. Chem.* 1996, 6, 1943-1949.

[80] Stark, J. V., Klabunde, K. *J. Chem. Mater.* 1996, 8, 1913-1918.

[81] Xu, B.-Q., Wei, J.-M., Wang, H.-Y., Sun, K.-Q., Zhu, Q.-M. *Catalysis Today* 2001, 68, 217-225.

[82] Makhluf, S., Dror, R., Nitzan, Y., Abramovich, Y., Jelinek, R., Gedanke, A. *Adv. Functional Materials* 2005, 15, 1708-1715.

[83] Stoimenov, P. K., Klinger, R. L., Marchin, G. L., Klabunde, K. J. *Langmuir* 2002, 18, 6679-6686.

[84] Ng, L. Y., Mohammad, A. W., Leo, C. P., Hilal, N. *Desalination* 2013, 308, 15-33.

[85] Zhao, D., Feng, J., Huo, Q., Melosh, N., Fredrickson, G. H., Chmelka, B. F., Stucky, G. D. *Science* 1998, 279, 548-552.

[86] Favvas, E. P., Stefanopoulos, K. L., Vairis, A., Nolan, J. W., Joensen, K. D., Mitropoulos, A. Ch. *Adsorption* 2013, 19, 331-338.

[87] Bareka, M. M. Sc. dissertation: *Advanced mesoporous materials: Preparation, properties and their applicability in catalysis and gas storage applications.* Postgraduate Program "Oil and Gas Technology" Dpt. of Petroleum and Mechanical

Engineering, School of Science, Eastern Macedonia and Trace Institute of Technology, February 2014.

[88] Oye, G., Sjöblom, J., Stöcker, M. *Advances Colloid Interf. Sci.* 2001, 89, 439-466.

[89] Vadia, N., Rajput, S. *Asian J. Pharmaceut. Clinic. Researc.* 2011, 4, 44-53.

[90] Sorribas, S., Zornoza, B., Téllez, C., Coronas, J. *J. Membr. Sci.* 452 (2014) 184-192.

[91] Foley, H. C. *J. Microporous Mater.* 1995, 4, 407-433.

[92] Vu, D. Q., Koros, W. J., Miller, S. J. *J. Membr. Sci.* 2003, 211, 311-334.

[93] Jones, C. W., Koros, W. J. *Carbon* 1994, 32, 1419-1425.

[94] Verma, S. K., Walker, J. R. P. L. *Carbon* 1992, 30, 829-836.

[95] Biggs, M. J., Buts, A., Williamson, D. *Langmuir* 2004, 20, 7123-7138.

[96] Vu, D. Q., Koros, W. J., Miller, S. J. *J. Membr. Sci.* 2003, 211, 335-348.

[97] Kresge, C. T., Leonowicz, M. E., Roth, W. T., Vartuli, J. C., Beck, J. S. *Nature* 1992, 359, 710-712.

[98] Kim, S.-S., Pinnavaia, T. J. *Chem. Commun.* 2001, 2418-2419.

[99] Xia, K., Gao, Q., Wu, C., Song, S., Ruan, M. *Carbon* 2007, 45, 1989-1996.

[100] Chang, H., Joo, S. H., Pak, C. *J. Mater. Chem.* 2007, 17, 3078-3088.

[101] Zhou, L., Liu, X., Li, J., Wang, N., Wang, Z., Zhou. Y. *Chem. Phys. Lett.* 2005, 413, 6-9.

[102] Grace W. R. & Co. *Enriching Lives, Everywhere: Zeolite Structure.* https://grace.com/en-us.

[103] http://en.wikipedia.org/wiki/Zeolite#cite_note-1.

[104] Cubillos Lobo, J. A. M. Sc. dissertation, title *Heterogeneous asymmetric epoxidation of cis-ethyl cinnamte over Jacobsen's catalyst immobilized in inorganic porous materials*, Aachen University, Germany, 2005.

[105] http://en.wikipedia.org/wiki/Clinoptilolite.

[106] http://en.wikipedia.org/wiki/Zeolite.

[107] Hudiono, Y. C., Carlisle, T. K., La Frate, A. L., Gin, D. L., Noble, R. D. *J. Membr. Sci.* 2011, 370, 141-148.

[108] Karatay, E., Kalipcilar, H., Yilmaz, L. *J. Membr. Sci.* 2010, 364, 75-81.

[109] Khan, A. L., Cano-Odena, A., Gutierrez, B., Minguillon, C., Vankelecom, I. F. J. *J. Membr. Sci.* 2010, 350, 340-346.

[110] Alexandre, M., Dubois, P. *Mater. Sci. Engin.* 2000, 28, 1-63.

[111] Vaia, R. A., Vasudevan, S., Krawiec, W., Scanlon, L. G., Giannelis, E. P. *Adv Mater.* 1995, 7, 154-156.

[112] Vaia, R. A., Giannelis, E. P. *Macromolecules* 1998, 30, 8000-8009.

[113] Giannelis, E. P., Krishnamoorti, R., Manias, E. *Adv. Polym. Sci.* 1999, 118, 108-147.

[114] Tsunoji, N., Ikeda, T., Sadakane, M., Sano, T. *J. Mater. Chem.* A 2014, 2, 3372-3380.

[115] Giannelis, E. P. *Advanced Materials* 1996, 8, 29-35.

[116] Kim, W.-G., Lee, J. S., Bucknall, D. G., Koros, J. W., Nair, S. *J. Membr. Sci.* 2013, 441, 129-136.

[117] Iijima S. *Nature* 1991, 354, 56-58.

[118] Aqel, A., Abou El-Nour, K. M. M., Ammar, R. A. A., Al-Warthan, A. *Arab. J. Chem.* 2010, 5, 1-23.

[119] Zhang, Y., Ali, S. F., Dervishi, E., Xu, Y., Li, Z., Casciano, D., Biris, A. S. *ACS Nano* 2010, 4, 3181-3186.

[120] Eddaoudi, M., Kim, J., Rosi, N., Vodak, D., Wachter, J., OÕKeeffe, M., Yagh, O. M. *Science* 2002, 295, 469-472.

[121] James, S.L. *Chem. Soc. Rev.* 2003, 32, 276-288.

[122] Kitagawa, S., Kitaura, R., Noro, S. *Angew. Chem.* 2004, 116, 2388-2430.

[123] Kitagawa, S., Kitaura, R., Noro, S. *Angew. Chem. Int. Ed.* 2004, 43, 2334-2375.

[124] Li, Y.-S., Liang, F.-Y., Bux, H., Feldhoff, A., Yang, W.-S., Caro J. *Angew. Chem.* 2010, 122, 558-561.

[125] Li, J.-R., Kuppler, R. J., Zhou, H.-C. *Chem. Soc. Rev.* 2009, 38, 1477-1504.

[126] Zacher, D., Shekhah, O., Wöll, C., Fischer, R. A. *Chem. Soc. Rev.* 2009, 38, 1418-1429.

[127] Civalleri, B., Napoli, F., Noël, Y., Roettia, C., Dovesi, R. *Cryst Eng Comm* 2006, 8, 364-371.

[128] Panella, B., Hirscher, M., Pütter, H., Müller, U. *Adv. Funct. Mater.* 2006, 16, 520-524.

[129] Bae, T. H., Lee, J. S., Qiu, W., Koros, W. J., Jones, C. W., Nair, S., *Angew. Chem. Int. Ed.* 2010, 49, 9863-9866.

[130] Caro, J. MOF membranes: Bright industrial future or a laboratory curiosity? *13th International Conference on Inorganic Membranes* (ICIM2014), 6-9 July (2014) Brisbane, Australia.

[131] Editorial. Virtual Issue: Graphene and Functionalized Graphene *J. Phys. Chem. C* 2011, 115, 3195-3197.

[132] Li, X., Cai, W., An, J., Kim, S., Nah, J., Yang, D., Piner, R., Velamakanni, A., Jung, I., Tutuc, E., Banerjee, S. K., Colombo, L., Ruoff, R. S. *Science* 2009, 324, 1312-1314.

[133] Dimitrakakis, G. K., Tylianakis, E., Froudakis, G. E. *Nano Lett.* 2008, 8, 3166-3170.

[134] Xue, C., Zou, J., Sun, Z., Wang, F., Han, K., Zhu, H. *Int. J. Hydrogen Energy* 2014, 39, 7931-7939.

[135] Du, H., Li, J., Zhang, J., Su, G., Li, X., Zhao, Y. *J. Phys. Chem. C* 2011, 115, 23261-23266.

[136] Han, Y., Xu, Z., Gao, C. *Adv. Funct. Mater.* 2013, 23, 3693-3700.

[137] Joshi, R. K., Carbone, P., Wang, F. C., Kravets, V. G., Su, Y., Grigorieva, I. V., Wu, H. A., Geim, A. K., Nair, R. R. *Science* 2014, 343, 752-754.

[138] He, H., Klinowski, J., Forster, M., Lerf, A. *Chem. Phys. Lett.* 1998, 287, 53-56.

[139] Szabo, T., Szeri, A., Dekany, I. *Carbon* 2005, 43, 87-94.

[140] Stankovich, S., Dikin, D. A., Dommett, G. H. B., Kohlhaas, K. M., Zimney, E. J., Stach, E. A., Piner, R. D., Nguyen, S. T., Ruoff, R. S. *Nature* 2006, 442, 282-286.

[141] Van der Bruggen, B. *ISRN Nanotechnology*, 2012, Article ID 693485, 17 pages, DOI: 10.5402/2012/693485.

[142] Cong, H., Zhang, J., Radosz, M., Shen, Y. *J. Membr. Sci.* 2007, 294, 178-185.

[143] Aroon, M. A., Ismail, A. F., Montazer-Rahmati, M. M., Matsuura, T. *J. Membr. Sci.* 2010, 364, 309-317.

[144] Haley, B., Frenkel, E. *Urologic Oncology* 2008, 26, 57-64.

[145] Sinha, N., Ma, J., Yeow, J. T. W. *J. Nanosci. Nanotechnol.* 2006, 6, 573-590.

[146] Varghese, O. K., Gong, D., Paulose, M., Ong, K. G., Grimes, C. A. *Sensors and Actuators* B 2003, 93, 338-344.

[147] Bernholc, J., Brenner, D., Nardelli, M. B., Meunier, V., Roland, C. *Annu. Rev. Mater. Sci.* 2002, 32, 347-375.

[148] Hone, J. *Carbon nanotubes: Thermal properties*, In Dekker Encyclopedia of Nanoscience and Nanotechnology, Marcel Dekker, New York, NY, USA, 2004.

[149] Che, R., Peng, L. M., Duan, X., Chen, Q., Liang, X. *Adv. Mater.* 2004, 16, 401-405.

[150] Kim, J., Van Der Bruggen, B. *Environ. Pollut.* 2010, 158, 2335-2349.

[151] Karkhanechi, H., Kazemian, H., Nazockdast, H., Mozdianfard, M. R., Bidoki, S. M. *Chem. Eng. Technol.* 2012, 35, 885-892.

[152] Moore, T. T., Mahajan, R., Vu, D.Q. *AIChE J.* 2004, 50, 311–321.

[153] Hinds, B. J., Chopra, N., Rantell, T., Andrews, R., Gavalas, V., Bachas, L. G. *Science* 2004, 303, 62-65.

[154] Weng, T. H., Tseng, H. H., Wey, M. Y. *Int. J. Hydrogen Energy* 2009, 34, 8707-8715.

[155] Goh, P. S., Ng, B. C., Ismail, A. F., Sanip, S. M., Aziz, M., Kassim, M. A. *Sep. Sci. Technol.* 2011, 46, 1250-1261.

[156] Ge, L., Zhu, Z., Rudolph, V. *Separ. Purif. Technol.* 2011, 78, 76-82.

[157] Ge, L., Zhu, Z., Li, F., Liu, S., Wang, L., Tang, X., Rudolph, V. *J. Phys. Chem. C.* 2011, 115, 6661-6670.

[158] Sieffert, D., Staudt, C. *Separ. Purif. Technol.* 2011, 77, 99-103.

[159] Goh, P. S., Ismail, A. F., Sanip, S. M., Ng, B. C., Aziz, M. *Sep. Purif. Technol.* 2011, 81, 243-264.

[160] Favvas, E. P., Papageorgiou, S. K., Stefanopoulos, K. L., Nolan, J. W., Mitropoulos, A. Ch. *J. Appl. Polym. Sci.* 2013, 130, 4490-4499.

[161] Favvas, E. P., Stefanopoulos, K. L., Nolan, J. W., Papageorgiou, S. K., Mitropoulos, A. Ch., Lairez, D. *Separ. Purif. Technol.* 2014, 132, 336-345.

[162] Chatzidaki, E. K., Favvas, E. P., Papageorgiou, S. K., Kanellopoulos, N. K., Theophilou, N. V. *Europ. Polym. J.* 2010, 43, 5010-5016.

[163] Stefopoulos, A. A., Chochos, C. L., Prato, M., Pistolis, G., Papagelis, K., Petraki, F., Kennou, S., Kallitsis, J. K. *Chemistry: A Europ. J.* 2008, 14, 8715-8724.

[164] Brûlet, A., Lairez, D., Lapp, A., Cotton, J. P. *J. Appl. Cryst.* 2007, 40, 165-177.

[165] Nitodas, S., Alexopoulos, N. D., Marioli-Riga, Z. An effective route for the functionalization/disperion of carbon nanotubes in polymer composites for transport applications. *6th International Congress for Composites* (*Composites in Automotive & Aerospace*), Munich, Germany, October, 2010.

[166] Srivastava, R., Banerjee, S., Jehnichen, D., Voit, B., Böhme, F. *Macromol. Mater. Eng.* 2009, 294, 96-102.

[167] Nunes, A., Amsharov, N., Guo, C., Van den Bossche, J., Santhosh, P., Karachalios, T. K., Nitodas, S. F., Burghard, M., Kostarelos, K., Al-Jamal, K. T. *Small* 2010, 6, 2281-2291.

[168] Raffa, V., Vittorio, O., Gherardini, L., Bardi, G., Ziaei, A., Pizzorusso, T., Riggio, C., Nitodas, S., Karachalios, T., Costa, M., Cuschieri, A. *Nanomedicine* 2011, 6, 1709-1718.

[169] Chen, P., Wu, X., Sun, X., Lin, J., Li, W., Tan, K. L. *Phys. Rev. Lett.* 1999, 82, 2548-2551.

[170] Zhang, H. B., Lin, G. D., Zhou, Z. H., Dong, X., Chen, T. *Carbon* 2002, 40, 2429-2436.

[171] Pimenta, M. A., Dresselhaus, G., Dresselhaus, M. S., Cancado, L. G., Jorio, A., Saito, R. *Phys. Chem. Chem. Phys.* 2007, 9, 1276-1291.

[172] Romanos, G. E., Likodimos, V., Marques, R. R. N., Steriotis, T. A., Papageorgiou, S. K., Faria, J. L., Figueiredo, J. L., Silva, A. M. T., Falaras, P. *J. Phys. Chem. C* 2011, 115, 8534-8546.

[173] Van't Hoff, J. P. *Wet spinning of polyethersulfone gas separation membranes*, Ph. D. Thesis, Twente University, 1988 (Chapter 2), pp. 25-50.

[174] Pereira, C. C., Nobrega, R., Borges, C. P. *Brazil. J. Chem. Eng.* 2000, 17, 599-605.

[175] Bonyadi, S., Chung, T. S., Krantz, W. B. *J. Membr. Sci.* 2007, 299, 200-210.

[176] Yu, D. G., Chou, W. L., Yang, M. C. *Sep. Purif. Technol.* 2006, 52, 380-387.

[177] Yu, D. G., Chou, W. L., Yang, M. C. *Sep. Purif. Technol.* 2006, 51, 1-9.

[178] Kuan, C. F., Chen, W. J., Li, Y. L., Chen, C. H., Kuan, H. C., Chiang, C. L. *J. Phys. Chem. Solids* 2010, 71, 539-543.

[179] Li, Y., Chung, T. S., Cao, C., Kulprathipanja, S. *J. Membr. Sci.* 2005, 260, 45-55.

[180] Ismail, A. F., Kusworo, T. D., Mustafa, A. *J. Membr. Sci.* 2008, 319, 306-312.

[181] Shao, L., Chung, T. S., Goh, S. H., Pramoda, K. P. *J Membr Sci.* 256 (2005) 46 –56.

[182] Q. Xiangyi, Chung, T. S. *AIChE* 2006, 52, 3462-3472.

[183] Shen, Y., Lua, A. C. *Chem. Engineer. J.* 2012, 188, 199-209.

[184] Pranzas, P. K., Knöchel, A., Kneifel, K., Kamusewitz, H., Weigel, T., Gehrke, R., Funari, S. S., Willumeit, R. *Anal. Bioanal. Chem.* 2003, 376, 602-607.

[185] Mergia, K., Stefanopoulos, K. L., Ordás, N., García-Rosales, C. *Micropor. Mesopor. Mater.* 2010, 134, 141-149.

[186] Bale, H. D., Schmidt, P. W. *Phys. Rev. Lett.* 1984, 53, 596-599.

[187] Porod, G. *General theory*. In: Small Angle X-ray Scattering, (Eds. Glatter O. and Kratky O.) pp. 17-51, Academic Press, London, 1982.

[188] Golosova, A. A., Adelsberger, J., Sepe, A., Niedermeier, M. A., Lindner, P., Funari, S. S., Jordan, R., Papadakis, C. M. *J. Phys. Chem.* C 2012, 116, 15765-15774.

[189] Miranda, S. M., Romanos, G. E., Likodimos, V., Marques, R. R. N., Favvas, E. P., Katsaros, F. K., Stefanopoulos, K. L., Vilar, V. J. P., Faria, J. L., Falaras, P. Silva, A. M. T. *Appl. Catal. B: Environmental* 2014, 147, 65-81.

[190] László, K., Geissler, E. *Carbon* 2006, 44, 2437-2444.

[191] Favvas, E. P., Stefanopoulos, K. L., Papageorgiou, S. K., Mitropoulos, A. C. *Adsorption* 2013, 19, 225-233.

[192] Salleh, W. N. W., Ismail, A. F. *J. Appl. Polym. Sci.* 2013, 127, 2840-2846.

[193] Le, N. L., Chung, T. S. *J. Membr. Sci.* 2014, 454, 62-73.

[194] Jiang, L. Y., Wang, Y., Chung, T. S., Qiao, X. Y., Lai, J. Y. *Prog. Polym. Sci.* 2009, 34, 1135-1160.

[195] Ayala, D., Lozano, A. E., de Abajo, J., Garcia-Perez, C., de la Campa, J. G., Peinemann, K. V., Freeman, B. D., Prabhakar, R. *J. Membr. Sci.* 2003, 215, 61-73.

[196] Dai, Y., Johnson, J. R., Karvan, O., Sholl, D. S., Koros, W. J. *J. Membr. Sci.* 2012, 401, 76-82.

[197] Zhang, Y., Musselman, I. H., Ferraris, J. P., Balkus Jr., K. J. *J. Membr. Sci.* 2008, 313, 170-181.

[198] Li, Y., Chung, T. S., Huang, Z., Kulprathipanja, S. *J. Membr. Sci.* 2006, 277, 28-37.

[199] Feng, C. Y., Khulbe, K. C., Matsuura, T., Ismail, A. F. *Separ. Purif. Technol.* 2013, 111, 43-71.

[200] Nagumo, R., Takaba, H., Nakao, S. *J. Phys. Chem.* B 2003, 107, 14422-14428.

[201] Scholes, C. A., Stevens, G. W., Kentish, S. E. *Fuel* 2012, 96, 15-28.

[202] Schrier, J. *J. Phys. Chem. Lett.* 2010, 1, 2284-2287.

[203] Bighane, N., Koros, W. J. *J. Membr. Sci.* 2011, 371, 254-262.

[204] Xianshe, F., Huang, R. Y. M. *J. Membr. Sci.* 1996, 118, 127-131.

[205] Dixon-Garrett, S. V., Nagai, K., Freeman, B. D. *J. Polym. Sci. Pol. Phys.* 2000, 38, 1461-1473.

[206] Lin, W. H., Chung, T. S. *J. Membr. Sci.* 2001, 186, 183-193.

[207] Bao, L., Dorgan, J.R., Knauss, D., Hait, S., Oliveira, N.S., Maruccho, I.M. *J. Membr. Sci.* 2006, 285, 166-172.

[208] Budd, P. M., McKeown, N. B., Ghanem, B. S., Msayib, K. J., Fritsch, D., Starannikova, L., Belov, N., Sanfirova, O., Yampolskii, Y., Shantarovich, V. *J. Membr. Sci.* 2008, 325, 851-860.

[209] Pinnau, I., He, Z. *J. Membr. Sci.* 2004, 244, 227-233.

[210] Robeson, L. *J. Membr. Sci.* 1991, 62, 165-185.

[211] Stern, S. A. *J. Membr. Sci.* 1994, 94, 1-65.

In: Innovations in Nanomaterials　　　　　ISBN: 978-1-63483-548-0
Editors: Al-N. Chowdhury, J. Shapter, A. B. Imran　© 2015 Nova Science Publishers, Inc.

Chapter 6

DETERMINING THE DISJOINING PRESSURE IN FOAM FILMS BY MEASURING THE CHARGE DEPTH PROFILES

C. Ridings and *G. Andersson*

Centre for NanoScale Science and Technology,
Flinders University, SA, Australia

ABSTRACT

Foam films consist of thin liquid films where the two surfaces come into close proximity of a few tens of nanometers or even less. The liquid films have a large surface area to bulk ratio which is energetically unfavorable and make the films unstable. In order to stabilize the foam films there must be an internal force preventing the film from collapsing. The main forces in a foam film are the forces resulting from the overlap of the electric double layer, and van-der-Waals forces. Both can be determined directly in the case that the concentration depth profiles of all species in the foam film are known. In this chapter a comprehensive description of the forces stabilizing a foam film is presented. The description presented is based on quantities which can be derived directly from measured concentration depth profiles. It is also described how the concentration depth profiles can be measured directly with the experimental method neutral impact collision ion scattering spectroscopy.

This ion scattering method is applied in a vacuum chamber, which requires a special preparation of the foam films in order to prevent too high evaporation of the solvent from the foam film. The structure of liquid surfaces is influenced by ion specific effects which also could affect the structure and forces in foam films. It is shown how neutral impact collision ion scattering spectroscopy can be used to identify the influence of ion specific effects on the structure of liquid surfaces and thus of foam films.

---

* Corresponding author: C. Ridings. Centre for NanoScale Science and Technology, Flinders University, SA, 5001, Australia. E-mail: chris.ridings@flinders.edu.au.

1. INTRODUCTION

In free-standing foam films two surfaces come into such close contact that each of the two surfaces of the foam film experiences a force due to the presence of the other surface. Free-standing foam films consist of a liquid core that is stabilised by the adsorption of surfactant molecules that cover the liquid/vapour interface. Foam films have a large surface to bulk ratio. Creating surfaces requires energy and thus is energetically unfavorable, especially in the cases where the surface to bulk ratio is large. As a consequence most foam films are not stable over a long time. In order to put a foam film at least in a metastable state, there must be an internal force which hinders collapsing of the foam film.

The stability of these films can be described using the concept of the disjoining pressure which is the sum of a long-range repulsive force due to the overlap of the electric double layer at interfaces (electrostatic forces), long-range attractive van der Waals forces, and short-range repulsive steric forces [1-3]. Quantifying both the repulsive force due to the overlapping double layer and the attractive van der Waals force requires knowing the structure of the foam films: the surface coverage with surfactant and solute molecules, the orientation of the molecules, and the distribution of the net charge along the surface normal. Measuring these structural properties is challenging because for the investigation of foam films a small section of the film has to be selected to keep the foam film stable over the time required for experimental measurements.

The method for evaluating the disjoining pressure in foam films described here is a first-principles approach. This approach is in particular useful in the case that the concentration depth profiles of the charges are known. Neutral impact collision ion scattering spectroscopy (NICISS) provides a technique that is uniquely suited for investigating the concentration depth profiles of the elements in thin foam films using a specialised setup [4].

By measuring the concentration depth profiles of the elements in foam films it is possible to determine the charge distribution and the composition in the film along the surface normal and the orientation of the molecules. In this approach all variables are measured directly and therefore comparison with the disjoining pressure as given by the thin film pressure balance (TFPB) method will allow for assessing the assumptions of the DLVO models used in the fitting. An example of this is the determination of the sign of the surface charge. Knowing the concentration depth profiles of the cation and anion will allow for the direct determination of the sign of the surface charge, something which cannot be determined directly in foam films using the TFPB, but instead must be inferred from corresponding wetting films [5].

Foams are used in technical processes such as mineral processing (froth flotation) [6-8] and fire fighting [9]. Thin liquid films, similar to foam films, play a role in wetting [10] and applications such as microfluidics and emulsions, the latter of which directly concerns the large industrial field of oil recovery [11]. In these cases, however, it is not two air/liquid interfaces that come into close proximity but a combination of air/liquid, liquid/solid or liquid/liquid interfaces.

The solution forming the foam film consists of surfactants and a solvent and in some cases also inorganic salts. Ionic species are present in the film from three different sources: i) from the surfactant in case an ionic surfactant is used, ii) from self-dissociation of the solute – plain water has a pH of 7 and provides a concentration of 10^{-7} M of OH$^-$ and H$_3$O$^+$ – and iii) from added inorganic salts [3].

The specific type of ion has been found to play a role in foam films. Examples are bubble coalescence and wetting films. Bubble coalescence describes the aggregation of gas bubbles in a liquid. Craig et al., found that bubble coalescence depends on the inorganic salts used as solute [12]. More specifically they found that there is a rule for the combination of ions used that leads to either bubble coalescence or inhibits bubble coalescence. Ions were divided into α and ß type of ions. The combination of like ions (α-α or ß-ß) inhibited bubble coalescence in aqueous solutions while the combination of unlike ions (α-ß) did not reduce bubble coalescence.

Craig et al., found a similar effect in non-aqueous solutions [13] and also could correlate the α and ß species to the partitioning of the ions at the surface [14]. Ion specific effects were also found for wetting films. Anions and cations influence in a different way the stability and thickness of the wetting film which is ascribed to the interaction of the air/liquid interface of the wetting film with the underlying substrate.

The wetting films were fabricated in a pH range which terminates the native oxide layer of the silicon with negative charges. The ionic radius of the halide and alkali ions increases with increasing atomic number. Increasing the ionic radius of the anion leads to an increase in stability and thickness of the wetting film while increasing cation size has the opposite effect.

Schelero et al., assign this effect to the interaction of the electric double layer of the liquid/air interface with the electric double layer at the liquid/solid interface. It is anticipated that the electric double layer in these experiments has the opposite distribution of charges along the surface normal in the case where cation ionic radius is varied as opposed to anion ionic radius. As a consequence it is assumed that the interaction of the electric double layer is different for both cases, leading to the observed differences is the thickness and stability of the wetting films.

Angarksa et al., studied the effects of various salts at a specific concentration (24 mM) on the thinning of sodium dodecylsulfate (SDS) foam films and stability of bulk foams [15]. The results were described in terms of ion polarisability, with the weakly polarisable ions such as Cl^- and NO_3^- leading to relatively unstable films compared to the more polarisable ions such as PO_4^{3-} and CO_3^{2-}.

The authors state that this film stability is related to the ions facilitating interconnection between the surfactant headgroups, which is further supported by the fact that changing from Na^+ to Mg^{2+} also increases film stability. For the anions it is thought that hydrogen bonding is the stabilising mechanism, while for the cations it is the direct electrostatic interactions of the cations with the negatively charged surfactant headgroups.

A similar study was performed by Pandey et al., (2003) who investigated the effect of various cations on the foamability of dodecyl sulphate using the so-called 'shaking method,' along with surface tension measurements [16]. They found that stability increased in the order $Li^+ < Na^+ < Cs^+$ for the monovalent cations, which they attribute to the varying ability of the cations to inhibit the micelle formation that stabilises these films. Coagulation [17], hydrodynamic slip [18], formation of vesicles [19] and micelles [20] are other areas where specific ion effects have been found when thin liquid films are formed.

The aim of this chapter is to give reconsideration to the description of forces in foam films in a way that it is described with parameters for the structure of foam films which can be measured experimentally. It then will be shown how the structural parameters can be measured directly.

2. EXPERIMENTAL

2.1. NICISS

NICISS is a method for determining the concentration profiles of elements in the interfacial region of soft matter with a depth resolution of ~0.2 nm [21]. In a NICISS experiment the target is bombarded with a pulsed beam of inert gas ions - commonly helium ions are used - with a kinetic energy of several keV. The energy of the projectiles backscattered from the atoms in the target is determined by their time of flight from the target to the detector. The projectiles lose energy during the backscattering process, and the energy transfer depends on the mass of the target atom, i.e., the atom from which the projectile is backscattered. This first type of energy loss is used to identify the element from which a projectile is backscattered. Furthermore, the projectiles lose energy on their trajectory through the bulk due to small angle scattering and electronic excitations of the molecules constituting the target (stopping power). This second type of energy loss is used to determine the depth of the atom from which a projectile is backscattered. The stopping power can be determined via Bragg's rule [22]. However, the data at low energies are too limited to make this a reliable approach [23]. A more reliable approach is measuring the stopping power directly in thin films [24]. In combination, these two types of energy loss are used to determine the concentration depth profiles of the elements. The zero mark in the depth scale is calibrated by gas phase experiments [25].

The energy and thus the depth resolution of a NICISS experiment is influenced by two factors. Firstly, the projectiles experience inelastic energy losses during the backscattering process which is due to charge exchange processes at the close encounter of projectile and target atom [26, 27] and can be determined from gas phase NICISS measurements [25]. Secondly, the stopping power of the projectiles is subject to statistical fluctuations which can be described with the Poisson statistic [28] and can be measured experimentally [29]. Knowing the energy resolution of the experiments, the NICIS spectra can be deconvoluted [29]. Due to the large number of variables the deconvolution process is complex. It has been shown that applying the Genetic algorithm for the deconvolution leads to reliable results [21]. The Genetic algorithm leads to reliable deconvolution results because the starting values for the fitting procedure are varied on the full scale of possible solutions [21]. The variation of starting parameters for the fitting is important in particular for cases where a large number of variables need to be fitted.

The separation of elements with similar atomic number is a challenge in a NICISS experiment. The separation becomes more difficult the smaller the ratio between the mass of the projectile and the target atom is. Separation of C, N and O is readily possible in a NICISS experiment.

The separation of P and Cl, however, is more difficult as can be seen in Figure 1. The figure shows measured and deconvoluted concentration depth profiles of the solution of the non-ionic surfactant dodecyldimethyl phosphineoxide ($C_{12}DMPO$) and NaCl in glycerol [30]. Cl appears at a position on the energy scale which can just be separated from the position of P and the presence of Cl⁻ in the spectrum is evident from the change in count rate around 1750 eV. The deconvolution of the measured concentration depth profiles reveals, that Cl⁻ adsorbs at the surface.

However, because it is difficult to fully separate the Cl and the P signal it cannot be quantified how much Cl⁻ adsorbs at the surface. Instead the qualitative conclusion can be drawn, that Cl⁻ adsorbs at the surface of the C12DMPO/NaCl glycerol solution.

A schematic of the NICISS apparatus is shown in Figure 2. The ion source produces an ion beam with kinetic energy of $1 - 5$ keV. Two sets of deflection plates are used to create a pulsed ion beam with a time length of each of the pulses of about $10 - 20$ ns. The pulsed ion beam is directed onto the target. Projectiles with an overall scattering angle of 168° enter the time of flight tube and are recorded with microchannel plates.

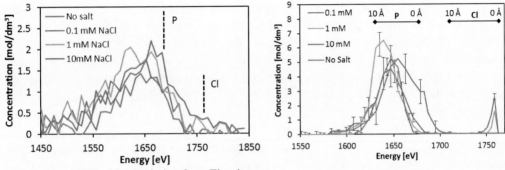

Reprinted from [30] with permission from Elsevier.

Figure 1. Measured concentration depth profiles of P and Cl in the solution of C12DMPO (left panel) and deconvoluted concentration depth profiles (right panel). In the left panel the Cl appears as a small contribution at a higher energy than the P contribution. In the deconvolution Cl appears as sharp peak which can only be interpreted as adsorption of the Cl at the surface.

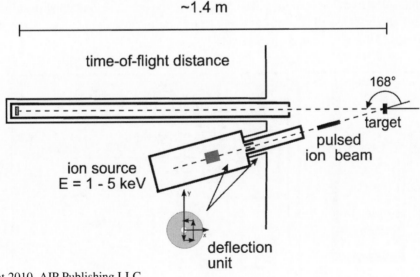

Figure 2. Schematic of the NICISS apparatus. Reprinted with permission from [31].

2.2. Foam Films

The facility that is used to create a foam film for the investigation with NICISS is similar to that used in a TFPB and is shown in Figure 3 and is described in detail in [31]. A porous plate with a hole of 2 mm and a pore size of 17 to 40 μm is filled with the liquid and connected to a capillary establishing a foam film over the hole. The disjoining pressure of the film is adjusted by varying the height difference (h) in the level of the liquid in the capillary and the centre of the foam film. It is possible to vary h in the vacuum chamber for about 0.5 m and thus the disjoining pressure for about 5000 Pa. This range is sufficient for the systems to be investigated. The main difference between the setup shown in Figure 3 and a typical TFPB equipment is that the setup shown in Figure 3 does not have a facility to measure the thickness of the foam films except in the case that profiles across an entire film can be measured. For the investigation of foam films with NICISS, the foam film has to be placed and equilibrated in a vacuum chamber because NICISS is applied under vacuum. However, in order to limit the evaporation of solvent from the foam film, the porous disc with the foam film may not be exposed directly to vacuum and is thus placed in a high pressure cell in the vacuum chamber of the NICISS apparatus. The high pressure cell is designed with a differential pumping stage allowing for pressures up to 1 mbar within the pressure cell. The pressure cell is required to reduce the evaporation from the foam film such that the evaporation from the film is negligible. This is achieved firstly through choosing a small aperture in the order of 1mm in diameter and secondly by covering the inside of the pressure cell with a porous material filled with the solvent ensuring that the partial pressure of the solvent in the pressure cell is very close to the vapour pressure of the solvent. The aperture in the closed housing unit has an entrance and aperture size of about 2 mm. The porous disc is placed 10 mm away from the apertures. The pressure in the pressure cell at the position of the porous disc is about 99.5% of that of the vapour pressure of the solvent and ensures that the evaporation rate of the solvent is negligible [31].

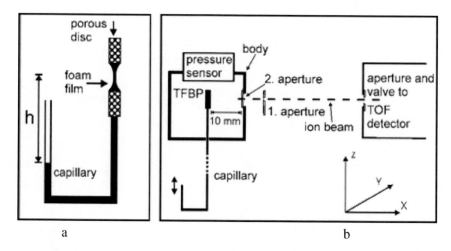

Figure 3. a) device to form a foam film for the investigation with NICISS (concept of the foam film generation in a) similar to Stubenrauch et al., [3]), b) the foam film is placed in a closed housing to reduce the evaporation of solvent. The foam film is investigated with the ion beam through apertures.

The setup also has a facility to fill the porous plate while the foam film holder is in the vacuum chamber.

3. FORCES IN FOAM FILMS

For quantifying the forces in foam films the fundamental equations are used here because all quantities used in the equations can be measured. There are approximations for the fundamental equations which are mainly used to model the forces in foam films by fitting TFPB data. At the end of this section we will discuss the approximations.

The forces in foam films are assumed to be due to a repulsive force of the overlapping electric double layers of the surfaces Π_{elec} and an attractive van-der-Waals force Π_{vdW} [1, 3]. Steric forces to be considered only for very thin films and can be omitted here.

$$\Pi = \Pi_{elec} + \Pi_{vdW} \tag{1}$$

In the presence of ions in the solution forming the foam film – see discussion above for the source of ions – it can be assumed that an electric double layer is formed at the surface [2]. In a foam film both surfaces reach close proximity and the electrostatic double layers interact. The close proximity potentially leads to an overlap of the double layers and as a consequence leads to an ion concentration in the middle of the foam film that is larger than the bulk concentration of the solution and thus to an osmotic pressure Π_{elec}.

For calculating the osmotic pressure the Gibbs free energy is used. The Gibbs free energy can readily be used here instead of the free energy [2]. For calculating the Gibbs free energy the electrochemical potential is used and related to the increase in pressure p_{rep} in the film. The pressure p_{rep} caused by the increase in repulsive potential energy can be determined from [2].

$$p_{rep}(h) = \Pi_{elec}(h) = \frac{\overline{\mu}(h_{eq}) - \overline{\mu}(h = \infty)}{v_{mol}} \tag{2}$$

where $\overline{\mu}(h_{eq})$ is the electrochemical potential at equilibrium film thickness, $\overline{\mu}(h = \infty)$ the electrochemical potential at infinite film thickness thus that of the bulk solution and v_{mol} the molar volume of the solvent.

The electrochemical potential can be determined from the measured concentration depth profiles. The procedure will be to calculate the change in the electrochemical potential $\overline{\mu}$. The electrochemical potential is given by

$$\overline{\mu} = \mu_0 + RT\ln(n) + zF\Phi \tag{3}$$

where μ_0 is a constant, R the ideal gas constant, T the temperature, n the concentration of the solute, z the charge of the ions of the solute, F the Faraday constant and Φ the electric potential.

From the change in electrochemical potential upon formation of the foam film (approaching of the two foam film surfaces) we can determine the change in energy in the foam film and thus of the pressure in the foam film through equation (3).

It must be emphasized that all quantities in equation (3) can be determined from measuring concentration depth profiles. The electric potential can be calculated from the concentration depth profile of the charges [32] provided that the dielectric constant at the surface is known [32]. The dielectric constant can be determined as described in [32].

The electrochemical potential is constant across the foam film and thus can be determined at the surface once the electric potential at the surface has been determined. $\mu_0 + RT \ln(n)$ is the chemical potential and can be determined from the surface excess through measuring the concentration depth profiles. This procedure has been shown to be applicable to liquid surfaces as described in [33, 34].

The van der Waals forces in foam films can be calculated with

$$\Pi_{vdW} = -\frac{A}{12\pi L^2}$$

$$(4)$$

where A is the Hamaker constant and is a material constant describing the van der Waals forces. The Hamaker constant depends on the elemental composition and density of a substance. $\frac{1}{12\pi L^2}$ considers the geometric structure of foam films with L being the film thickness. For accurate calculations A should not only consider the static polarisability but also its frequency dependence. Equation (4) is a first approximation for the dispersion force and gives only a rough estimate [2].

For the full calculation of the van der Waals forces, retardation effects and the composition of the film have to be taken into account [35]. For the latter the change of composition along the surface normal has to be taken into account. The composition of the foam film at the liquid/gas interface is different to the composition inside the film and thus the Hamaker constant describing the material properties changes across the foam film.

4. MEASUREMENTS OF CONCENTRATION DEPTH PROFILES IN FOAM FILMS

Ridings et al., have investigated foam films of hexadecyltrimethylammonium bromide ($C_{16}TAB$) in glycerol. The solvent glycerol was chosen for two reasons. Firstly, glycerol has a low vapour pressure allowing formation of stable films under vacuum conditions (10^{-5} mbar).

Secondly, the self-dissociation constant of glycerol is about 4 orders of magnitude lower than that of water [36] thus, as opposed to aqueous foam films, the concentration of ions in the bulk and at the surface due to self-dissociation of the solvent can be neglected.

In Figure 4 the concentration depth profiles of the alkyl chain and the anion Br⁻ are shown. The concentration depth profile of the alkyl chain is derived from the profile of carbon. Carbon is a constituent in both the solvent and the surfactant, and both contributions need to be separated to obtain the profile of the carbon in the alkyl chain only.

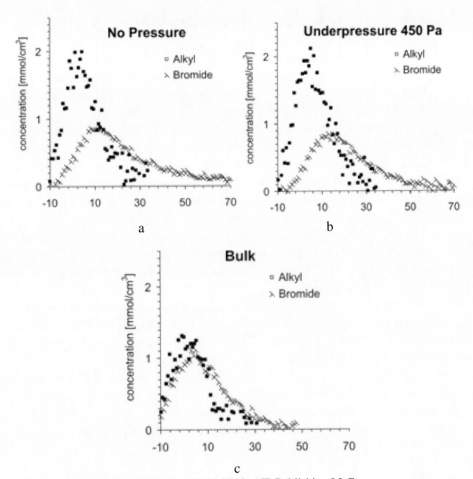

Figure 4. Concentration depth profiles of the anion bromide and the alkyl chain of the surfactant $C_{16}TAB$ in glycerol. In a) the concentration depth profiles of the foam film formed without external pressure are shown, in b) those of the foam films formed at an underpressure of 450 Pa and c) the concentration depth profiles of the bulk surface.

Oxygen is a constituent of the solvent only. Because the glycerol molecule is rather small, the oxygen profile can be considered as representing the profile of the carbon in the glycerol. The difference between the profile of the total carbon and the oxygen then results in the profile of the carbon in the alkyl chain [31]. Figure 4 shows that the alkyl chain and the anion separate upon formation of the foam film.

Unfortunately the carbon cannot be considered as representing the exact concentration depth profile of the cation because the nitrogen is located at one side of the cation and its profile could differ from that of the alkyl chain. The position of the cation could be determined with the profile of nitrogen.

However, the signal of projectiles backscattered from nitrogen is too weak and cannot be evaluated. For this reason measurements are on the way where the nitrogen is replaced with phosphorous. The measurements on hexadecyltrimethylphosphonium bromide solutions in glycerol will then allow determining the charge distribution directly.

5. SPECIFIC ION EFFECTS DETERMINED WITH NICISS

NICISS measurements on various tetra-N-butylonium halides (tetra-N-butylammonium iodide (Bu$_4$NI), (tetra-N-butylammonium bromide (Bu$_4$NBr), (tetra-N-butylphosphonium bromide (Bu$_4$PBr)) allowed the determination of the position of two different anions (Br$^-$ and I$^-$) relative to the liquid/vapour interface of formamide solutions [25] while keeping the cation the same. In the left panel of Figure 5 the profiles of Br$^-$ and I$^-$ of 0.25 M solutions of Bu$_4$NBr and Bu$_4$NI in formamide show slightly different positions with the I$^-$ located closer to the surface than the Br$^-$. It can also be seen that the surface coverage, and as a consequence the surface excess, with the I$^-$ surfactant is larger than that of the Br$^-$ surfactant. The same trend for the surface excess and the position of the anion was found by Schulze et al., for the same series of surfactants but also including a surfactant with Cl$^-$ as the anion.

The same trend for the surface excess was also reported by Holmberg et al., who concluded from ARXPS data that the surface excess increases in the order Bu$_4$NCl < Bu$_4$NBr < Bu$_4$NI [37]. It is important to note that the ionic radius of the anions increases in the order of Cl$^-$ < Br$^-$ < I$^-$.

Wang et al., found a similar trend also for the cations of anionic surfactants. The solutions of sodium dodecyl sulfate (SDS) and cesium dodecyl sulfate (CDS) in formamide have a higher surface excess for the cesium version of the surfactant compared to the sodium version [38-40] with the Cs$^+$ being located closer to the surface than Na$^+$.

For the ionic radius it holds Cs$^+$ > Na$^+$. The preferential adsorption of I$^-$ at the surface of LiI solutions in formamide was found with NICISS while Cl$^-$ did show neither desorption nor preferential adsorption at the surface (see Figure 6) [41]. It is still unclear whether the ionic radius and the size of the solvation sphere are correlated to preferential adsorption of ions at a liquid surface [12, 13, 17, 20, 42-51].

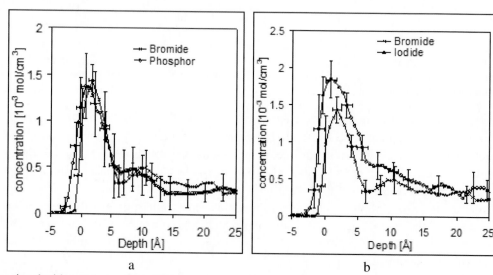

a   b

Reprinted with permission from [25].

Figure 5. a) concentration depth profiles of cation and anion of Bu$_4$PBr and b) concentration depth profiles of the anions of Bu$_4$NBr and of Bu$_4$NI in 0.25 M solutions in formamide.

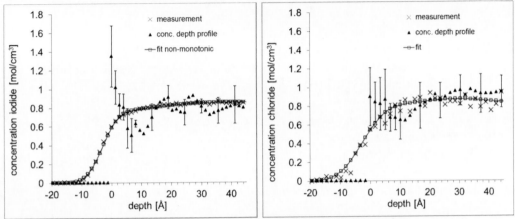

Figure 6. on the left side the concentration depth profile of iodide of 0.85 M solutions of LiI in formamide and on the right side the concentration depth profile of chloride of 0.85 M solutions of LiCl in formamide are shown. The measured concentration depth profiles have been deconvoluted.

It has been found for aqueous solutions by means of computer simulations that the ionic radius in solution increases with the atomic number both for the halide ions [52, 53] and the alkali ions [54]. The simulations also show that the larger ions seem to have a less ordered solvation sphere in aqueous solutions [53, 54].

Despite the fact that ion specific adsorption at interfaces has been demonstrated with computer simulations and calculations [17, 46, 50, 51], other authors have raised questions whether the potentials used for calculations and simulations are appropriate [49].

CONCLUSION

Foam films are stabilized by the disjoining pressure, where the main contributors to this are the overlap of electric double layers, and van der Waals forces. Both these contributions can be directly determined once the concentration depth profiles of all species in the film are known. Neutral impact collision ion scattering spectroscopy provides a technique for determining concentration depth profiles of elements at soft matter surfaces, and has been thoroughly applied to the investigations of bulk liquid surfaces. Recently this technique has been adapted to accommodate foam films for measurement, which have led to the first experimentally obtained concentration depth profiles for foam films. Investigations into solutions of glycerol/C_{16}TAB have shown that the surfactant has a different orientation at the surface of a foam film compared to the surface of the corresponding bulk solution, indicating the uncertainty in correlating the adsorption behavior of surfactants in foam films directly from the corresponding bulk liquid surface. Additionally, investigations into foam films and bulk liquid surfaces have demonstrated NICISS is capable of determining the concentration depth profiles of ions to such a degree that specific ion effects in the preferential adsorption of ions at surfaces can be identified. Finally, a description is given for how the disjoining pressure can be calculated using the electrochemical potential, once the concentration depth profiles of all species are known.

REFERENCES

[1]   V. Bergeron, *Journal of Physics: Condensed Matter*, 1999, 11, R215-R238.

[2]   R. J. Hunter, *Foundation of Colloid Science Oxford University Press*, Oxford, 1989.

[3]   C. Stubenrauch and R. V. Klitzing, *Journal of Physics: Condensed Matter*, 2003, 15, R1197-R1232.

[4]   C. Ridings, V. Lockett and G. Andersson, *Colloids and Surfaces A: Physicochemical and Engineering Aspects*.

[5]   K. Hanni-Ciunel, N. Schelero and R. von Klitzing, *Faraday Discussions*, 2009, 141, 41-53.

[6]   E. D. Manev and A. V. Nguyen, *International Journal of Mineral Processing*, 2005, 77, 1-45.

[7]   D. W. Fuerstenau and Pradip, *Advances in Colloid and Interface Science*, 2005, 114-115, 9-26.

[8]   M. C. Fuerstenau and K. N. Han, *Journal of Colloid and Interface Science*, 2002, 256, 175-182.

[9]   T. J. Martin, In: *Foam Engineering: Fundamentals and Applications*, ed. P. Stevenson, Wiley-Blackwell, Chichester, UK 2012, pp. 411-457.

[10]   J. Ralston, *Australian Journal of Chemistry*, 2005, 58, 644-654.

[11]   P. Brown, A. Bushmelev, C. P. Butts, J. Cheng, J. Eastoe, I. Grillo, R. K. Heenan and A. M. Schmidt, *Angewandte Chemie International Edition*, 2012, 51, 2414-2416.

[12]   V. S. J. Craig, *Current Opinion in Colloid and Interface Science*, 2004, 9, 178-184.

[13]   C. L. Henry and V. S. J. Craig, *Langmuir*, 2008, 24, 7979-7985.

[14]   C. L. Henry and V. S. J. Craig, *Langmuir*, 2010, 26, 6478-6483.

[15]   J. K. Angarska, K. D. Tachev, P. A. Kralchevsky, A. Mehreteab and G. Broze, *Journal of Colloid and Interface Science*, 1998, 200, 31-45.

[16]   S. Pandey, R. P. Bagwe and D. O. Shah, *Journal of Colloid and Interface Science*, 2003, 267, 160-166.

[17]   P. dos Santos and Y. Levin, *Physical Review Letters*, 2011, 106, 167801.

[18]   D. M. Huang, C. Cottin-Bizonne, C. Ybert and L. Bocquet, *Langmuir*, 2007, 24, 1442-1450.

[19]   W. Sun, Y. Shen and J. Hao, *Langmuir*, 2010, 27, 1675-1682.

[20]   S. Murgia, M. Monduzzi and G. Palazzo, *Langmuir*, 2011, 28, 1283-1289.

[21]   G. Andersson and H. Morgner, *Surface Science*, 2000, 445, 89-99.

[22]   W. H. Bragg and R. Kleeman, *Philosophical Magazine Series 6*, 1905, 10, 318-340.

[23]   J. F. Ziegler and J. M. Manoyan, *Nuclear Instruments and Methods in Physics Research Section B: Beam Interactions with Materials and Atoms*, 1988, 35, 215-228.

[24]   G. Andersson and H. Morgner, *Nuclear Instruments and Methods in Physics Research Section B: Beam Interactions with Materials and Atoms*, 1999, 155, 357-368.

[25]   G. Andersson, H. Morgner and K. D. Schulze, *Nuclear Instruments and Methods in Physics Research Section B: Beam Interactions with Materials and Atoms*, 2002, 190, 222-225.

[26]   H. H. Brongersma, M. Draxler, M. de Ridder and P. Bauer, *Surface Science Reports*, 2007, 62, 63-109.

[27] G. G. Andersson, In: *Reference Module in Chemistry, Molecular Sciences and Chemical Engineering*, Elsevier 2013.

[28] G. Andersson, H. Morgner and H. Pohl, *Physical Review A*, 2008, 78, 032904.

[29] G. Andersson, *Physical Review A*, 2007, 75, 032901.

[30] C. Ridings and G. G. Andersson, *Nuclear Instruments and Methods in Physics Research Section B: Beam Interactions with Materials and Atoms*, accepted for publication 24th of July 2014.

[31] C. Ridings and G. G. Andersson, *Review of Scientific Instruments*, 2010, 81, 113907-113908.

[32] K. D. Schulze and H. Morgner, *Journal of Physics: Condensed Matter*, 2006, 18, 9823-9839.

[33] G. Andersson, T. Krebs and H. Morgner, *Physical Chemistry Chemical Physics*, 2005, 7, 136-142.

[34] T. Krebs, G. Andersson and H. Morgner, *The Journal of Physical Chemistry B*, 2006, 110, 24015-24020.

[35] J. N. Israelachvili, *Intermolecular and Surface Forces*, Academic Press, Amsterdam, 1991.

[36] A. Manfredi, E. Ranucci, M. Suardi and P. Ferruti, *Journal of Bioactive and Compatible Polymers*, 2007, 22, 219-231.

[37] S. Holmberg, Z. C. Yuan, R. Moberg and H. Siegbahn, *Journal of Electron Spectroscopy and Related Phenomena*, 1988, 47, 27-38.

[38] C. Wang and H. Morgner, *Colloid Polym. Sci.*, 2011, 289, 967-970.

[39] C. Wang and H. Morgner, *Langmuir*, 2009, 26, 3121-3125.

[40] C. Wang and H. Morgner, *Physical Chemistry Chemical Physics*, 2011, 13, 3881-3885.

[41] G. Andersson, H. Morgner, L. Cwiklik and P. Jungwirth, *The Journal of Physical Chemistry C*, 2007, 111, 4379-4387.

[42] P. Creux, J. Lachaise, A. Graciaa and J. K. Beattie, *The Journal of Physical Chemistry C*, 2007, 111, 3753-3755.

[43] E. M. Knipping, M. J. Lakin, K. L. Foster, P. Jungwirth, D. J. Tobias, R. B. Gerber, D. Dabdub and B. J. Finlayson-Pitts, *Science*, 2000, 288, 301-306.

[44] S. Ghosal, J. C. Hemminger, H. Bluhm, B. S. Mun, E. L. D. Hebenstreit, G. Ketteler, D. F. Ogletree, F. G. Requejo and M. Salmeron, *Science*, 2005, 307, 563-566.

[45] P. Jungwirth and D. J. Tobias, *Journal of Physical Chemistry B*, 2001, 105, 10468-10472.

[46] P. Jungwirth and D. J. Tobias, *Chemical Reviews*, 2006, 106, 1259-1281.

[47] R. I. Slavchov and J. K. Novev, *Journal of Colloid and Interface Science*, 2012, 387, 234-243.

[48] N. Schelero, G. Hedicke, P. Linse and R. V. Klitzing, *The Journal of Physical Chemistry B*, 2010, 114, 15523-15529.

[49] B. W. Ninham, T. T. Duignan and D. F. Parsons, *Current Opinion in Colloid and Interface Science*, 2011, 16, 612-617.

[50] A. Martín-Molina, J. G. Ibarra-Armenta and M. Quesada-Pérez, *The Journal of Physical Chemistry B*, 2009, 113, 2414-2421.

[51] P. Koelsch, P. Viswanath, H. Motschmann, V. L. Shapovalov, G. Brezesinski, H. Mohwald, D. Horinek, R. R. Netz, K. Giewekemeyer, T. Salditt, H. Schollmeyer, R.

von Klitzing, J. Daillant and P. Guenoun, *Colloids and Surfaces A: Physicochemical and Engineering Aspects*, 2007, 303, 110-136.

[52] F. Bresme, E. Chacón, P. Tarazona and A. Wynveen, *The Journal of Chemical Physics*, 2012, 137, -.

[53] V. Migliorati, F. Sessa, G. Aquilanti and P. D'Angelo, *The Journal of Chemical Physics*, 2014, 141, -.

[54] T. Ikeda, M. Boero and K. Terakura, *The Journal of Chemical Physics*, 2007, 126, -.

In: Innovations in Nanomaterials
Editors: Al-N. Chowdhury, J. Shapter, A. B. Imran

ISBN: 978-1-63483-548-0
© 2015 Nova Science Publishers, Inc.

Chapter 7

Transport Properties of Colloids in Bulk and in Confinement at Nanoscale

Gaganpreet[*], Sunita Srivastava and K. Tankeshwar

Institute of Nano Science and Technology, Sector-64, Ph-X, Mohali (Punjab), India
Department of Physics, Panjab University, Chandigarh, India

Abstract

Fluid like water and fluid with Gaussian core model (GCM) potential show anomalous behaviour of transport coefficients. Different behaviour is seen in fluid with stable colloidal dispersion of nanoparticles in it and also when fluid is confined to nano-geometries. In the present work, theoretical results have been presented to explain the anomalous behaviour of nanofluids and soft core fluids subject to confinement. Statistical mechanism, which include aggregation of nanoparticles and the concept of formation of interfacial layer around the nanoparticles have been incorporated to explain the cause of anomalous increase in thermal conductivity and viscosity of nanofluids. This chapter also includes soft GCM fluid which behaves differently and shows anomalous behaviour even in the bulk. Particularly, work presented here is the study of velocity autocorrelation function and self-diffusion coefficient of fluid by employing the Mori-Zwanzig memory function formalism. Bulk values of contributions to the second and the fourth frequency sum rules of velocity autocorrelation function have been modified for the confined system. These modified sum rules and an ansatz for the memory function have been used to study diffusion for different densities and temperatures as a function of distance from the wall.

1. Introduction

Study of fluid confined to nanoscale is an emerging field in the area of applied sciences as well as technology having interdisciplinary features. Foremost unifying theme is the control of matter on a scale smaller than 1 μm, normally 1-100nm as well as fabrication of

---

[*] E-mail: gagan@inst.ac.in.

devices on a small length scale. Nanoscale endows the properties and behaviour that are different from bulk material and the interaction with the walls that confine the sample inside some channels. Transport of fluids with colloidal dispersion of nanoparticles or in nanoscale conduits has gained a lot of technological relevance in the last few decades like in the area of biological and chemical separations [1, 2], energy storage [3] and biosensing [3, 4], nanodrug delivery, cancer therapeutics [5, 6] and water desalination [7]. In 1993, investigation by Choi [8] of Argonne National lab, USA on advanced fluids culminated in the invention of nanofluids and subsequently, anomalous increase of thermal conductivity in metallic-based nanofluidswas reported in 2001 [9]. This special class of fluids with the colloidal dispersion of mesoscopically few hundred nanometre particles in conventional thermal base fluids such as ethylene glycol (EG), water, polyolephin oil, etc. aretermed as nanofluids [10]. In most of the cold regions of the world, use of water as coolant has been replaced by ethylene glycol due its low freezing point under normal conditions. Added nanoparticles may be oxide ceramics (e.g., CuO, Al_2O_3, SiO_2), metallic (e.g., Al, Cu, Si, Fe), metal nitrides (e.g., SiN, AIN) or carbon nanotubes (CNT), fibers which change the transport properties and heat transfer characteristics of the base fluids. These nanofluids have demonstrated significant increase in their critical heat flux in boiling heat transfer and in thermal conductivity compared with base fluids. They also exhibit strong dependence on temperature, size, shape, concentration of nanoparticles and surfactants. Nanoparticles have unique features such as high particle mobility and large surface to volume ratio. Properly engineered colloidal nano dispersions possess the following unique features over conventional solid-liquid suspensions [10]

1) More heat transfer surface between the particles and the fluids.
2) Reduced pumping power of nanofluids
3) More dispersion stability and predominant Brownian motion of particles.
4) Less chances of clogging as compared to the conventional slurries in microchannels.

Nanofluid properties including thermal conductivity, viscosity, diffusion and surface wettability, optical properties, etc. can be adjusted by varying particle size, particle volume fraction, shape and other fluid parameters to meet the requirements of various applications. Significant research on nanofluids for thermal properties and other applications has been carried out.

Excellent potential of the nanofluids explored by the researchers led both industries and universities to launch research activities in nanofluid technology. This effort has increased manifold over the past few years with emphasis on further development and removal of some key barriers in the nanofluid technology. Industrial cooling applications are one of the major technical challenges facing many diverse industries including transportation, microelectronics and most of these demands [11-13] shall be taken care of by nanofluids. In the case of the automotive industry, nanofluids with engine oils as base fluids find application at low pressure and engine operating temperatures [14]. The most desirable heat transfer base fluids for the cooling systems are EG, water and mineral oils. Use of nanofluids as coolant in industry has the potential to conserve energy and reduce the emission of gases [15]. Besides, these nanofluids have found uses in many other applications [15].

Some of the applications of nanofluids are circulation for thermal management of weapons armed with pulsed systems, mass transfer, gas absorption [14, 16-17], Plasmon resonance [18-20] to produce optical filters or lasers. ZnO Nanofluids exhibited antibacterial properties against E. coli DH5a bacteria [21]. By adding nanosized particles like TiO_2, the dielectric strength of the mineral oils was found to improve [22] and also enhanced the breakdown voltage. Recent research demonstrated that nanofluids provide a promising path in increasing the efficiency of solar collection systems [18, 23, 24], find an extensive use in medical applications like treatment of hyperthermia, magnetic-targeted drug delivery and flow in micro-nanofluidic devices [25, 26]. Nanofluids have specific heat capacity which is nearly 50% more than the base fluids and hence have better efficiency in thermal energy storage applications compared to the [27] cooling applications. To explain the anomalous increase in thermo-physical properties [28-31] such as viscosity, specific heat, density and mass transfer, number of theoretical attempts have been made in this direction by including various factors such as aggregation, interfacial layer, Brownian motion, nano-convection. Heat transfer coefficient was of prime interest in the first decade of nanofluid research. However, comprehensive theoretical models for nanofluid that into account all the main factors are lacking and hence require more systematic efforts in this direction.

Extensive work has been reported in the literature covering different types of fluids and fluid mixtures subject to nano-confinement. However, in this chapter we have focussed chiefly on diffusion properties of Gaussian core model (GCM) fluids under confinement.

The chapter is organised as follows. In sections 2 and 3, thermal conductivity and shear viscosity of nanofluids are discussed. In section 4, GCM fluid has been introduced and diffusion has been studied in confined geometry. The work has been concluded in Section 5.

2. THERMAL CONDUCTIVITY OF NANOFLUIDS

Thermal conductivity is an imperative property of nanofluids. Recent report by Sundar et al. [33] showed enhancement in thermal conductivity of hybrid composite of nano-diamond and nickel (ND-Ni) nanoparticles based nanofluid for a 3.03% wt. of ND-Ni nanoparticles dispersed in water and EG of 21% and 13%, respectively. Usowicz et al. [34] explained enhanced conductivity for oxide nanofluids based on the physical statistical model. Choi et al. [35] reported an enhancement up to 160% in thermal conductivity for multi walled carbon nanotubes dispersed in synthetic poly (α-olefin) oil nanofluids. Masuda et al. [36] reported an enhancement up to 30% in thermal conductivity at volume fractions of less than 4.3% in water based nanofluids of oxide nanoparticles such as Al_2O_3. Eapen et al. [37] studied enhancement as a function of aggregation state of nanoparticles. Lee et al. [38] investigated CuO-water/EG and Al_2O_3-water/ EG systems and found an increase of about 20% in thermal conductivity at volume fraction of 4%. Eastman et al. [39] found such an enhancement for CuO-water, Al_2O_3-water and Cu-Oil nanofluids using the transient hot wire method. These results were much higher than those accounted for by classical models. Various factors also contribute to the thermal conductivity enhancement which includes: particle volume fraction, size, shape, spatial distribution, temperature, pH value as well as type of the base fluid [34-39].

There exist many experimental determination of thermal conductivity [31-33, 39-44] and also classical models to explain the enhancement in the thermal conductivity of nanofluids [45-47]. Classical theories like Maxwell's model showed that the thermal conductivity of nanofluids depend only on volume fraction of dispersed nanoparticles when the volume concentration is low. However, from the experimental reports discussed above, it is clear that thermal conductivity of nanofluids also depends on other parameters like particle shape, size, temperature etc. This implies that classical theories have not been able to explain the fundamental mechanism responsible for enhancement of thermal conductivity. The mechanisms proposed for the explanation of thermal conductivity typically fall into two categories: static [48-53] and dynamic mechanisms.

2.1. Interfacial Layer

There is a formation of a semi solid ordered layer of liquid molecules around nanoparticles when they are dispersed in a fluid due to adsorption of molecules of base fluid on the surface of the nanoparticle, which has been confirmed through molecular dynamics simulation and TEM imaging [54, 55]. Concept of interfacial layers was first introduced by Keblinski et al. [56] and Yu and Choi [57] to analyse the enhanced thermal conductivity. Lin et al. [54] reported the thickness of the interfacial layer to be 0.75 nm around the copper nanoparticle in EG. Similarly, from TEM images, Oh et al. [55] showed that ordering of liquid molecules exists on the interface between liquid aluminium and sapphire. Liang and Tsai [58], using non-equilibrium molecular dynamics simulation, showed that the thermal conductivity of Au-Ar system becomes 1.6 to 2.5 times higher than that of base fluid when the thickness of the interfacial layer is taken as 1nm.

The varied simulation results reveal that initially when the materials such as oxides, sulphides and insoluble salts are immersed in aqueous solution, solid and liquid materials are confined in their own areas and the value of density within each area is approximately constant. However, with time, liquid density near the interface continues to increase resulting in the formation of the interfacial layer [58]. Even the change in volume concentration of nanoparticles in fluid affects the thickness of the interfacial layer. This layer at the solid-liquid interface is expected to exhibit properties intermediate between the base fluid and the bare particle.

The interfacial layer around a nanoparticle may also be considered as a monolayer formed by the adsorption of liquid molecules on the nanoparticle surface. Wang et al. [52] estimated the adsorption thickness h from the carrier fluid property as

$$h = \frac{1}{\sqrt{3}} \left(\frac{4M}{\rho_f N_A} \right)^{\frac{1}{3}}. \tag{1}$$

where M is the molecular weight of the liquid, ρ_f the density of the base fluid and N_A the Avogadro's number. This nanolayer effectively acts as a thermal bridge between a solid nanoparticle and the bulk liquid surrounding this. Accordingly, the nanoparticle in liquid suspension could be viewed as a three phase composite, in which each solid nanoparticle is surrounded by a interfacial layer having different thermal conductivity from particles and base fluids. Depending on its nature, the layer could also act as a thermal barrier for heat transfer and, therefore, it is a key structure based mechanism is enhancing thermal conductivity of nanofluids. It is pertinent to note that the interfacial layer can acquire surface charges [59], which attract and bind counter ions with opposite polarity in the liquid. In view of the fact that the ions are often hydrated, association of such ions with water molecules at the interface is responsible for the hypothetical charge-induced interfacial layer. As an extension of this, we can say that electric double layer (EDL) [59-61] consists of two parallel layers of ions. Depending on the particle surface first layer can be either positive or negative known as Stern layer and the other layer is in the fluid which electrically screens the first layer. Second layer is formed due to free ions in the fluid under the influence of electric attraction and thermal motion. This layer is known as diffuse layer which is loosely associated with the object and resulting in decrease in potential. Typical interfacial width is only of the order of few distances, i.e., 1nm while in simulations this width goes up to 2 nm [62]. The reason of such a big difference between simulation and the experiments is still unknown. Hence, this aspect of the problem also deserves more endeavours. The thickness, h, of interfacial layer of the EDL is expressed as

$$h = C(dk)^{-1},$$

(2)

and it is often scaled as reciprocal Debye-Huckel parameter $(dk)^{-1}$ and C is a constant [52-61]

$$d\kappa^{-1} = \sqrt{\left(\varepsilon_0 \varepsilon_r R_0 T / 2000 F^2 I\right)}$$

(3)

where F is the Faraday constant, ε_o is permittivity of vacuum and R_0 is the gas constant. The remaining parameters, viz., dielectric constant ε_r, ionic strength I and temperature, T all pertain to the solution.

In the next section of this chapter, we have discussed the concept of effective medium theory and aggregation models which is mainly responsible for thermal enhancement in nanofluids.

2.2. Aggregation Based Thermal Conductivity Models

There are number of thermal conductivity studies on the basis of effective medium approximations (EMA) which have been derived from continuum formulations and the concept of nanoparticle clusters. EMA is used to explain the thermal conductivity of the nanofluids without taking in to consideration the size, geometry, distribution or the motion of dispersed nanoparticles, while only particle volume fraction and the thermal conductivities of particles and base fluids are included. EMAs are physical models that describe the macroscopic properties of a medium based on relative fractions and the properties of its components. There are two approaches commonly used in EMA to explain the effective transport coefficient of mixture and composites: the Maxwell-Garnett (MG) self-consistent approximation [63] and the Bruggeman approach [64]. Numerous theoretical studies for particle-fluid mixtures have been conducted based on the classical work of Maxwell [63]. Maxwell considered a heterogeneous mixture of suspension of spherical particles in which the interaction between the particles was ignored in the dilute limit.

Based on effective medium theory and fractal theory, Maxwell gave the expression of effective thermal conductivity for dispersed particles [63] as

$$K_{eff} = \left[\frac{k_{agg} + (n-1)k_f - (n-1)\phi_{agg}(k_f - k_{agg})}{k_{agg} + (n-1)k_f + \phi_{agg}(k_f - k_{agg})} \right] k_f.$$

$$(4)$$

Here, k_{agg} is the thermal conductivity of aggregates, k_f is the thermal conductivity of base fluid, ϕ_{agg} is the volume fraction of aggregates and

$$n = 3/\psi$$

where ψ is the measure of sphericity and has value 3 for spherical particle. Wang et al. presented a fractal model to account for the observed thermal conductivity of nanofluids [52]. According to them, the effective thermal conductivity of nanofluids in terms of fractal model is expressed as

$$K_{eff} = \left[\frac{(1-\phi) + 3\phi \int_0^\infty \dfrac{k_{cl}(a_{cl})n(a_{cl})}{k_{cl}(a_{cl}) + 2k_f} da_{cl}}{(1-\phi) + 3\phi \int_0^\infty \dfrac{k_{bf}(a_{cl})n(a_{cl})}{k_{cl}(a_{cl}) + 2k_f} da_{cl}} \right] k_f.$$

$$(5)$$

Here, ϕ is the volume fraction of nanoparticles, a_{cl} is the equivalent radius of nanoparticle cluster and depends on the fractal dimension of the cluster structure, $k_{cl}(a_{cl})$ is the effective thermal conductivity of nanoparticle cluster and $n(a_{cl})$ is the radial distribution function. It explains well the thermal conductivity of CuO-water nanofluids when the adsorption is included in the calculations but it under predicts the increase when the adsorption effect is excluded. This model explains the thermal conductivity well for nonmetallic nanofluids. Xuan et al. [65] performed simulations on random motion and the aggregation process of the nanoparticles by using the theory of brownian motion and diffusion-limited aggregation model. Their expression for thermal conductivity at temperature T can be written as

$$K_{eff} = \left[\frac{k_p + 2k_f + 2(k_p - k_f)\phi}{k_p + 2k_f - (k_p - k_f)\phi} \right] k_f + \frac{\rho_p \phi c_p}{2} \sqrt{\frac{K_B T}{3\pi a_{cl} \mu}}.$$

$$(6)$$

Here, ρ_p, c_p are the density and specific heat of nanoparticle, k_p is the thermal conductivity of nanoparticle, K_B is the Boltzmann constant μ and T are the viscosity and temperature of fluid. The first term in Equation (6) corresponds to Maxwell model, while the second term gives contribution due to brownian motion of suspended nanoparticle clusters. Prasher et al. [66] studied the aggregation effect based on aggregation kinetics of nanofluids. They showed that effective thermal conductivity is affected by the fractal like aggregates through which heat transfer depends on zeta potential, pH, Hamaker Constant, ion concentration and temperature. Evans et al. [67] analyzed and carried out simulation studies on the effect of interfacial thermal resistance and aggregation on the effective thermal conductivity of nanofluids and nanocomposites. Keblinski et al. [56] assumed that clustered nanoparticles provide local percolation-like paths for rapid heat transport and increase the effective volume fraction. Gaganpreet and Srivastava [31] explained the substantial high thermal conductivity in oxide nanoparticle suspensions by taking 1-D thermal electrical analogue to find heat conduction through aggregate of nanoparticles.

In this chapter, substantially high thermal conductivity of nanofluids is analysed by making use of the involved structural mechanisms such as liquid layering and the concept of nanoparticle aggregation which is termed as clusters. Particle in nanofluids may aggregate with each other and may resemble solid like structure. This analysis is based on the assumption that there are two paths of heat flow through nanofluids: one through medium and the other through aggregate particles. Nanoparticles inside the base fluid no longer retain their original shape due to agglomeration, particle adhesion to the walls and coagulations [68]. To reduce the mathematical complexity, the deviated shape of nanoparticles has been taken to be a simple prolate spheroid. The overall heat transfer of the system for one dimensional heat flow may be expressed as

$$q = q_{md} + q_{il} \tag{7}$$

where the subscripts md and il denote respectively the quantities for medium, and for clusters of particles with interfacial layer. The nanoparticles soon after dispersion form an interfacial layer around them and hence hereafter are termed as equivalent nanoparticles [57]. The thermal conductivity of the equivalent particles can be analysed using the concept of effective medium theory [31, 57, 69] as:

$$k_{pe} = \frac{\left[2(1-\sigma)+(1+\delta_{maj})(1+\delta_{min})^2(1+2\sigma)\right]\sigma k_p}{\left[(\sigma-1)+(1+\delta_{maj})(1+\delta_{min})^2(1+2\sigma)\right]}, \tag{8}$$

where $\sigma = k_{layer}/k_p$ is the ratio of thermal conductivities of interfacial layer to that of the nanoparticle. Due to these equivalent nanoparticles, volume fraction ϕ of nanoparticles also gets modified and results in an equivalent volume fraction given by

$$\phi_{mod} = \phi(1+\delta_{maj})(1+\delta_{min})^2, \tag{9}$$

Here,

$$\delta_{maj} = \frac{h}{a}, \delta_{min} = \frac{h}{b}. \tag{10}$$

where h is the nanolayer thickness, a and b are the semi major and semi minor axes of prolate spheroid nanoparticles. The effective thermal conductivity for the medium due to these equivalent nanoparticles for low volume concentration [57] is obtained to be

$$K_m = \frac{k_{pe} + 2k_f + 2(k_{pe} - k_f)\phi_{mod}}{k_{pe} + 2k_l - (k_{pe} - k_f)\phi_{mod}} k_f.$$

(11)

Following Feng et al. [70], a theoretical model has been developed for the prolate shaped nanoparticles based on the fact that thermal conductivity increases because of aggregating nanoparticles [31-32, 70]. Since some nanoparticles may aggregate to form clusters, while others remain well dispersed. Therefore, the effective thermal conductivity of medium and aggregating particles is given by adding the two terms by employing a parameter α which measures the contribution of interfacial layer to the effective thermal conductivity of nanofluids

$$K_{eff} = (1 - \alpha)K_m + \alpha K_{il}.$$

(12)

Dimensionless thermal conductivity is given by

$$K_{eff}^* = K_m^* + K_{il}^*,$$

(13)

where

$$K_{il}^* = \frac{6\phi_{mod} E(e)(a+h)}{\pi\gamma\sqrt{1-\gamma^2}}\left[\frac{1}{\gamma}\ln\frac{1}{1-\gamma} - 1\right],$$

(14)

and

$$K_m^* = (1 - \phi_m)\left[\frac{k_{pe} + 2k_f + 2(k_{pe} - K_f)\phi_{mod}}{k_{pe} + 2k_f - (k_{pe} - K_f)\phi_{mod}}\right].$$

(15)

where K_m, K_{il}, k_{pe} and k_f are thermal conductivities of medium, interfacial layer, equivalent nanoparticle (i.e., nanoparticle surrounded by nanolayer of thickness, h and base fluid, respectively. For calculations, $\alpha = \phi_{mod}$ is taken where ϕ_{mod} is the modified volume fraction.

Here, circumference of ellipse with eccentricity E(e) is given as

$$E(e) = \frac{\pi}{2}\left\{ 1 - \left(\frac{1}{2}\right)^2 e^2 - \left(\frac{1.3}{2.4}\right)^2 \frac{e^4}{4} - \left(\frac{1.3.5}{2.4.6}\right)^2 \frac{e^6}{5} - \ldots\ldots - \left(\frac{(2n-1)!!}{2n!!}\right)^2 \frac{e^{2n}}{2n-1} \right\}$$

$$= \frac{\pi}{2}\left[1 - \sum_{n=1}^{\infty} \frac{e^{2n}}{2n-1} \prod_{m=1}^{n} \frac{\pi}{m}\left(\frac{2m-1}{2m}\right)^2 \right]$$

(16)

Obviously, for a circle, eccentricity $e = \sqrt{1 - \frac{(b+h)^2}{(a+h)^2}}$ is zero as a + h = b + h and $E(e) = \frac{\pi}{2}$.

Results based on the model developed here have been compared with the experimental data on oxide nanofluids available in the literature. This clearly brings out the fact that the thermal conductivity enhancement of nanofluids is influenced by the volume fraction of nanoparticles, interfacial layer, eccentricity, thermal conductivity of nanoparticle and base fluid, aggregation and size of the dispersed nanoparticles. Calculations have been carried out to study the effect of particle shape on thermal conductivity of nanofluids. For this analysis was performed by varying values of eccentricity. However, only those values of eccentricity have been reported for which the thermal conductivity values match well with the experimental values. For example in Figure 1(a) the choice of eccentricity e = 0.1 is found to give results for Al$_2$O$_3$-EG system whereas e = 0.9 appears to be better choice for Al$_2$O$_3$-H$_2$O system as shown in Figure 1(e). It is observed that e = 0.9 lies closer to the experimental values as compared to other values of eccentricity. Our model has been built by modifying the model given by Feng et al. [70] and hence comparison has been carried out only with the values reported in [70]. Other works follow different approach and gives result which deviate from experiment data.

The calculations for Al$_2$O$_3$, CuO nanoparticles in C$_2$H$_5$OH and H$_2$O base fluids have been carried out using the values k$_p$(Al$_2$O$_3$) = 46.0 W/m-K, k$_p$(CuO) = 69.0 W/m-K, k$_f$(C$_2$H$_5$OH) = 0.258 W/m-K and k$_f$(H$_2$O) = 0.604 W/m-K [71], for different values of eccentricity and interlayer thickness and the results are plotted in Figures 1 and 2. We have taken the thermal conductivity of interfacial layer to be 2 times that of base fluid. Obviously, thermal conductivity increases monotonically with the volume fraction. A perusal of Figures 1(a) to 1(e) for Al$_2$O$_3$ in EG and water as a base fluid reveals that effective thermal conductivity increases nonlinearly with volume fraction which may correspond to the fact that suspensions are more prone to aggregation and deviation from sphericity with an increase in particle concentration. Formed clusters are of porous natures which actively participate in the heat transfer across the temperature gradient, by increasing the surface area.

It is interesting to note that the nanofluids of both the oxide particles of the same size in EG have somewhat higher thermal conductivity than in water although the latter has larger thermal conductivity value. A possible explanation for this seems to be that the interfacial layer formed with EG is more densely packed than that in the case of water.

3. VISCOSITY OF NANOFLUIDS

Flow of nanofluids in bulk and inside channels requires the knowledge of viscosity behaviour and it is reported that viscosity of nanofluids is much higher than the conventional dispersions at equal volume concentration of dispersed particles [73-74]. There are certain studies which showed that viscosity of nanofluids increases with decrease in nanoparticle size [74]. However, opposite results have also been reported in [75, 76]. Like particle size, there are other contributing factors to the change in behaviour of viscosity of nanofluids like particle shape, particle concentration, temperature, PH value of the base fluid, aggregation, uniformity of size distribution, type of materials, sonication time etc. [31, 77,78]. There are various experimental as well as theoretical studies of viscosity and few of them are listed below:

Einstein [79] established the relation for suspension viscosity by assuming that there is no interaction among the particles since they are far apart of each other and the hydrodynamic equations of motion with inertia neglected (the Stokes equations) were used to describe the flow around each particle valid for ($\phi << 1$). The relation is expressed as

$$\mu_r = 1 + 2.5\phi. \tag{17a}$$

where ϕ is the particle volume fraction and μ_r is relative viscosity. H. Brinkman [80] explained behaviour of relative viscosity μ_r for moderate particle concentrations, by assuming the effect of the addition of one solute-molecule to an existing solution, which is considered as a continuous medium. The expression is given by

$$\mu_r = \frac{1}{1 + 2.5\phi}. \tag{17b}$$

Frankel and Acrivos [81] presented a relation given as

$$\mu_r = \frac{9}{8}\left[\frac{(\phi/\phi_m)^{\frac{1}{3}}}{1 - (\phi/\phi_m)^{\frac{1}{3}}}\right], \tag{18}$$

where, ϕ_m is the maximum packing fraction. Lundgren [82] proposed the following equation in the form of a Taylor series as given by

$$\mu_r = 1 + 2.5\phi + \frac{25}{4}\phi^2 + f(\phi^3). \tag{19}$$

Considering the effects of variable packing fraction within the aggregate structure, Chen et al. modified K-D equation [83] and named it as modified Krieger and Dougherty [84] equation given by

$$\mu_r = \left(1 - \frac{\phi_a}{\phi_m}\right)^{-2.5\phi_m} \tag{20}$$

where

$$\phi_a = \phi \left(\frac{r_a}{r} \right)^{3-D_f}.$$

Equation r and r_a, are the radii of primary particles and of aggregates, respectively. D_f is the fractal index having a typical value of 1.8 for nanofluids. Shear viscosity has been explained by Equation (20) for semi-dilute nanofluids with aggregation of nanoparticles, and there is no obvious shear-thinning behaviour.

Masoumi et al. presented a new equation to find the effective viscosity of nanofluids by considering the Brownian motion of nanoparticles [77]. This is done theoretically by considering two abstracted contributions; one from the base fluid μ_{bf} and other due to nanoparticles which was termed as apparent viscosity μ_{app}. While developing a model Correction factor, C_1 was introduced to calculate the shear stress and the expression for viscosity is

$$\mu_r = 1 + \frac{\rho_p V_B d_p^{\,2}}{72 C_1 \delta}.$$

(21)

where ρ_p is the density of particle, V_B is the brownian velocity of suspended nanoparticle, d_p is diameter of suspended particle in volume element with dimension δ.

Gaganpreet and Srivastava [85] reported relative viscosity of nanofluids by taking in to account the effect of fractal aggregates and particle shape as

$$\mu_r = \left(1 - \frac{\phi_{ag}}{\phi_m} \right)^{-[\eta]\phi_m}.$$

(22)

Here, $[\eta]$ is the intrinsic viscosity for prolate shaped nanoparticles. Intrinsic viscosity for spheroids is expressed as [85, 88]

$$[\eta] = \left(\frac{4}{15} \right)(C_1 + C_2 - C_3) + \frac{2}{3}[C_3] + \frac{1}{3}[C_4] + \frac{C_5}{15}[C_6]$$

(23)

The constants appearing in the above expression are given as

(24)

$$C_4 = \frac{\left(1-e^2\right)^{0.5}}{J_\alpha'}, C_2 = \frac{C_4}{2}, C_1 = C_2 \frac{J_\alpha'''}{J_\beta'''}, C_3 = \frac{\left(1-e^2\right)^{1.5}}{\left(1-\frac{e^2}{2}\right)J_\beta'},$$

$$C_6 = \frac{6\left(1-e^2\right)^{1.5} e^2}{J_\alpha + \left(1-e^2\right)J_\beta}, C_5 = \frac{e^2}{2-e^2}.$$

The different Jeffrey integrals (J_α's) appearing in Equations (24) are expressed in terms of eccentricity (e) by the following relations

$$J_\alpha = \frac{\left(1-e^2\right)^{1.5}}{b^3} \int_0^\infty \frac{dy}{(y+1)\left(y\left(1-e^2\right)+1\right)^{1.5}},$$ (25)

$$J_\beta = \frac{\left(1-e^2\right)^{0.5}}{b^3} \int_0^\infty \frac{dy}{(y+1)^2 \left(y\left(1-e^2\right)+1\right)^{0.5}},$$ (26)

$$J_\alpha' = \frac{\left(1-e^2\right)^{0.5}}{b^5} \int_0^\infty \frac{dy}{(y+1)^3 \left(y\left(1-e^2\right)+1\right)^{0.5}},$$ (27)

$$J_\beta' = \frac{\left(1-e^2\right)^{1.5}}{b^5} \int_0^\infty \frac{dy}{(y+1)^2 \left(y\left(1-e^2\right)+1\right)^{1.5}},$$ (28)

$$J_\alpha'' = \frac{\left(1-e^2\right)^{0.5}}{b^3} \int_0^\infty \frac{dy}{(y+1)^3 \left(y\left(1-e^2\right)+1\right)^{0.5}},$$ (29)

$$J_\beta'' = \frac{\left(1-e^2\right)^{1.5}}{b^3} \int_0^\infty \frac{ydy}{(y+1)^2 \left(y\left(1-e^2\right)+1\right)^{1.5}},$$ (30)

Here, ϕ_m is an asymptotic value for prolate particles corresponding to maximum particle loading. From experiments and simulations, the maximum possible volume fraction ϕ_m for prolate spheroids has been found to be $\phi_m = 0.68$ to 0.71 and approach to $\phi_m \approx 0.74$ for ellipse spheroids [88]. The volume fraction of aggregates, ϕ_{ag} is expressed as

$$\phi_{ag} = \phi_{mod} \left(\frac{r_a}{r}\right)^{3-d_f},$$ (31)

where d_f is the fractal index representing the extent of changes in the packing fraction from the centre to the edge of aggregates. The average value of d_f is 1.8 for nano-sized alumina suspensions, guided by the results of simulation and measurements and by the experimental findings on aggregation [87-88]. Molecular dynamics simulations, it was shown that mechanism responsible for increase in viscosity of nano suspensions is non equilibrium microfluctuations of density and velocity of the carrier medium, which are induced by nanoparticles [89]. Rudyak et al. [89] showed that the variation in viscosity with particle is mass, radius and concentration through correlations it is explained the behaviour by considering that force acting on the nanoparticle is not stationary in comparison to Stokes force [89]. Using Equation (22) we have plotted the variation of relative viscosity with particle volume fraction for SiO_2-H_2O and Al_2O_3-H_2O nanofluid system as shown in Figures 3 and 4. Parameters used to evaluate the viscosity of nanofluid are $d_f = 1.8$, maximum packing fraction $\phi_m = 0.605$ (spherical particles) and $\phi_m = 0.74$ (prolate particles). In order to determine the effect of aggregates and particle shape on the viscosity of nanofluids, the effective radii of aggregates were first evaluated for spherical particles using the concept of equivalent volume fraction for spheres and the modified K-D equation as shown in Figure (3a). From Figure 3a, it is observed that there is increase in viscosity with particle concentration and it matches well with the experimental data [90] for $r_a = 2r$ and hence keeping the value of $r_a = 2r$ fixed, effect of particle shape in terms of eccentricity has been studied as depicted in Figure 3b. From Figure 4a for Al_2O_3-H_2O system, calculated results are close to experimental data [91] for $r_a = r$. It is perceived that there are less chances of particle aggregation as the particle size increases as in the case of $d_p = 40nm$. With the increase in fractal aggregate size, viscosity of the nanofluid increases which implies that there is strong bonding among the aggregated nanoparticles.

4. SOFT CORE FLUIDS AND CONFINEMENT OF FLUIDS

Most of the theoretical attempts to study confined fluids are made in the direction for simple fluids which are interacting by Leenard-Jones potential [92, 93], binary liquids [94, 95] and fluids with orientational degrees of freedom [96, 97]. There exist very few studies dedicated to confined soft core fluids. The interactions among the soft core fluids are generally governed by Yukawa potentials, GCM potential [98-102]. GCM has meanwhile turned out to be reasonably reliable model system for colloids like polymer coils, dendimer solution [101], micelles dispersed in good solvent [100], effective interactions of micellar aggregates of ionic surfactants [103] and certain glasses [104]. The non-conventional behaviour of GCM has been actively investigated, showing anomaly in their structural and thermo-dynamic characteristics [105-107]. Typical properties include self-diffusion anomaly (mobility of Gaussian-Core (GC) particles increases with density and temperature) [108], viscosity anomaly (decrease in viscosity with increase in density), density anomaly (expansion upon isobaric cooling). GCM fluid is solely repulsive potential which explains the phase transition and structure in simple or molecular fluids. The GCM potential used to examine the equilibrium properties of a fluid of soft interacting particles is defined as

$$U(r) = \varepsilon \exp\left(-\left(\frac{r_{ij}}{\sigma}\right)^2\right),$$

(32)

where r_{ij} is the distance between two particles, $\varepsilon > 0$ and σ is a length scale determining the size of the particles. This potential shows weak (logarithmic) divergence as r approaches to 0 unlike hard-sphere and inverse power potentials. The other available bounded potential is the penetrable sphere model (PSM) [109] similar to GCM but the phenomenological expression and phase diagram are quite different from GCM.

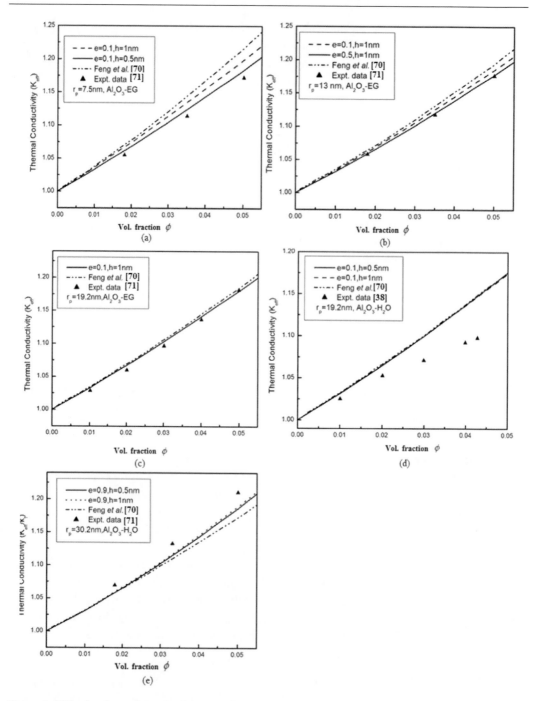

Figure 1. Effective thermal conductivity as a function of particle volume fraction for Al_2O_3 in ethylene glycol (EG) and water as base fluid for different sizes and eccentricity. (a) Al_2O_3-EG (r_p = 7.5nm) (b) Al_2O_3-EG (r_p = 13nm) system (c) Al_2O_3-EG (r_p = 19.2nm) system (d) Al_2O_3-H_2O (r_p = 19.2 nm) system and (e) Al_2O_3-H_2O (r_p = 30.2 nm) system. Dash lines and solid lines are theoretical results for different values of h. Dash dot dot (-...) and (▲) are results of Feng et al. and experimental data [71], respectively.

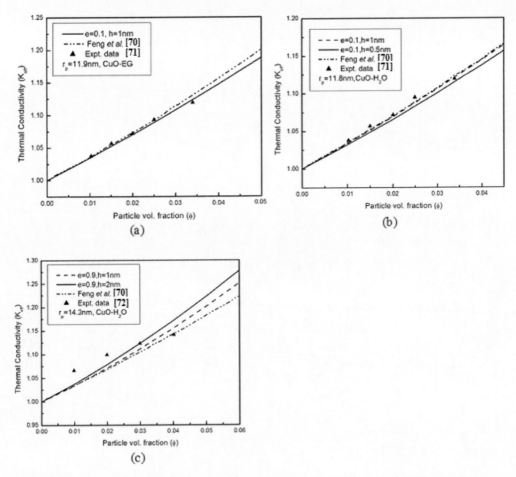

Figure 2. Effective thermal conductivity for CuO nanoparticle in ethylene glycol (EG) and water as base fluid for different size of nanparticles. (a) CuO-EG (r_p = 11.9 nm) (b) CuO-H$_2$O (r_p = 11.8nm) system (c) CuO- H$_2$O (r_p = 14.3 nm) system. Dash lines and solid lines are theoretical results for different values of h. Dash dot dot (-...) and (▲) are results of Feng et al. [70] and experimental data [72], respectively.

There are various approaches available to study the behaviour of GCM fluids such as density functional theory (DFT) [110] and memory function approach [108]. Shall and Egorov explained the diffusion anomaly by mode coupling (MC) theory and obtained structural input from integral equation theory. However, in comparison with MC approach, they reported that their ansatz for memory function gives better result [110]. Mausbach and May explained the diffusion behaviour by molecular dynamics simulations and reported that diffusion anomaly exist for densities higher than approximately 0.33 [106]. Gaganpreet et al. reported the dynamics of GCM fluid using the Mori Zwanzig formalism [108] and it is noted that triplet contribution to the fourth frequency sum rules of velocity autocorrelation function becomes negative for higher densities which is responsible for diffusion anomaly. There is still much to understand in GCM fluids, particularly when they are confined to nanochannels.

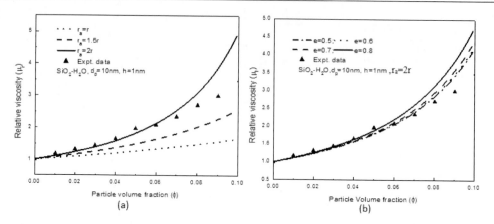

Figure 3. Variationof relative viscosity with particle volume fraction for SiO_2-H_2O nanofluid.(a)For sphere (b) Prolate shape particles at different values of eccentricities.

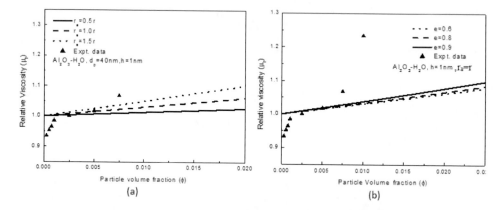

Figure 4. Variationof relative viscosity with particle volume fraction for Al_2O_3-H_2O nanofluid. (a)For sphere (b) Prolate shape particles at different values of eccentricities.

Lab on chip concept and miniaturization of electronic devices appealed researcher's interest towards fluidics area. Confinement of colloids, polymeric solutions, nanofluids etc. provide applications in many areas of science. When fluid is confined, it exhibits much stronger inhomogeneity in local density and pressure which further results in structure change of the equilibrium phases, dynamic, kinetic and phase transitions properties. Due to nano confinement and the complex liquid-solid interactions, most liquid molecules and ions confined in nanocavities behaves differently. Fluids can be confined inside nanoporous materials (Zeolites and nanoporous carbons), nanotubes (CNTs and boron nitrides), slits and biological channels etc. System performance can be altered by controlling various materials parameters (e.g., pore size and geometry, liquid phase, solid phase, and surface treatment) and system parameters (e.g., temperature, electric field, and mechanical load). As length reduces to nano-scale, dynamical behaviour of fluids gets affected as the density of the system becomes a function of position and time of fluid particles. The physical properties gets affected due to irregular density distribution hence changes the dynamical behaviour of fluids. Generally, for macroscopic sample of fluid, two independent variables, like temperature and pressure define the state of system, however for nanoconfined fluids state of the system are function of geometry, internal structure, wall effects [109]. Chen et al. [111] reported the effect of nanoconfinement on molecular interactions within mesoporous silica and showed that confinement enhances the interaction. Better understanding of confined systems leads to use in many real world physical and biological phenomena like blood flow inside arteries, [112-114], energy absorption actuation. Theoretically, it was shown that when simple fluids is confined between two wall with small separation it becomes ordered in to layers that have lateral internal ordering as transverse ordering between layers. And transport properties like diffusion can be controlled by the width of the layers formed inside channel due to wall effect and particle interactions [92-93]. In next section, diffusion dynamics of GCM fluids is mainly discussed using frequency sum rules up to fourth order.

4.1. Diffusion

Diffusion coefficient is associated with mass transport which is independent of space and time in the classical continuum theory. However, in confining conduits transport depends on the fluctuation statistics of molecules and available phase space to explain their dynamics. Generally, to compute transport properties linear response theory provides the general framework in which transport coefficients are related to auto correlations functions and these relations are known as Green Kubo formulae. For example one of the Green Kubo formulae relates the self-diffusion coefficient to velocity auto correlation function. The macroscopic law that describes phenomena of diffusion is known as Fick's law.

$$D = \frac{K_B T}{m} \int_0^\infty dt V(t),$$ (33)

where $V(t)$ is velocity autocorrelation (VAC)function, $\langle v(t)v(0)\rangle$ and is a measure of the projection of the particle velocity $v(t)$ at time t on to its initial value, averaged over all the initial conditions. In order to calculate the time evolution of the VAC function, Mori-Zwanzig formalism is coupled with sum rules of the VAC function. The short time expansion of the normalised VAC function is given by

$$V(t) = 1 - V_2 \frac{t^2}{2!} + V_4 \frac{t^4}{4!} - V_6 \frac{t^6}{6!} + \ldots\ldots,$$ (34)

where V_2, V_4 and V_6 are called the second, fourth and sixth frequency sum rules of the VACF. There exists number of theoretical [115-117] and simulation studies [115, 118-120,] for studying the behaviour of fluids in confined channels limiting to micro and nanochannels. There also exist simulation studies [107, 121-122] for fluids where GCM fluid is confined in pair of parallel walls. Theoretical studies [123] have been made to understand the dynamics of the fluid confined by two parallel walls. However this has not been used for fluid like GCM fluid where in anomalous behaviour of VACF and self diffusion coefficient is observed. Effect of confinement of simple fluid has been studied by the dynamical model proposed by Tankeshwar and Srivastava [93] which showed that diffusion perpendicular to the walls decrease as one approaches the confining wall. The model [93] is based on the following assumptions

- Configuration space of the many body system is divided into a number of cells which are characterized by a fixed configuration associated with local minima on the potential energy hyper surface of the system.
- System jumps between the cells with a fixed frequency or a spectrum of frequencies. Let the fluid be confined in x direction and thus affecting the motion as shown in Figure 5.

Due to nano scale confinement, the particles find themselves in a compression-like situation. To study the effect of confinement on the frequency of liquid configuration executing motion in the x direction, consider

$$x(t) = A\sin(\omega t),$$
(35)

where A is the amplitude when the fluid is not confined. Let a particle in a given cell experience compression due to confinement which results in decrease in its amplitude say by A_1 along x direction. If at time $t = t_1$ the particle reaches the confining wall along x direction with

$$x(t_1) = A - A_1 \sin(\omega t),$$
(36)

This provides [93]

$$t_1 = \frac{1}{\omega}\sin^{-1}\left(1 - \frac{A_1}{A}\right),$$
(37)

This change in amplitude results is a change in frequency of the particle and hence its velocity. Let the new frequency be represented by at Ω_n. The velocity in x direction at $\Omega_n t_1 = \pi/2$ is given as

$$v_x(t_1) = A\Omega_n \cos(\Omega_n t_1) = 0,$$
(38)

This provides the relation between new and old frequency

$$\left(\frac{\Omega_n}{\omega}\right) = \left(\frac{\pi}{2\sin^{-1}\left(1 - \frac{A_1}{A}\right)}\right),$$
(39)

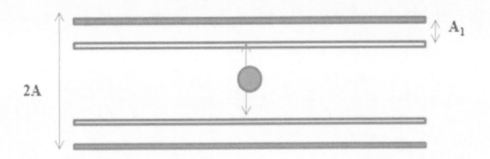

Figure 5. Particle in confined situation.

Equation (39), shows that if compression (A_1/A) is zero, the frequency remains the same and for positive compression, it increases. Flow of fluid within micro/nano channel experience compression like situation which varies with distance from the walls. For such a situation A_1/A is to be replaced by continuous function of x say by c (l, x). Here $2l$ is the width of the channel and x is the distance measured from the middle of channel. T he expression for the frequency ratio is then modified to

$$\left(\frac{\Omega_n}{\omega}\right) = \left(\frac{\pi}{2\sin^{-1}\left(1-\frac{A_1}{A}\right)}\right),$$

(40)

Dynamics of the system described by velocity autocorrelation function $V(t)$ and is related to first order memory function $M_1(t)$. The phenomenological form of Memory function is chosen as

$$M_1(t) = \delta_1 \sec h(b_1 t)\cos(\Omega t),$$

(41)

where b_n, Ω and δ_1 are real and positive. Parameters b and Ω can be obtained either by fitting or can be determined through the comparison of short time expansion of Equation (41) and is given as

$$b_1 = \sqrt{\frac{\delta_3 - \delta_2}{4}}, \Omega = \sqrt{\frac{5\delta_2 - \delta_3}{4}},$$

(42)

where $\delta_k' s$ are known as Mori's coefficient and δ_k is zero time value of kth stage memory function, i.e., $m_k(0)$.$\sqrt{\delta_1}$ also called as einstein frequency. If this frequency is assumed to modify in confined situation in a similar way as that of oscillator then δ_1 gets modified to

$\delta_1\left(\dfrac{\Omega_n}{\omega}\right)^2$ and correspondingly change are

$$V_2(n) = V_2\left(\dfrac{\pi}{2\sin^{-1}\left(1 - \dfrac{A_1}{A}\right)}\right)^2,$$

(43)

Due to which δ_2 also gets modified.
The modified form of memory function is then given by

$$M_{eff} = \delta_1\left(\dfrac{\Omega_n}{\omega}\right)^2 \sec h\left(b_1' t\right)\cos\left(\Omega' t\right),$$

(44)

with new values of b' and Ω' by using modified δ_2 and V_2 given by Equation (44). The analytic expression for self-diffusion coefficient now become function of x and is given by

$$D(x) = \dfrac{K_B T}{m} \dfrac{2b\cosh\left(\pi\Omega/2b\right)}{\delta_1\pi\left(\dfrac{\Omega_n}{\omega}\right)^2}.$$

(45)

In order to investigate the effect of the confinement on the transport processes in one component fluid, we use the above discussed model. In the following, we shall use reduced thermodynamic quantities namely reduced density $\rho^* = n\sigma^3$, reduced temperature $T^* = K_B T/\varepsilon$, reduced time $t^* = t(\varepsilon/m\sigma^2)^{1/2}$ and reduced diffusion $D^* = D(m/\varepsilon\sigma^2)^{1/2}$. In order to calculate the self-diffusion coefficient, we require Mori coefficients for the calculation *of b_1 and Ω in Equation (45), the values obtained for the fluid is not subject to any confinement. The values for these were obtained from the second and fourth sum rules of velocity auto correlation function (VACF). The width of the nano channel is taken to be $2l = 40$, with wall at $x = \pm 20$ where x is the distance from the middle of geometry. Since the effect of wall increases as fluid particle moves closer to it and at the centre shall depend on how distant, i.e., from the wall. By assuming, it to be exponentially decaying analogous to damping, for the decrease of the effect of the wall on atomic motion then $c(l, x)$ taken to be $exp(-(l-x))$ for the calculation D^*. Results for the self diffusion coefficient D^* for GCM fluid are plotted in Figure 6. In the diffusion graphs solid line correspond to density $\rho^* = 0.4$, dash dot line $\rho^* = 1.0$ and dotted line shows $\rho^* = 1.2$ respectively. From the obtained results we concluded that diffusion of fluid particle decreases significantly closer to the walls and diffusion anomalies observed in bulk fluid persist even in the confined fluid.

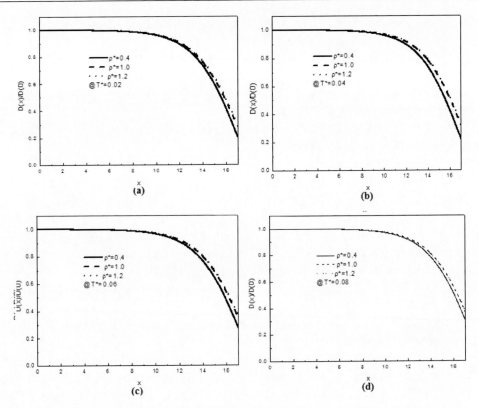

Figure 6. Self-diffusion coefficient as function of x, i.e., distance from the middle of the geometry at different temperatures. The solid line represents diffusion at $\rho^* = 0.4$, dash line at $\rho^* = 1.0$ and the dotted line corresponds to $\rho^* = 1.2$.

CONCLUSION

In this chapter, thermal conductivity and shear viscosity of nanofluids have been discussed. The model for the thermal conductivity is based on the effective medium theory and ellipsoidal shape of molecules. The result obtained has been compared with experimental observation and it is observed that different eccentricity parameters are required to get best comparison. It is also found that same nanoparticles provide more values of thermal conductivity in EG fluid than in water. Results of the shear viscosity of nanofluids by using modified KD expression has also provided better results than other models as judged by comparison with experimental observation. The results of self– diffusion coefficient of GCM fluid in confined state has been studied by assuming that frequency sum rules are modified in perpendicular direction to wall. It is observed that diffusion fall significantly near the walls.

REFERENCES

[1]   Han, J., Craighead, H. G. *Science* 2000, 288, 1026-1029.
[2]   Zeng, Y., Harrison, D. *Anal. Chem.* 2007, 79, 2289-2295.

[3]   Siria, A., Poncharal, P., Biance, Anne-Laure, Fulcrand, R., Blase, X., Purcell, S. T.,Bocquet, L. *Nature* 2013, 494, 455-458.

[4]   Fan, R., Karnik, R., Yue, M., Li, D., Majumdar, A., Yang, P. *Nano Lett.* 2005, 5, 1633-1637.

[5]   Shawgo, R. S., Grayson, A. C. R., Li, Y., Cima, M. J. *Curr. Opinion Solid State Materials Science* 2002, 6, 329-334.

[6]   Chiang, P.-C., Hung, D.-S., Wang, J.-W., Ho, C.-S., Yao, Y.-D., *IEEE Trans. on Magnetics* 2007, 43, 2445-2447.

[7]   Kim, S., Ko, S., Kang, K., Han, J. *Nature Nanotech.* 2010, 5, 297-301.

[8]   Choi, S. U. S., *ASME Fluid Eng.* 1993, 231, 99-105.

[9]   Choi, S. U. S., Zhang, Z. G., Yu, W., Lockwood, F. E., Grulke, E. A. *Appl. Phys. Lett.* 2001, 79, 2252-2254.

[10]   Das, S. K., Choi, S. U. S., Yu, W., Pradeep, T. *Nanofluids-Science and Technology* John Wiley & Sons, Inc. Hoboken, NJ, USA, 397 pages Newyork, (2007).

[11]   Han, Z. H., Cao, F. Y., Yang, B. *Appl. Phys. Lett.* 2008, 92, 243104-3.

[12]   Kim, S. J., Bang, I. C., Buongiorno, J., Hu, L. W. *Bull. Polish Academy Sci. Tech. Sci.* 2007, 55, 211-216.

[13]   Boungiorno, J., Hu, L. W., Kim, S. J., Hannink, R., Truong, B., Forrest, E., *Nuclear Technology,* 2008, 162, 80-91.

[14]   Cussler, E. L., *Diffusion Mass Transfer in Fluid Systems*, Cambridge Series in Chemical Engineering, 2nd Ed. p. 580 (Cambridge University Press, New York 2009).

[15]   Wong, K. V., Leon, O. D. *Adv. Mech. Eng.* 2010, Article ID 519659, 11 pages.

[16]   Kulkarni, D. P., Vajjja, R. S., Das, D. K., Oliva, D. Appl. Therm. Eng. 2008, 28, 1774-1781.

[17]   Kim, W. G., Kang, H. U., Jung, K. M., and Kim, S. H., *Sep. Sci. Tech.* 2008, 43, 3036-3055.

[18]   Otanicar, T. P., Phelan, P. E., Taylor, R. A., Tyagi, H. *J. Sol. Energy Eng.* 2011, 133, 024501.

[19]   Taylor, R. A., Phelan, P. E., Otanicar, T. P., Walker, C. A., Nguyen, M., Trimble, S., Prasher, R. *J. Renewable Sustainable Energy* 2011, 3, 023104.

[20]   Sani, E., Mercatelli, L., Barison, S., Pagura, C., Agresti, F., Colla, L., Sansoni, P. *Sol. Energy Mater. Sol. Cells* 2011, 95, 2994-3000.

[21]   Zhang Yu, Ding Y., Povey M., York D. *Progress in Nat. Sci.* 2008,18,939-944

[22]   Mock, J. J., Barbic, M., Smith, D. R., Schultz, D. A., Schultz, S. S. *J. Chem. Phys.* 2002, 116, 6755-6759.

[23]   Nelayah, J., Kociak, M., Stephan, O., Garcia de Abajo, F. J., Tence, M., Henrard, L., Taverna, D., Pastoriza-Santos, I., Liz-Marzan, L. M., Colliex, C. Nat. Phys. 2007, 3, 348-353.

[24]   Liu, Z., Hou, W., Pavaskar, P., Aykol, M., Cronin, S. B. *Nano Lett.* 2011, 11, 1111-6.

[25]   Saidur, R., Leong, K. Y., Mohammad, H. A. Renewable Sustainable Energy Rev. 2011, 15, 1646-1668.

[26]   Patel, R., *Phys. Rev. E* 2012, 85, 026316.

[27]   Gupta, H. K., Agrawal, G. D., Mathur, J. *Int. J. Env. Sci.* 2012, 3, 433-440.

[28]   Putnam, S. A., Cahill, D. G., Braun, P. V., Ge, Z., Shimmin, R. G. *J. Appl. Phys.* 2006, 99, 084308.

[29] Venerus, D. C., Kabadi, M. S., Lee, S., Perez-Luna, V. *J. Appl. Phys.* 2006, 100, 094310-1-5.

[30] Shaikh, S., Lafdi, K., Silverman, E. *J. of Appl. Phys.* 2007, 101, 064302 -7.

[31] Gaganpreet, Srivastava, S. *Appl. Nanosci.* 2012, 2,325-331.

[32] Gaganpreet, Srivastava, S. *AIP Conf. Proc.* 2011, 1349, 407-408.

[33] Sundar, L. S., Singh, M. K., Ramana, E. V., Singh, B., Gracio, J., Sousa, A. C. M. *Scientific reports* 2014, 4, 4039 (14 pages).

[34] Usowicz, B., Usowicz, J. B., Usowicz, L. B. *J. Nanomaterials* 2014, Article ID 756765.

[35] Choi, S. U. S., Zhang, Z. G., Yu, W., Lockwood, F. E., Grulke, E. A. *Appl. Phys. Lett.* 2001, 79, 2252-2254.

[36] Masuda, H., Ebata, A., Teramea, K., Hishinuma, N. *Netsu Bussei* 1993, 4, 227-233.

[37] Eapen, J., Rusconi, R., Piazza, R.,Yip, S. *J. Heat Transfer* 2010, 132, 102402-1-14

[38] Lee, S., Choi, S. U. S., Li, S., Eastman, J. A. J. Heat Trans.1999, 121,280-289.

[39] Eastman, J. A., Choi, S. U. S., Li, S., Thompson, L. J., Lee, S. *MRS Res. Soc. Symp. Proc.* 1996, 457, 3-11.

[40] Wang, X. W., Xu, X. F., Choi, S. U. S. *J. Thermophys. Heat Trans.* 1999, 13, 474-480.

[41] Li, C. H., Peterson, G. P. *J. Appl. Phys.* 2006, 99, 084314-084321.

[42] Li, C. H., Peterson, G. P. *J. Appl. Phys.* 2006, 99, 084314-084321.

[43] Murshed, S. M. S., Leong, K. C., Yang, C. *Appl. Therm. Eng.* 2009, 29, 2477.

[44] Zhu, D., Li, X., Wang, N., Wang, X., Gao, J. W., Li, H. *Curr. Appl. Phys.* 2009, 9, 131-139.

[45] Pak. B.C., Cho, Y. I. *Exp. Heat Transfer* 1998, 11, 151-170.

[46] Kleinstreuer, C., Feng, Y. *Nanoscale Res. Lett.* 2011, 6, 229-241.

[47] Rizvi, I. H., Ayush, J., Ghosh, S. Kr., Mukherjee, P. S. *Heat Mass Trans.* 2013 49, 595-600.

[48] Chen, H., Ding, Y., He Y., Tan, C. *Chem. Phys. Lett.* 2007, 444, 333-337.

[49] Hong, K. S., Hong T. K., Yang, H. S. *Appl. Phys. Lett.* 2006, 88, 03190.

[50] Keblinski, P., Prasher, R., Eapen, J. *J. Nanopart. Res.* 2008, 10, 1089.

[51] Nan, C. W., Shi, Z., Lin, Y. *Chem. Phys. Lett.* 2003, 375, 666-669.

[52] Wang, B. X., Zhou, L. P., Peng, X. F. *Int. J. Heat. Mass Transfer* 2003, 46, 2665-2672.

[53] Teng, T. P., Hung, Y. H., Teng, T. C., Chen, J. H. *Nanoscale. Res. Lett.* 2011, 6, 488.

[54] Lin, Y. S., Hsiao, P. Y., Chieng, C. C. *Appl. Phys. Lett.* 2011, 98, 153105.

[55] Oh, S. H., Kauffmann, Y., Scheu, C., Kaplan, W. D., Ruhle, M. *Science* 2005, 310, 661-663.

[56] Keblinski, P., Phillpot, S. R., Choi, S. U. S. Eastman, J. A. *Int. J. Heat Mass Trans.* 2002, 45, 855-863.

[57] Yu, W., Choi, S. U. S. *J. Nanopart. Res.* 2004, 6, 355-361.

[58] Liang, Z., Tsai, H.-L. *Phys. Rev. E* 2011, 83,041602.

[59] Lee, D., Kim, J. W., Kim, B. G., *J. Phys. Chem.* B 2006, 110, 4323-8.

[60] Hunter, R. J. *Foundation of Colloid Science*, I[st] Edition, (Clanderon Press Oxford, 1987).

[61] Lee, D. *Langmuir* 2007, 23, 6011-6018.

[62] Xue, L., Keblinski, P., Phillpot, S. R., Choi, S. U. S., Eastman, J. A. *Int. J. Heat Mass Trans.* 2004, 47, 4277-4284.

[63] Maxwell, J. C., *A Treatise on Electricity and Magnetism*, 3rd edition Oxford: (Clarendon Press 1891).

[64] Bruggeman, D. A. G. *Annalen der Physik* (Leipzig) 1935, 24, 636-664.

[65] Xuan, Y., Li, Q., Hu, W. *AIChE J.* 2003, 49, 1038-1043.

[66] Prasher, R., William, E., Meakin, P., Jacob, F., Patrick, P., Keblinski, P. *Appl. Phys. Lett.* 2006, 89, 143119.

[67] Evans, W., Prasher, R., Fish, J., Meakin, P., Phelan, P., Keblinski, P. *Int. J. Heat Mass Trans.* 2008, 51, 1431-1438.

[68] Lee, J.-H., Hwang, K. S., Jang, S. P., Lee, B. H., Kim, J. H., Choi, S. U. S., Choi, C. J. *Int. J. Heat and Mass Trans.* 2008, 51, 2651-2656.

[69] Schwartz, L. M., Garboczi, E. J. and Bentz, D.P., *J. Appl. Phys.* 1995, 78, 5898-5908.

[70] Feng, B., Yu, P., Xu and M. Zou, *J. Phys. D: Appl. Phys.* 2007, 40, 3164-3171.

[71] Xie, H. Q., Wang, J. C., Xi, T. G., Liu, Y., Ai, F., Wu, Q. R. *J. Appl. Phys.* 2002, 91, 4568-4572.

[72] Das, S. K., Putra, N., Thiesen, P., Roetzel, W. *J. Heat Transfer* 2003, 125, 567-574.

[73] Mahbubul, I. M., Saidur, R., Amalina, M. A. *Int. J. Heat Mass Transf.* 2012, 55 874-885.

[74] Rudyak, V. Y., *Adv. Nanopart.* 2013, 2, 266-279.

[75] He, Y., Jin, Y., Chen, H., Ding, Y., Cang, D., Lu, H., *Int. J. Heat Mass Trans.*2007,50, 2272-2281

[76] Nguyen, C. T., Desgranges, F., Roy, G., Galanis, N., Maré, T., Boucher, S., Mintsa, H. A. *Int. J. Heat and Fluid flow* 2007, 28, 1492-1506.

[77] Masoumi, N., Sohrabi, N., Behzadmehr, A. *J. Phys. D: Appl. Phys.* 2009, 42, 055501-055506.

[78] Sadril, R., Ahmadi, G., Togun, H., Dahari, M., Kazi, S. N., Sadeghinezhad, E., Zubir, N. Nanoscale Res. Lett. 2014, 9, 151 (16 pages).

[79] Einstein A., Eineneuebestimmung der moleküldimensionen. *Ann. Phys.* 1905, 324, 289-306.

[80] Brinkman, H., *J. Chem. Phys.*1952, 20:571-581.

[81] Frankel N, Acrivos A. *Chem. Eng. Sci.*1967, 22, 847-853.

[82] Lundgren, T, S., *J. Fluid Mech.*1972, 5, 273-299.

[83] Krieger, I. M., Dougherty, T. J. *Trans. Soc. Rheol.* 1959, 3, 137-152.

[84] Chen H., Ding Y., Tan C., Rheological behavior of Nanofluids. *New J. Phys.* 2007, 9, 367.

[85] Gaganpreet, S. Srivastava, *Phys. And Chem. Liquids* 2014, 53, 174-186

[86] Scheraga, H. A., *J. Chem. Phys.* 1955, 23, 1526-1531.

[87] Donev, A., Cisse, I., Sachs, D., Variano, E. A., Stillinger, F. H., Connelly, R., Torquato, S., Chaikin, P. M. *Science* 2004, 303, 990-993.

[88] Goodwin, J. W., Hughes, R. W. *Rheology for Chemists: An Introduction* (London: The Royal Society of Chemistry), 2000.

[89] Rudyak, V. Y., Belkin, A., Egorov, V. V., *Tech. Phys.* 2009, 54, 1102-1109.

[90] Jamshidi, N., Farhadi, M., Ganji, D. D., Sedighi, K. *IJE Transactions B: Applications*, 25, 2012 201-209.

[91] Azari, A., Kalbasi, M., Rahimi, M. Brazilian *J. Chem Eng.* 2014, 31, 469-481.

[92] Goyal, I., Srivastava, Sunita, Tankeshwar, K. *Nano-Micro Letters* 2012, 4,154-157.

[93] Tankeshwar, K., Srivastava, S. *Nanotechnology* 2007, 18, 485714-18.

[94] Scholl-Paschinger, E., Levesque, D., Weis, J. J., Kahl, G., *Phys. Rev.* E 2001, 64, 011502.

[95] Fernaud, M. J., Lomba, E., Weis, J. J. *Phys. Rev.* E 2001, 64, 051501.

[96] Fernaud, M. J., Lomba, E., Martin, C., Levesque, D., and Weis, J. J. *J. Chem. Phys.* 2003,119, 364.

[97] Spoler, C., Klapp, S. H. L. *J. Chem. Phys.* 2003,118, 3628.

[98] Flory P. J., Krigbaum, W. R. *J. Chem. Phys.* 1950, 18, 1086.

[99] Stillinger, F. H. *J. Chem. Phys.* 1976, 65, 3968-3974.

[100] Baeurle, S. A., Kroener, J. *J. Math. Chem.* 2004, 36,409.

[101] Likos, C. N., Schmidt, M., Löwen, H., Ballauff, M., Pötschke, D., Lindner, P., *Macromolecules* 2001,34, 2914.

[102] Götze, I. O., Harreis, H. M., Likos, C. N. *J. Chem. Phys.* 2004, 120, 7761-7771.

[103] Likos C. N., Rosenfeldt S., Dingenouts N., Ballauff M., Lindner P., Werner N. and Vögtle F., *J. Chem. Phys.* 117, 1869 (2002).

[104] Ikeda, A., Miyazaki, K., *Phys. Rev. Lett.* 2011,106, 015701.

[105] Stillinger, F. H., Stillinger, D. K., *Physica* A 1997, 244, 358-369.

[106] Mausbach, P., May, H.-O. *Fluid Phase Equilib.* 2006, 249, 17-23.

[107] Archer, A. J., Evans, R., *Phys. Rev.* E 2001, 64, 041501 p 1-12.

[108] Gaganpreet, Srivastava, S., Tankeshwar, K. *Chem. Phys.* 2012,405, 60-66.

[109] Vakili-Nezhaad, G.R., Mansoori, G.A. *J. Comp. Theor. Nanosci.* 2004, 1, 233-235.

[110] Shall, L. A., Egorov, S. A. *J. Chem. Phys* 2010, 132, 184504.

[111] Chen, Y., Wang, S., Ye, J.., Li, D., Liu, D., Wu, X. Nanoscale2014, 6, 9563-9567.

[112] Stepanek, F., Soos, M., Rajniak, P. *Colloid Surf.* A 2007, 300, 11-20.

[113] Henrich B., Cupelli,C., Santer, M., M. Moseler, *New J. Phys.* 2008,10, 113022.

[114] Krishnamurthy, V., Cornell, B. *Protoplasma* 2012, 249, 3-9.

[115] Krishnan, S. H., Ayappa, K. G. *J. Chem. Phys.* 2004, 121, 3197-3205.

[116] Devi, R., Sood, J., Srivastava, S., Tankeshwar, K. *Microfluid Nanofluid* 2010, 9, 737-747.

[117] Klein, J., Kumacheva, E. *J. Chem. Phys.* 1998, 108, 6996-7009.

[118] Kim, H., Kim, C., Lee, E. K., Talkner, P., Ha¨nggi P., *Phys Rev* E 2008, 77, 031202-1-031202-8.

[119] Bhatia, S. K., Nicholson, D. *J. Phys. Chem.* 2011, 115, 11700-11711.

[120] Cosentino, C., Amato, F., Walczak, R., Boiarski, A., Ferrari, M. *J. Phys. Chem.* B 2005,109, 7358-7364.

[121] Almenar, L., Rauscher, M. *J. Phys.: Condens. Matter* 2011, 23,184115.

[122] Schwanzer, D. F., Coslovich, D., Kurzidim, J. Kahl, G. *Mol. Phys.* 2009, 107, 433-441.

[123] Terao, T, *Soft Mater* 2010, 8, 63.

In: Innovations in Nanomaterials
Editors: Al-N. Chowdhury, J. Shapter, A. B. Imran

ISBN: 978-1-63483-548-0
© 2015 Nova Science Publishers, Inc.

Chapter 8

MICROEMULSIONS AS TEMPLATE FOR SYNTHESIZING NANOPARTICLES WITH TUNABLE ANTIBACTERIAL AND OPTICAL PROPERTIES: CURRENT TRENDS AND FUTURE PERSPECTIVE

Shazia Sharmin Satter[1], Mahfuzul Hoque[1],
M. Muhibur Rahman[2], M. Yousuf A. Mollah[1,2]
and Md. Abu Bin Hasan Susan[1,]*
[1]Department of Chemistry, University of Dhaka, Dhaka, Bangladesh
[2]University Grants Commission of Bangladesh, Dhaka, Bangladesh

ABSTRACT

Nanoscience is the forerunner in all technological sectors because of the neoteric properties exhibited by all types of nanoparticles due to the quantum size effect which is absent for the bulk particles. Bottom-up processes have been the pioneering strategies opted for by chemists for synthesizing nanoparticles and the microemulsion template method promises to be a versatile bottom-up method for controlling dimension, shape and size of the particles. In this chapter, we address the factors for synthesizing nanoparticles in water-in-oil microemulsions with controllable size and morphology and correlate various parameters to tune their antibacterial and optical properties. The long-term objective is to invoke the demand for novel ideas and intuition amongst the nanochemists in this particular approach.

[*] Corresponding author: Md. Abu Bin Hasan Susan. Department of Chemistry, University of Dhaka, Dhaka 1000, Bangladesh. E-mail: susan@du.ac.bd.

1. INTRODUCTION

Nanoscience is the science of today and tomorrow. From time immemorial, life, lifestyle, science and technology, all have direct influences from this fascinating area of science; however it is only the modern development that brought this interesting field into the light. If thought sensibly, it is the most ancient science that existed; but has been unveiled only in the recent days. Hemoglobin may be considered; it is only 5.5 nm in diameter but has the ability to carry oxygen throughout the body letting humans breathe. DNA, the building block of human life, is another example. With strand being only 2 nm in size it is the most powerful entity in the human body storing all the information. "Small is not only beautiful but also powerful" thus complements Feynman's lecture on "There's plenty of room in the bottom" [1a]. Not only our body but our lives have strong impact from nanoscience with innovations such as stain-resistant fabrics inspired by nanoscale features found on lotus plants and computer hard drives storing information on magnetic strips that are just 20 nm thick. Hence nanoparticles (NPs) are the most important and simplest forms and have been striding their ways in the varied fields of engineering and science [1b].

In nanotechnology, NP is defined as a small object that behaves as a whole unit in terms of its transport properties and potential applications. The heart of today's smart devices in physical, chemical, biomedical and pharmaceutical applications [2-4] are novel nanodevices composed of either simple or composite NPs. The diversity in the structure, shape and geometry of the NPs has been bewildering and quintessential for many years from both scientific and technological perspectives owing to their robust properties which *inter alia* include unusual electronic, optical, magnetic and chemical properties that can hardly be obtained from conventional bulk materials. At this miniaturized size, crystallites are influenced by scalable effects stemming from increase in the number of surface atoms and quantum effects due to the quantum confinement of the electronic states. These two factors dictate penultimate properties of NPs.

Amongst many physical and chemical techniques, the one involving microemulsions (mE) is a bottom-up method which enables control of properties of particles such as size, geometry, morphology, homogeneity and surface area [5, 6]. There have been extensive reviews on different aspects of mEs [7-13]. Relatively low energy consumption and ease of implementation of wet routes for synthesis of NPs make the mE based method a preferred choice over dry routes. This chapter aims to address the past and present successes of mEs as a bottom-up nanosynthetic approach for the synthesis of various NPs. It also aims at the correlation between the microscopic properties of mE and tunable optical and anti-bacterial properties of the synthesized NPs. Future perspectives of mE and strategic development to create new horizon in the innovation of nanomaterials will also be discussed.

2. MICROEMULSIONS

The word "microemulsion" was originally proposed by Schulman et al., (1959). Homogeneity on the macroscale but heterogeneity on the microscale make mEs a unique bottom-up approach for the synthesis of nanostructured materials and since the early 1980s [14, 15] there has been a tremendous upsurge in the interest for synthesizing nanomaterials via mE based method. MEs are isotropic, thermodynamically stable dispersion of nanodroplets within the range of 20-400 nm in diameter stabilized by an interfacial film of surface active species e.g., surfactants separating polar (usually water) and non-polar (usually oil) domains. MEs are generally made up of an oil phase, a polar phase, a surfactant which has both 'water loving' and 'water hating' part and a co-surfactant. In case of non-ionic surfactants, use of co-surfactant may more often be avoided.

Surfactants are amphiphilic substances with a hydrophilic head group, hydrophobic group constituting tail and a counterion which can form aggregates at certain concentrations [16-22, 89, 105, 176, 177] and depending on the nature of polar, hydrophilic head, surfactants in general used for preparation of microemulsions can be classified as shown in Figure 1. Since the key component of the microemulsions is the surfactant, the nature of the surfactant governs the formation of mEs [23, 24].

The nature of the interfacial layers composed of surfactants and occasionally that of co-surfactants vary which can reversibly change from one kind to another. In addition, the composition of various components, hydrophilic-lipophilic balance (HLB) of the surfactant and the nature of the co-surfactant play an important role in determining the type of mEs formed.

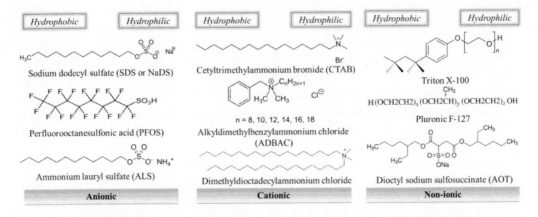

Figure 1. Types of Surfactants depending on their varying polar head.

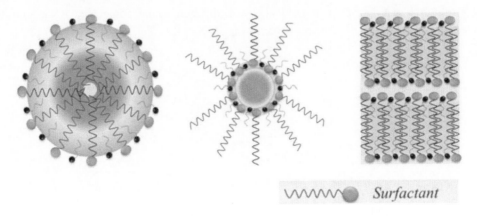

Figure 2. Different types of mEs: o/w, w/o and BI.

The mEs, in general, can be classified as water-in-oil (w/o), oil-in-water (o/w), and intermediate bicontinuous (BI) as shown in Figure 2. It is the w/o mEs also termed as reverse mEs which have been receiving increasing attention as templates for synthesis of NPs with controllable size.

3. MICROEMULSION BASED NANOREACTORS USED IN SYNTHESIS OF NANOPARTICLES

In w/o mEs, the dispersed oil phase consists of monodisperse droplets in the size range of 5-100 nm in diameter where thermodynamically driven surfactant self-assembly generates aggregates known as reverse micelles (RMs) [25]. Such RMs have the ability to solubilize water in the polar core to form water-pools of nanometer size.

The water trapped inside the reverse micelles is said to exhibit major changes from bulk behavior when water to surfactant molar ratio, $W_o < 20$ [26] and therefore, the hydrophilic core, containing water behaves as a medium different from ordinary water. Reduction of metal salts can thus be carried out efficiently in the hydrophilic core of the RMs which yields metal NPs. Since the metal NPs are synthesized in a confined medium, excessive growth of particles is not observed. Henceforth the size and dispersity of the particles can be controlled and the water pools serve as 'nanoreactors' as shown in Figure 3.

One of the first NPs to be synthesized in these nanoreactors was reported by Boutonnet et al., where Pt, Rh and Ir NPs were prepared in the RMs of cetyltrimethylammonium bromide (CTAB)/water/octanol [14] and this was considered to be a breakthrough in this field. This concept was then further applied for synthesis of various metal NPs such as Au, Ag, Cu [15]. Ag NPs were synthesized in RMs of Triton X-100 (TX-100)/cyclohexane/butanol/water [27] by reduction of silver nitrate with $NaBH_4$.

Copper NPs have also been synthesized in the water-pools of CTAB/1-butanol/ cyclohexane/water by the reduction of $CuCl_2$ with $NaBH_4$ [28]. Ag and Cu NPs have been successfully synthesized in anionic w/o mE of sodium dodecyl sulphate (SDS)/1-pentanol/cyclohexane/water [29]. Hossain et al., [29] prepared Ag NPs with antibacterial activity in the water pools of the reverse mEs of SDS/butanol/water/cyclohexane. Reverse micellar solutions of CTAB/1-pentanol/water with different water content have successfully been used as a medium for the electrodeposition of Ni NPs [30, 31].

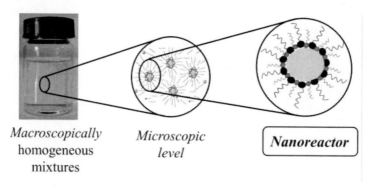

Macroscopically homogeneous mixtures   *Microscopic level*   **Nanoreactor**

Figure 3. Water-in-oil microemulsion envisaged as nanoreactors in microscopic level.

Petit et al., synthesized CdS NPs in reverse micelles of AOT (Dioctyl sodium sulfosuccinate) and triton [33]. Semiconductor NPs such as CdS, PbS, CuS, Cu$_2$S and CdSe [33-36] and metal oxide NPs such as ZrO$_2$, TiO$_2$, SiO$_2$, GeO$_2$ and Fe$_2$O$_3$ [37-40] could be synthesized in the water pools of w/o mE. Basic hydrolysis of zinc nitrate in w/o mE of TX-100 resulted in synthesis of ZnO NPs with controlled size and shape and growth kinetics of such NPs was also studied in the water-pools of the nanoreactors of SDS/pentanol/water and CTAB/pentanol/water [41]. The approach was then further applied for synthesis of nanocomposites.

Nanocomposites are composite materials made up of two types of particles in which at least one of the phases shows dimensions in the nanometer range. Solvent casting, novel combustion, and sol gel methods have so far been used for synthesizing different types of polymer based nanocomposites and composites such as CdS-ZnS [42], CdS/PbS, CdS/HgS [43, 44], CdS/Ag$_2$S [45], CdSe/CdSe, CdSe/ZnS [46, 47]. RMs have also played a great role in the synthesis of CdS-ZnS, CdS-Ag$_2$S and CdSe-ZnS, CdSe-ZnSe [45, 46] composite NPs.

Additionally, mEs have also been successfully employed in the synthesis of core/shell NPs (CSNPs). Such NPs are receiving an upsurge of interest, since they have emerged at the frontier between materials chemistry and many other fields, such as electronics, biomedical, pharmaceutical, optics, and catalysis. In particular in the biomedical field, the majority of these particles are used for bioimaging, controlled drug release, targeted drug delivery, cell labeling and tissue engineering applications [48-56]. Goikolea et al., [56] studied the magnetic behavior of Ag/Fe$_2$O$_3$ NPs synthesized in w/o mEs of CTAB/octane/n-butanol/water and Pileni et al., [57] synthesized NPs in mEs where co-precipitation in w/o mEs enables multiple steps reactions and composite NPs are formed. Magnetic MnZn-ferrite NPs with a narrow size distribution were synthesized in water-CTAB-hexanol mEs using Pileni's concept [57]. A wide range of CSNPs such as CdS/Ag$_2$S [58, 59], Fe/Au [60], Fe$_3$O$_4$/SiO$_2$ [61], PbTe/CdTe [62], CaO/Fe$_2$O$_3$ [63], CaO/Fe$_2$O$_3$, MgO/Fe$_2$O$_3$, and ZnO/Fe$_2$O$_3$ [64] were all synthesized using mEs. Recently, Satter et al., [32] synthesized ZnO/Ag CSNPs via reverse mEs of TX-100/hexanol/water/cyclohexane.

4. PARAMETERS CONTROLLING SIZE OF MES

Before describing the parameters that control the size of NPs, the mechanism of the formation of the NPs in w/o mEs is described briefly.

W/o mEs are dynamic systems — RMs undergo random *Brownian motion* and coalesce to form dimers, which may exchange contents then break apart again [65, 66]. Clearly, any inorganic reagents encapsulated inside the micelles will be mixed.

This exchange process is fundamental to the synthesis of NPs inside reversed micellar *'templates,'* allowing different reactants solubilized in separate micellar solutions to react upon mixing. RMs in these systems described as *"nanoreactors,"* provide a suitable environment (cage like effect) for controlled nucleation and growth.

By controlling several parameters of mEs and the operation, nanoreactors can be used to produce tailor-made products down to a nanoscale. Various parameters which may be controlled during preparation of mEs used to synthesize NPs of controllable size are shown in Table 1.

Table 1. Various parameters for preparation of mEs and NPs with controllable size in such mEs

Microemulsion parameters	Operating parameters
Water to surfactant molar ratio (W_o)	
Polar volume fraction	Reactant addition scheme
Precursor salts concentration	Mixing time
Oil phase	Temperature
Types of surfactants	pH
Types of co-surfactant	

Optimization of these parameters can be governed by varying the feed composition, constituents and reaction conditions during synthesis. As the nanoreactors can demonstrate numerous features microscopically including rigidity/flexibility of interfacial film; large/small in size; more/less stability; wide/narrow width of size distribution etc.; they can be called *versatile nanoreactors*. As a result size of the NPs can be effectively controlled.

4.1. Effect of Surfactant

MEs are found in fascinating and useful variety of shapes, sizes, and forms. These are related to the flexibility of the amphiphilic interface to change its geometry which is known as surfactant film (SF). This film can be characterized by three phenomenological constants: *tension*, *bending rigidity*, and *spontaneous curvature*. Their relative importance depends on the constraints felt by the film. It is important to understand how these parameters are related to interfacial stability since surfactant films determine the static and dynamic properties of mEs. These include phase behavior and stability, structure, droplet exchange behavior and solubilization capacity.

General classification of the types of surfactants has already been made in Section 2. Among those, non-ionic surfactant based RMs are smaller in size compared to those of ionic ones since the polar heads are not charged and hydrophobic part of a non-ionic surfactant is in general smaller than the hydrophilic part. In case of ionic surfactants, charged head groups repel each other making the hydrodynamic radii of the water pools larger in size. This phenomenon is shown in Scheme 1 where the non-ionic TX-100 based mE is smaller in size compared to those of ionic ones say, CTAB and SDS.

In case of ionic surfactants, co-surfactants are necessary as they anchor the two surfactant species to form the surfactant film at the oil-water interface whereas non-ionic surfactants do not necessarily need co-surfactants. As the number of tail beads of the surfactant increases, the energy cost of bending the membrane for monolayers with a given area per molecule is expected to increase. This is also true if one compares the different chain lengths at the same interfacial tension rather than interfacial density. Hoque et al., [178] showed the effect of type of surfactant on the hydrodynamic radii of the water pools using TX-100, CTAB and SDS and then synthesized Fe/Au CSNPs in these water pools. Nanoreactors of SDS based mEs were larger in size due to high HLB value and low local mean curvature. Due to decrease in curvature and presence of co-surfactant, the surfactant film flexibility (SFF) increases. Therefore, when the RMs undergoes colascence, RMs of larger dimensions are formed.

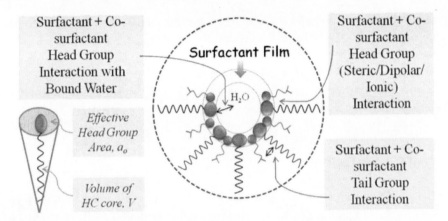

Figure 4. Different types of hydrophobic and hydrophilic interactions within the SF of RM in reverse mEs.

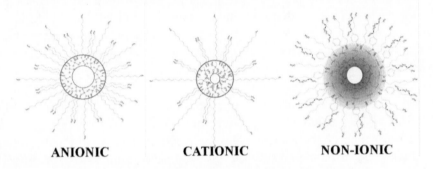

ANIONIC   **CATIONIC**   **NON-IONIC**

Scheme 1. Effect of type of surfactants on hydrodynamic radii of the water-pool.

Use of CTAB results in the formation of flexible film which gives rise to high exchange dynamics of the micelles. Porta et al., [67] obtained smaller Au NPs using CTAB as a surfactant. ZrO_2-Y_2O_3 NPs were synthesized in a CTAB/hexanol/water mEs by Fang and Yang [68]. Such mEs were used to produce other materials like spinel $ZnAl_2O_4$ [69], perovskite nLaMnO$_3$ [69], bioceramics hydroxyapatite [70], cerium oxide [71], and coated materials like, SiO_2-coated Pt, Pd and Pt/Ag [72].

Different kinds of NPs such as CeO_2, Ce_{1-x}-Zr_xO_2, Ce-Tb mixed oxides, Al_2O_3, Y_2O_3:Eu^{3+}, TiO_2 and silver halides [73-77] were synthesized in w/o microemulsions using non-ionic TX-100 surfactant. Synthesis of Pt-Ru bimetallic NPs was carried out in w/o mEs of water/TX-100/propanol/cyclohexane [90]. Pt-Co NPs were also synthesized in the same media [91]. By varying the type of surfactants, silica NPs were coated with shell of iron using different precipitating agents (NH_4OH or NaOH) [92]. Ultra-small particles (< 5 nm in diameter) aggregated in different morphologies and depending on the experimental conditions, the extent of adsorption of surfactant on surface of the particle varied. In the case of Brij-97, more ordered structure was observed due to its long hydrophobic chain compared to Triton X and Igepal. This resulted in ordered aggregation of particles due to strong hydrophobic-hydrophobic interactions between the oleyl groups attached to adjacent NPs. Tartaj and Serna [78] found that the nature of the surfactant (Igepal CO-720 or TX-100) did not significantly affect the microstructure of the prepared iron oxide NPs using cyclohexane as oil and *n*-hexanol as co-surfactant.

NPs such as hydroxyapatite and metallic Bi were synthesized using poly(oxyethylene)-5-nonylphenol ether (NP-5), poly (oxyethylene)-9-nonylphenol ether (NP-9) and poly (oxyethylene)-12-nonylphenol ether (NP-12) [79, 80] and their catalytic activity was also studied [81]. Rh NPs of different sizes were also synthesized in NP-5/cyclohexane microemulsion using different reducing agents [82] and bimetallic Pt-Ru NPs were synthesized using different non-ionic surfactants [81].

4.2. Effect of Co-Surfactants

Co-surfactants are molecules with weak surface active properties and the effect of replacing long surfactant molecules with short ones is larger than the effect of reducing the chain length of all molecules. They are usually alcohols or amines ranging from C4 to C10 and help in the formation and stabilization of micelles/mEs. It provides a "dilution effect" in addition to that of the surfactant and causes a further decrease of the interfacial tension. When salt is added to the solution, the surface potential is partly neutralized. This results in decrease of the coulombic repulsion between adjacent head groups and allows the formation of larger micelles. The short hydrophobic chain and terminal hydroxyl groups of co-surfactants are known to enhance the interaction with surfactant monolayers at the interface.

Increase in chain length lowers the diameter of the water pools of the RMs. In case of butanol based mEs the average size of the nanoreactors are larger than those for hexanol based mEs. This means that the micro structural properties of the SF are heavily influenced. Increase in chain length of the alcohol raises their hydrophobicity. The polar head group position of the alcohol from the bound water inside a nanoreactor thereby increases which causes an increase in SF thickness. As a result hydrophobic interaction between the surfactant and co-surfactant tail group increases which results in the removal of surfactant from the SF to the oil phase and lowers the size of a nanoreactor as shown in Scheme 2 [178].

The size distribution is narrowed down which is also contributed by the decrease in interfacial film tension and increase in local mean curvature. Hoque et al., [178] also showed that a branched chain polymeric alcohol of pentanol, 2-methyl-1-butanol showed marked decrease in hydrodynamic radius of water pools (Scheme 3).

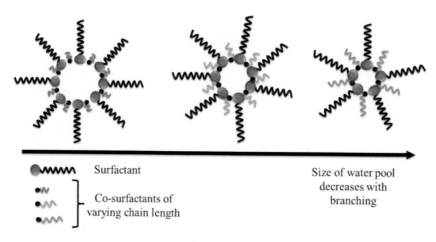

Scheme 2. Effect of co-surfactants based on their chain length over the size of the RMs.

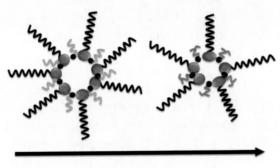

Size of water pool decreases with
branching of co-surfactant

Scheme 3. Effect of co-surfactants based on their chain branching over the size of the RMs.

The branched alcohol increases the steric congestion at the interface which screens the dense packing of surfactants to result in lower water uptake which in turn decreases the size of RMs [84, 85].

4.3. Effect of Solvent

Particle size is affected by the type of solvent [85]. The effect of the solvent on the size of NPs depends on the type of surfactant [86, 87] used in mEs. Viscosity is one of the important parameters for controlling the mechanism of particle formation in mEs. It controls the intramicellar exchange dynamics indirectly. Any increase in the intermicellar exchange rate will result in higher rate of growth in contrast with the nucleation resulting in systems with lower polydispersity. For example, when cyclohexane (viscosity = 0.85 cP at 25 °C) is used as the continuous phase instead of isooctane (viscosity = 0.32 cP at 25 °C), reduction in the rate of intermicellar exchange can be observed; therefore, although the size of the water pools remains constant, because of less intermicellar exchange the ultimate particle size is reduced in the case of cyclohexane [88]. The decrease in particle size due to the use of cyclohexane instead of isooctane (all other parameters kept constant) can be explained in terms of intramicellar nucleation and growth mechanism, and lower aggregation for cyclohexane system. Recently, Eastoe et al., [89] studied the extent of solvent penetration in mEs stabilized by di-chained surfactants, using small angle neutron scattering (SANS) and selective deuteration. Results suggested that oil penetration is a subtle effect, which depends on the chemical structures of both surfactant and oil. In particular, unequal surfactant chain length or presence of C = C bonds results in a more disordered surfactant/oil interface, thereby providing a region of enhanced oil mixing.

4.4. Water to Surfactant Molar Ratio

Both w/o droplet structures have a polar core region surrounded by an interfacial film. The value of water-to-surfactant molar ratio, W_o can be varied by increasing/decreasing surfactant concentration by keeping water content constant and vice versa.

It is evident that the particle size initially increases with increasing water to surfactant molar ratio, W_o as shown in Scheme 4. It shows that although the size reaches a plateau [88], the polydispersity still continues to increase.

Pileni and co-workers [180, 93] explained that at low water content, the number of water molecules per surfactant species is too few to hydrate the counterions and the surfactant headgroups. This induces strong interactions between the water molecules and the polar headgroups, so that the water molecules are present as *"bound"* water.

With a progressive increase in the water content, there is a change from *"bound"* to *"free"* water in the water pool; as a result, the particle size also increases. But whenever it reaches a plateau any further increase in W_o does not effectively increase the particle size since excess water can no longer contribute to the hydration of new ions. Another trend is also evident from the literature. It shows that initially the particle size increases with increasing W_o until a certain value is reached, beyond which particle size again starts to decrease [94]. This may be due to the decrease in the rate of effective collision and intermicellar transfer after a certain W_o value is attained.

Another effect of changing W_o is to vary the effective concentration of reagents inside the RMs, if the overall reagent is kept constant. The final particle size is dependent on the initial W_o [95-97]. Unfortunately in many cases, the same effect is not seen [98, 99]. Kithcens et al., [98] observed that at any given value of W_o, NPs of the same size can be synthesized if left for sufficient time for the completion of reaction. They proposed that the rate of NP growth is affected by varying W_o. Surfactants are fully hydrated at given water content and increase as the water content induces formation of a thermodynamically stable emulsion made of several phase with interconnected cylinders trapped inside an onion phase and filling the interstices between spherulites [7]. With the water bound, the micelle interface is said to be *"rigid,"* lowering intermicellar exchange and thus the growth rates. As W_o is raised, the film becomes more fluid, so the rate of growth increases, until it reaches a point when all extra water is just added to bulk water (at around $W_o = 10\text{-}15$). The extra water added merely dilutes the reagents decreasing reduction rates; so any increase in rate of intermicellar exchange is negated and in some cases a decrease in particle size is observed [99].

These contrast with few other works reported [95, 100] which suggested that particle size can indeed be controlled at any composition of mEs by changing W_o.

4.5. Effect of Reagent

With increasing reactant concentration, occupancy per reverse micelle increases for the individual species. The occupancy number is defined as the number of reactant species in a RM. This results in increased rate of intramicellar nucleation and growth; as a result, the ultimate particle size decreases [101, 102].

In case of CSNPs, shell formation occurs around the core via *heterogenous nucleation.* Instead of forming new nuclei in the bulk, the embryo of the shell materials gets deposited directly on surface of the core and then continues formation of nuclei and hence growth occurs. Slow reaction favors formation of a uniform coating and low reactant concentration is favorable for synthesis of CSNPs [103, 104].

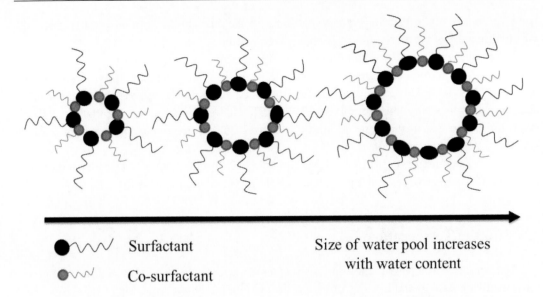

Surfactant

Co-surfactant

Size of water pool increases
with water content

Scheme 4. Change in size of the hydrodynamic radius of the water-pools with change in W_o value.

Normally the reducing agents have no effect on controlling the particle size but weak reducing agents increase the polydispersity of NPs due to slow rate of reaction [93, 105]. When rate of reaction is fast, *intra* and *intermicellar* exchange rates increase and so nuclei concentration increases and size decreases. CdS NPs were synthesized by both a single and double microemulsion addition scheme where H_2S gas was used as a sulfide precursor for the single mE method, whereas in the case of the double mE method aqueous Na_2S was used as the sulfide precursor [106]. Particle size was greater in the former case due to increased concentration of sulfide ion. Literature reports support the view that particle size and polydispersity do not depend on the nature of the reactant [102, 107-109].

4.6. Polar Volume Fraction

In a w/o mE with increasing polar volume fraction, the population of RMs or nanoreactors increase in the continuous oil phase. Therefore, the collision between particles in populated RMs also increases, which in turn leads to an increase in the probability of aggregation. Another reason for increasing size with increasing polar fraction may be concomitant increase in reactant concentration so that the nucleation and growth mechanism change from intramicellar to intermicellar and hence ultimately particle size increases [101, 102, 107, 108].

4.7. Effect of Mixing

During particle formation, reaction steps involve mass transfer which necessitates the study of the effect of mixing. Two means are commonly used for synthesis of NPs. In *single microemulsion reactant addition* systems aggregation may lead to increase in particle size.

However, ultrasound sonications effectively improve the reactant distribution within the RMs, which ultimately gives more uniformly small sized particles.

In *double microemulsion addition* systems, mixing results in narrow particle size distribution. It improves the rate of nanoparticle formation and lowers the time required for uptake of NPs [107]. However, there are reports which show no observable effect on particle size and polydispersity due to mixing [94, 110]. No explanation was available on the controlling step since the rate of mass transfer and surface area for reaction varied simultaneously during reaction.

In general, an increasing [110-112], decreasing [113, 114] or both [115] trends on particle size were found to be dependent on the types of particles and the medium for synthesis.

4.8. Effect of Temperature

Temperature is one of the important factors which governs the stability of the micromeulsion system and kinetics of the NP formation. Increased temperature increases rate of reaction and influences rate of mass transfer. Nucleation depends on diffusion of the atoms or embryo from the bulk to the nuclei surface which in turn depends on temperature. In case of endothermic reactions, concentration of product in bulk is more thereby, diffusion increases. The opposite phenomenon is observed in case of an exothermic reaction. However, if rate of nucleation is very high, the overall reaction becomes very fast and growth rate depends on diffusion only. In general, a low temperature is preferred for the formation of CSNPs in order to maintain unsaturated shell material formation within the reaction mixture which influences appropriate coating on the core surface rather than the formation of separate nuclei. Decreased intermixing of core and shell atoms occur at the interfaces due to low temperature. Surface modification occurs due to adsorption of appropriate surface active agent on surface of core so that oppositely charged shell material gets deposited on surface. At high temperatures, the adsorption density of the surface active agent on the core surface is reduced; achieving selective uniform coating becomes a difficult task. So, an optimum temperature is maintained throughout the reaction so that both the parameters are taken into consideration. In general, a low temperature is preferred for core/shell particle synthesis in the presence of a surface modifier [116-120]. However, if the shell material can be coated onto the core surface without any surface modifier then normally the temperature is maintained according to requirements of the reaction [121-123].

4.9. Effect of pH

Effect of pH is visible on size of particles when H^+ and OH^- ions are directly related to the reaction mechanism. Types of reactions such as reduction, precipitation and redox reactions are highly affected by pH. Influence of pH can also be extended to the topic of surface modification. Control of pH controls surface modification and in case of RMs with hydrophilic cores, surface charge can be adjusted by controlling pH.

5. MECHANISM OF FORMATION OF NANOPARTICLES IN MICROEMULSIONS

The RMs in the w/o mEs are in continuous Brownian motion and undergo process like fusion-fission. The RMs containing precursor salts undergo collisions which lead to opening of surfactant bilayer and hence reaction occurs inside the confined media. Formation of particles occurs by a nucleation process which is followed by other processes as described below.

5.1. Nucleation of Monomers

Nucleation is the process by which atoms (or ions), which are free in solution come to closer approach to produce a thermodynamically stable cluster and it is one of the fastest processes alongside chemical reaction.

In mEs, sufficient number of atoms needs to reach local saturation inside the core of the micelle in order to overcome the surface energy required to form a solid nucleus. This means that, for the nucleation to be effective, one micelle must carry enough material for the nuclei to reach a critical size. Below this size, nuclei are unstable and redissolve. The critical size of the nuclei depends on the reagent concentration and the surface energy of the solid–liquid interface, according to the equation:

$$\Delta G_{nucl} = S\gamma + V\Delta G_{cryst}...\tag{1}$$

where S and V are the surface and volume of the nucleus, γ the interfacial energy, and ΔG_{cryst} is the free energy of crystallization. The ΔG_{cryst} depends on the reactant concentration and becomes negative as saturation is reached.

Due to the very small size of micelles, local concentrations can be very high, favoring nucleation. Moreover, the interfacial energy is lowered by the adsorption of surfactant to the nuclei; that is, nucleation can be induced at the droplet interface by using surfactants/co-surfactants that promotes nucleation. RM thus constitutes environment in which *critical nucleus size* (the minimum size of a stable nucleus, which depends on the specific material), C^* is smaller than that associated with precipitation in simple aqueous solutions. A small value of C^* implies that a high number of nucleus seeds appear quickly; that is, the nucleation rate is faster (higher reduction potential) as the critical nucleus is smaller.

Therefore, the metal with a higher nucleation rate will give rise to the majority of seeds from which NPs will be formed. These seeds will grow giving rise to a monometallic core, on which the slower metal is deposited. As a consequence, the bimetallic structure of NPs is directly related to the difference between the nucleation rates of both metals: a large difference in the nucleation rates of the two metals results mainly in a core/shell structure, and a small difference in the nucleation rates leads to an alloy one. Ion-displacement and heterogeneous nucleation are considered to be the shell formation mechanisms in case of CSNPs. The phenomenon of finite homogeneous nucleation also occurs and leads to the formation of NPs of only shell-forming material.

It is likely that every molecule of the shell deposited on the core particle surface blocks the core surface molecules available for the displacement reaction.

5.2. Growth of Nanoparticles

NP growth can take place via reaction on an existing aggregate (growth by autocatalysis) or via ripening (growth by ripening). The metal precursor and reducing agent have to be in the same droplet to react, so the interdroplet reactant exchange controls the autocatalytic growth. On the other hand, growth by ripening is determined by the SFF, which controls the maximum size of exchanged growing particles [124].

An increase in the SFF implies a quicker exchange of reactants and the exchange of larger aggregates of products, favoring the growth by ripening. Two consequences arise from this increase in the ripening contribution: First, as the synthesis proceeds the growing particles are located in fewer droplets, and more droplets become empty. Second, the whole process becomes longer because ripening is possible for a longer time (more time is needed to reach the maximum particle size which cannot be exchanged), and because collisions between two nucleated droplets is a less probable event. Many studies have been reported for different growing, aging, or ripening mechanisms in bulk. A particular coarsening process that has to be considered in mEs is *Ostwald ripening*. The particle size changes by solubilization and condensation of material. *Ostwald ripening* assumes that the largest particles will grow by condensation of material, coming from the smallest particles that solubilize more readily than larger ones. The process of *Ostwald ripening* occurs because of the variation of the solubility of particles with the particle size. In this process, the smaller particles dissolve in the bulk fluid phase due to higher solubility and this dissolved material gets deposited on the larger particles which are relatively less soluble in the bulk phase. This difference in solubility is due to the *Gibbs-Thomson effect* [125].

5.3. Stabilization of Nanoparticles

Internal contents of the mEs droplets are known to exchange, typically on the millisecond time scale [126-128]. They diffuse and undergo collisions. If collisions are sufficiently vigorous, then the SF may rupture thereby facilitate droplet exchange that is the droplets are kinetically unstable. However emulsions with small droplets (< 500 nm in diameter) have less tendency to coalesce due to the energy barrier meaning that the droplet exchange process will be activation controlled (E_a, barrier to fusion) process [126]. The system will remain dispersed and transparent for a long period of time (months) [127]. Such an emulsion is said to be kinetically stable [128].

Therefore, further growth and agglomeration of CSNPs are prevented by the surfactant protective layer, and the concentration of the NPs and their size distribution becomes time-dependent. Particles growing beyond that range precipitate under gravity effect. Aforementioned features concerning the mechanism of NPs formation are depicted in Scheme 5.

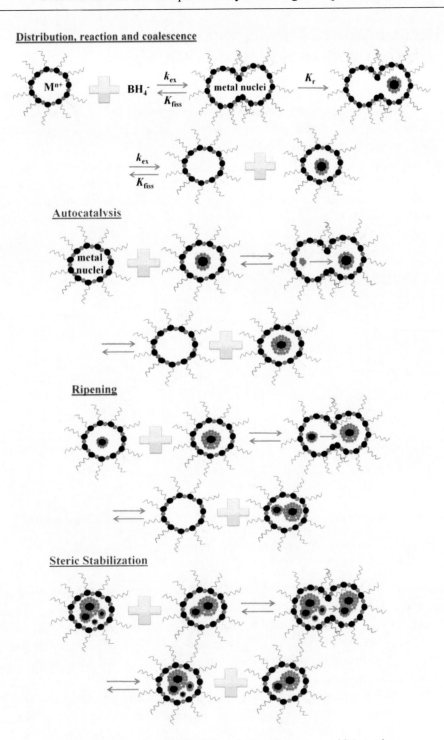

Scheme 5. Mechanism of the formation of CSNPs by *single mE reactant addition* scheme.

6. PREPARATION OF METAL, METAL OXIDE AND CORE/SHELL NANOPARTICLES IN WATER-IN-OIL MICROEMULSION

Although several methodologies have been reported for the synthesis of NPs, the mE mediated method is becoming quite popular due to its increased ability to control the size of NPs. Synthesis of Ag and Cu NPs were reported by Hossain et al., [29] in water-in-oil (w/o) mEs of SDS/1-pentanol/cyclohexane/water.

With increase in W_o, the size of the NPs also increases as shown in Figure 5. An increase in water content refers to an increase in the water content in the hydrophilic core of reverse micelles which in turn results in slow nucleation. Rigidity of the surfactant layer decreases which lowers the particle size.

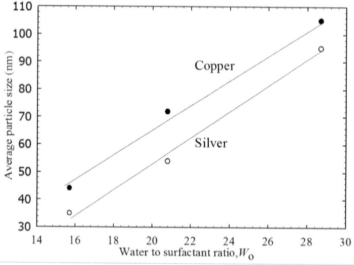

Reproduced from Reference [29] with permission of Journal of Bangladesh Chemical Society.

Figure 5. Average particle size of Ag and Cu NPs in SDS/1-pentanol/cyclohexane/water microemulsions as a function of water-to-surfactant molar ratio, W_o.

Synthesis of ZnO NPs in w/o mE of TX-100 was recently achieved by the basic hydrolysis of zinc nitrate precursor salt in the reverse micelles followed by successful coating of Ag around the ZnO NPs at $W_o = 13.34$ [32]. Although the crystalline structure of ZnO is hexagonal cubic packed (HCP) and Ag is face centred cubic (FCC), the crystalline mismatch was overcome in the confined region of the water pools of w/o mE and hence uniform coating could be observed as shown in Scheme 6.

In these confined water pools of the reverse micelles which have a 'cage-like' effect, Gibbs energy required is lower for Ag NPs to undergo heterogeneous nucleation on the surface of ZnO NPs. The surface of the ZnO NPs might have numerous defects and are thermodynamically unstable. Surface reconstruction would further decrease their energy and thus provide active sites for heterogenous nucleation and growth of Ag NPs. Hence, successful uniform coating of Ag around ZnO NPs has been achieved in this highly unique media. The size of the core and the thickness of the shell may be effectively controlled by systematic variation of different parameters to influence the properties of the mEs as nanoreactors.

For instance, in case of synthesizing ZnO/Ag CSNPs, an optimized W_o value was chosen for the w/o mE. Other parameters, such as type of surfactant, co-surfactant and reagent concentration can also be widely varied in order to have a marked effect on the size of water pools of RMs which in turn have an effect on size and distribution of NPs synthesized in the water pool. This concept has also been applied to synthesize Fe/Au CSNPs. The change in parameters shows noticeable change in size of the core as well as the shell [178]. TX-100 showed the best results in terms of size of the CSNPs and non-ionic surfactants may be preferred for the synthesis [178]. Variation in W_o values shows marked effect on the size of RMs which can be used to efficiently control the size of the Fe/Au CSNPs.

Branching of co-surfactant also shows marked influence on the size of RMs as well as the CSNPs. The CSNPs are basically synthesized inside a confined region and the interfacial film (IF) of the RMs act as a barrier and prevents excess growth of NPs.

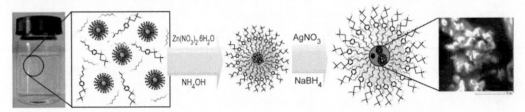

Reproduced from Reference [32] with permission of The Royal Society of Chemistry.

Scheme 6. Synthesis of ZnO/Ag CSNPs synthesized in the water pool of w/o mEs.

Variation in salt concentration also affects the size of core and thickness of shell. This phenomenon is prominent in the surface plasmon resonance (SPR) of the core and the CSNPs as discussed in the next section.

7. OPTICAL PROPERTY

The electronic energy levels and quantum confinement determine the optical properties of materials as well as their functionalities. The need to fine-tune different properties of NPs to make them suitable for specific applications has sparked a large number of worldwide research efforts aimed at their tailoring. Closely related to size-induced changes in the electronic structure are the optical properties of NPs. Novel optical properties of nanomaterials are both linear and non-linear types shown in Figure 6 usually investigated by spectroscopy based techniques. Spectroscopic methods probe the energy differences between two states for allowed transitions as well as the lifetimes of excited states and their respective energy relaxation channels using time-resolved techniques.

Since the "Middle Ages" exhibition of colorful colloidal solutions of noble metal NPs baffled scientists but they did not know the origin. It was Faraday who first recognized this phenomenon, and Mie [129] was able to explain it theoretically in 1908 by solving Maxwell's equations. The physical origin of the strong light absorption by noble metal NPs is the coherent oscillation of the conduction band electrons induced by interaction with an electromagnetic field. These resonances are known as surface plasmons (SPs) and correspond to a small particle as well as a surface effect because they are absent in the individual atoms as well as in the bulk. Their extinction coefficient scales with the volume of the particles and can reach values several orders of magnitude larger compared to common organic dye molecules [130]. In addition to this, the weak intrinsic fluorescence of noble bulk metals resulting from the electronic *interband transition* was discovered in 1969 by Mooradian [131]. Recent research has shown that NPs have enhanced fluorescence (FL) emission over the bulk, particularly in small clusters [132]. Nanorods also have enhanced emission over bulk metal and nanospheres, due to the large enhancement of the longitudinal plasmon resonance [130, 133-136]. Clusters and nanorods have an emission that shifts wavelengths as the size or aspect ratio increases. In the case of NPs with a diameter much smaller than the wavelength of light, the origin of the second harmonic (SH) signal is, therefore, of electric quadrupole origin and is thus expected to be rather weak.

However, it is possible to enhance this nonlinear optical response through local field resonances corresponding to the surface plasmon excitations of the particles [137]. Second harmonic generation (SHG) (frequency doubling) is forbidden for centrosymmetric media within the dipole approximation, and it is therefore strongly sensitive to the symmetry of the system.

Since centrosymmetry is inherently broken at an interface, SHG has been applied extensively as a probe to study processes on bulk surfaces or at interfaces between materials.

NPs with a metal nanoshell have recently been developed. They are composed of a dielectric core, with a nano range metallic layer, which is typically gold. Similar to the size-dependent color of pure gold NPs, the optical response of gold nanoshells depend dramatically on the relative size of the core NP as well as the thickness of the gold shell [138]. By varying the relative core and shell thickness, the color of such gold nanoshell can be varied across broad range of the optical spectrum spanning the visible and the near-infrared (NIR) spectral regions. A plot of the core/shell ratio versus resonance wavelength for a silica core gold shell NP displayed [138] extreme agile *"tunability"* of the optical resonance.

This is a property unique to nanoshells; no other molecular or NP structures can induce the resonances of the optical absorption properties and thereby do not to have the prospect to be systematically "designed." The result is that the gold nanoshells may prove useful for biomedical imaging in future applications. Gold coating on any particles enhances many physical properties, such as chemical stability by protecting the core material from oxidation and corrosion, biocompatibility, bio-affinity through functionalization of amine/thiol terminal groups, and optical properties.

Other shell metals such as nickel [139], cobalt [140], palladium [141, 142], platinum [143-146] and copper [147] are also important for some specific applications in the field of catalysis, solar energy absorption, permanent magnetic properties, etc. The absorption band of Ni/Au NPs was reported to be red shifted at 600 nm and broadened as compared with the band of 15 nm gold NPs at 522 nm as shown in Figure 7 [148].

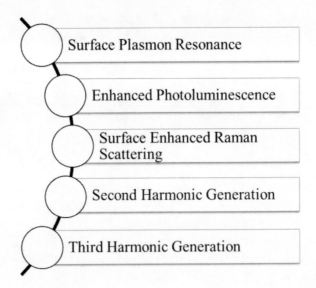

Figure 6. Optical properties of nanoparticles.

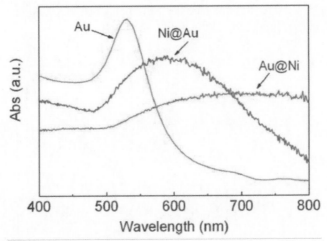

Figure 7. UV-vis absorption spectra of Au, Ni/Au and Au/Ni NPs.

This red shift is closely related to the nanoshell effects. The overall trend is that the plasmon resonance red-shifts when the thickness of the shell is decreased and the size of the core is increased as shown schematically in Figure 8. The red-shift of the SPR band of gold was also experimentally observed in NPs having a core of magnetic metals, such as Fe/Au [149], Ni/Au [148, 150] and FePt/Au [151] NPs.

The theoretical calculation on the Ag/Au core/shell structure also demonstrated a red-shift for Au-coated metallic cores [64], although no detailed calculations were performed on Ni/Au NPs. The increase in band width may be explained by inhomogeneous polarization of the gold shell in the electromagnetic field as well as multipole excitation effects, since the geometry of the shell is unlike that of a sphere which is completely symmetrical and has only one plasmon resonance.

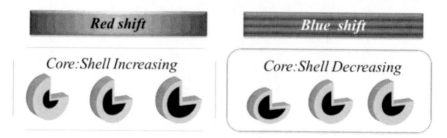

Figure 8. Shifting of SPR as a function of the core to shell ratio.

The absorption spectra of Au/Ni NPs were also quite different from that of gold NPs. The SPR band belonging to the gold was greatly damped and replaced by a slow increasing jump extending to the NIR region. This damping effect was due to the coating of non-plasmonic nickel. Unlike noble metals which have a strong visible-light plasmon resonance, most transition metals only have a broad and poorly resolved absorption band in the UV region of UV-vis absorption spectrum.

This is attributed to the strong coupling between the *plasmon transition* and the d-d *interband excitation*. Therefore, the plasmon energy is lost by excitation of the single electron interband transition [152]. The damping of the SPR band of gold was also observed in the case of Au/Ni nanocrystals prepared via the polyol method [153].

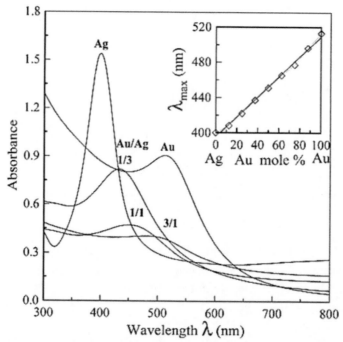

Reproduced from Reference [154] with permission of The Royal Society of Chemistry.

Figure 9. UV-vis absorption spectra of Au-Ag bimetallic NPs at various molar ratios. The inset shows the dependence of the absorption maximum of the plasmon band on the composition. [metal salts] = 0.1 M; W_o = 4; [AOT] = 0.1 M; reaction time = 3 h.

Figure 9 shows the UV-vis absorption spectra for Au, Ag and three typical Au–Ag bimetallic systems. Only one plasmon band was observed for each bimetallic system and the plasmon maximum was red-shifted almost linearly from 400 to 520 nm with increasing Au content as shown in the inset in Figure 9 to reveal the formation of Au–Ag alloy. The absorbance of Au–Ag bimetallic NPs decreased with increasing Au content and they all were lower than those for Au and Ag. This is in agreement with the calculated spectra of Au–Ag alloy NPs using the Mie equations [129]. According to the suggestion of Mulvaney et al., [155] one monolayer of Au should be sufficient to mask the Ag plasmon resonance band completely.

So, the above phenomenon could be due to the fact that the increased Au content on the particle surface resulted in the damping of the underlying Ag surface plasmon band.

The SPR of ZnO/Ag core/shell NPs was shifted to longer wavelength compared to the plasmon resonance of Ag NPs showing an absorption maximum at 408 nm as shown in Figure 10 [32]. The ZnO@Ag CSNPs were prepared in a restricted region of the RMs which did not allow excessive growth and kept the size within a controlled range. The core NPs were synthesized in nanorange which resulted in increased band gap of ZnO. The red shift observed in the UV-visible spectrum is explained using energy band structure of silver and ZnO. The work function of ZnO is more than that of silver as shown in Scheme 7.

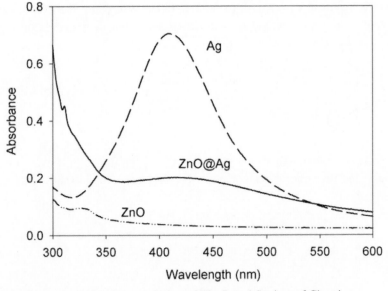

Reproduced from Reference [32] with permission of The Royal Society of Chemistry.

Figure 10. Absorbance spectra of (a) ZnO, (b) Ag and (c) ZnO/Ag CSNPs synthesized in w/o microemulsion of TX-100/hexanol/cyclohexane/water at $W_o = 13.34$.

The Fermi level of Ag $[E_f(Ag)]$ is greater than that of ZnO $[E_f(ZnO)]$ which results in the transfer of electrons from silver to ZnO until a new Fermi energy level (E_f) is formed in order to attain equilibration. The electron transfer from silver to ZnO enhances charge separation to cause deficiency of electrons on silver nanocrystals.

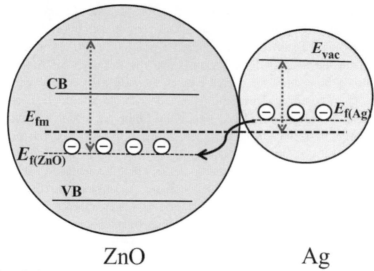

ZnO   Ag

Scheme 7. Energy band structure of Ag and ZnO showing formation of new Fermi level by back-donation of electrons.

Hence there is a shift in wavelength towards a higher value. Interfacial coupling between Ag NPs may cause broadening and red shift of the surface plasmon absorption [156]. These can be explained theoretically as given below.

For coated NPs, plasmon peak position, λ_{peak} is determined by [157]

$$\frac{\lambda_{peak}^2}{\lambda^2} = \epsilon^\infty + 2\,\epsilon_m + \frac{2g(\epsilon_s - \epsilon_m)}{3} \tag{2}$$

where, ϵ_s is the dielectric function of shell layer, ϵ_m is the dielectric function of medium, i.e., cyclohexane (the w/o microemulsion mostly comprises cyclohexane), g is the volume fraction of shell layer.

The refractive index and dielectric function is related by, $n_s = \epsilon_s^{1/2}$

$$\lambda_{peak} = \lambda[\epsilon^\infty + 2n_{cyclohexane}^2 + \frac{2g(2n_{Ag}^2 - 2n_{cyclohexane}^2)}{3}]^{1/2} \tag{3}$$

where bulk plasma wavelength, $\lambda = [\frac{4\pi^2 c^2 m_{eff}\epsilon_o}{Ne^2}]^{1/2}$
m_{eff} is the effective mass of free electron of the metal, N is the electron density of the metal.

$$n_{Ag} = 1.07 < n_{cyclohexane} = 1.43248$$

Although refractive index of Ag is less than that of cyclohexane, calculation shows that λ_{peak} is greater than λ to result in a red shift.

If the particles are well separated (have a thick coating), the dipole-dipole coupling is fully suppressed and the plasmon band is located nearly at the same position of the individual metal particles. In this case, it can be presumed that a thin shell of Ag has been formed. The ZnO NPs synthesized in w/o microemulsion were kept overnight and Ag core was formed after one day. Therefore, the polydispersity of the ZnO NPs might have increased thus showing such a broad peak for the CSNPs formed which is near the SPR of Ag NPs.

Recently, Hoque et al., [178] have shown that by tuning different parameters such as varying type of surfactant, co-surfactant, precursor salt concentration and composition, mEs can be prepared with varying dimensions of the water pool which causes systematic variation in the size of Fe/Au CSNPs. The variation of size of core and thickness of shell ultimately results in marked variation in SPR in UV-vis spectrum.

Increase in core to shell ratio results in red shift and decrease in core to shell ratio of Fe/Au CSNPs results in blue shift as shown in Scheme 7. This serves as a guide to explain the red/blue shift of SPR of heterogeneous CSNPs synthesized in reverse mEs of different types for variation in core to shell ratio of CSNPs.

Besides SPR, photoluminescence (PL) behavior of CSNPs, due to radiative combination of electron-hole pair may also be explained. Stokes shift is observed when the energy of the emitted photons is usually lower than the energy of the incident excitation photons. This is observed in case of Fe/Au CSNPs which eventually have the capability of exhibiting interesting SHG.

At a particular W_o value, Fe/Au CSNPs were synthesized [178] and excitation and emission spectra of the NPs prepared by changing the chain length of the co-surfactant were studied. Selection of excited wavelengths was based on the SPR response of the corresponding Fe/Au CSNPs obtained from UV-visible absorption spectra. In resemblance to the SPR, blue shift was also observed in case of fluorescence response of Fe/Au CSNPs.

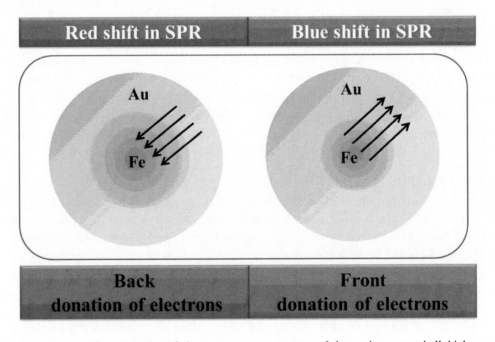

Scheme 8. Back and front donation of electrons as a consequence of change in core to shell thickness ratio.

The second harmonic (SH) signal was also blue shifted showing correlation with the fluorescence and SPR which was governed by the dimension of the Fe/Au CSNPs specifically core to shell ratio of CSNPs. Change in precursor salt concentration also influences luminescence behavior. In general, normalized luminescence intensity for both fluorescence and second harmonic generation of Fe/Au CSNPs increase proportionally with precursor salt concentration of Au NPs. Due to the increase in precursor salt concentration, the number of ions inside the droplets increases to speed up the *intermicellar coalescence* exchange rate.

Eventually the thickness of the Au shell increases. Due to increase in the number of Au NPs as shell in the Fe/Au CSNPs there was a gradual increase of intensity. On the other hand, enhancement of SH signal was also observed due to the continuation of breaking of symmetry stemming from homogeneous nucleation of Au NPs over the pre synthesized Fe/Au CSNPs at a particular mole ratio. The assumption is based on the fact that as nucleation rate of Au is very fast and it will be less uniform over the former Au shell of Fe/Au CSNPs. So as the number of Au layers increases due to the gradual addition, the shell becomes less spherical resulting in increase in the SH intensity. Such observations were indeed very striking and needs further attention at the atomic scale.

Also, there is a possibility of aggregation of Fe/Au CSNPs which may contribute to the SH response as well. Furthermore, a Stokes shift was also observed with tunable excitation and emission wavelength. It may be worth mentioning that such findings are new in case of metal based CSNPs and the origin of this behavior is rather vague. The Stokes shift was tunable and governed by the core to shell ratio of Fe/Au CSNPs as well. This invokes new inquisition in the field of plasmonics.

8. ANTIBACTERIAL PROPERTY

The antimicrobial properties of silver have been known to cultures all around the world for many centuries. The Phonecians stored water and other liquids in silver coated bottles to discourage contamination by microbes. Silver dollars used to be put into milk bottles to keep milk fresh, and water tanks of ships and airplanes that are "silvered" are able to render water potable for months. In 1884, it became a common practice to administer drops of aqueous silver nitrate to newborn's eyes to prevent the transmission of *Neisseria gonorrhoeae* from infected mothers to children during childbirth [158].

In 1893, the antibacterial effectiveness of various metals was noted and this property was named the oligodynamic effect. It was later found that out of all the metals with antimicrobial properties, silver has the most effective antibacterial action and the least toxicity to animal cells [159]. Silver became commonly used in medical treatments, such as those of wounded soldiers in World War I, to deter microbial growth.

The prevention of microbial surface contamination has become most crucial in today's health care system, and in food and pharmaceutical production. Human beings are often infected by microorganisms such as bacteria, viruses, and molds that often infect living environments. Ag NPs or Ag ions have been known to have strong inhibitory and bactericidal effects [160-167]. Metallic NPs are highly promising as they show good antibacterial properties due to their large surface area to volume ratio. They are coming up as the current interest among the researchers due to the growing microbial resistance against metal ions, antibiotics and the development of resistant strains [167].

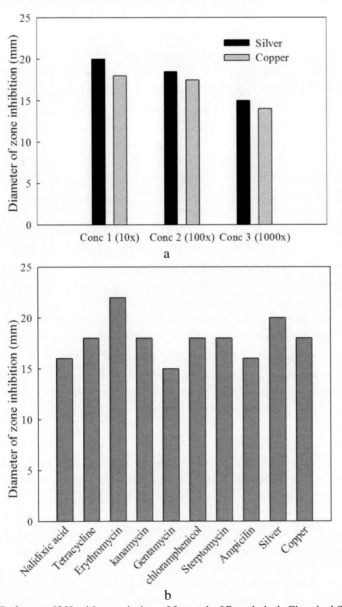

Reproduced from Reference [29] with permission of Journal of Bangladesh Chemical Society.

Figure 11. (a) Diameter of inhibited zone against *E. coli* for different concentration of Ag and Cu NPs in ethanol suspensions (b) A comparison of antibacterial sensitivity Ag and Cu NPs with conventional antibiotics.

Hossain et al., [29] tested the antibacterial sensitivity of Ag and Cu against *E. coli*, a gram negative bacterium commonly found in the lower intestine of warm-blooded organisms (endotherms) using zone inhibition method. The diameters of inhibited zone by Ag and Cu NPs at different dilutions are shown in Figure 11a. Silver showed a better antibacterial activity against *E. coli* (zone inhibition 20 mm) than that of copper (zone inhibition 18 mm). The antibacterial activity decreases as concentration decreases although to a smaller extent. From this, it can be inferred that even at high dilution limit, NPs synthesized in w/o mE will show antibacterial activity.

Figure 11b demonstrates antibacterial sensitivity of Ag and Cu NPs compared with conventional antibiotics. Ag NPs showed better bactericidal efficiency compared to some conventional antibiotics such as ampicilin, tetracycline, gentamycin, kanamycin, streptomycin etc. for bactericidal effects on *E. coli*. Cu NPs show almost equivalent sensitivity compared to conventional antibiotics. CSNPs such as ZnO/Ag core/shell NPs show synergistic antibacterial effect by acting upon gram-positive and gram-negative bacteria which is highly useful for water purification [170]. ZnO/Ag NPs have shown different types of antibacterial behavior depending on their synthesis methods [169-171]. Supermagnetic CSNPs of Fe_2O_3/Ag NPs synthesized in the w/o mEs of showed good antibacterial performance against *Escherichia coli* (gram-negative bacteria), *Staphylococcus epidermidis* (gram-positive bacteria) and *Bacillus subtilis* (spore bacteria) [165]. Antibacterial effect was also shown by Ni NPs synthesized in the RMs solution of TX-100 [172]. Construction of a sandwich-like antibacterial reagent (Ag/Halloysite nanotubes (HNTs)/rGO) has also been reported. Synthesis has been carried out through the direct growth of Ag NPs on the surface graphene-based HNTs nanosheets. Various nanomaterials were combined by adhesion effect of L-3, 4-dihydroxyphenylala-nine, DOPA after self-polymerization. Ag/HNTs/rGO possesses enhanced antibacterial ability against *E. coli* and *S. aureus* compared with individual Ag NPs, rGOnanosheets or their nanocomposites [173]. Recently, new strategy has been developed to improve the antibacterial potential of Ag NPs by their surface modification with the surface corona of biologically active polyoxometalates (POMs) [174].

Smart textiles are next generation textiles which will be free from bacteria and odor and have a range of potential applications from clothing – putting an end to smelly socks - to sporting gear and uniforms.

It has been found that organic materials with semi-conductor properties can have superior antibacterial effects over metal salts of silver which are already known for their antibacterial properties and to test this concept, nanowires have been grown on fabric which confirmed the antibacterial properties of Ag-tetracyanoquinodimethane [175].

There are very few reports on antibacterial effect of NPs synthesized in mEs. Hossain et al., [29] synthesized metal NPs in w/o mE and were successful in showing antibacterial activity. If different parameters are changed to control the dimension of the core of mEs, the size of NPs prepared from these mEs can be tuned.

Therefore it is envisioned that, the NPs synthesized in such templates can be used to show enhanced antibacterial activity in a controllable fashion. Even core size and shell thickness of CSNPs synthesized in such medium can be tuned to show improved tunable antibacterial activity.

CONCLUSION

MEs are liquid structures which can be controlled by changing different parameters over a wide range. Water pools of the w/o mEs can be controlled and in this way, different types of

NPs can be synthesized with controlled dimensions. Hence, NPs of desired and controlled size, shape and morphology can be synthesized which can eventually be used in task specific applications. Metal, metal oxide and core/shell NPs can be synthesized and the size of the core and the shell can be tuned which will allow control of optical and antibacterial activities in NPs.

Although separation of the NPs from the mEs and recyclability of the media is still a challenge, major applications focus on the use of NPs without separation from the mEs. However, conventional separation techniques such as ultracentrifugation, solvent evaporation, addition of a suitable solvent to cause phase separation or precipitation (ethanol, acetone, water) have been applied to recover NPs from the w/o mE system. The stability of mEs, easy access of chemicals and cost-effectiveness make mE a very novel medium for synthesis of NPs.

ACKNOWLEDGMENTS

The authors gratefully acknowledge financial support for a sub-project (CP-231) from the Higher Education Quality Enhancement Project of the University Grants Commission of Bangladesh financed by World Bank and the Government of Bangladesh. The collaborative research findings with the Center of Advanced Research for Sciences (CARS), University of Dhaka have also been used. MABHS also acknowledges the research grant (UNESCO FR: 3240262724) TWAS, the academy of science for developing world.

REFERENCES

[1]   Feynmann, R. P.; *Caltech Eng. Sci.*, 1960, 23, 22-36. (b) Sounderya, N.; Zhang, Y. *J. Biomed. Eng.* 2008, 1, 34-42.

[2]   Chan, W. C. W.; Maxwell, D. J.; Gao X.; Bailey, R. E.; Han, M.; Nie, S. *Curr. Opin. Biotechnol.* 2002, 13, 40-46.

[3]   Brigger, I.; Dubernet, C.; Couvreur, P. *Adv. Drug Deliver. Rev.* 2002, 54, 631-651.

[4]   Sondi, I.; Siiman, O.; Matijevic, E. *Langmuir* 2000, 16, 3107-3118.

[5]   Hu, A.; Yao, Z.; Yu, X. *J. Appl. Polym. Sci.* 2009, 113, 2202-2208.

[6]   Pileni, M. P. *C. R. Chimie.* 2003, 6, 965-978.

[7]   Pileni, M. P. *Langmuir* 2001, 17, 7476-7487.

[8]   Pileni, M. P. *J. Phys. Chem. C* 2007, 111, 9019-9038.

[9]   Pileni, M. P. *Acc. Chem. Res.* 2008, 41, 1799-1809.

[10] Lopez-Quintela, M. A.; Tojo, C.; Blanco, M. C.; Garcia Rio, L.; Leis, J. R.; *Curr. Opin. Colloid. Interf. Sci.* 2004, 9, 264-278.

[11] Cushing, B. L.; Kolesnichenko, V. L.; Connor, C. J. O. *Chem. Rev.* 2004, 104, 3893-3946.

[12] Shervani, Z.; Ikushima, Y.; Hakuta, Y.; Kunieda, H.; Aramaki, K. *Colloid. Surf. A: Physicochem. Eng. Asp.* 2006, 289, 229-232.

[13] Holmes, J. D.; Lyons, D. M.; Ziegler, K. J. *Chem. Eur. J.* 2003, 9, 2144-2150.

[14] Boutonnet, M.; Kitzling, J.; Stenius, P. *Colloids Surf.* 1982, 5, 209-225.

[15] Capek, I. *Adv. Colloid. Interf. Sci.* 2004, 110, 49-74.

[16] Kabir, A. M. R.; Susan, M. A. B. H. *J. Saudi Chem. Soc.* 2008, 12, 543-554.

[17] Shahed, S. M. F.; Islam, M. J.; Choudhuri, M. M. R.; Susan, M. A. B. H. *J. Bangladesh Chem. Soc.* 2009, 22, 123-130.

[18] Mahmud, I.; Samed, A. J. F.; Haque, M. A.; Susan, M. A. B. H. *J. Saudi Chem. Soc.* 2011, 15, 203-208.

[19] Haque, M. A.; Rahman, M. M.; Susan, M. A. B. H. *J. Solution Chem.* 2011, 40, 861-875.

[20] Mredha, T. I.; Roy, C. K.; Mollah, M. Y. A.; Susan, M. A. B. H. *Electrochim. Acta* 2013, 97, 231-237.

[21] Rahman, M. M.; Rahman, M. M.; Mollah, M. Y. A.; Susan, M. A. B. H. *ISRN Electrochemistry*, 2013, Article ID 839498, 10 pages.

[22] Arzuman, L.; Karobi, S. N.; Islam, M. J.; Ara, G.; Mollah, M. Y. A.; Rahman, M. M.; Susan, M. A. B. H. *Synth. Reactivity Inorg. Metal-Org. Nano-Metal Chem.* 2014, in press.

[23] Begum, F.; Mollah, M. Y. A.; Rahman, M. M.; Susan, M. A. B. H. *J. Bangladesh Chem. Soc.*, 2011, 24, 173-184.

[24] Haque, M. A.; Rahman, M. M.; Susan, M. A. B. H. *J. Solution Chem.* 2012, 41, 447-457.

[25] Ekwall, P.; Mandell, L.; Solyom, P. *J. Colloid Interf. Sci.* 1970, 35, 266-272.

[26] Piletic, I. R.; Moilanen, D. E.; Spry, D. B.; Levinger, N. E.; Fayer, M. D. *J. Phys. Chem. A*, 2006, 110, 4985-4999.

[27] Akhter, S. S. *4th year project report submitted for the partial fulfillment of the requirement for the degree of Bachelor of Science in Chemistry*, 2011.

[28] Siddiki, N. A, *4th year project report submitted for the partial fulfillment of the requirement for the degree of Bachelor of Science in Chemistry*, 2010.

[29] Hossain, S.; Fatema, U. K.; Mollah, M. Y. A.; Rahman, M. M.; Susan, M. A. B. H. *J. Bang. Chem. Soc.* 2012, 25, 71-79.

[30] Saha, S.; Sultana, S.; Islam, M. M.; Rahman, M. M.; Mollah, M. Y. A.; Susan, M. A. B. H. *Ionics* 2014, 20, 1175-1181.

[31] Sultana, S.; Saha, S.; Islam, M. M.; Rahman, M. M.; Mollah, M. Y. A.; Susan, M. A. B. H. *J. Electrochem. Soc.* 2013, 160, D524-D529.

[32] Satter, S. S.; Hoque, M.; Rahman, M. M.; Rahman, Mollah, M. Y. A.; Susan, M. A. B. H. *RSC Adv.* 2014, 4, 20612-20615.

[33] Petit, C.; Ixon, L.; Pileni, M. P. *J. Phys. Chem.* 1990, 94, 1598-1603.

[34] Eastoe, J.; Warne, M. *Curr. Opin. Colloid. Interf. Sci.* 1996, 1, 800-805.

[35] Robinson, B. H.; Towey, T. F.; Zourab, S.; Visser, A. J. W. G.; Van Hoek, A. *Colloid. Surf.* 1991, 61, 175-188.

[36] Haram, S. K.; Mahadeshwar, A. R.; Dixit, S. G. *J. Phys. Chem.* 1996, 100, 5868-5873.

[37] Chang, S. Y.; Liu, L.; Asher, S. A. *J. Am. Chem. Soc.* 1994, 116, 6739-6744.

[38] Esquena, J.; Tadros, T. F.; Kostareios, K.; Solans, C. *Langmuir* 1997, 13, 6400-6406.

[39] Lopez-Perez, J. A.; Lopez-Quintela, M. A.; Mira, J.; Rivas, J.; Charles, S. W. *J. Phys. Chem. B* 1997, 101, 8045-8047.

[40] Chen, M.; Wu, Y.; Zhou, S.; Wu, L. *J. Phys. Chem. B* 2008, 112, 6536-6541.

[41] Ahmed, P.; Miran, M. S.; Susan, M. A. B. H.; Mollah, M. Y. A. *J. Bang. Chem. Soc.* 2013, 26, 20-29.

[42] Hirai, T.; Shinojiri, S.; Komasawa, I. *J. Chem. Eng. Jpn*. 1994, 27, 590-597.

[43] Kamalov, V. F.; Little, R.; Logunov, S. L.; El-Sayed, M. A. *J. Phys. Chem*. 1996, 100, 6381-6384.

[44] Hota, G.; Idage, S. B.; Khilar, K. C. *Colloid. Surf. A: Physicochem. Eng. Asp*. 2007, 293, 5-12.

[45] Han, M. Y.; Huang, W.; Chew, C. H.; Gan, L. M.; Zhang, X. J.; Ji, W. *J. Phys. Chem. B* 1998, 102, 1884-1887.

[46] Peng, X.; Sclamp, M. C.; Kadavanich, A. V.; Alivisatos, A. P.; Epitaxial, A. P. *J. Am. Chem. Soc*. 1997, 119, 7019-7029.

[47] Kortan, A. R.; Hull, R.; Opilla, R. L.; Bawendi, M. G.; Steigerwald, M. L.; Carroll, P. J.; Bres, L. E. *J. Am. Chem. Soc*. 1990, 112, 1227-1332.

[48] Laurent, S.; Forge, D.; Port, M.; Roch, A. C.; Elst, L. V.; Muller, R. N. *Chem. Rev*. 2008, 108, 2064-2110.

[49] Babes, L.; Denizot, B.; Tanguy, G.; Le Jeune, J. J.; Ballet, P.; *J. Colloid Interf. Sci*. 1999, 212, 474-482.

[50] Gupta, A. K.; Gupta, M. *Biomaterials* 2005, 26, 3995-4021.

[51] Dresco, P. A.; Zaitsev, V. S.; Gambino, R. J.; Chu, B. *Langmuir* 1999, 15, 1945-1951.

[52] Yan, E.; Ding, Y.; Chen, C.; Li, R.; Hu, Y.; Jiang, X. *Chem. Commun*. 2009, 19, 2718-2720.

[53] Jaiswal, J. K.; Mattoussi, H.; Mauro, J. M.; Simon, S. M. *Nat. Biotechnol*. 2003, 21, 47-51.

[54] Michalet, X.; Pinaud, F. F.; Bentolila, L. A.; Tasy, J. M.; Doose, S.; Li, J. J.; Sundaresan, G.; Wu, A. M.; Gambhir, S. S.; Weiss, S. *Science* 2005, 307, 538-544.

[55] De, M.; Ghosh, P. S.; Rotello, V. M. *Adv. Mater*. 2008, 20, 4225-4241.

[56] Goikolea, E.; Insausti, M.; Lezama, L.; Gil de Muro, I.; Garitaonandia, J. S. *J. Non-Cryst. Solids* 2008, 354, 5216-5218.

[57] Pileni, M. P. *J. Phys. Chem*. 1993, 97, 6961-6974.

[58] Hota, G.; Idage, S. B.; Khilar, K. C. *Colloids Surf. A* 2007, 293, 5-12.

[59] Hota, G.; Jain, S.; Khilar, K. C. *Colloids Surf. A* 2004, 232, 119-127.

[60] Carpenter, E. E.; Sangregorio, C.; O'Connor, C. *J. IEEE Trans. Magn*. 1999, 35, 3496-3498.

[61] Santra, S.; Tapec, R.; Theodoropoulou, N.; Dobson, J.; Hebard, A.; Tan, W. *Langmuir* 2001, 17, 2900-2906.

[62] Lambert, K.; Geyter, B. D.; Moreels, I.; Hens, Z. *Chem. Mater*. 2009, 21, 778-780.

[63] Decker, S.; Lagadic, I.; Klabunde, K. J.; Moscovici, J.; Michalowicz, A. *Chem. Mater*. 1998, 10, 674-678.

[64] Carnes, C. L.; Klabunde, K. *J. Chem. Mater*. 2002, 14, 1806-1811.

[65] Bommarius, A. S.; Holzwarth, J. F.; Wang, D. I. C.; Hatton, T. A. *J. Phys. Chem*. 1990, 94, 7232-7239.

[66] Fletcher, P. D. I.; Howe, A. M.; Robinson, B. H. *J. Chem. Soc. Faraday Trans. I* 1987, 83, 985-1006.

[67] Porta, F.; Prati, L.; Rossi, M.; Scari, G. *Colloid. Surf. A: Physicochem. Eng. Asp*. 2002, 211, 43-48.

[68] Fang, X.; Yang, C. *J. Colloid. Interf. Sci*. 1999, 212, 242-251.

[69] Giannakas, A. E.; Vaimakis, T. C.; Ladavos, A. K.; Trikalitis, P. N. *J. Colloid. Interf. Sci*. 2003, 259, 244-253.

[70] Koumoudilis, G. C.; Katsouldilis, A. P.; Ladavos, A. K.; Pomonis, P. J.; Trapalis, A. T.; Vaimakis, T. C. *J. Colloid. Interf. Sci.* 2003, 259, 254-260.

[71] Wu, Z.; Zhang, J.; Benfield, R. E.; Ding, Y.; Grandjean, D.; Zhang, Z.; Ju, X. *J. Colloid. Interf. Sci.* 2002, 106, 4569-4577.

[72] Yu, K. M. K.; Yeung, C. M. Y.; Thompsett, D.; Tsang, S. C. *J. Phys. Chem. B* 2003, 107, 4515-4526.

[73] Rodriguez, J. A.; Hanson, J. C.; Kim, J. Y.; Liu, G. *J. Phys. Chem. B* 2003, 107, 3535-3543.

[74] Hungria, A. B.; Martinez-Arias, A.; Fernandez-Garcia, M.; Iglesias-Juez, A.; Guerrero-Ruiz, A.; Calvino, C. C.; Conesa J. C.; Soria, J. *Chem. Mater.* 2003, 15, 4309-4316.

[75] Pang, Y. X.; Bao, X. *J. Mater. Chem.* 2002, 12, 3699-3704.

[76] Pang, Q.; Shi, J.; Liu, Y.; Xing, D.; Gong, M.; Xu, N. *Mater. Sci. Eng. B* 2003, 103, 57-61.

[77] Xu, S.; Li, Y. *J. Mater. Chem.* 2003, 13, 163-165.

[78] Tartaj, P.; Serna, C. J. *Chem. Mater.* 2002, 14, 4396-4402.

[79] Bose, S.; Saha, S. K. *Chem. Mater.* 2003, 15, 4464-4469.

[80] Fang, J.; Stokes, K. I.; Wiemann, J. A.; Zhou, W. L.; Dai, J.; Chen, F.; Connor, C. J. O. *Mater. Sci. Eng. B* 2001, 83, 254-257.

[81] Liu, Z.; Lee, J. Y.; Han, M.; Chen, W.; Gan, L. M. *J. Mater. Chem.* 2002, 12, 2453-2458.

[82] Althues, M.; Sterlund, L.; Ljungstro, S.; Palmqvist, A. *Langmuir* 2002, 18, 7428-7435.

[83] Bose, S.; Saha, S. K. *Chem. Mater.* 2003, 15, 4464-4469.

[84] Marchand, K. E.; Tarret, M.; Lechaire, J. P.; Normand, L.; Kasztelan, S.; Cseri, T. *Colloid Surf. A: Physicochem. Eng. Asp.* 2003, 214, 239-248.

[85] Bagwe, R. P.; Khillar, K. C. *Langmuir* 1997, 13, 6432-6438.

[86] Petit, C.; Lixon, P.; Pileni, M. P. *J. Phys. Chem.* 1993, 97, 12974-12983.

[87] Ahmad, T.; Chopra, R.; Ramanujachary, K. V.; Lofland, S. E.; Ganguli, A. K. *Solid State Sci.* 2005, 7, 891-895.

[88] Lisiecki, I.; Pileni, M. P. *J. Phys. Chem.* 1995, 99, 5077-5082.

[89] Marium, M.; Rahman, M. M.; Mollah, M. Y. A.; Susan, M. A. B. H. *RSC Adv.* 2015, 5, 19907-19913.

[90] Zhang, X.; Chan, K. Y. *Chem. Mater.* 2003, 15, 451-459.

[91] Zhang, X.; Chan, K. Y. *J. Mater. Chem.* 2002, 12, 1203-1206.

[92] Santra, S.; Tapec, R.; Theodoropoulou, N.; Dobson, J.; Hebard, A.; Tan, W. *Langmuir* 2001, 17, 2900-2906.

[93] Pileni, M. P. *J. Phys. Chem.* 1993, 97, 6961-6973.

[94] Sugih, A. K.; Shukla, D.; Heeres, H. J.; Mehra, A. *Nanotechnology* 2007, 18, 35607(1-9).

[95] Nanni, A.; Dei, L. *Langmuir* 2003, 19, 933-938.

[96] Berkovich, Y.; Aserin, A.; Walchtel, E.; Garti, N. *J. Colloid. Interf. Sci.* 2002, 245, 58-67.

[97] Lemyre, J. L.; Ritcey, A. M. *Chem. Mater.* 2005, 17, 3040-3043.

[98] Kitchens, C. L.; Mcleod, M. C.; Roberts, C. B. *J. Phys. Chem. B* 2003, 107, 11331-11338.

[99] Krauel, K.; Davies, N. M.; Hook, S.; Rades, S. S. *J. Control Release* 2005, 106, 76-87.

[100] Kimijama, K.; Sugimoto, T. J. *Colloid. Interf. Sci.* 2005, 286, 520-525.

[101] Husein, M. M.; Rodli, E.; Vera, J. H. *J. Colloid Interf. Sci.* 2004, 273, 426-434.

[102] Husein, M. M.; Rodli, E.; Vera, J. H. *J. Nanopart. Res.* 2007, 9, 787-796.

[103] Kumar, S.; Zou, S. *Langmuir* 2007, 23, 7365-7371.

[104] Li, J. F.; Yang, Z. L.; Ren, B.; Liu, G. K.; Fang, P. P.; Jiang, Y. X.; Wu, D. Y.; Tian, Z. Q. *Langmuir* 2006, 22, 10372-10379.

[105] Afrin, T; Mafy, N. N.; Rahman, M. M.; Mollah, M. Y. A.; Susan, M. A. B. H. *RSC Adv.* 2014, 4, 50906-50913.

[106] Pileni, M. P.; Motte, L.; Petit, C. *Chem. Mater.* 1992, 4, 338-345.

[107] Husein, M. M.; Rodli, E.; Vera, J. H. *J. Colloid Interf. Sci.* 2005, 288, 457-467.

[108] Husein, M. M.; Rodli, E.; Vera, J. H. *J. Nanopart. Res.* 2006, 9, 787-796.

[109] Li, Q.; Li, H.; Pol, V. G.; Bruckental, I.; Koltypin, Y.; Calderon- Moreno, J.; Nowik, I.; Gedanken, A. *New J. Chem.*, 2003, 27, 1194-1199.

[110] Husein, M. M.; Rodil, E.; Vera, J. H. *Langmuir* 2006, 22, 2264-2272.

[111] Hu, Z.; Ramirez, D. J. E.; Cervera, B. E. H.; Oskam, G.; Searson, P. C. *J. Phys. Chem. B* 2005, 109, 11209-11214.

[112] Furedi-Milhofer, H.; Babic-Ivancic, V.; Brecevic, L.; Filipovic-Vincekovic, N.; Kralj, D.; Komunjer, L.; Markovic, M.; Skrtic, D. *Colloids Surf.* 1990, 48, 219-230.

[113] Pal, T.; Sau, T. K.; Jana, N. R. *Langmuir* 1997, 13, 1481-1485.

[114] Pawar, M. J.; Chaure, S. S. *Chalcogenide Lett.* 2009, 6, 689-693.

[115] Ghosh Chaudhuri, R.; Paria, S. *J. Colloid Interf. Sci.* 2010, 343, 439-446.

[116] Zanella, R.; Sandoval, A.; Santiago P.; Basiuk, V. A.; Saniger, J. M. *J. Phys. Chem. B* 2006, 110, 8559-8565.

[117] Liu, Y. F.; Yu, J. S. *J. Colloid Interf. Sci.* 2009, 333, 690-698.

[118] Park, J. J.; Lacerda, S. H. D. P.; Stanley, S. K.; Vogel, B. M.; Kim, S.; Douglas, J. F.; Rahgavan, D.; Karim, A. *Langmuir* 2009, 25, 443-450.

[119] Ni, K. F.; Sheibat-othman, N.; Shan, G. R.; Fevotte, G.; Bourgeat-lami, E. E. *Macromolecules* 2005, 38, 9100-9109.

[120] Madani, A.; Nessark, B.; Brayner, R.; Elaissari, H.; Jouini, M.; Mangeney, C.; Chemini, M. M. *Polymer* 2010, 51, 2825-2835.

[121] Wang, L.; Luo, J.; Fan, Q.; Suzuki, M.; Suzuki, I. S.; Engelhard, M. H.; Lin, Y.; Kim, N.; Wang, J. Q.; Zhong, C. J. *J. Phys. Chem. B* 2005, 109, 21593-21601.

[122] He, Q.; Zhang, Z.; Xiong, J.; Xiong, Y.; Xiao, H. *Opt. Mater.* 2008, 31, 380-384.

[123] Chen, Y. C.; Zhous, S. X.; Yang, H. H.; Wu, L. M. *J. Sol-Gel Sci. Technol.* 2006, 37, 39-47.

[124] Tojo, C.; de Dios, M.; Lopez-Quintela, M. A. *J. Phys. Chem. C* 2009, 113, 19145-19154.

[125] Shukla, D.; Mehra, A. *Langmuir* 2006, 22, 9500-9506.

[126] Fletcher, P. D. I.; Howe, A. M.; Robinson, B. H. *J. Chem. Soc. Faraday Trans.* 1987, 83, 985-1006.

[127] Fletcher, P. D. I.; Clarke, S.; Ye, X. *Langmuir* 1990, 6, 1301-1309.

[128] Friberg, S.; Mandell, L.; Larson, M. *J. Colloid Interf. Sci.* 1969, 29, 155-156.

[129] Mie, G. *Ann. Phys.* 1908, 25, 377-445.

[130] Link, S.; El-Sayed, M. A. *Annu. Rev. Phys. Chem.* 2003, 54, 331-366.

[131] Mooradian, A. *Phys. Rev. Lett.* 1969, 22, 185-187.

[132] Zheng, J.; Zhang, C.; Dickson, R. M. *Phys. Rev. Lett.* 2004, 93, 077402 (1)-077402(4).

[133] Link, S.; El-Sayed, M. A. *J. Phys. Chem. B* 1999, 103, 8410-8426.

[134] Link S.; El-Sayed, M. A. *Int. Rev. Phys. Chem.* 2000, 19, 409-453.

[135] Eustis, S.; El-Sayed, M. A. *J. Phys. Chem. B* 2005, 109, 16350-16356.

[136] Mohamed, M. B.; Volkov, V.; Link, S.; El-Sayed, M. A. *Chem. Phys. Lett.* 2000, 317, 517-523.

[137] Ghosh, S. K.; Pal, T. *Chem. Rev.* 2007, 107, 4797-4862.

[138] Loo, C.; Lin, A.; Hirsch, L.; Lee, M.; Barton, J.; Halas, N.; West, J.; Drezek, R. *Technol. Cancer Res. Treat.* 2004, 3, 33-40.

[139] Bao, F.; Li, J. F.; Ren, B.; Yao, J. L.; Gu, R. A.; Tian, Z. Q. *J. Phys. Chem. C* 2008, 112, 345-350.

[140] Kumar, S.; Zou, S. *Langmuir* 2007, 23, 7365-7371.

[141] Li, J. F.; Yang, Z. L.; Ren, B.; Liu, G. K.; Fang, P. P.; Jiang, Y. X.; Wu, D. Y.; Tian, Z. Q. *Langmuir* 2006, 22, 10372-10379.

[142] Zhang, P.; Chen, Y. X.; Cai, J.; Liang, S. Z.; Li, J. F.; Wang, A.; Ren, B.; Tian, Z. Q. *J. Phys. Chem. C* 2009, 113, 17518-17526.

[143] Hu, J. W.; Li, J. F.; Ren, B.; Wu, D. Y.; Sun, S. G.; Tian, Z. Q. *J. Phys. Chem. C* 2007, 111, 1105-1112.

[144] Yang, Z.; Li, Y.; Li, Z.; Wu, D.; Kang, J.; Xu, H.; Sun, M. *J. Chem. Phys.* 2009, 130, 234705(1-7).

[145] Zhang, X. B.; Yan, J. M.; Han, S.; Shioyama, H.; Xu, Q. *J. Am. Chem. Soc.* 2009, 131, 2778-2779.

[146] Feng, L.; Wu, X.; Ren, L.; Xiang, Y.; He, W.; Zhang, K.; Zhou, W.; Xie, S. *Chem-Eur. J.* 2008, 14, 9764-9771.

[147] Zhang, N.; Gao, Y.; Zhang, H.; Feng, X.; Cai, H.; Liu, Y. *Colloid Surf. B* 2010, 81, 537-543.

[148] She, H.; Chen, Y.; Chen, X.; Kun, Z.; Peng, D. L. *J. Mater. Chem.* 2012, 22, 2757-2765.

[149] Ban, Z.; Barnakov, Y.; Li, F.; Golub, V.; O'Connor, C. *J. Mater. Chem.* 2005, 15, 4660-4666.

[150] Chen, D.; Li, J.; Shi, C.; Du, X.; Zhao, N.; Sheng, J.; Liu, S. *Chem. Mater.* 2007, 19, 3399-3405.

[151] Tsuji, M.; Yamaguchi, D.; Matsunaga, M.; Ikedo, K. *Cryst. Growth Des.* 2011, 11, 1995-2005.

[152] Mulvaney, P. *Langmuir* 1996, 12, 788-800.

[153] Xie, X. L.; Li, R. K. Y.; Liu, Q. X.; Mai, Y. W. *Polymer* 2004, 45, 6665-6673.

[154] Chen, D. H.; Chen, C. J. *J. Mater. Chem.* 2002, 12, 1557-1562.

[155] Mulvaney, P.; Giersig, M.; Henglein, A. *J. Phys. Chem.* 1993, 97, 7061-7064.

[156] Nasakumar, N.; Rayappan, J. B. B. *J. Appl. Sci.,* 2012, 12, 1758-1761.

[157] Templeton, A. C.; Pietron, J. J.; Murray, R. W.; Mulvaney, P. *J. Phys. Chem. B,* 2000, 104, 564-570.

[158] Guggenbichler, J. P.; Boswald, M.; Lugauer, S.; Krall, T. *Infection.* 1999, 27, S16-23.

[159] Song, K. C.; Lee, S. M.; Park, T. S.; Lee, B. S. *Kor. J. Chem. Eng.* 2009, 26, 153-155.

[160] Kang, H. Y.; Jung, M. J.; Jeong, Y. K. *Kor. J. Biotechnol. Bioeng.* 2000, 15, 521-524.

[161] Burrell, R. E. A. *Ostomy/Wound Manage.* 2003, 49, 19-24.

[162] Feng, Q. L.; Wu, J.; Chen, G. Q.; Cui, F. Z.; Kim, T. N.; Kim, J. O. *J. Biomed. Mater. Res.* 2000, 52, 662-668.

[163] Hollinger, M. A. *Crit. Rev. Toxicol.* 1996, 26, 255-260.

[164] Ratte, H. T. *Environ. Toxicol. Chem.* 1999, 18, 89-108.

[165] Gong, P.; Li, H.; He, X.; Wang, K.; Hu, J.; Tan, W.; Zhang, S.; Yang, X. *Nanotechnology* 2007, 18, 285604. 7 pages.

[166] Navarro, E.; Piccapietra, F.; Wagner, B.; Marconi, F.; Kaegi, R.; Odzak, N.; Sigg, L.; Behra, R. *Environ. Sci. Technol.* 2008, 42, 8959-8564.

[167] Loher, S.; Schneider, O. D.; Maienfisch, T.; Bokorny, S.; Stark, W. J. *Small* 2008, 4, 824-832.

[168] Ghosh, S.; Goudar, V. S.; Padmalekha, K. G.; Bhat, S. V.; Indi, S. S.; Vasan, H. N. *RSC Adv.* 2012, 2, 930-940.

[169] Lu, W.; Liu, G.; Gao, S.; Xing, S. Wang, J. *Nanotechnology*, 2008, 19, 445711-445732.

[170] Koga, H.; Kitaoka, T.; Wariishi, H. *J. Mater. Chem.* 2009, 19, 2135-2140.

[171] Lee, L-H.; Deng, J. C.; Deng, H. R.; Liu, Z. L.; Li, X. L. *Chem. Eng. J.* 2010, 160, 378-382.

[172] Kumar, H.; Renu, Rani, R.; Salar, R. *Advances in Control, Chemical Engineering, Civil Engineering and Mechanical Engineering*, 2010, 88-94.

[173] Yu, L.; Zhang, Y.; Zhang, B.; Liu, J. *Sci. Rep.* 4, 2014, 4551.

[174] Daima, H. K.; Selvakannan, P. R.; Kandjani, A. E.; Shukla, R.; Bhargavaa, S. K.; Bansal, V. *Nanoscale*, 2014, 6, 758-765.

[175] Davoudi, Z. M.; Kandjani, A. E.; Bhatt, A. I.; Kyratzis, I. L.; O'Mullane, A. P.; Bansal, V. *Adv. Funct. Mater.* 2014, 24, 1047-1053.

[176] Afrin, T.; Karabi, S. N.; Rahman, M. M.; Mollah, M. Y. A. M.; Susan, M. A. B. H. *J. Solution Chem.* 2013, 42, 1488-1499.

[177] Keya, J. J.; Islam, M. M.; Rahman, M. M.; Rahman, M. Y. A. M.; Susan, M. A. B. H. *J. Electroanal. Chem.* 2014, 712, 161-166.

[178] Hoque, M.; Shazia, S. S.; Rahman, M. M.; Mollah, M. Y. A.; Susan, M. A. B. H. unpublished results.

In: Innovations in Nanomaterials   ISBN: 978-1-63483-548-0
Editors: Al-N. Chowdhury, J. Shapter, A. B. Imran   © 2015 Nova Science Publishers, Inc.

Chapter 9

PLASMA MODIFICATION OF SP² CARBON NANOMATERIALS

J. S. Quinton[1,], A. J. Barlow[2] and L. Velleman[3]*

[1]Flinders Centre for NanoScale Science and Nanotechnology, School of
Chemical and Physical Sciences, Flinders University, Adelaide, Australia
[2]National EPSRC XPS Users' Service, School of Mechanical and Systems
Engineering, Newcastle University, Newcastle upon Tyne, UK
[3]Department of Chemistry, Imperial College London,
South Kensington Campus, London, UK

ABSTRACT

With carbon being one material at the very heart of the 'nano' revolution, the discovery of nanotubes, Buckminsterfullerenes and graphene have spear-headed a significant number of advances in nanomaterials in recent times. With potential applications of these materials are used as supports for catalysis, sensing architectures, charge transport in photovoltaic devices and integrated circuits, through to providing mechanical strength in nanocomposites, it is no surprise that they have become ubiquitous. The ability to control their surface chemistry, however, is critical to the fabrication and performance of smart devices that utilise them in their architecture. Modification with plasma offers the ability to alter and control the surface chemistry with high efficiency in extremely reactive environments. We will highlight key advances in this area of research.

* Corresponding author: J. S. Quinton. Flinders Centre for NanoScale Science and Nanotechnology, School of Chemical and Physical Sciences, Flinders University, GPO Box 2100 Adelaide, Australia, SA 5001. E-mail: Jamie.Quinton@flinders.edu.au.

INTRODUCTION

Carbon Nanomaterials

A 1991 article by Sumio Iijima, and subsequent publications in 1993 by Iijima and Donald Bethune, spurred a rapid growth of research into single and multi-walled carbon nanotubes (CNTs) [1-3]. The reports of these nanoscale structures, generated using arc-discharge plasma, gave nanoscience and nanotechnology an enormous boost leading into the 21st century and still continuing today. Carbon nanomaterials have since become fundamental architectural building blocks of choice for novel device fabrication. Iijima was recognised for his contribution to this burgeoning field of science in 2008 when he was awarded the inaugural Kavli Prize alongside Louis Brus. When one considers the properties that are exhibited by carbon nanotubes in particular, it becomes clear why they have garnered so much attention from research groups around the world.

Single-walled carbon nanotubes;

- can adopt different electrical characteristics depending on the physical structure of the sidewall, i.e., their electronic structure can vary from semiconducting to metallic depending on the arrangement of carbon atoms in the nanotube [4, 5];
- display ballistic electron transport, whereby electrons can travel the length of the nanotube without scattering and heat losses, resulting in very high current carrying capacity [6, 7];
- can have incredible mechanical resilience, with Young's moduli much greater than steel, and can be repeatedly buckled and deformed through large angles without failure [8, 9].

Further sustaining the nanomaterial 'revolution,' the laboratory production of graphene, an individual layer of the hexagonal close-packed sp^2 carbon atoms that form the multi-layered structure in graphite, has established itself as the nanomaterial of choice at the moment and has been called a 'miracle material' by Konstantin Novoselov (2010 Physics Nobel Laureate for his work on graphene) [10]. Graphene in its own right is an enabling material for a number of current and potential future applications [11], but both it and highly oriented pyrolytic graphite (HOPG) serve as ideal, model surfaces for studies of atomic and molecular interactions with nanotubes, Buckminsterfullerenes (aka Buckyballs or Fullerenes) and nanocones. Some of the astounding properties of carbon nanotubes are also true for graphene, but in addition, graphene:

- also has incredible mechanical properties, with an ultimate tensile strength of 130 GPa (c.f. 400MPa for A36 structural steel) and a Young's modulus of 1TPa (c.f. 200GPa for steel);
- has an electron mobility at room temperature which is also very close to predicted limits, and can sustain very high current densities;
- displays very high thermal conductivity;
- is impermeable to gases through the material [11], despite being one atomic layer thick.

Another important property of both nanotubes and graphene is that they can be chemically functionalised. While this may seem obvious, it provides the means through which these materials are incorporated into devices or surface architectures.

Applications

In recent times there has been a push to reduce the energy and material costs involved in manufacturing [12]. To successfully do so will require the replacement of pervasive materials upon which devices are made such as silicon, which in itself is costly to purify and process, and other expensive inorganic rare-earth materials such as indium; with disruptive carbon-based material technologies where possible.

This is not to be confused with so-called 'low-carbon technologies,' which alleviate energetically costly manufacturing technologies by reducing the total CO_2 production involved. Successful materials replacement however, will only be achievable once the ability to manufacture macroscale structures that exploit the properties of each of the allotropic forms of carbon, particularly the 1D nanotube and 2D graphene varieties, has been mastered.

Carbon nanomaterials have found ubiquitous application in next generation transistors and supercapacitors; field emission sources; energy storage devices such as fuel cells and hydrogen storage materials; high-performance catalysts; solar cells; lithium-ion batteries; biomedical devices and implants; cancer treatments; nano-enabled drug delivery; antimicrobial agents and thin films; tissue fabrication; neurogenesis; and even in food packaging and cosmetics [13].

Plasma Processing

Along with the solid, liquid and gas phases, there exists the fourth state of matter known as plasma. Neglecting exotic forms of matter, most of the observable universe in stellar interiors and nebulae is in the plasma state [14, 15].

On Earth however, there are a few examples of plasma that one may encounter on a daily basis, both from natural and anthropological sources [14]. In addition to fire, the intense glow produced during a lightning strike is possibly the most impactful, with the sudden extremely high voltage producing such intense electric fields as to ionise the surrounding air. While far more benign, the ionosphere surrounding our planet is plasma and in combination with the magnetosphere, shields the surface of plasma solar wind to create aurorae, which are also plasma [14]. Synthetic examples of plasma include the discharge within fluorescent tubes or neon signs, and the generation of light and colour within plasma displays [14, 15]. Needless to say, one is inevitably exposed to plasma in some form on a day-to-day basis. Over the past few decades the plasma state has also established itself as one of the most important industrial tools used for large-scale materials processing and cutting.

It could be argued that the semiconductor industry would not have reached its current capabilities without the use of plasma processing methods. Much silicon-based technology relies on this highly reactive and energetic state to develop intricate integrated circuit components. For example, trenches narrow enough to fit onto a single human hair hundreds of times across can be plasma-etched into silicon to produce circuitry [16].

The process can be used to deposit various metals in precisely controlled quantities, as well as to grow oxides or films of specific thicknesses. Ion implantation can also be performed, which is an extremely important process for the production of doped p- or n- type silicon components [16]. Each of these processes can be used in tandem to produce an integrated circuit of incredibly small dimensions with extreme intricacy. While this industry is a very large and important one today, its use of plasma processing methods is not unique. The medical technologies and food industries also use plasma to sanitise surfaces.

In medicine, new operating tools are being developed to sanitise local areas of tissue such as open wounds, by using a plasma-based torch generated from an inert gas [17, 18]. The food industry uses plasma to sanitise surfaces and materials used in the packaging of consumables [19]. Due to the very high energies required for the fusion of deuterium atoms into helium, plasma has of course been at the heart of controlled nuclear fusion research. This research has existed since the 1950's and continues today toward ever increasing scales [14]. In possibly the grandest challenge, plasma is even being investigated as a method of propulsion in space, with the development of ion thrusters [20-22]. Simply put, plasma is the ionised form of a gas. To quote Chen; "A plasma is a quasi-neutral gas of charged and neutral particles which exhibits collective behaviour" [14].

There are several ways to produce plasma;

- The ionisation of atoms can be enacted by the application of strong electric or electromagnetic fields [16, 23]. In these systems the atoms are ionised through either the dissociation of an electron from the atom, resulting in two charged species, or the collision of an atom with an electron that has been accelerated by fields. In either case, charged species result and act to continue the ionisation of the gas and provided that energy is continuously delivered into the system, the gas will remain in the plasma state. The degree of ionisation can be controlled by the magnitude of the applied fields (among other factors), i.e., more energy delivered to the charged species results in a greater number of collisions and thus more ionisation. Typically, lab-based plasma have a degree of ionisation below 1%, and are commonly referred to as 'cold' plasma [16].

- To initiate ionisation, some plasma systems utilise a static high voltage (~kV) electric field setup between two parallel plates immersed within the gas. The sample stage often takes the place of one electrode. These systems can be generally referred to as dc (direct current) plasma sources. Examples include sputter ion coaters and glow discharge tubes. These systems suffer from very high directionality since the electric field is static. Charge flow becomes unidirectional with electrons and negative ions flowing towards the cathode and positive ions towards the anode. This is desirable in some situations, such as an ion sputter coater where the intention is to sputter a given target material to subsequently deposit a coating of that material onto a substrate of choice. However sputtering of a material may also be an inconvenience, for instance when chemical functionalisation of the surface is the desired outcome.

- To alleviate this issue to some extent, plasma can also be generated using ac (alternating current) electric fields. A myriad of frequencies can be chosen ranging from kHz to GHz, but a commonly used frequency is 13.56 MHz, corresponding to a special band that is set aside for scientific purposes. Plasma sources generated in this manner are commonly referred to as radio frequency (rf) plasma sources. Ionisation in these systems occurs in much the same way as their dc counterpart, however the electric field from the source is not static and thus the system ideally displays little to no directionality. In practice however, this is not strictly true as through other means the source can develop what is known as a 'self-bias' voltage which can also lead to sputtering of material from surfaces.

While plasma methods have been applied to materials synthesis, engineering and processing for a number of decades, the demand for these processes to be amenable at the nanoscale has only become critical since the growth of modern nanoscience and nanotechnology. The ignition of a gaseous species into the plasma state produces a reactive environment that is capable of modifying surfaces and structures [16, 23].

Plasma affords the use of low reactant volumes compared to many wet-chemical techniques, with the additional benefit of a clean, low-pressure regime. The relative ease of scalability to industrial sizes means that such processes could enable the large-scale production of future nanoscale devices. Increased acceptance and utilisation of plasma-based fabrication and processing techniques during the early 21st century has ultimately led to a sudden boost in research publications in the area. This is due, in part, to the desire to have well-controlled and up-scalable processing techniques for devices that might be fabricated from novel materials. Furthermore plasma techniques have evolved from somewhat of a 'black-box' process, where a particular protocol is arrived upon through trial-and-error, to one where the process is designed from first-principles via an in-depth understanding of the underlying plasma physics and chemistry.

Challenges and the Need for Further Research

In all of the applications for which sp^2 carbon nanostructures are used, each requires the careful design and fabrication of specific molecular architectures in order to function. Furthermore, the significant manufacturing challenge will be to produce these architectures reproducibly and in large quantities, thus requiring methods such as plasma surface modification in their fabrication. To achieve this aim there is considerable importance in understanding the nature of the plasma, the atomic and molecular level processes that take place at the surface, and the fundamental science that underpins the surface modification process so that they can be optimised to suit each particular application. For this approach to be truly successful, however, atomic-scale control requires the ability to preferentially alter, remove or attach species to pre-existing surface-mounted nanostructures, usually in multiple steps and treatments, and this is where much of the ongoing research in plasma surface modification is focussed.

PLASMA SURFACE MODIFICATION OF CARBON NANOMATERIALS

One of the greatest benefits of using plasma to perform chemical functionalisation of surfaces and carbon nanomaterials is that it can be achieved at room temperature. This is a benefit of the method through which plasma is generated in typical lab-based dc, or capacitively/inductively coupled plasma reactors. Consequently, these plasma are usually non-equilibrium, meaning that the electron temperature (typically quantified in units of electron-Volts) is very much greater than the ion or neutral temperatures. Under conditions such as these, minimal energy is transferred to any surface exposed to the plasma, and the target material remains near the ambient temperature of the laboratory. The implications of this capability are that in fabricating nanostructured surfaces or devices that require multiple fabrication steps, plasma processing is ideal for situations where surface structures might not typically survive chemical functionalisation at elevated temperatures without unwanted and uncontrollable degradation.

Plasma processes can be operated in a wide range of pressure regimes, from atmospheric pressure to high vacuum [24]. Processing gas volumes can therefore be greatly reduced while still maintaining high levels of reactivity at the surface. Also due to the high reactivity of plasma environments, processing times can be reduced, lowering overall energy expenditure while also minimising reactant usage. Plasma processing technologies are an inherently scalable, since relatively simple and low-power technology can be used to generate high reactivity environments and maintain high throughput for the target material. This is well-demonstrated by the food packaging industry, where the sterilisation of packaging surfaces needs to be performed continuously over a large area of material and in a matter of seconds [19].

MODEL STUDIES ON HIGHLY ORIENTED PYROLYTIC GRAPHITE (HOPG)

The nature of how hydrogen and methane plasma interacts with HOPG surfaces has been studied with Time of Flight Secondary Ion Mass Spectrometry (ToF-SIMS), Scanning Tunnelling Microscopy (STM), X-ray Photoelectron Spectroscopy (XPS), Elastic Recoil Detection Analysis (ERDA) and Rutherford Backscattering Spectroscopy (RBS) [25-27].

Typical ToF-SIMS spectra, collected under static total ion fluence conditions to prevent ion beam damage to the surface, for untreated, hydrogen and methane plasma treated HOPG are shown in Figure 1. Upon inspection the spectra possess some subtle differences, but apart from these they are quite similar, exhibiting the expected peak families of hydrocarbon C_xH_y species often seen in mass spectra of carbonaceous materials.

To mine more deeply into the available data, multivariate principal component analysis (PCA) was adopted and when combined with the ToF-SIMS technique, can powerfully discriminate properties of the sample depending upon the type of plasma treatment used. As an example, the technique was applied to the case of methane plasma-modified HOPG, to examine subtle variations in the obtained ToF-SIMS spectra for a range of applied plasma rf coupling power, during plasma treatment. This is shown in Figure 2. In the figure, principal component 1 (PC1) is dominated by the hydrogen peak at 1 m/z, but PC2 is more (albeit negatively) influenced by the presence (absence) of the peak at 27 m/z, corresponding to the $C_2H_3^+$ ion species that is liberated from the surface film.

The scores plot in (a) indicates that the relative presence of this particular peakdecreases with increasing applied power.

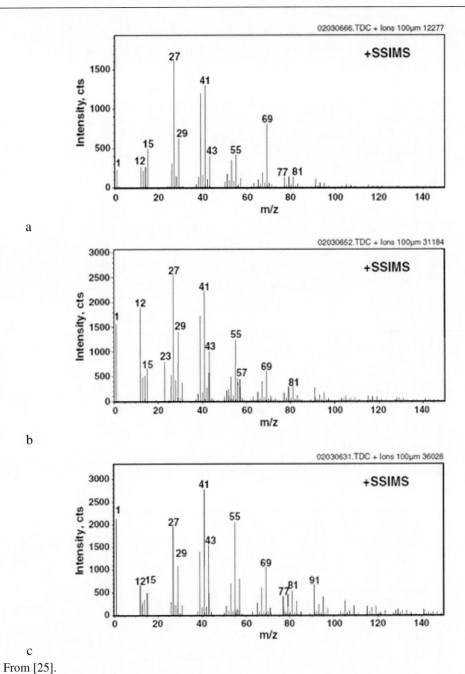

From [25].

Figure 1. Positive ion Static SIMS spectra of (a) untreated; (b) hydrogen plasma treated; and (c) methane plasma treated HOPG.

Furthermore, the scores plot illustrate how against these two components, the samples can be discriminated from one another and thus, one has some control over the chemical and molecular structure of the modified surface through careful selection of the plasma coupling power. STM (and ERDA) studies of the same plasma treatments reveal that the nature of the interaction between HOPG and hydrogen- and methane- plasma are quite different to one another.

Hydrogen plasma etches the HOPG surface, abstracting sp^2 bonds and hydrogenating the surface at the resultant dangling bonds of the newly transformed sp^3 carbon atoms, and does so incredibly rapidly and randomly.

On the other hand, methane plasma treatment results in the island growth of sp^3 diamond-like carbon films upon the HOPG surface [25]. These are illustrated in Figure 3.

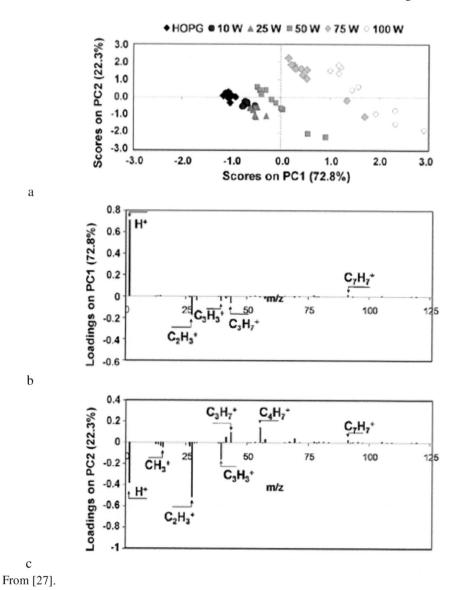

a

b

c

From [27].

Figure 2. Principal Component Analysis of methane plasma treated HOPG for a range of coupled plasma power values. (a) Scores plot against PC1 and PC2; (b) loadings of spectral peak intensities on PC1; (c) loadings on PC2.

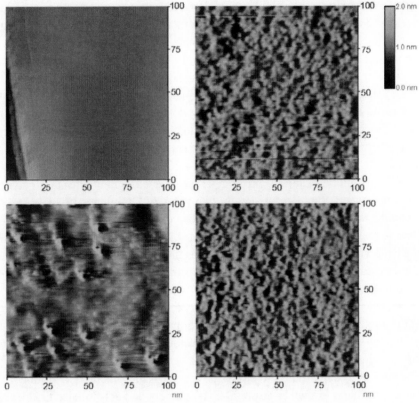

From [25].

Figure 3. 100x100 nm^2 STM images of plasma treated HOPG. Top left: clean HOPG; Bottom left: HOPG exposed to CH$_4$ plasma for 60 min; Top right: HOPG exposed to H$_2$ plasma for 1 min; Bottom right: HOPG exposed to H$_2$ plasma for 60 min.

CARBON NANOTUBES

Reactivity of Carbon Nanotubes

Graphite is a particularly stable form of carbon. It follows that pristine carbon nanotubes are also quite resistant to chemical attack [28]. That said, they are not inert structures and some level of reactivity exists due to the nanotube's inherent curvature [29]. Multiple reviews on the covalent surface chemistry of nanotubes have been published with extensive literature surveys [28-34]. Given that the nanotube can be thought to be formed by the rolling of a graphene sheet, it is somewhat intuitive to realise that it should have surface properties that are similar to graphene and HOPG, except that the bonds between the sp^2 hybridised carbon atoms must undergo some level of strain due to the curvature of the tube. In effect, this forces the carbon atoms away from the ideal trigonal planar arrangement that is found in graphite, and results in curvature-induced 'pyramidalisation' of the carbon atoms, an effect that is illustrated in Figure 4. By curving the graphene lattice, strain on the σ bonds increases the angle relative to the π orbital from the ideal value of 90° as it is in graphite, to values as great as 110° [34].

It follows that nanotubes of smaller diameter will experience greater pyramidalisation and thus greater bond strain. Ultimately, reactions at these sites become energetically favourable since the breaking of bonds within the nanotube structure allows for some of this strain to be relieved. Moreover, these reactions would typically proceed through conversion to sp^3 hybridisation, or by introducing vacancies and decreasing the spatial constraint of the carbon atoms [34]. The bonds within the caps typically experience even greater strain in single-walled nanotubes with closed ends. Thus the ends become preferential sites for chemical attack, and consequently chemical functionalisation is commonly expected to be more favourable at the nanotube ends than along the sidewalls [34].

The curvature of the nanotube also gives rise to another aspect to their reactivity. Irrespective of the nanotube's chirality, from a strain-induced pyramidalisation perspective it is possible to consider that there are generally two types of C-C bonds in a CNT; those that are aligned parallel with the nanotube axis, and those that are oriented at some angle to the axis. In Figure 4 these two cases are highlighted in blue and green respectively, where the arrows indicate the viewpoint for the extracted atoms on the left and right respectively. Recall that in the sp^2 hybridised carbon atoms making up the CNT structure, π orbitals protrude normal to nanotube axis.

In the case where the C-C bond is parallel to the tube axis (blue) the π orbitals are aligned, i.e., the angle between them is zero as shown in Figure 4. However, for the C-C bond that is oriented at an angle to the tube axis, the curvature induces a twist along the bond axis and as a result the angle between the π orbitals is *not* zero as shown in Figure 4.

This misalignment and strain increases as the curvature increases, or the nanotube diameter decreases. This is known as π orbital misalignment and acts to raise the local surface energy and reduce the stability of the nanotube structure. Any reaction that may reduce this misalignment becomes energetically favourable, giving rise to further chemical reactivity [34]. The discussion to this point has only considered an ideal lattice structure. Defects that may be present in the sidewall structure of carbon nanotubes are also expected to exhibit greater reactivity compared with pristine sites [31].

This defect-induced reactivity can be caused by vacancies, dopants and imperfect lattice structures such as pentagonal and heptagonal carbon rings [29]. A unique defect caused by these imperfect structures is the so-called Stone-Wales defect, where 2 pentagons and 2 heptagons are fused together within the nanotube sidewall [35, 36].

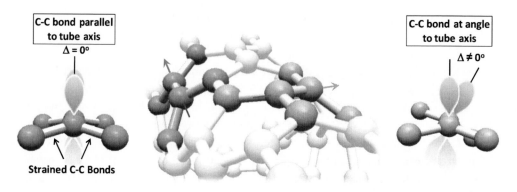

Figure 4. Pictorial representation of curvature induced pyramidalisation and π orbital misalignment that leads to CNT reactivity.

Lu et al., showed through density functional theoretical calculations that while the centre of this defect was no more reactive than a pristine site, the periphery of the defect, i.e., where it meets with the perfect nanotube structure, results in a significant increase in reactivity [36].

Plasma Processing of Carbon Nanotubes (AB)

Due to how widely applicable the plasma state is to the processing of carbon nanotubes, there are a great number of publications and reviews surrounding the topic. There are some areas that are more heavily focused on than others however, such as the growth of carbon nanostructures and the sidewall functionalisation of these structures. It is these that we will focus on here.

Plasma-Enhanced Growth of Nanotubes

Thermal chemical vapour deposition (CVD) processes have been established as a successful method towards large-scale production of nanotubes. An issue with the typical thermal CVD process however is that processing temperatures in excess of 800°C are generally required to result in single-walled nanotube (SWCNT) growth [37]. Below this temperature, multi-walled growth is predominant. When SWCNTs are desired, such high temperatures limit the applicability of the CVD process to only those substrates that are particularly resilient. The high temperature is required largely to initiate the decomposition of the feedstock into components suitable for growth.

By using the plasma state, much of this energy can be supplied by the collisions of electrons at high energy, and thus decomposition can occur without the need for the gas to be heated significantly [38]. This is the principle of plasma-enhanced CVD (PECVD), which in many ways is the same as thermal CVD, except that it is performed at reduced pressure where the feedstock gas can be ignited into plasma to facilitate decomposition. Over recent years, multiple extensive reviews of the current state of knowledge of plasma nano-science have been performed with particular focus on PECVD growth of nanostructures [37-41].

Using PECVD, SWCNT growth has been achieved with substrate temperatures as low as 450°C, and MWCNT growth at room temperature [39]. Although very desirable, low temperature synthesis is not the only advantage to PECVD over thermal CVD.

Due to the formation of a plasma sheath above a growth substrate (a potential and electric field results from electrons having larger transit speeds and reaching earthed chamber surfaces faster than ions), nanotubes grown via PECVD tend to have a natural vertical alignment. Upright forests of CNTs have been produced via CVD, as have arrays of vertical columns, however detailed inspection finds that the nanotubes within these structures are not necessarily well aligned, and appear more like the traditional nanotube 'spaghetti' that has been pushed upward from the surface. PECVD on the other hand has been used to generate well-aligned freestanding vertical CNT arrays of very small dimensions [38].

This level of precise control over morphology is paramount for device development, especially field emission [42], photovoltaics, sensing and filtration, where vertical alignment is highly desirable.

The fact that PECVD is performed under vacuum also gives the process intrinsic cleanliness, since atmospheric contaminants become negligible. Furthermore, in PECVD the sample can be easily cooled under either a pure inert atmosphere or vacuum.

Following growth via thermal CVD, the atmosphere must be precisely controlled as the surface is cooled to ensure that oxidation from atmospheric species does not occur.

Great control over resulting nanotube structures grown by PECVD has been demonstrated [43]. Using a radio-frequency inductively coupled plasma system [24], Bissett et al., demonstrated growth of both multi-walled and single-walled nanotubes from a feedstock mixture of argon and methane. Growth was achieved on 5 nm iron catalyst films patterned onto silicon substrates via shadow masking from a copper grid, as seen in the electron micrographs in Figure 5(a). In the reactor, the substrates were heated to temperatures between 450 and 650°C during the plasma treatment. Using a very low power (10 W), it was found that a transition from predominantly single-walled to multi-walled growth occurred, depending on the amount of exposure to the growth plasma. At short exposure times (10 minutes), nanotube growth of single-walled structures with diameters around 1.5 nm, as evidenced by the radial breathing mode (RBM) position in Raman spectroscopy, and very low D-band to G-band ratio, a measure of the disorder in the graphitic structure, was demonstrated. This is illustrated in Figure 5(b) where Raman spectra were collected on and off the patterned catalyst areas.

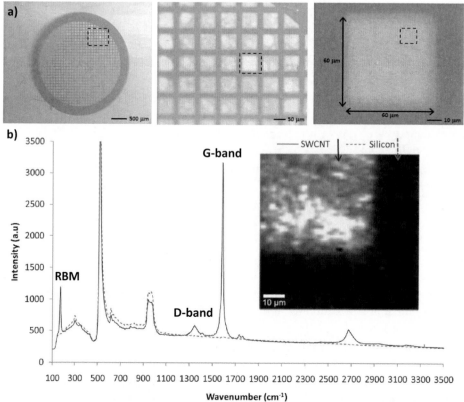

From [43].

Figure 5. Patterned PECVD growth of single-walled nanotubes on silicon. Scanning electron micrographs in (a) show the nanotube growth on iron catalyst patterned via shadow masking with a copper grid and (b) Raman spectroscopy from the patterned region showing the RBM confirming the presence of single-walled structures, and the excellent ratio of D-band to G-band intensity, suggesting very pure growth.

The regions where the catalyst was deposited show very sharp RBM features, and intense G-band signal. Extended exposure shifted the measured diameter of the structures until the material was no longer considered single-walled, and the D-band intensity increased significantly indicating poor quality growth.

This same technique was used to compare chemically attached SWCNTs with thermally and plasma grown CNTs on the performance of nanotube-based photovoltaic devices [44].

Conclusions drawn from photo-response measurements shown in Figure 6 were that while chemically attached nanotubes were superior to either of the two growth techniques, thermal or plasma assisted, the PECVD electrodes outperformed the thermal CVD electrodes.

This was concluded to be a result of the much greater quality of nanotubes produced by the PECVD technique, not to mention the fact that the surface was mostly covered in single-walled nanotubes, with more desirable electrical properties than that of the multi-walled nanotubes grown from thermal CVD.

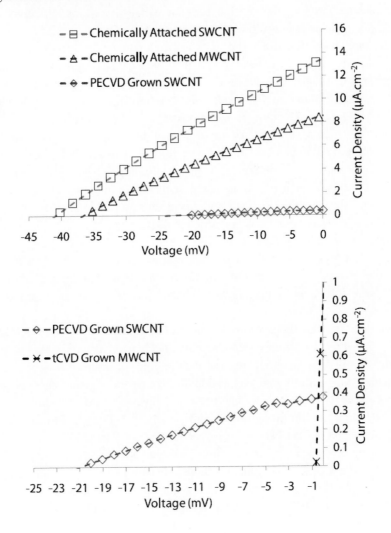

From [44].

Figure 6. I-V curves for nanotube-based solar cells. Chemically attached single and multi-walled nanotubes compared with PECVD grown nanotubes (top) and PECVD versus thermal CVD grown nanotubes (bottom).

Chemical Functionalisation

While ideally nanotubes would be engineered at the production phase to match a given application, most often the properties of the structures need to be tailored retroactively through chemical modification. The goal is to create a set of building blocks that can be predictably used to fabricate devices at the nanoscale, with each type of building block being defined by the type of chemical functionality present.

There are a number of pathways for producing carbon nanotubes with desirable functionality, from fluorination using F_2, CF_4 and SF_6, oxygenation via ozonolysis, and hydrogenation and alkylation through exposure to organics, to name just a few [45].

Due to the anticipated inertness of nanotubes, a by-product of graphite chemistry, fluorination was an early contender for sidewall functionalisation of these materials and as such has received significant attention [29, 31], so much so that it will discussed separately in this chapter. This section however will focus on other sidewall chemical functionalities produced via plasma. Though this is far from an exhaustive study, it demonstrates the flexibility available from covalent sidewall functionalisation of nanotubes.

Nanotubes have been functionalised with oxygen-containing groups in an effort to enhance the solubility of the structures in polar solvents, and improve their inclusion in polymer composite materials. Plasma of O_2 and CO_2 have been used to form hydroxyl, carboxyl and carbonyl groups on CNT sidewalls, as confirmed by XPS measurements by Bubert et al., though limited selectivity was demonstrated in this case [46].

Okpalugo et al., demonstrated very rapid formation of O - C = O bonding on the sidewalls of multi-walled nanotubes, with saturation being reached in just seconds after exposure to an atmospheric pressure dielectric barrier discharge plasma in air [47]. Treatment of nanotubes with radio frequency O_2 plasma can enhance the number of sites available for interaction between the nanotube surface and metal clusters, allowing for some degree of tunability in the density of decorated metal particles [48].

Self-assembly has been enhanced through the thiolation of the ends of bundles of nanotubes [49]. Mixtures of SWCNTs and elemental sulfur were exposed to Ar/H_2 plasma and heated to 120°C, producing a material with around 5% thiol functionality at the nanotube surface. These were subsequently dispersed in ethanol and drop-cast across micro-patterned gold electrodes.

It was found that the nanotubes showed a preference to align across the electrodes, rather than along the length, suggesting a self-assembly mechanism that was dominated by the interaction between the thiol functional groups and the gold substrate.

Nitrogen-containing groups can be grafted to nanotubes via plasma exposure in the presence of N_2 and NH_3 [50-52]. This has been demonstrated as an effective route toward control over the field emission properties of multi-walled CNTs [50]. Primary amines have been selectively attached to SWCNTs in an Ar/N_2 microwave plasma [52].

This was achieved through careful control of the atmosphere around the nanotube material after the initial plasma treatment, where the addition of H_2 greatly enhanced the preferential formation of –NH_2 functionality at the sidewalls. Using a glow-discharge in NF_3, superhydrophobic mats of CNTs were produced in just minutes with contact angles of almost 160° using water, polyethylene glycol and glycerol [53].

Amination can be achieved via an intermediate fluorination step, and allows the formation of both primary and secondary amines to be produced at the sidewalls of the nanotubes [54, 55].

Fluorination

Sidewall fluorination is an extremely effective modification of carbon nanotubes, and enables further derivatisation reactions to be performed, allowing for a broad range of complex, yet tunable nanotube chemistries to be achieved. A figure presented by Banerjee, and reproduced in part in Figure 7, showed that from pristine nanotube material the addition of fluorine to the sidewalls enables multiple reaction pathways to produce materials with amine, hydroxyl and alkyl functionalisation [56]. Fluorination has also been used as a first step towards incorporation into polymer composites [57]. The strain induced by the intrinsic curvature in fullerenes results a relative weakening of C-F bonds (5.33 eV in $C_{60}F_{60}$ vs 6.2 eV in CF_4) [58-60] and this in turn enables these addition reactions to take place.

By tailoring the amount of fluorine attached to the nanotubes to begin with, the level of further functionalisation could be controlled. Exposure to stoichiometric quantities of hydrazine affords this control through 'defluorination' and gives a significant level of control over the amount of fluorine left attached to nanotubes [61].

While it is important to note the broad functionality achievable through fluorination, the initial process itself also gives rise to interesting phenomena. For example, the addition of active fluorine sites to nanotube sidewalls makes them more receptive to hydrogen bonding. Mickelson et al., showed that after fluorination nanotubes could be more readily solubilised via sonication in a wide range of alcohol solvents. The solutions were found to be metastable, with the nanotubes remaining in solution for up to a week [62].

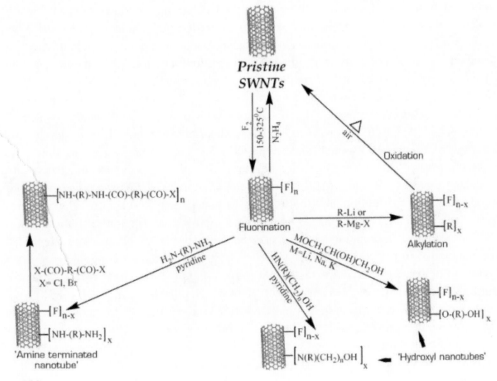

From [56].

Figure 7. The various functional nanotubes possible via an initial fluorination step.

Dissolution of nanotubes is an important step towards their further use since Van der Waals forces resulting from the delocalised π orbitals promote the formation of nanotube 'bundles' in the raw material, similar to the interlayer interaction within crystalline graphite [63]. Much of the work surrounding the fluorination of nanotubes has made use of fluorine (F_2) gas at elevated temperature to achieve the desired functionality. Fluorine gas is extremely reactive, as evidenced by fluorine attachment to nanotubes being achieved at room temperature [64], albeit from an extreme exposure time [65]. Elemental fluorine is considerably toxic which complicates its handling during an experiment [66]. Reactants capable of providing fluorination without the toxicity of F_2 are therefore desired.

More recently, plasma-based processes have been identified as a very useful tool for controllable fluorination. Multiple research groups have investigated the use of this reactive state using fluorine rich gases such as tetrafluoromethane (CF_4) and sulfur hexafluoride (SF_6) [57, 67-78]. Each of these species has significantly greater stability than F_2. In fact SF_6 has been used extensively as an insulating gas in high voltage switch-gear due to its very high breakdown potential [79]. Furthermore, the toxicity is greatly reduced with these substances, particularly with SF_6 which is commonly used to display the opposite effect to breathing in helium; its significant density lowers ones voice. The improved stability of these molecules requires reactivity to be induced through ionisation into the plasma state where it produces radicals and ions that can functionalise surfaces.

In 2004 Tressaud et al., performed a direct comparison between F_2 gas and CF_4 plasma exposure on graphite samples [68]. It was found that while the high temperature fluorine gas exposure had an increased level of C-F bonding, the CF_4 plasma provided a greater level of covalency to those bonds (as determined via XPS) [68]. Thus the choice between these two techniques, at least on graphite, appeared to be a balance between quality and quantity.

Plank et al., in 2004 performed a comparison of the fluorination of nanotubes using either plasma generated within CF_4 or the more fluorine-rich molecule SF_6 [69].

While each of the exposures resulted in similar levels of fluorine attachment over the bias range studied, the sample exposed to the SF_6 plasma resulted in essentially purely covalent C-F bonding, while the CF_4 treatment gave mixed covalent/ionic bonding. This was despite the fact that the SF_6 flow rate was half that of the CF_4. The authors suggested that this was a result of the greater atomic fluorine pressure in the SF_6 plasma, a conclusion based on a plasma diagnostic study. Thus, for half the flow rate, SF_6 produced a greater concentration of fluorine ions than CF_4. Additionally the use of CF_4 plasma may result in the deposition of carbonaceous films and therefore SF_6 is potentially a more viable solution for fluorination of CNTs. Interestingly, fluorination via exposure to SF_6 plasma occurs without significant contamination by sulfur containing species, which is curious considering there must be energetic sulfur containing species in the plasma itself. Lab-based XPS results from SF_6 fluorinated nanotube surfaces have shown no detectable S2p presence in spectra [80]. It was not until advanced synchrotron-based photoelectron spectroscopy was applied to these surfaces, as seen in Figure 8, that it was shown that indeed sulfur is present on these surfaces, but at very low concentrations.

It is also worthwhile to note in Figure 8 the difference in level of functionalisation between two different types of nanotubes; ones produced via arc discharge, and ones produced via high pressure carbon monoxide disproportionation (HiPCO). Recalling what we have already discussed on the reactivity of nanotubes, bond strain and π orbital misalignment plays a key role.

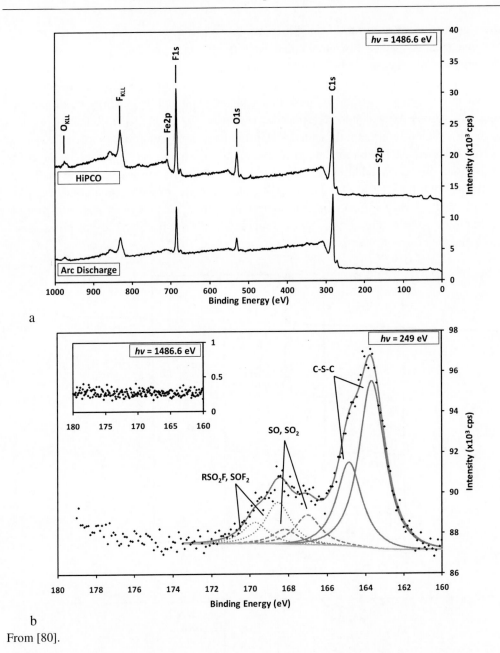

a

b

From [80].

Figure 8. XPS spectra from nanotubes that were fluorine functionalised using radio-frequency SF$_6$ plasma. Survey spectra in (a) show the presence of strong F1s signal after fluorination yet appear to lack any detectable sulfur signal. High resolution scans of the S2p region at much lower excitation energy in (b) however shows that indeed sulfur is present and has complex chemistry.

The HiPCO nanotubes are much smaller in diameter when compared with the arc discharge type, and thus the reactivity is expected to be greater due to the greater strain and misalignment in these sidewalls, resulting in greater fluorine attachment to the HiPCO type nanotubes. Further control over the degree of fluorination achieved after exposure to fluorine containing plasma can be gained through addition of oxygen either into the plasma itself, or directly onto the target surface as a pre-treatment [81].

The fluorination of a surface from SF_6 plasma relies on the breakdown of the parent molecule into smaller species, liberating a reactive fluorine ion or radical at each step.

The addition of oxygen enables further reaction pathways to be exploited during the breakdown of the SF_6 molecule, forming SO_xF_y species with a fluorine ion produced for each breakdown [82]. Figure 9(a) shows how this can be exploited by adding small partial pressures of gaseous oxygen into SF_6 plasma, via either O_2 or H_2O [81] resulting in the degree of fluorination being greatly enhanced. In the case of O_2 addition, increased oxygen presence is detected on the nanotube surface after plasma treatment, however when H_2O vapour is used this extra addition of oxygen at the surface vanishes, resulting in pure enhancement of the fluorine level. Another protocol is to increase the oxygen content at the material surface first, via pre-treatment with oxygen containing plasma, or by chemical means such as exposure to acid. Figure 9(b) and (c) present XPS data from a study of pre-treated nanotubes and the effect on fluorination [81].

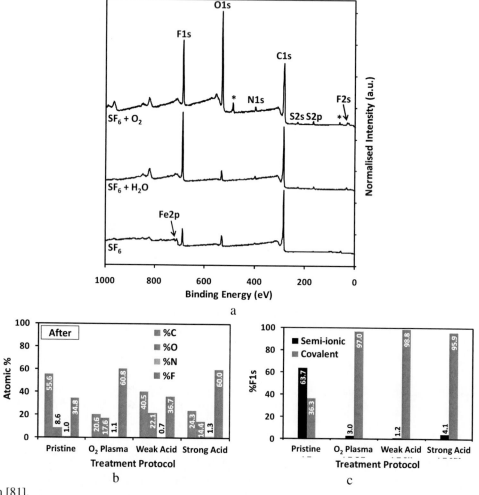

From [81].

Figure 9. The fluorination of nanotubes using SF_6 plasma in the presence of oxygen (a) in the plasma, where O_2 and H_2O vapour was added to the plasma during treatment and (b) at the surface of the nanotubes via pre-treatment prior to SF_6 plasma exposure.

Increasing the amount of oxygen present at the surface prior to exposure to SF$_6$ plasma not only increases the level of fluorination, but also markedly improves the extent of covalently versus ionically bound C-F species that result. Control over the bonding at the nanotube surface is an important factor when synthesising materials that are resistant to degradation.

BIOMEDICAL APPLICATIONS

Nanostructured carbon has been explored over the past decade in a number of biomedical areas ranging from implant biocompatibility [83], transport (delivery) systems, sensors, tissue engineering and biological agents (for example antimicrobials). Fullerenes in particular have shown much promise for transfecting energy or charge, particularly in biomedical contexts ranging from high performance contrasting agents for X-Ray and MRI imaging, drug and gene delivery, and found to aid in photodynamic therapy by enhancing the delivery of therapeutic X-ray energy into tumorous and cancerous cells [84]. A comparison of the applicability and the inherent implementation challenges of the various carbon nanoforms into these areas is summarised by Kovoselov et al., [85]. Indeed, plasma processing is frequently adopted in many of the fabrication steps of the nanostructured surface architectures that are required for each of the aforementioned areas.

CONCLUSION

The future of materials will increasingly entail carbon-based alternatives as replacements for expensive inorganic materials. As this occurs we will see an expansion in the number of novel carbon nanoforms that will be be developed and utilised [86]. With the discovery of exotic carbon nanomaterials such as nanocones and hybrid yarns [87] for example, the need for us to deepen our knowledge of these materials, how to process them and tailor their surface properties to suit each specific application will become increasingly important. For treatment of large quantities of materials at industrial scales, plasma processing will be a key method used to alter the surface physicochemical, optical and electronic properties and will do so with a widening range of active molecular species. Consequently, a thorough knowledge of the underlying science behind plasma-surface interactions will remain crucial to their success.

REFERENCES

[1]   Iijima, S., Helical Microtubules of Graphitic Carbon. *Nature*, 1991. 354: p. 56-58.
[2]   Bethune, D. S., et al., Cobalt-catalysed growth of carbon nanotubes with single-atomic-layer walls. *Nature*, 1993. 363(6430): p. 605-607.
[3]   Iijima, S. and T. Ichihashi, Single-shell carbon nanotubes of 1-nm diameter. *Nature*, 1993. 363(6430): p. 603-605.

[4]   Hamada, N., S.-i. Sawada and A. Oshiyama, New One-Dimensional Conductors: Graphitic Microtubules. *Phys. Rev. Lett.*, 1992. 68(10): p. 1579-1581.

[5]   Mintmire, J. W., B. I. Dunlap and C. T. White, Are fullerene tubules metallic? *Physical Review Letters*, 1992. 68(5): p. 631.

[6]   Yao, Z., C. L. Kane and C. Dekker, High-Field Electrical Transport in Single-Wall Carbon Nanotubes. *Physical Review Letters*, 2000. 84(13): p. 2941.

[7]   Radosavljevic, M., J. Lefebvre and A. T. Johnson, High-field electrical transport and breakdown in bundles of single-wall carbon nanotubes. *Physical Review B*, 2001. 64 (24): p. 241307.

[8]   Treacy, M. M. J., T. W. Ebbesen and J. M. Gibson, Exceptionally high Young's modulus observed for individual carbon nanotubes. *Nature*, 1996. 381(6584): p. 678-680.

[9]   Falvo, M. R., et al., Bending and buckling of carbon nanotubes under large strain. *Nature*, 1997. 389(6651): p. 582-584.

[10] Novoselov, K. S., et al., Electric Field Effect in Atomically Thin Carbon Films. *Science*, 2004. 306(5696): p. 666-669.

[11] Novoselov, K. S., et al., A roadmap for graphene. *Nature*, 2012. 490 (7419): p. 192-200.

[12] *Additive Manufacturing: Pursuing the Promise.* August 2012 DOE/EE-0776: US Department of Energy Efficiency and Renewable Energy, Advanced Manufacturing Office.

[13] Sharon, M. and M. Sharon, *Carbon Nano Forms and Applications.* 2009: McGraw Hill.

[14] Chen, F. F., *Introduction to Plasma Physics and Controlled Fusion.* Vol. 1 Plasma Physics. 1984: Plenum Press: New York.

[15] Fridman, A., *Plasma Chemistry.* 2008, Cambridge: Cambridge University Press.

[16] Lieberman, M. A. and A. J. Lichtenberg, *Principles of Plasma Discharges and Materials Processing.* Second ed. ed. 2005, New Jersey: John Wiley and Sons.

[17] Laroussi, M. and X. Lu, Room-temperature atmospheric pressure plasma plume for biomedical applications. *Applied Physics Letters*, 2005. 87(11): p. -.

[18] Laroussi, M., et al., Inactivation of Bacteria by the Plasma Pencil. *Plasma Processes and Polymers*, 2006. 3(6-7): p. 470-473.

[19] Schneider, J., et al., Investigation of the practicability of low-pressure microwave plasmas in the sterilisation of food packaging materials at industrial level. *Surface and Coatings Technology*, 2005. 200(1-4): p. 962-966.

[20] Aanesland, A., A. Meige and P. Chabert, Electric propulsion using ion-ion plasmas. *Journal of Physics: Conference Series*, 2009. 162(1): p. 012009.

[21] Frisbee, R. H., Advanced Space Propulsion for the 21st Century. *Journal of Propulsion and Power*, 2003. 19(6): p. 1129-1154.

[22] Mitterauer, J., Micropropulsion for small spacecraft: a new challenge for field effect electric propulsion and microstructured liquid metal ion sources. *Surface and Interface Analysis*, 2004. 36(5-6): p. 380-386.

[23] Bonizzoni, G. and E. Vassallo, Plasma physics and technology; industrial applications. *Vacuum*, 2002. 64(3-4): p. 327-336.

[24] Barlow, A. J., A. Deslandes and J. S. Quinton, Langmuir probe characterization of low-frequency oscillations in a radio-frequency SF 6 plasma. *Plasma Sources Science and Technology*, 2011. 20(6): p. 065011.

[25] Quinton, J. S., et al., RF Plasma Functionalized Carbon Surfaces for Supporting Sensor Architectures. *Current Applied Physics*, 2008. 8: p. 376-379.

[26] Deslandes, A., et al., ToF-SIMS Characterisation of Methane- and Hydrogen- Plasma-Modified Graphite using Principal Component Analysis. *Surface and Interface Analysis*, 2009. 41: p. 216-224.

[27] Deslandes, A., et al., Hydrogenation of sp2-bonded carbon surfaces using methane plasma. *Applied Surface Science*, 2010. 256 p. 1888-1894.

[28] Kuzmany, H., et al., Functionalization of carbon nanotubes. *Synthetic Metals*, 2004. 141 (1-2): p. 113-122.

[29] Banerjee, S., T. Hemraj-Benny and S. S. Wong, Covalent Surface Chemistry of Single-Walled Carbon Nanotubes. *Adv. Mater.*, 2005. 17(1).

[30] Bahr, J. L. and J. M. Tour, Covalent chemistry of single-wall carbon nanotubes. *J. Mater. Chem.*, 2002. 12: p. 1952-1958.

[31] Hirsch, A., Functionalisation of Single-Walled Carbon Nanotubes. *Angew. Chem. Int. Ed.*, 2002. 41(11): p. 1853-1859.

[32] Hirsch, A. and O. Vostrowsky, Functionalization of Carbon Nanotubes. *Top Curr. Chem.*, 2005. 245: p. 193-237.

[33] Liu, M., et al., Chemical modification of single-walled carbon nanotubes with peroxytrifluoroacetic acid. *Carbon*, 2005. 43: p. 1470-1478.

[34] Niyogi, S., et al., Chemistry of Single-Walled Carbon Nanotubes. *Acc. Chem. Res.*, 2002. 35: p. 1105-1113.

[35] Lu, X. and Z. Chen, Curved Pi-Conjugation, Aromaticity, and the Related Chemistry of Small Fullerenes (< C60) and Single-Walled Carbon Nanotubes. *Chem. Rev.*, 2005. 105: p. 3643-3696.

[36] Lu, X., Z. Chen and P. v. R. Schleyer, Are Stone-Wales Defect Sites Always More Reactive Than Perfect Sites in the Sidewalls of Single-Wall Carbon Nanotubes? *J. Am. Chem. Soc.*, 2005. 127(1): p. 20-21.

[37] Meyyappan, M., Plasma nanotechnology: past, present and future. *Journal of Physics D: Applied Physics*, 2011. 44(17): p. 174002.

[38] Meyyappan, M., A review of plasma enhanced chemical vapour deposition of carbon nanotubes. *Journal of Physics D: Applied Physics*, 2009. 42(21): p. 213001.

[39] Lim, S., et al., Plasma-Assisted Synthesis of Carbon Nanotubes. *Nanoscale Research Letters*, 2010. 5(9): p. 1377-1386.

[40] Meyyappan, M., et al., Carbon nanotube growth by PECVD: a review. *Plasma Sources Science and Technology*, 2003. 12(2): p. 205.

[41] Ostrikov, K., Plasma nanoscience: setting directions, tackling grand challenges. *Journal of Physics D: Applied Physics*, 2011. 44(17): p. 174001.

[42] Shearer, C., et al., Highly resilient field emission from aligned single-walled carbon nanotube arrays chemically attached to n-type silicon. *Journal of Materials Chemistry*, 2008. 18(47): p. 5753-5760.

[43] Bissett, M. A., et al., Transition from single to multi-walled carbon nanotubes grown by inductively coupled plasma enhanced chemical vapor deposition. *Journal of Applied Physics*, 2011. 110: p. 034301.

[44] Bissett, M., et al., Comparison of carbon nanotube modified electrodes for photovoltaic devices. *Carbon*, 2012. 50(7): p. 2431-2441.

[45] Hirsch, A. and O. Vostrowsky, *Functionalization of Carbon Nanotubes*, in *Functional Molecular Nanostructures*, A. D. Schlüter, Editor. 2005, Springer Berlin Heidelberg. p. 193-237.

[46] Bubert, H., et al., Characterization of the uppermost layer of plasma-treated carbon nanotubes. *Diamond and Related Materials*, 2003. 12(3-7): p. 811-815.

[47] Okpalugo, T. I. T., et al., Oxidative functionalization of carbon nanotubes in atmospheric pressure filamentary dielectric barrier discharge (APDBD). *Carbon*, 2005. 43(14): p. 2951-2959.

[48] Felten, A., et al., Nucleation of metal clusters on plasma treated multi wall carbon nanotubes. *Carbon*, 2007. 45(1): p. 110-116.

[49] Plank, N. O. V., R. Cheung and R. J. Andrews, Thiolation of single-wall carbon nanotubes and their self-assembly. *Appl. Phys. Lett.*, 2004. 85(15): p. 3229-3231.

[50] Gohel, A., et al., Field emission properties of N2 and Ar plasma-treated multi-wall carbon nanotubes. *Carbon*, 2005. 43(12): p. 2530-2535.

[51] Bystrzejewski, M., et al., Functionalizing Single-Wall Carbon Nanotubes in Hollow Cathode Glow Discharges. *Plasma Chemistry and Plasma Processing*, 2009. 29(2): p. 79-90.

[52] Ruelle, B., et al., Selective Grafting of Primary Amines onto Carbon Nanotubes via Free-Radical Treatment in Microwave Plasma Post-Discharge. *Polymers*, 2012. 4(1): p. 296-315.

[53] Hong, Y. C., et al., Surface transformation of carbon nanotube powder into super-hydrophobic and measurement of wettability. *Chemical Physics Letters*, 2006. 427(4-6): p. 390-393.

[54] Plank, N. O. V., et al., Electronic properties of n-type carbon nanotubes prepared by CF_4 plasma fluorination and amino functionalisation. *J. Phys. Chem. Lett.* B, 2005. 109: p. 22096-22101.

[55] Stevens, J. L., et al., Sidewall Amino-Functionalization of Single-Walled Carbon Nanotubes through Fluorination and Subsequent Reactions with Terminal Diamines. *Nano Lett.*, 2003. 3(3): p. 331-336.

[56] Banerjee, S., T. Hemraj-Benny and S. S. Wong, Covalent surface chemistry of single-walled carbon nanotubes. *Advanced Materials*, 2005. 17(1): p. 17-29.

[57] Valentini, L., et al., Use of plasma fluorinated single-walled carbon nanotubes for the preparation of nanocomposites with epoxy matrix. *Composites Science and Technology*, 2008. 68(3-4): p. 1008-1014.

[58] Khabashesku, V. N., W. E. Billups and J. L. Margrave, Fluorination of Single-Wall Carbon Nanotubes and Subsequent Derivatization Reactions. *Acc. Chem. Res.*, 2002. 35 (12): p. 1087-1095.

[59] Taylor, R., Progress in fullerene fluorination. *Russian Chemical Bulletin*, 1998. 47(5): p. 823-832.

[60] Dunlap, B. I., et al., Geometric and electronic structures of C60H60, C60F60, and C60H36. *The Journal of Physical Chemistry*, 1991. 95(15): p. 5763-5768.

[61] Mickelson, E. T., et al., Fluorination of single-wall carbon nanotubes. *Chemical Physics Letters*, 1998. 296(1-2): p. 188-194.

[62] Mickelson, E. T., et al., Solvation of Fluorinated Single-Wall Carbon Nanotubes in Alcohol Solvents. *The Journal of Physical Chemistry B*, 1999. 103(21): p. 4318-4322.

[63] Cranford, S., et al., A single degree of freedom 'lollipop' model for carbon nanotube bundle formation. *Journal of the Mechanics and Physics of Solids*, 2010. 58(3): p. 409-427.

[64] Claves, D., et al., An unusual weak bonding mode of fluorine to single-walled carbon nanotubes. *Carbon*, 2009. 47(11): p. 2557-2562.

[65] Kawasaki, S., et al., Fluorination of open- and closed-end single-walled carbon nanotubes. *Physical Chemistry Chemical Physics*, 2004. 6(8): p. 1769-1772.

[66] Ricca, P. M., A Survey of the Acute Toxicity of Elemental Fluorine. *American Industrial Hygiene Association Journal*, 1970. 31(1): p. 22-29.

[67] Ho, K. K. C., A. F. Lee and A. Bismarck, Fluorination of carbon fibres in atmospheric plasma. *Carbon*, 2007. 45(4): p. 775-784.

[68] Tressaud, A., E. Durand and C. Labrugere, Surface modification of several carbon-based materials: comparison between CF4 rf plasma and direct F-2-gas fluorination routes. *Journal of Fluorine Chemistry*, 2004. 125(11): p. 1639-1648.

[69] Hou, Z., et al., Ar, O-2, CHF3, and SF6 plasma treatments of screen-printed carbon nanotube films for electrode applications. *Carbon*, 2008. 46(3): p. 405-413.

[70] Kalita, G., et al., Fluorination of multi-walled carbon nanotubes (MWNTs) via surface wave microwave (SW-MW) plasma treatment. *Physica E: Low-dimensional Systems and Nanostructures*, 2008. 41(2): p. 299-303.

[71] Bishun, N. K., W. Patrick and M. Meyyappan, The fluorination of single wall carbon nanotubes using microwave plasma. *Nanotechnology*, 2004. 15(11): p. 1650.

[72] Plank, N. O. V. and R. Cheung, Functionalisation of carbon nanotubes for molecular electronics. *Microelectronic Engineering*, 2004. 73-4: p. 578-582.

[73] Plank, N. O. V., et al., Electronic Properties of n-Type Carbon Nanotubes Prepared by CF4 Plasma Fluorination and Amino Functionalization. *The Journal of Physical Chemistry B*, 2005. 109(47): p. 22096-22101.

[74] Plank, N. O. V., L. D. Jiang and R. Cheung, Fluorination of carbon nanotubes in CF4 plasma. *Applied Physics Letters*, 2003. 83(12): p. 2426-2428.

[75] Shoda, K., et al., Feasibility study for sidewall fluorination of SWCNTs in CF4 plasma. *Journal of Applied Physics*, 2008. 104(11): p. -.

[76] Kaoru, S. and T. Seiji, Systematic Characterization of Carbon Nanotubes Functionalized in CF 4 Plasma. *Japanese Journal of Applied Physics*, 2007. 46(12R): p. 7977.

[77] Shoda, K. and S. Takeda, Transmission Electron Microscopy Study on the Surface Properties of CNTs and Fullerites Exposed to CF4 Plasma. *Mater. Res. Soc. Symp. Proc.*, 2007: p. 1018.

[78] Valentini, L., et al., Sidewall functionalization of single-walled carbon nanotubes through CF4 plasma treatment and subsequent reaction with aliphatic amines. *Chemical Physics Letters*, 2005. 403(4-6): p. 385-389.

[79] Malik, N. H. and A. H. Qureshi, A Review of Electrical Breakdown in Mixtures of SF6 and Other Gases. *Electrical Insulation, IEEE Transactions on*, 1979. EI-14(1): p. 1-13.

[80] Barlow, A. J. and J. S. Quinton. Radio frequency SF6 plasma modified single-walled carbon nanotubes: Synchrotron spectroscopy and plasma characterisation studies. In: *2010 International Conference on Nanoscience and Nanotechnology (ICONN)*. 2010. Sydney, Australia.

[81] Barlow, A. J., et al., *A Synchrotron X-ray Photoelectron Spectroscopy Study of the Controlled Fluorination of Single-Walled Carbon Nanotubes in Sulfur Hexafluoride Plasma.* In Preparation, 2015.

[82] Ryan, K. R. and I. C. Plumb, Gas-phase reactions in plasmas of SF6 with O2 in He. *Plasma Chemistry and Plasma Processing*, 1988. 8(3): p. 263-280.

[83] Grabarczyk, J. and I. Kotela, Plasma modification of medical implants by carbon coatings depositions. *Journal of Achievements in Materials and Manufacturing Engineering*, 2009. 37(2): p. 277-281.

[84] Lalwani, G. and B. Sitharaman Multifunctional fullerene and metallofullerene based nanobiomaterials. *NanoLIFE* 08/2013. 1342003.

[85] Kostarelos, K. and Novoselov, K. S., Graphene Devices for Life. *Nature Nanotechnology*, 2014. 9: p. 744-745.

[86] Cazorla-Amoros, D., Grand Challenges in Carbon-Based Materials Research. *Frontiers in Materials*, 2014. 1: p. 6.

[87] Foroughi, J., et al., Highly Conductive Carbon Nanotube-Graphene Hybrid Yarn. *Advanced Functional Materials*, 2014. 24: p. 5859-5865.

In: Innovations in Nanomaterials
Editors: Al-N. Chowdhury, J. Shapter, A. B. Imran

ISBN: 978-1-63483-548-0
© 2015 Nova Science Publishers, Inc.

Chapter 10

GRAPHENE OXIDE MEMBRANES A NEW PARADIGM IN MOLECULAR AND IONIC SEPARATIONS

*Luke J. Sweetman and Amanda V. Ellis**

Flinders Centre for Nanoscale Science and Technology,
School of Chemical and Physical Sciences, Flinders
University, Bedford Park, SA, Australia

ABSTRACT

Membranes made from the carbon-based material graphene oxide are capable of separating ions from water, making these materials potentially useful in a variety of applications. In recent years, membranes composed of graphene and graphene oxide derivatives have been prepared using a variety of techniques such as vacuum filtration and spin coating. These membranes have shown potential towards desalination, separation of metal ions and the selective transport of gases. This chapter will present the current progress in the fabrication and transport properties of graphene oxide membranes. The ability of these membranes to exclude molecules and ions will also be explored via recent theoretical simulations (molecular dynamic) and experimental reports.

INTRODUCTION

Membrane-based methods have become an integral feature of the industrial separations sector. This is due to their low cost compared to other techniques, ease of scale-up, low impact on the environment, and flexibility [1, 2]. Perhaps the most important use of membrane technology is in helping to secure access to clean, safe water sources, particularly in light of an ever-growing population.

---

* Corresponding author: Amanda V Ellis. Flinders Centre for Nanoscale Science and Technology, School of Chemical and Physical Sciences, Flinders University, Sturt Road, Bedford Park, SA 5042, Australia. E-mail: Amanda.ellis@flinders.edu.au.

Despite the success of current membrane methods, there has been a growing interest to find new materials which may help to overcome problems associated with existing membrane materials, including poor selectivity, fouling, limited service lifetimes and high energy costs [3]. One material that is rapidly emerging as a promising membrane material is graphene, and in particular its highly oxidized form graphene oxide (GO).

This unique allotrope of carbon possesses unprecedented mechanical strength and flexibility, and can be easily fabricated into films which are highly suited to existing membrane processes [4, 5]. These membranes can consist of either a single-layer, or multiple graphene layers stacked together to form a free-standing or substrate supported film. Aspects of both forms will be discussed in more detail in subsequent sections.

Multi-layer GO membranes are of great interest due to their ease of preparation and their remarkable transport properties. For example, GO membranes have already demonstrated the ability to act as a barrier film for gas leakage testing due to complete impermeability to even the smallest gases such as helium, yet aqueous solutions can pass freely across its surface [6]. Consequently, a number of studies have emerged showing the potential of these membranes for ion separation, water purification and the separation of small molecules [7].

PRODUCTION OF GRAPHENE OXIDE MEMBRANES

There are a number of methodologies which can be used to prepare multi-layer GO membranes, however, all employ suspensions of GO which are converted to a film. Many approaches can be used to obtain GO dispersed in solution, however, the Hummers' method has emerged as, by far, the most common wet chemical route to produce sufficient quantities of GO [8-10]. In general, the Hummers' method utilizes graphite powder, fibers or flakes which are oxidized to form graphite oxide.

Graphite oxide is a layered structure containing functional groups such as epoxides, ketones, hydroxyls and carboxylic acids spread across the graphene basal plane sheets [11, 12]. These functional groups increase the interlayer spacing between adjacent graphene sheets and weaken the van der Waals forces between layers. Thus, unlike graphene which is hydrophobic and has low solubility in most solvents, graphite oxide can be easily exfoliated into individual GO sheets via sonication in water or various organic solvents (e.g., DMF, THF, and NMP). In this process, GO is stabilized by the solvent molecules through hydrophilic, hydrophobic and electrostatic interactions. In this form it can remain stable for several weeks, depending on the choice of solvent [13, 14].

In a typical Hummer's method synthesis $KMnO_4$ is added to a mixture of H_2SO_4 and $NaNO_3$ at 0 °C, which is stirred for an extended period. As the reaction proceeds a brown-grey suspension is produced which is further treated with hydrogen peroxide to reduce any residual permanganate to soluble manganese sulphate [8]. To remove the manganese sulfate and other residues, the suspension is filtered and washed several times with water to obtain a yellow-brown graphite oxide product. This product can then be suspended in an appropriate solvent with the aid of ultrasonication.

Two main methods have been used to produce multi-layer GO membranes from suspensions obtained *via* the Hummers' method, namely vacuum filtration or spin (or spray) coating.

The most common approach is simple vacuum filtration of a GO dispersion onto a porous membrane support [15, 16]. This process allows for the layered deposition of GO laminates, which can be removed from the underlying substrate upon drying. Alternatively, the obtained membrane can instead be left on the substrate to provide additional mechanical reinforcement. The thickness of the deposited membranes can be easily controlled by varying the quantity of GO suspension being filtered, but is typically between 0.5-10 μm [15]. Once dry, these membranes can then be transferred to either a pressurized, or diffusion driven, transport cell in order to measure the passage of an analyte. Typically, these are held in place with the aid of additional reinforcing materials such as copper foil and are sealed in place *via* the use of rubber o-rings [15, 16].

Spin or spray-coating have also been used to produce a GO membrane, which are of the order of 0.1-10 μm in thickness [17]. Both procedures utilize GO suspensions that are either applied directly onto a rotating stage or sprayed as an aerosol onto a substrate. To increase the deposition rate, the underlying substrate is often heated to approximately 50°C [6].

Unlike vacuum filtration, the substrate used for this method does not have to be porous and so provides a greater degree of flexibility. In order to reveal the porous GO membrane, it is possible to selectively etch it away and leave the desired geometry intact. For example, Nair et al., used a copper substrate which was chemically etched using solution of nitric acid [6]. To ensure that only the desired portion of copper was removed during etching a polymer mask was deposited onto the substrate to protect regions of the copper prior to applying nitric acid solution (Figure 1). These membranes were cleaned, dried and then sealed in an aluminium alloy cell for subsequent transport experiments.

Although the majority of GO membrane research has focused on the deposition of homogeneous GO suspensions, the preparation of variable-layer GO membranes has also been explored. This so-called layer-by-layer (lbl) approach involves the alternate deposition of graphene nanosheets containing complementary charges or by incorporation of cross-linking agents between adjacent graphene layers [18, 19].

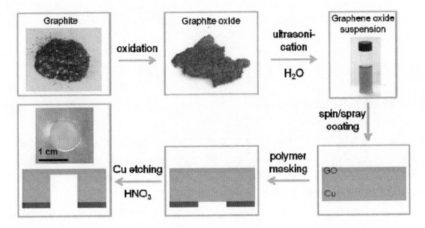

From Nair, R. R.; Wu, H. A.; Jayaram, P. N.; Grigorieva, I. V.; Geim, A. K., *Science*, 2012, 335, 442-444 (supplementary material). Reprinted with permission from AAAS.

Figure 1. Fabrication procedure used to produce GO membranes via spin/spray coating procedure.

The former method developed by Park et al., involves the amidation of carboxyl groups on GO, which were activated using thionyl chloride and subsequently reaction with ethylenediamine [19]. These amine functionalized nanosheets possessed a positive charge and thus were able to strongly electrostatically interact with the negatively charged GO nanosheets. Deposition onto a quartz substrate involved sequentially applying alternate solutions of the positively charged amine functionalized GO containing with the negatively charged GO *via* the use of a slide stainer [19]. However, due to the use of a nonporous substrate these membranes have not been explored for membrane filtration applications.

A later method developed by Hu and co-workers involved the deposition of alternating layers of GO and 1,3,5-benzenetricarbonyl trichloride (trimesoyl chloride) (TMC) onto a polydopamine coated porous polysulfone substrate (Figure 2) [18]. The TMC provided a means of covalently cross-linking adjacent GO sheets by the reaction of its acid chloride groups with the carboxyl groups on the GO. By using this process it was possible to control the layered structure of GO to between 5-50 layers and due to the porous polysulfone substrate it was possible to examine the membrane performance of this material directly.

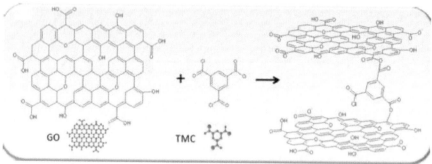

Adapted with permission from Hu, M.; Mi, B. X., *Environ. Sci. Technol.*, 2013, 47, 3715-3723. Copyright 2013 American Chemical Society.

Figure 2. Scheme showing the mechanism of the reaction between graphene oxide (GO) and 1,3,5-benzenetricarbonyl trichloride (TMC) producing a covalent cross-linking between adjacent nanosheets.

MOLECULAR TRANSPORT PROPERTIES OF GRAPHENE

In the past decade, significant advances have been made in the understanding of the fluid transport properties of graphene.

Despite this progress, pristine graphene nanosheets composed of just a single atomic monolayer are still generally considered one of the most impermeable materials known. Research in this area has primarily focused on the theory associated with the transport of fluids across graphene via the aid of computer-based simulations.

The reason attributed to the lack of permeability is the highly sp^2 hybridized C-C structure within graphene, giving rise to a large, delocalised π−electron cloud capable of repelling molecules from the surface.

Notwithstanding this, if a molecule could breach these strong electrostatic interactions, the geometric pore diameter of 0.064 nm which lies between aromatic rings, is simply far too small for any molecule to pass through a graphene sheet (Figure 3) [20].

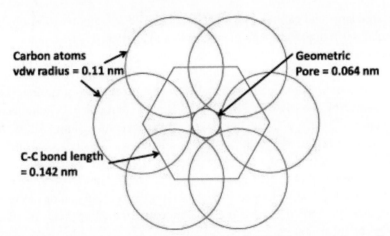

Figure 3. Schematic representation of the repeating molecular structure of graphene, illustrating the geometric pore diameter between aromatic rings is only 0.064 nm, too small for the passage of any molecule.

Consequently, the use of single-layer graphene membranes has sparked considerable interest for use as protective coatings to prevent corrosion and as barriers to protect against diffusion of hazardous substances [21-23]. Although the inherent impermeability of single-layer graphene has been widely accepted, a recent report by Tsetseris et al., has emerged which challenges this belief [24]. In this report the energy barriers and mechanisms associated with the transport of selected small elements were determined using a new computation technique, known as nudged elastic band method [24].

Energy barriers for hydrogen, oxygen and nitrogen were found to be extremely large (4.2 eV, 5.5 eV and 3.2 eV, respectively), however, in comparison, boron only required 1.3 eV to pass through the membrane. This energy barrier can be breached with the aid of only moderate annealing between 150-200 °C, thus the selectivity of these single-layered graphene membranes, for certain molecules, may be controlled by manipulating the temperature [24].

MOLECULAR AND IONIC TRANSPORT THROUGH GRAPHENE NANOPORES

In order to facilitate the passage of fluid across the largely impermeable single-layer graphene, nanopores can also be engineered. Water permeation through graphene membranes possessing pores ranging from 0.75-2.75 nm in diameter was modelled by Suk et al., and compared to analogous carbon nanotube (CNT) membranes [25]. The report found that for membranes containing very small pores, where single-file water transport took place, CNT membranes outperformed graphene. The results were due to frequent disturbances in the hydrogen bonding network and defects in dipole orientation (so-called L/D defects) which were observed throughout the graphene nanopores. This required reorientation of water molecules, slowing down their passage across the graphene nanopore.

In the case of larger pores, however, where single-file passage was not observed, flux rates were instead much higher through graphene, due to a reduced permeation energy barrier at the pore entrance and more bulk-like water behaviour [25].

Aside from water alone, the flux of a number of analytes such as DNA, gases and ionic species across nanoporous graphene membranes has been extensively explored [26-32]. For example, the passage of Li^+, Na^+, K^+, F^-, Cl^- and Br^- ions through F-N terminated and H-terminated nanopores (5 Å in diameter) have been investigated by Sint et al., via molecular dynamic simulations (Figure 4) [29]. In these simulations, a single solvated ion was accelerated via an electric field running perpendicular to a graphene pore and the transport measured. It was found that, in the presence of this large electric field (0.1 V/m), only cations were able to cross the negatively charged F-N pore. Furthermore, only anions were capable of crossing the positively charged H-pore due to Coulombic coupling between ions and the functional groups on the nanopore surface.

Interestingly, smaller radii of ions did not correlate to an increased passage across the nanopores in the graphene membrane. In fact the opposite trend was observed. This was reportedly due to the larger coupling of water in the hydrated shells of the smaller ions.

This finding is in good agreement with a more recent study by Suk et al., who investigated the passage of Na^+ and Cl^- ions across sub 5 nm graphene nanopores [28]. They found that as the pore diameter became smaller and approached the hydration diameter, ion concentrations within the nanopore reduced significantly from the bulk due to the free energy barrier [28]. The observed ion partitioning in these uncharged graphene nanopores was also primarily attributed to this steric exclusion and dehydration effect [28].

The potential for use of nanoporous graphene membranes in desalination applications has shown promise in a number of studies [26, 27, 33]. Cohen-Tanugi et al., modelled the desalination performance of nanoporous graphene (diameters of 1.5-62 Å) functionalized with either hydroxyl groups (hydrophilic) or hydrogen atoms (hydrophobic) and found that graphene possessed exceptional water permeability of between 0-129 L/cm^2/day/MPa [27].

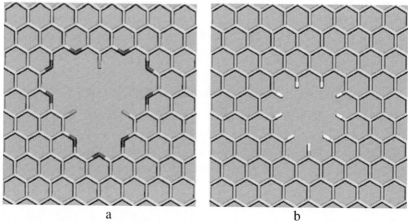

a   b

Reprinted with permission from Sint, K.; Wang, B.; Král, P., *J. Amer. Chem. Soc.*, 2008, 130, 16448-16449. Copyright 2008 American Chemical Society.

Figure 4. Molecular representation of functionalized graphene nanopores. (a) The F-N terminated nanopore, (b) The H-terminated nanopore.

This permeability rate increased with pore size and hydrophilicity of the pore surface. Salt rejection by the graphene mono-layer was strongly dictated by pore diameter, with optimal rejection achieved at pore diameters < 5.5 Å due to size exclusion [27].

For the larger pores, however, salt rejection decreased with increasing pressure and was significantly lower for hydrophilic membranes [27]. This dependence on hydrophilicity was attributed to the OH groups forming hydrogen-bonding with salt ions, lowering the barrier energy for ion passage across the nanopore [27].

WATER TRANSPORT THROUGH MULTI-LAYERED GRAPHENE OXIDE

Larger macroscopic membranes containing multiple graphene oxide layers have begun to emerge due to their ease of synthesis and reproducibility, as described previously. Nair et al., postulated the transport mechanism of water across a graphene oxide (GO) membrane made of multiple GO crystallites stacked on top of each other [6].

In these multi-layer membranes the functionalized regions of the GO are responsible for keeping a large spacing between the GO sheets (~5 Å), which is sufficient to allow passage of a single mono-layer of water (Figure 5).

Due to the strongly hydrophobic nature of the pristine 'nano-capillary' regions, which form between graphene sheets, it is believed that a low-friction flow of mono-layer water similar to that observed through small diameter CNTs would be achieved [34]. This hypothesis was corroborated using molecular dynamics simulations whereby the passage of a monolayer of water for 6 Å $\leq$ d $\leq$ 10 Å was modelled through pristine regions of GO membranes, revealing a capillary pressure of the order of 1000 bars [34].

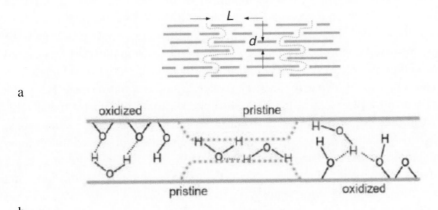

a

b

From Nair, R. R.; Wu, H. A.; Jayaram, P. N.; Grigorieva, I. V.; Geim, A. K., *Science*, 2012, 335, 442-444. Reprinted with permission from AAAS.

Figure 5. An illustration of water transport through multi-layer graphene oxide membranes as proposed by Nair et al., (a) [6]. The enlargement (b) shows the narrowing of the nano-capillaries due to the lowering of humidity, resulting in insufficient van der Waals distance between adjacent graphene oxide sheets to accommodate water transport.

However, this mono-layer water flux was shown to be strongly impeded in the functionalized regions of GO due to hydrogen bonding and a narrower space for diffusion [34].

PERFORMANCE OF MULTI-LAYERED GRAPHENE OXIDE MEMBRANES

In light of the unprecedented permeability of graphene shown via computational methods, researchers have begun to examine the molecular transport properties of these membranes experimentally. Studies involving these materials have focused primarily on multi-layer graphene oxide membranes due to the difficultly in producing large-scale high-quality, single layered graphene. The primary focus in this regard has been in the development of multi-layered membranes made from GO or GO derivatives, due to the ability to easily process aqueous suspensions of this more hydrophilic material. At present, work in this area has focused on either diffusion or pressure driven transport of a number of species including gases, water, ions and selected organic molecules across GO membranes.

A recent study by Joshi et al., examined the diffusion of ions and molecular solutes across a 1cm diameter GO membrane mounted onto copper foil within a 2.5 cm diameter U-shaped tube [15]. Although these membranes were completely impermeable to helium, transport of water could be observed from the permeate (water only) to feed solution when a feed solution of 1 M sucrose was applied. The large sucrose molecule was unable to diffuse across the membrane resulting in a flow rate of water of c.a. $0.2 \, L \, m^{-2} \, h^{-1}$, which was further increased upon increased sucrose concentration [15].

Passage of water was facilitated by an osmotic pressure differential of ~ 25 bar, which in turn was driven by a large capillary pressure of the order of 1000 bar. The diffusion of selected ions, molecular solutes and solvents across these GO was also explored (Figure 6) [15]. It was found that in order for these species to freely diffuse across GO membranes the hydrated radius of the analyte must be less than 4.5 Å [15]. Transport in this case was caused by an interlayer swelling between GO sheets (13 ± 1 Å) which occurred when submerged in solution during testing, thus allowing for multiple water layers and ion passage [15]. The rapid permeation was also attributed to the high adsorption affinity of GO for metal salts, resulting in an ionic pressure of > 50 bar (determined by molecular dynamics simulations) [15]. Similar work by Sun et al., using a drop-casting technique rather than vacuum filtration revealed comparable results when using GO membranes [35, 36]. The passage of a larger Rhodamine B dye molecule was completely blocked due to a combination of steric and electrostatic interactions with the GO nanosheets, however, alkali cations were able to permeate across the membrane freely. The permeation of a given cation was influenced by the identity of the accompanied anion, in particular the chemical reactivity of the anion with the GO nanosheets. For example, the rate of passage of four Na^+ salts containing different anions the permeability was determined to be $NaOH > NaHSO_4 > NaCl > NaHCO_3$. The rapid transport of NaOH was attributed to the reaction of OH- with the acidic functional groups on GO sheets (i.e., carboxyl and hydroxyl groups). This resulted in an increase in the number of negative charge sites present on the GO nanosheets, thus increasing the interlayer distance and consequently rate of flux.

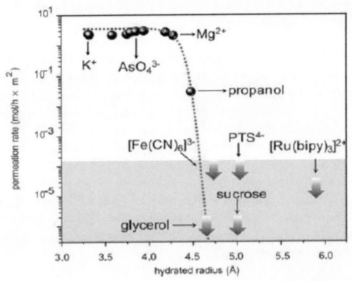

From Joshi, R. K.; Carbone, P.; Wang, F. C.; Kravets, V. G.; Su, Y.; Grigorieva, I. V.; Wu, H. A.; Geim, A. K.; Nair, R. R., *Science*, 2014, 343, 752-754. Reprinted with permission from AAAS.

Figure 6. The permeation rate of selected molecular and ionic species diffusing across multilayer graphene oxide membranes as a function of hydrated radius of the transported analyte.

For $NaHSO_4$, the H^+ ions generated from this salt inhibited the deprotonation of the GO which was observed with NaOH and the interlayer distance remained unchanged yielding a lower rate of flux. On the other hand, although $NaHCO_3$ is also basic in nature it had the lowest flux of all due to generation of CO_2 through the reaction with HCO_3^- ions resulting in a reversed compression of the sheets impeding ion permeability.

By far, the identity of the cation dictates the permeability of metal salts across the GO membrane. The passage of alkaline earth cations was significantly greater than that of alkali metal cations, which was attributed to their ability to form strong cation-π interactions with the GO [35, 36]. This was caused, in part, by the increased screening of the cation charge of the alkaline earth metals to the GO nanosheets due to their larger hydration shell than that of alkali metals. Consequently, the magnitude of cation-π interactions between the hydrated alkali cations and GO sheets was considerably larger, resulting in a more impeded flow. Transition metal cations on the other hand had a significantly lower ability to form cation-π interactions but instead can form coordinate complexes between adjacent GO layers. Differences in permeability between transition metals was strongly related to the confirmation and overall binding energy associated with the metal and the various oxygen containing functional groups on GO.

Pressure driven transport across GO membranes prepared *via* vacuum filtration by Huang et al., revealed these materials possessed a water flux of 71 L m^{-2} h^{-1} bar^{-1} and displayed a 85% rejection of Evans blue (EB) dye molecules [37]. The flux and rejection rates were strongly influenced by the pH, applied pressure and the presence of ionic species. The addition of NaCl resulted in a sharp reduction in flux across GO membranes due to the screening of negative charges between adjacent GO layers due to the increased electrolyte concentration.

In turn, this caused the nanochannels between the GO sheets to narrow, resulting in a reduced flux. At low pH the water flux was significantly reduced while the rejection of EB was enhanced. This occurred as a result of the protonation of carboxyl groups on GO which resulted in a reduction in the electrostatic repulsion between neighbouring GO sheets, thus lowering the inter-sheet spacing impeding the passage of water and EB.

As the pH was raised an increase in flux and subsequent decrease in EB rejection was observed due to the deprotonation of these carboxyl sites on GO, enhancing electrostatic repulsion between adjacent sheets. This broadened the nanochannels within the GO membrane and allowed for an increase in the rate of passage. However, when the pH was raised above 9 the spacing of GO layers begins to contract. This was attributed to an increase in screening between the negatively charged GO sheets due to the greater ion concentration present, resulting in a large decrease in flux and an increase in EB rejection. Finally, increased pressure was also shown to decrease membrane flux. This was believed to be due to the collapse of the corrugated GO sheets under the higher pressure, reducing the size of the nanochannels and hence lower permeability.

TRANSPORT PROPERTIES OF CHEMICALLY AND THERMALLY REDUCED GRAPHENE OXIDE MEMBRANES

The behaviour of modified GO membranes, in particular, the development of chemically converted graphene obtained by thermal or chemical reduction of GO has recently been explored. Nair et al., observed that after thermal reduction, graphene membranes became completely impermeable to water, organic solvents and gas molecules [6]. However, Han et al., produced graphene membranes via filtration of dilute base-refluxing reduced GO (brGO), resulting in membranes which were highly permeable towards water (up to 21.8 L m^{-2} h^{-1} bar^{-1} depending on brGO loading) [38]. This permeation was attributed to the low friction between water and the hydrophobic, unfunctionalized carbon on the brGO and also the formation of ordered hydrogen bonds formed by the single-file of water molecules across the membrane. In addition, the permeation of a number of other more hydrophobic solvents such as ethanol, hexane and toluene was examined and were found to permeate much more. All of these species were found to have significantly lower rate of flux across the graphene membrane. This was primarily due to the increased level of interaction with the more hydrophobic liquids and the graphene walls. Han et al., also studied the permeation of various analytes, such as organic dyes and selected ions, across the graphene membranes [38]. Solutions containing dye molecules displayed a > 99% exclusion for all dyes in the permeate, with the subsequent permeation of water still maintaining nearly 90% of that the flux of pure water alone.

This significant level of rejection was attributed mainly to a physical sieving by the membrane nanochannels. It was also proposed that some of the rejection was due to an electrostatic interaction between the negatively charged graphene membrane (zeta potential = -43 mV) and anionic dyes. The retention of ions was also shown to be between 10-60% depending on the salt solution being filtered. The extent of which was governed by the ratio of valency of the anion to cation charge, as per the Donnan exclusion theory [38].

Although there have been a number of reports which have explored the use of GO or modified GO membranes, as described previously, there is still some contention regarding the long-term stability of such membranes for aqueous based separation. Consequently, a limited number of studies have been reported describing methods to improve the longevity of these membranes by cross-linking the extremely hydrophilic nanosheets. This was performed using the lbl method [18, 19]. Permeability of these cross-linked membranes towards water was shown to vary between 80-276 L $m^{-2} h^{-1}$ MPa^{-1} depending on the number of graphene layers deposited. The separation performance of these membranes was also investigated using two salts (NaCl and Na_2SO_4) and two organic dyes (Methylene Blue; MB and Rodamine-WT; R-WT). In the case of NaCl and Na_2SO_4 the rejection rates were measured to be 6-19% and 26-46%, respectively, depending on the number of GO layers present. On the other hand, the rejection of MB and R-WT varied between 46-66% and 93-95%, respectively. Contrary to the report by Han et al., the primary mode of removal of the dyes (within the first hour) was attributed to the inherent adsorption capacity of GO and not size exclusion or charge effects as previously believed [38].

CONCLUSION

Graphene oxide has been shown as a promising membrane material towards the selective transport of a myriad of ionic and molecular species. These ultrathin multilayer membranes display exceptional flux to aqueous solution and are consequently ideal candidates for water-based separation processes. Despite this, there is still a great deal more to learn about these unique membrane structures and how they interact with analytes.

At present, the transport or rejection across these materials is primarily attributed to size exclusion mechanisms; however, adsorption and electrostatic interactions with the membrane are likely to play important roles and need to be better understood.

This may help to open the door for the future use of these membranes in the growing fields of desalination and water purification. Finally, the long-term stability of these membranes is certainly a point of concern, particularly in light of the extremely hydrophilic nature of GO nanosheets. Despite conflicting reports, there is no question that an improvement in their mechanical properties in order to survive the turbulent and harsh conditions they will be applied to is warranted. The use of the lbl method represents one approach that may be useful in this regard, however, its ability to be employed on a large scale is yet to be explored. Alternatively, the use of covalent GO cross-linking methods may hold the answer if a suitable procedure can be found that will firmly lock these sheets in place.

REFERENCES

[1]   Ulbricht, M. *Polymer*; 2006, 47, 2217-2262.
[2]   Nunes, S.; Peinemann, K. V. *Membrane Technology in the Chemical Industry*; Second Edition; Wiley-VCH; Weinheim; Germany; 2006.
[3]   Mulder, M.; *Basic Principles of Membrane Technology*; Second Edition; Kluwer Academic Publishers; Netherlands; 1996.

[4] Novoselov, S. K.; Fal'ko, V. I.; Colombo, L.; Gellert, P. R.; Schwab, M. G.; Kim, K. *Nature* 2012, 490, 192-200.

[5] Li, D.; Müller, M. B.; Gilje, S.; Kaner, R. B.; Wallace, G. G. *Nat. Nanotechnol.* 2008, 3, 101-105.

[6] Nair, R. R.; Wu, H. A.; Jayaram, P. N.; Grigorieva, I. V.; Geim, A. K. *Science* 2012, 335, 442-444.

[7] Mi, B. *Science*, 2014, 343, 740-742.

[8] Hummers, W. S.; Offeman, R. E. *J. Am. Chem. Soc.* 1958, 80, 1339-1339.

[9] Dreyer, D. R.; Park, S.; Bielawski, C. W.; Ruoff, R. S. *Chem. Soc. Rev.* 2010, 39, 228-240.

[10] Allen, M. J.; Tung, V. C.; Kaner, R. B. *Chem. Rev.* 2009, 110, 132-145.

[11] Dreyer, D. R.; Park, S.; Bielawski, C. W.; Ruoff, R. S. *Chem. Soc. Rev.* 2010, 39, 228-240.

[12] Gao, W.; Alemany, L. B.; Ci, L.; Ajayan, P. M. *Nature Chem.* 2009, 1, 403-408.

[13] Li, D.; Müller, M. B.; Gilje, S.; Kaner, R. B.; Wallace, G. G. *Nat. Nanotechnol.* 2008, 3, 101-105.

[14] Paredes, J. I.; Villar-Rodil, S.; Martínez-Alonso, A.; Tascón, J. M. D. *Langmuir* 2008, 24, 10560-10564.

[15] Joshi, R. K.; Carbone, P.; Wang, F. C.; Kravets, V. G.; Su, Y.; Grigorieva, I. V.; Wu, H. A.; Geim, A. K.; Nair, R. R. *Science* 2014, 343, 752-754.

[16] Dikin, D. A.; Stankovich, S.; Zimney, E. J.; Piner, R. D.; Dommett, G. H. B.; Evmenenko, G.; Nguyen, S. T.; Ruoff, R. S. *Nature* 2007, 448, 457-460.

[17] Yeh, T.-M.; Wang, Z.; Mahajan, D.; Hsiao, B. S.; Chu, B. *J. Mater. Chem. A*, 2013, 1, 12998-13003.

[18] Hu, M.; Mi, B. X. *Environ. Sci. Technol.* 2013, 47, 3715-3723.

[19] Park, J. S.; Cho, S. M.; Kim, W.-J.; Park, J.; Yoo, P. J. *ACS Appl. Mater. Interfaces* 2011, 3, 360-368.

[20] Berry, V. *Carbon* 2013, 62, 1-10.

[21] Zhao, Y.; Xie, Y.; Hui, Y. Y.; Tang, L.; Jie, W.; Jiang, Y.; Xu, L.; Lau, S. P.; Chai, Y. *J. Mater. Chem. C* 2013, 1, 4956-4961.

[22] Guo, F.; Silverberg, G.; Bowers, S.; Kim, S.-P.; Datta, D.; Shenoy, V.; Hurt, R. H. *Environ. Sci. Technol.* 2012, 46, 7717-7724.

[23] Kirkland, N. T.; Schiller, T.; Medhekar, N.; Birbilis, N. *Corros. Sci.* 2012, 56, 1-4.

[24] Tsetseris, L.; Pantelides, S. T. *Carbon* 2014, 67, 58-63.

[25] Suk, M. E.; Aluru, N. R. *J. Phys. Chem. Lett.* 2010, 1, 1590-1594.

[26] Zhao, S.; Xue, J.; Kang, W. *J. Chem. Phys.* 2013, 139, 114702.

[27] Cohen-Tanugi, D.; Grossman, J. C. *Nano Lett.* 2012, 12, 3602-3608.

[28] Suk, M. E.; Aluru, N. R. *J. Chem. Phys.* 2014, 140, 084707.

[29] Sint, K.; Wang, B.; Král, P. *J. Am. Chem. Soc.* 2008, 130, 16448-16449.

[30] Hu, G.; Mao, M.; Ghosal, S., *Nanotechnology*, 2012, 23, 395501.

[31] Garaj, S.; Liu, S.; Golovchenko, J. A.; Branton, D., *PNAS*, 2013, 110, 12192-12196.

[32] Sun, C.; Boutilier, M. S. H.; Au, H.; Poesio, P.; Bai, B.; Karnik, R.; Hadjiconstantinou, N. G., *Langmuir*, 2013, 30, 675-682.

[33] Wang, E. N.; Karnik, R., *Nat. Nanotechnol.*, 2012, 7, 552-554.

[34] Holt, J. K.; Park, H. G.; Wang, Y.; Stadermann, M.; Artyukhin, A. B.; Grigoropoulos, C. P.; Noy, A.; Bakajin, O., *Science*, 2006, 312, 1034-1037.

[35] Sun, P. Z.; Zheng, F.; Zhu, M.; Song, Z. G.; Wang, K. L.; Zhong, M. L.; Wu, D. H.; Little, R. B.; Xu, Z. P.; Zhu, H. W., *ACS Nano*, 2014, 8, 850-859.

[36] Sun, P. Z.; Zhu, M.; Wang, K. L.; Zhong, M. L.; Wei, J. Q.; Wu, D. H.; Xu, Z. P.; Zhu, H. W., *ACS Nano*, 2013, 7, 428-437.

[37] Huang, H.; Mao, Y.; Ying, Y.; Liu, Y.; Sun, L.; Peng, X., *ChemComm*, 2013, 49, 5963-5965.

[38] Han, Y.; Xu, Z.; Gao, C., *Adv. Funct. Mater.*, 2013, 23, 3693-3700.

In: Innovations in Nanomaterials
Editors: Al-N. Chowdhury, J. Shapter, A. B. Imran

ISBN: 978-1-63483-548-0
© 2015 Nova Science Publishers, Inc.

Chapter 11

GRAPHENE SUPPORTED NOBLE METAL NANOSTRUCTURES: SYNTHESIS AND ELECTROCHEMICAL APPLICATION

Subash Chandra Sahu, Aneeya K. Samantara
and Bikash Kumar Jena[*]
CSIR- Institute of Minerals and Materials Technology, Academy of Scientific &
Innovation Research (AcSIR), Bhubaneswar, India

ABSTRACT

Nanostructured materials of noble metals such as Pt, Pd, Au and Ag have been gaining tremendous research interest in the current decade owing to their unique physiochemical properties. These nanostructured materials have fascinated the scientists because of their wide range applications including the fuel cell, biosensor and sensor, electronics and so forth. Further, these metal nanostructures have been loaded on conducting support to enhance their catalytic activity and stability. Recently, graphene has been accepted as the ideal supporting material, because of its large surface area and high electrical and thermal conductivity. Therefore, many scientific reports have been documented on the synthesis of graphene-noble metal hybrid nanomaterials and their diverse application. In this chapter, synthesis of graphene supported noble metal nanostructures and their applications have been discussed briefly. In particular, the interesting application towards the development of fuel cell and the electrochemical sensor is highlighted.

Keywords: noble metals, graphene, graphene oxide, fuel cell, methanol, formic acid, electro-oxidation, oxygen Reduction, electrochemical sensors

---

[*] Corresponding author: Email: bikash.immt@gmail.com.

1. INTRODUCTION

Materials in the nanoscale have been attracted because of their potential and technological application in various fields [1]. In general, nanomaterials have dimensions in the range of 1-100 nm [2]. These materials possess interesting physical and chemical properties which are extraordinarily different from their bulk counterparts. A large number of nanomaterials of novel properties have been investigated with the advancement of material research. Out of various nanomaterials, the metals attract special attention because they cover two third of the total elements present in the periodic table. Further, most of the metals crystallize in the face centred cubic lattice that helps their facile characterization [3]. In particular, nanomaterials of noble metals such as Au, Ag, Pd and Pt have been the frontier of nanoscience for their wide uses in catalysis, electrocatalysis, bio-medical applications and so forth [4]. For example, Pt, a precious metal has been extensively applied as a unique and efficient electrocatalyst for fuel cell applications [4 a,b]. It also plays an important role in catalytic reduction of pollutants, hydrogenation reactions of organic molecules, petroleum refining industry and so forth [4c]. On the other hand, palladium metal has been attracted for its unique application in catalysis, hydrogen storage and sensing [5]. It has also been accepted as an alternative material for platinum as electrocatalyst owing to its low cost and high resistance towards surface poisoning [6]. Nanostructures of Au have been received huge research interest because of their potential applications in catalysis, development of chemical sensor and biosensors, effective surface enhanced Raman scattering (SERS) and so forth [7]. The biocompatible nature of Au nanostructures enables them to apply in biomedical areas such as phototherapy, cancer therapy, drug delivery, bioimaging (in-vivo and in-vitro), etc [8]. Similarly, the nanostructures of silver are being used as a potential nanomaterial due to their exciting applications such as catalyst, electrocatalyst, antibacterial and antimicrobial agents [9]. Ag nanoparticles owing its unique optical property have especially been applied as efficient SERS active metallic substrate for detection of bio-molecules and analytes [9].

Nanomaterials of carbon have been attracting special research interest because of their vast and potential utility in various fields [10, 11]. Carbon, the second most abundant element on the earth has been fascinating scientific world for its existence in different materials. Its name has been originated from Latin word "*carbo*" which means "*coal*". Its unique self-linking property called "*catenation*" has made it as the basic element of 80% of existing organic compounds. It was discovered in the prehistoric age and was known in the form of soot particle and charcoal during the earliest human civilisations. In 1789, Antoine Lavoisier for the first time listed the name "carbon" as an element in his textbook. Carbon exists in different elemental forms called allotropes such as charcoal, graphite, diamond, coal, etc. It is present in three hybridization states such as sp hybridised carbon (carbon dioxide, acetylene), sp^2 hybridised carbon (graphite, graphene, ethylene) and sp^3 hybridised carbon (diamond, methane). Based on the structural arrangement and bonding of carbon atom, these allotropes have been classified into two categories such as amorphous and crystalline carbon [12].

Out of various carbon-based nanomaterials, graphene has been the ever rising nanomaterial since its discovery in 2004 [13]. It comprises of a single layer of sp^2 bonded carbon atoms those are closely packed to form a honey-comb like lattice. Moreover, it has been regarded as the basic structural unit of important allotropes such as graphite, carbon nanotube and fullerene (Figure 1).

Further, graphene possesses unique optical, electronic, thermal, mechanical and electrical properties due to which it has been employed for many technological applications. The novel properties such as large surface area (2,630 m^2g^{-1}), high conductivity (106 Scm^{-1}) and excellent thermal and chemical stability have made graphene as unique and attractive material among the others [13]. Moreover, it has been treated as an ideal catalytic support for loading of metal/metal oxide nanoparticles owing to large surface area and good conductivity which find interesting applications in various fields [14]. In the subsequent sections, synthesis of graphene and graphene-based noble metal nanostructures, novel properties and their electrocatalytic and analytic applications have been reviewed.

2. CARBON ALLOTROPES AND GRAPHENE

2.1. History

It was believed that carbon was discovered in prehistorical time and commonly known in the form of soot particles and charcoals to the earliest human civilisation. At the beginning of 2500 BC, the diamond was found in China and charcoal was made during Roman times. Antoine Lavoisier documented carbon as an element in his 1789 textbook and thereby different allotropes of carbon was discovered. In 1985, Smalley et al. at Rice University synthesized first fullerene molecule called Buckminsterfullerene (C_{60}), which bestowed the Nobel Prize in Chemistry in 1996 [15]. For the first time, L. V. Radushkevich and V. M. Lukyanovich observed the clear images of carbon tubes in 1952. But Sumio Iijima of NEC, Japan in 1991 discovered the hollow, nanometre-size carbon nanotubes comprising of graphitic carbon atoms [16]. During the 4th millennium B.C., graphite was applied as material for making ceramic paints to decorate pottery in south-eastern Europe. The word "*graphite*" was given by Abraham Gottlob Werner in 1789 which means writing stone. The graphite sheets were being applied to make the mark on instruments in the middle ages as well as the present use of graphite in pencil today. It was also used as solid dry lubricants along with other carbon materials. In contrast, the term "graphene", which is a single sheet of graphite, was first observed in 1987 as one of the major components of graphite intercalation compounds. The scientific motivation towards graphene was highly intensified after A. Geim and K. Novoselov reported the successful separation of a single-atom-thick layer from bulk graphite by a simple scotch tape technique in 2004. This discovery of graphene, as new member of carbon material, got the Nobel Prize in Physics for the year 2010 [17]. Consequently, various research groups, all over the world, are actively involved to explore the fundamental properties and technological application of graphene.

2.2. Carbon Allotropes

The major elemental forms, i.e., allotropes of carbon are fullerene (0D, zero dimensional carbon material), carbon nanotubes (1D, one dimensional carbon material), graphene (2D, two dimensional carbon material), graphite (3D, three dimensional carbon material) and diamond (3D, three dimensional carbon material). All the above allotropic forms have common sp^2 hybridised carbon atoms except diamond that has sp^3 hybridised carbon atom.

2.2.1. Fullerene (0D)

Fullerene is a large spherical molecule composed of mostly sp^2 hybridised carbon atoms and its shape resembles with the soccer ball and hence the name buckyball. The first fullerene molecule known as Buckminsterfullerene (C_{60}) was prepared by R. Smalley, R. Curl, J. Heath, S. O'Brien and H. Kroto of Rice University in 1985 [18]. Fullerene molecules consist of both hexagonal and pentagonal rings of sp^2 hybridised carbon that are arranged in a specific manner to produce the shape of a hollow sphere.

2.2.2. Carbon nanotube (1D)

Carbon nanotube (CNT) is cylindrical in shape consisting of various carbon rings (hexagonal, pentagonal and heptagonal). It is structurally similar to fullerene but differ in shape. CNT is mainly available as single-walled nanotube (SWNT) and multi-walled nanotube (MWNT). The SWNT is long and hollow structure with the single wall formed by one-atom-thick sheets of carbon. But MWNT is the form when individual single walled CNTs naturally align themselves with Van der Waals forces and pi-stacking [19].

2.2.3. Graphene (2D)

Graphene is a single layer of sp^2 bonded carbon atoms which are packed in honeycomb lattice as regular hexagonal ring. The bond length of carbon-carbon (C-C) bond is nearly 0.142 nm. The graphene sheets stack themselves to form graphite with an interplanar distance of 0.335 nm. Further, graphene is the most stable two dimensional carbon materials, which is the basic building block of graphite, carbon nanotubes and fullerene. Moreover, graphene possesses extraordinary physio-chemical properties such as higher electrical conductivity, better chemical and thermal stability and larger surface area [20].

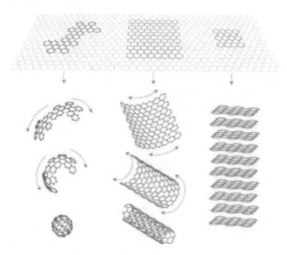

Figure1. Schematic presentation of graphene as building block for Fullerene, Carbon Nanotube and Graphite. Reproduced with permission from reference [13].

2.2.4. Graphite (3D)

It is a three dimensional carbon material which consists of a large number of one atom thick graphene sheets. The graphene sheets are stacked together by weak Van der Waal's forces. The name "*graphite*" was given by Abraham Gottlob Werner in 1789. It has derived from the ancient Greek word, which means "to draw/write" for its use in pencils and commonly named as lead. Each layer of graphite consists of honeycomb pattern carbon atoms with a C-C bond distance of 0.142 nm and interlayers distance of 0.335 nm. The carbon atoms present in graphite is sp^2 hybridised and linked to each other by covalent bonds whereas the unhybridized p-orbitals form π bonds forming the delocalised π-electron clouds, which is responsible for its higher electrical conductivity [21].

2.2.5. Diamond (3D)

Diamond is a highly transparent crystal in which sp^3 hybridised carbon atoms are arranged in a covalent network lattice with tetrahedral structures. It possesses extraordinary physical properties such as extreme hardness, high thermal conductivity as well as large optical dispersion. It is a highly dense material which leads to unique hardness but the carbon-carbon (C-C) bond strength is weaker than that of graphite [22]. In diamond, the bonds arrange themselves in such a manner resulting inflexible three-dimensional lattice.

3. ATOMIC STRUCTURE AND CHEMISTRY OF GRAPHENE

Graphene consists of a large number of six-membered rings of sp^2 hybridised carbon atom [20]. Each carbon atom forms three sp^2 hybrid orbitals and an unhybridized p-orbital. The hybrid orbitals of one carbon atom form three strong covalent sigma bonds whereas the overlapping of unhybridized p-orbitals forms the weak covalent pi- bond. The C-C-C bond angle of the ring is 120°, resulting a planar structure. The filled π-bonding orbital and empty π^* orbitals formed are called valence band and conduction band. Further, the pi bonds in six membered ring as well as graphene network are in conjugation with each other by resonance. The planner structure and large extended and conjugated π network make graphene as a unique material. Due to peculiar structural features, graphene can anchor various chemical compounds/molecule/ions on its surface through π electrons. This results in the enhanced activity of the foreign compounds by the synergistic behaviour. Further, graphene surface can be functionalized with different functional groups and doped with heteroatoms in its network by resulting in modified graphene with enhanced properties. Although, the atoms and planes of graphene are exposed, graphene does not react and is fairly stable thus, exhibiting high chemical inertness property. But, graphene reacts with chemicals under harsh reaction conditions. Moreover, the edge atoms of graphene have been found to be more reactive as compared to surface atoms [21].

4. PROPERTIES OF GRAPHENE

Graphene possesses distinctive optical, electronics, electrical and mechanical properties because of peculiar structure than other carbon-based materials. These extraordinary properties of graphene have exploited for their potential and technological applications. Graphene possesses unique optical properties. Although it is one atom thin sheet of carbon, it absorbs as high 2.3% of light passing through it because of its distinct electronic structure. Further, the optical transmittance of a single layer graphene is very high ~ 97.7% [23]. The optical transparency of graphene depends upon the number of layers of graphene present [24]. The refractive index of single layer graphene varies in the range of 2.0-1.1 in visible light.

Graphene has the distinct electronic property which is rarely observed in conventional two or three-dimensional materials [25]. This is attributed to the inherent fundamental properties such as Quantum Hall effect and massless charge carriers called Dirac fermions [26]. Further, the electronic property depends upon the nature and density of charge carriers. Moreover, it shows ballistic charge transport mobility $15000 \, cm^2 \, (Vs)^{-1}$ at room temperature. Various novel electrical properties of graphene have been reported because of its charge carriers act as massless Dirac fermions. The electrical properties of graphene are dependent on size, curvature of the surface and interaction with surrounding medium. Owing to Quantum Hall effect and unique band structure, it exhibits highest electrical conductivity. Another interesting property of graphene is its high mechanical strength. Pure and high quality graphene sheet possesses Young's modulus of 1.0 T and fracture strength of 130 GPa. It is known to be the hardest material even more than diamond and around 300 times stronger than steel. The mechanical strength of graphene depends upon structure, defect and size. It is theoretically predicted that fracture strength of zigzag graphene ribbons are higher than that of armchair graphene ribbons [27]. Despite the high mechanical strength, graphene is very flexible, and it can be stretched up to 20% of its original length in any direction.

5. Synthesis of Graphene

The synthetic protocol for graphene has been developed by various research groups in large number since its historic discovery by A. Geim and K. Novoselov in 2004. The synthesis of graphene has been broadly classified into two categories such as (1) Top down approach and (2) Bottom-up approach. The top-down approach involves various synthetic routes such as mechanical cleavage by scotch tape method [28], exfoliation of graphite by ultrasonication [29], ball- milling technique [30] and graphene oxide mediated technique [31]. The bottom-up approach also involves various synthetic routes such as epitaxial growth on silicon carbide [32], chemical vapour deposition of graphene from various carbon sources on metal substrates like ruthenium, iridium, copper and nickel [33], growth of graphene from metal-carbon melts [34], reduction of gaseous CO_2 or dry ice [35], total organic synthesis of graphene [36] and unzipping of carbon nanotube by control oxidation [37].

Among the various methods, graphene oxide mediated technique has been widely applied for the large-scale synthesis of graphene. This method involves two steps for synthesis of graphene from graphite i.e, (1) Oxidation of graphite to graphene oxide by a mixture of concentrated acid and oxidising agent followed by thermal exfoliation and (2) Reduction of graphene oxide to graphene by the use of reducing agents. In 1859, Brodie et al. demonstrated for the first time, the synthesis of GO by adding $KClO_3$ to a slurry of graphite in the presence of fuming HNO_3 [38]. In 1898, Staudenmaier et al. made some modification to Brodie's method by using both concentrated H_2SO_4 and fuming HNO_3 and adding the

$KClO_3$ at different intervals over the course of the reaction. It results in the production of GO at higher extent in one step process [39]. A new method, is widely adopted for the oxidation of graphite to graphene oxides, was reported by Hummers et al. in 1958 [40]. In this method, graphite is oxidized by $KMnO_4$ and $NaNO_3$ in the presence of concentrated H_2SO_4 without the use of HNO_3.

The drawback of the above mentioned graphene oxide synthetic protocols is the generation of very toxic gases (NO_2, N_2O_4, and ClO_2), and also the chemical $NaNO_3$ being an explosive. Therefore, efforts have been made for developing various synthetic routes for graphene oxide avoiding toxic chemicals like HNO_3 and $KClO_3$ [41]. In spite of this, Hummer's method has been found to be a safer, quicker and more efficient process for the synthesis of graphene oxide that is widely applied, often with some modifications. Marcano et al. modified the Hummer's method by using more amount of $KMnO_4$ [41b]. The method has been improved by excluding the use of $NaNO_3$ and carried out the reaction in a 9:1 mixture of H_2SO_4/H_3PO_4 with increasing amount of $KMnO_4$. The improved method shows higher extent oxidation of graphite and produces more hydrophilic and oxidised graphene oxide as compared to the Hummer's method. In another synthetic protocol, $K_2S_2O_8$ and P_2O_5 have been utilised for the synthesis of graphene oxide in combination with H_2SO_4 and $KMnO_4$ [42].

Graphene oxide is obtained as an intermediate product during the preparation of graphene from graphite by chemical oxidation-reduction process. It is a single layer graphene with various functional groups like carboxylic, aldehyde/ketone and alcoholic/epoxy on its surface and edges. There are a large number of methods documented so far for the reduction of graphene oxide to graphene. The methods, being used for preparing graphene by the reduction of GO, involve thermal reduction [43], photochemical reduction [44], electrochemical reduction [45], microwave-assisted reduction [44, [46], hydrothermal reduction [47] and chemical reduction using several reducing agents [48]. Among various reduction techniques of GO, the chemical method is very versatile, which utilizes reducing agents like $NaBH_4$ [49], hydrazine hydrate [50, 51], hydroquinone [52], urea [53, 54], vitamin C [55-57], thiophene [58], aluminium Powder [59], amino acid [60, 61], pyrrole [62, 63], alcohols [64], carbon monoxide [65], $LiAlH_4$ [66], oxalic acid [67] etc. Although most of the functional groups, present in the intermediate graphene oxide are removed, the π-conjugation of graphene surface is restored after reduction. But the chemical reduction processes for the development of reduced graphene oxide (rGO) lower the electrical conductivity properties compared to intrinsic graphene. Hence, the development of new techniques for the production of high quality graphene in bulk amount is a very challenging task.

6. GRAPHENE-NOBLE METAL HYBRID NANOSTRUCTURES

Owing to large surface area and extraordinary properties, graphene has been attracted as the ideal catalytic support for loading of metal nanoparticles [13]. The graphene-supported metal nanoparticle has explored as the new generation of the hybrid catalyst because of better dispersion of metal nanoparticles on the support and fast electron transfer [68]. In particular, graphene-noble metal nanoparticle has been extensively studied and applied in many research fields such as electrocatalysis, catalysis, surface-enhanced Raman scattering (SERS), and electrochemical sensing [69]. In this chapter, we provide brief information about different methods for synthesis of graphene-noble metal nanocomposites and their versatile electrochemical applications towards the development of fuel cell and the fabrication of sensors.

6.1. Synthetic Approaches

Several strategies have been developed to prepare graphene-noble metal hybrid nanomaterials like chemical reduction, electrochemical reduction, microwave reduction, hydrothermal reduction, chemical vapour deposition method, ultrasonication technique and photochemical reduction etc. Generally, the fabrication methods have been classified based on the preparation method as ex-situ and in-situ approaches (Figure 2).

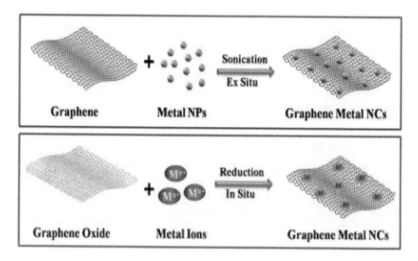

Figure 2. Schematic illustration for synthesis of graphene-metal nanocomposites by ex-situ and in-situ approaches.

6.1.1. Ex-situ approach

In the ex-situ approach, the as-prepared graphene sheets and pre-synthesized or commercially available metal nanoparticles are mixed in solution. Usually, in this process, the metal nanocrystals and /or graphene are modified with different functionalized molecules before their mixing. So the hybrid materials are stabilised by a strong non-covalent interaction or chemical bonding. For example, bovine serum albumin (BSA) protein, an amphiphilic bio-polymer, has been utilised to functionalize the surface of rGO via the π-π interaction which serves as universal binder molecule to absorb Au, Ag, Pd and Pt nanoparticles [70]. Similarly, Jin et al. successfully prepared graphene-Pd nanoparticle hybrid from the mixture of as synthesized Pd nanoparticle in hexane solution and graphene in dimethyl formamide solution [71]. This method provided the uniform distribution of Pd nanoparticle on graphene with high density in comparison to commercially available Pd-C.

6.1.2. *In-situ approach*

The ex-situ approach can preselect nanostructures with desired functionalities. However, it sometimes produces low coverage of nanostructures on the graphene surfaces [72]. In contrast; the in-situ approach can produce high and uniform surface coverage of nanocrystals on graphene through controlled growth on graphene via surface functionalization. It gives a continuous and well dispersion of nanoparticles on graphene surfaces. Further, the in-situ approach involves various methods like chemical reduction, hydrothermal reduction and microwave reduction etc.

6.1.2.1. Chemical reduction method

Chemical reduction method has been followed as the most reliable strategy for metal nanostructures synthesis. Both GO and noble metal precursors, such as $HAuCl_4$, $AgNO_3$, K_2PtCl_4 and H_2PdCl_6, can be simultaneously reduced by various reducing agents. For example, Chen et al. documented the synthesis of graphene–Au nanoparticle composites by the simultaneous reduction by $NaBH_4$ [73]. In another strategy, dense coverage of nanoparticle modified reduced graphene–Au nanoparticle composites have been developed by the in-situ reduction of 3,4,9,10-perylene tetracarboxylic acid (PTCA) modified reduced graphene sheets and $HAuCl_4$ [74]. Metal nanostructures of different shapes like nanorods and snowflakes supported on GO/rGO surface have also been synthesized via chemical reduction [75]. For example, Jasuja and Berry successfully prepared snowflakes like Au nanostructures on graphene oxide by seed mediated growth approach (Figure 3) [75b]. Further, the bimetallic nanostructures Pt-on-Pd has also been developed on graphene support by surface functionalization and seed-mediated method [76]. The spherical Pd nanoparticles on reduced graphene support are synthesized by the reduction of H_2PdCl_4 with HCOOH, and then, Pt-Pd nanodendrites are obtained by reducing K_2PtCl_4 with ascorbic acid on the Pd seeds [76]. Liu et al. reported the formation of Pd-Ag nanoring supported on graphene in a two step method. At first, graphene anchored Ag nanoparticle is prepared by in-situ reduction of graphene oxides and silver ion by sodium citrate and it follows the galvanic displacement by using Palladium (II) nitrate [77].

6.1.2.2. Photochemical reduction method

The reduction approach by photochemical has been followed for development of graphene-metal hybrid materials of interest. For example, Huang et al. has developed a new method to modify the GO/rGO surfaces with tiny fluorescent Au nanodots with particle size < 2nm via in-situ reduction of $HAuCl_4$ by photochemical irradiation technique [78]. They observed an interesting phenomenon from this process. If the rGO surface is pre-modified with octadecylthiol (ODT), the in-situ reduction process leads to the development of small Au nanodots and the Au nanodots self-assemble into a short chain-like structure on the rGO surface. Interestingly, the self-assembly of Au nanodots occurs along the <100> direction of a pure graphitic lattice structure. They explained that this interesting pattern formation is assisted by the specific assembly of ODT on the rGO surface. The pre-assembled thiol groups assist as a secondary template for formation and self-assembly of Au nanodots into this specific pattern.

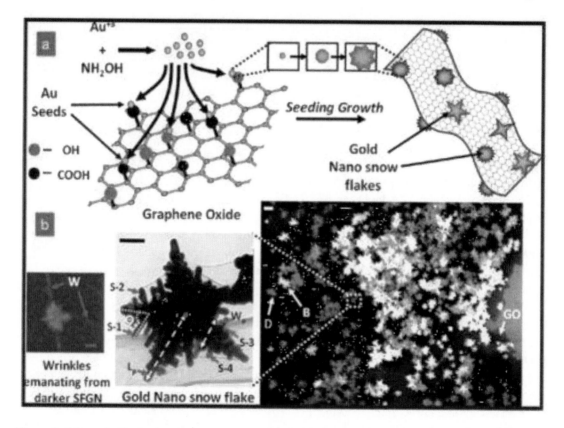

Figure 3. Schematic illustration for growth of snowflake shaped Au nanostructures on graphene oxide and (b) Corresponding TEM images. Reproduced with permission from reference [75b].

6.1.2.3. Microwave reduction method

Microwave irradiation method is a facile process for the synthesis of nanostructures. It has been utilised for the simultaneous reduction of GO and metal precursor. Many reports have been documented for the development of nanoparticle supported graphene by this approach [79]. For instance, Hassan et al. presented the microwave-assisted process for the synthesis of graphene decorated with Pd, Cu, Ag, Au and PdCu nanoparticle in aqueous as well as organic solvent [79b]. However, the synthesis of a controlled size and uniform distribution of nanoparticles on graphene surface is quite challenging with technique.

6.1.2.4. Hydrothermal reduction method

The hydrothermal method has been used mostly for the synthesis of inorganic nanoparticles. This method is operated at an elevated temperature and high pressure. This is a one-pot synthetic process to produce nanostructures of high crystallinity avoiding the post synthesis high temperature annealing process. This method has been used for the in-situ development of nanoparticle supported graphene. For example, Hu et al. developed well dispersed ternary Pt/PdCu nanoboxes on graphene (3D) by the hydrothermal process (Figure 4) [80].

6.1.2.5. Chemical vapour deposition method

The chemical vapour deposition (CVD) is the most reliable approach for the growth of high quality graphene. Further, the CVD process has been extended by several groups to produce metal nanoparticle decorated graphene. For instant, Giovanni et al. reported the formation of graphene supported Au, Ag, Pt and Pd nanoparticles by this technique [81].

6.1.2.6. Ultra-sonication Method

The substantial research effort has been made to avoid the high usages of external reagents for the development of nanoparticles. The sonochemical method has been attracted as the facile method for development of nanomaterials with minimal use of reducing reagents. This method has been employed for the development of reduced graphene oxide supported metal nanostructures. This process provides advantage for proper exfoliation of graphene sheets and the homogeneous dispersion of metal precursors on its surface [82]. For example, Anandan et al. reported the preparation of graphene supported Pt-Sn bimetallic hybrid material by the sonochemical method [82c].

6.1.2.7. Electrochemical deposition method

The electrochemical deposition of metal nanocrystals on graphene substrate without the use of reducing agent and surfactant, etc is an attractive and green approach to prepare the graphene-metal hybrid material [83]. For example, Zhou et al. successfully synthesized graphene-Pt nanoparticle by electrochemical deposition method [83b]. They reported that the electrochemical reduction of GO occurs in the potential range of -0.75 V to -1.2 V vs. SCE and at higher negative potential -1.5 V, high quality graphene can be prepared. They developed one step electrodeposition method for the synthesis of Pt NPs@G with average particle size of 10 nm by the simultaneous reduction of GO and Pt ions at -1.5 V vs. SCE.

6.2. Electrocatalytic and Electroanalytical Application

The synergetic properties of graphene and noble metal nanohybrids establish their promising application in many fields. The unique feature of graphene to support the nanostructures of desired properties drives for special applications. The high specific surface area of graphene not only plays the best support in loading nanostructures but triggers their activities. The enhancement of the properties of hybrid materials occurs due to the synergetic effect of both the graphene and nanostructures supported on it. As catalytic support, graphene possesses unique properties such as large surface area, high inertness towards the chemical environment and better thermal as well as chemical stability. These properties of graphene not only enhance the performance of the catalyst but also provide sufficient stability to the catalyst that finds extensive utilisation in the field of electrocatalytic and electroanalytic applications.

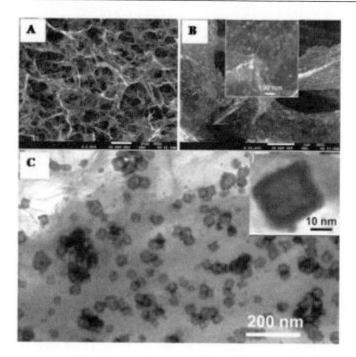

Figure 4. SEM images of Pt/PdCu nanoboxes supported of 3D graphene framework synthesized by hydrothermal method (A,B) and corresponding TEM image(C) (Inset of C shows TEM images of a single nanobox). Reproduced with permission from reference [80].

6.2.1. Electrocatalytic application

In the recent years, substantial interest has been devoted to develop the energy conversion devices to meet the need of huge energy demands. The fuel cell is a major device in which the electrical energy is produced by oxidizing fuel on electrodes using catalyst materials of interest [84]. The low-temperature fuel cells have been attracting because of the important advantage as a portable power sources. Usually, the low-temperature fuel cell is designed to work at the temperature below 200°C [85]. The efficiency of fuel cells mainly depends on catalysts on the electrode surface. Platinum and its alloys have been widely used as high efficient electrocatalysts for low-temperature fuel cells. However, the cost of platinum and high surface poisoning limits its practical application. Therefore, platinum free electrocatalysts of other noble metals such Pd, Ag, Au and their bimetallic/trimetallic alloy or core-shell nanostructures have been developed by several research groups for better activity and durability [86].

Despite the huge advancement in unsupported noble metal nanostructures as electrocatalyst, the low performance, poor stability and high usage of noble metal prevents their practical applications. This has motivated the researchers to enhance the efficiency of noble metal based electrocatalysts. The best process adopted to enhance the efficiency of noble metal electrocatalysts is to develop an ideal catalytic support that anchors the catalysts on its surface, increase the activity and provide the stability. Therefore, efforts have been made to utilise carbon based materials such as carbon black, carbon nanotube, fullerene, carbon nanohorns as catalytic support. In most of the cases, the catalytic support corrodes in the fuel cell operation conditions and thereby decreasing the performance of the electrocatalyst. In this aspect, graphene has been treated as the ideal support owing to its low cost production, large surface area for wide dispersion of metal nanoparticles and high corrosion resistance [87].

6.2.1.1. Graphene-Pt nanostructures

It has been reported that graphene-Pt metal nanoparticle shows the excellent electrocatalytic activity and durability towards methanol oxidation. For example, Honma and group revealed the excellent activity and durability of graphene supported Pt nanocatalysts compared to the commercially available CB–Pt catalyst towards methanol oxidation reaction (MOR) [88a]. Further, Wang et al. checked the performance of graphene supported platinum (Pt/G) towards methanol oxidation and compared with CNT supported platinum (Pt/CNTs). They have observed a significant improvement in the catalytic efficiency of Pt/G compared to Pt/CNTs (dimension of Pt = 3nm for both the support) [88b].

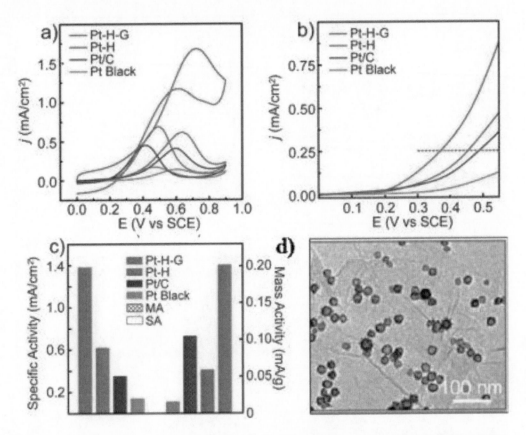

Figure 5. (a) Cyclic voltammograms and (b) linear-sweep voltammograms curves for MOR catalysed by Pt–H–G (red), Pt–H (green), commercial Pt/C (blue) and Pt black (cyan) in aqueous solution containing 0.5 M H$_2$SO$_4$ and 0.5 M methanol. (c) Comparison of MOR area specific activity (left) and mass activity (right) of the four catalysts, calculated from MOR current in CV curves obtained at 0.65 V (vs. SCE) and ECSA and mass of the used catalysts, respectively. (d) TEM images of hollow Pt on graphene. Reproduced with permission from reference [89].

The use of Pt/G also has been verified towards the oxygen reduction reaction (ORR) for fuel cell interest. It proves a significant improvement in the catalytic activity as observed in the case of methanol oxidation reactions (MOR). It is proposed that the high electrochemically accessible surface area (ECSA) of Pt/G promotes the efficiency of ORR and MOR activity. Liu et al. verified the effect of the specific surface area of functionalized graphene sheets supported Pt nanoparticles (Pt-FGS) towards electrochemical oxygen reduction reaction (ORR), and compared with commercially available Pt nanoparticles on carbon black (E-TEK) [88c]. Interestingly, they observed that the Pt-FGS possessed high ECSA and improved ORR activity. Also, Pt-FGS catalyst is stable beyond the cyclic voltammetry (CV) cycles of 5000. Further, the shape and morphology of Pt nanoparticle on graphene greatly affect the electrocatalytic performance. For example, Wan and co-workers reported the enhanced activity of hollow Pt nanostructures supported on graphene utilising the hollow structure and synergistic effect of graphene Pt interaction (Figure 5) [89].

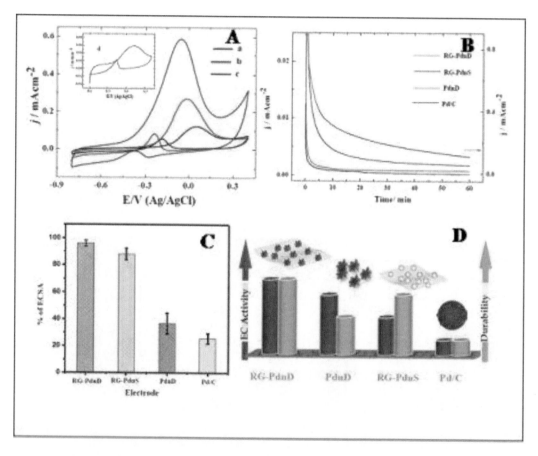

Figure 6. (A) Cyclic Voltammogram measurements for (a) RG-PdnD, (b) PdnD, (c) RG-PdnS and (d) Pd/C modified electrodes towards oxidation of methanol (0.25M) in 0.1 M KOH, respectively. Scan rate: 10 mV/s. (B) corresponding Chronoamperometric profiles of electrode matrerials. (C) The change in ESCA value of electrodes after continuous cycles (100) in absence of methanol (D) summarized comparative electrocatalytic and durability performance of RG-PdnD, PdnD, RG-PdnS and Pd/C modified electrodes based on CV and CA measurements methanol electrooxidation (The plot bars are not drawn to scale). Reproduced with permission from reference [92].

One dimensional Pt nanowires supported on S-doped graphene exhibited 2-3 times the higher catalytic performance as compared to Pt/C towards methanol oxidation and oxygen reduction owing to its anisotropic nature [90]. Graphene supported other anisotropic Pt nanostructures such as dendritic, flower shaped, and branched structures have been reported as an excellent electrocatalysts. For example, Sahu et al. reported the in-situ synthesis of branched Pt nanostructure-graphene hybrid as excellent bifunctional electrocatalyst for MOR and ORR in 0.5 M H_2SO_4 [91].

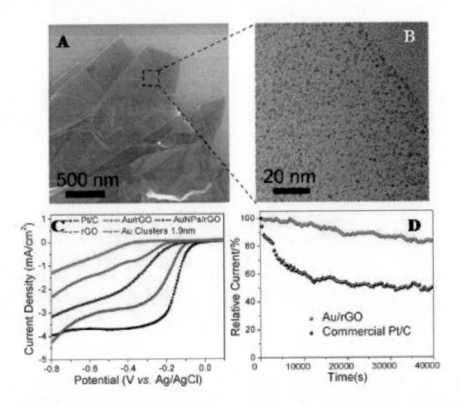

Figure 7. TEM images of Au nanocluster supported graphene (A,B) and RDE curves of commercial Pt/C, Au/rGO hybrids, Au NP/rGO hybrids, rGO sheets, Au clusters in O_2-saturated 0.1M KOH at a scanning rate of 50 mV/s at 1600 rpm; (B) Current-time (i-t) chronoamperometric responses for ORR at the Au/rGO hybrids and commercial Pt/C electrodes in an O_2-saturated 0.1 M KOH solution at -0.2 V vs Ag/AgCl. Reproduced with permission from reference [97a].

6.2.1.2. Graphene-Pd nanostructures

Graphene-Pd nanostructures have been attracting as the efficient electrocatalysts because of their high performance and stability resulting from low cost and high surface poisoning resistance of Pd and the outstanding properties of graphene. For example, Sahu et al. have synthesized Pd nanodendrite-graphene hybrid nanocomposite by in-situ approach with excellent performance towards MOR activity and high durability (Figure 6) [92]. Zhao et al. reported the synthesis of Pd nanoparticle-graphene nanosheet by using copper as the sacrificial template with the high electrocatalytic performance for oxidation of formic acid [93]. Wang and co-workers observed that the much enhanced electrocatalytic performance of Pd nanoparticles on three dimensional reduced graphene oxide as compared to carbon black supported Pd nanoparticles with the same amount of Pd loading. This higher catalytic performance was believed to be due to the porous structures of graphene and well distribution of Pd nanoparticles on its surface [94]. In another work, Fan and co-workers reported the high electrocatalytic performance of graphene-Pd nanocomposites towards ethanol oxidation compared to the commercially available Vulcan XC-72R carbon supported Pd nanoparticles [95]. Further, graphene supported Pd nanostructures are also attractive as cathode catalysts for the oxygen reduction reaction. For instance, Carrera-Cerritos et al. reported the graphene supported Pd nanoparticle as high performance ORR catalyst [96].

6.2.1.3. Graphene-Au and Ag nanostructures

The coinage metals such as Au and Ag are not electrochemically active in bulk form. But, the nanostructured Au and Ag metals exhibit relatively high performance. Moreover, the catalytic performance of these nanomaterials has been improved by supporting on conducting support like graphene [97]. For example, Yin et al. prepared Au cluster/graphene hybrid by the surfactant-free method, and reported comparable ORR activity and superior stability as compared to commercial Pt/C (Figure 7) [97a]. Davis et al. reported the silver-graphene nanoribbon composite as high performance electrocatalyst for ORR activity in alkaline media [97b].

6.2.1.4. Graphene supported Pt and Pd-based bimetallic/ trimetallic nanostructures

The Pt-based bimetallic or trimetallic nanostructures have been developed to improve the catalytic performance, durability, and to minimise the utility of Pt. Although, a great advancement has been achieved in the synthesis of these electrocatalysts, the unsupported nanostructures are vulnerable to surface poisoning. Therefore, attempts have been made to improve their performance by making composite with conductive support such as graphene. A large number of graphene supported Pt-based bimetallic nanostructures like PtRu/GNs, PtPd/GNs, PtNi/GNs and PtFe/GNs, etc. have been synthesized with enhanced electrocatalytic performances [98]. Among all the graphene supported bimetallic catalysts, PtRu/GNs was found to exhibit better performance. It effectively reduces the surface poisoning through the removal of CO like reaction intermediates by the hydroxyl ions adsorbed on Ru surface [98c]. For example, the graphene supported PtRu nanoparticles possess improved catalytic activity towards methanol oxidation compared to the commercial available PtRu/C counterpart. This is believed due to the extraordinary properties of graphene like high electronic conductivity and large specific surface area [99].

Despite the advances in the development of Pt-based supported and unsupported nanomaterials, the cost effectiveness of Pt-based electrocatalysts limits their practical applications [100]. Therefore, researchers have devoted more efforts towards the development of low cost platinum free electrocatalysts with enhanced performance and durability. In this respect, graphene supported Pd based electrocatalysts have been attracting much interest. The synergistic effect of graphene support, high surface resistance of Pd and low cost with less Pd content and platinum free has made graphene supported Pd based bi/trimetallic nanostructures as efficient electrocatalysts. For example, Niu et al. reported the ionic liquid mediated synthesis of graphene supported hollow Au/Pd alloy nanostructures. They observed the enhanced electrocatalytic activity towards the formic acid oxidation reaction that is attributed to the distinct electronic conductivity of graphene and hollow morphology of Au/Pd alloy nanoparticles [101].

6.2.2. Electroanalytical application

Graphene possesses large surface to volume ratio and fascinating electrochemical properties [102]. It has the large electrochemical potential window in the buffer solution which enables the detection of molecules oxidised/reduced at a higher potential. The edges and defects on graphene surface facilitate the high electron transfer rate. Further, the electron transfer rate can be enhanced by direct conducting wiring between the electrode surface and redox enzyme centre. It has been reported that noble metal nanostructures supported on graphene are excellent hybrid material for sensing biomolecules/ions/chemicals and gases. The graphene-based nanomaterial can improve the performance by high sensitivity and selective detection of analytes through the synergistic effect. These graphene based materials have been utilised as enzymatic and non-enzymatic sensors for detection of analytes. The details about the graphene-noble metal nanostructure based sensors have been discussed in the subsequent section.

6.2.2.1. Graphene-noble metal nanoparticles based enzymatic sensor

Graphene-noble metal nanostructures based enzymatic sensors have been attractive for their unique sensing property utilising the large surface for immobilisation of enzyme and well dispersion of nanoparticles. For example, Zhou et al. fabricated microperoxidase-11 based RGO-gold nanoparticles for detection of H_2O_2. They reported that the synergetic effect of gold nanoparticles and RGO sheets enhanced the catalytic activity and increased charge transfer kinetics [103]. Dey et al. designed a cholesterol biosensor by immobilizing cholesterol oxidase and cholesterol esterase enzymes on the surface of Pt nanoparticle-graphene hybrid material (Figure 8) [104]. Guo and co-workers reported aptasensor for detection of L-histidine by immobilising structure changing DNAzymes on Au NPs-graphene hybrid [105]. Yang et al. reported the electrochemical synthesis of graphene sheets-AuPd alloy nanoparticles for enzymatic sensing of glucose using glucose oxidase [106].

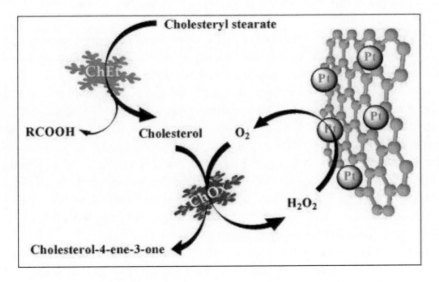

Figure 8. Scheme illustrating the Biosensing of Cholesterol Ester with the GNS-nPt based Biosensor. Reproduced with permission from reference [104].

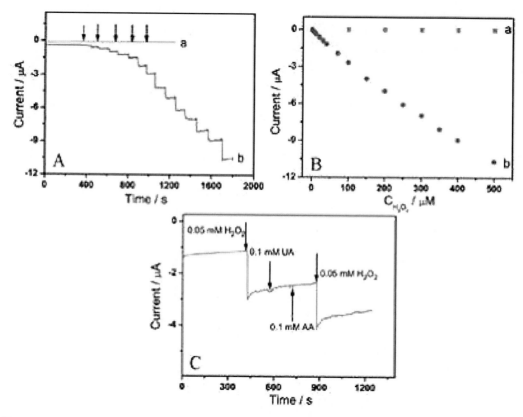

Figure 9. (A) Current-time recordings for successive additions of H_2O_2 at the GNs/GCE (a) and PNEGHNs/GCE (b) measured at 0 V. (B) The plot of current of H_2O_2 *versus* its concentrations. (C) Current responses of the PNEGHNs/GCE to the sequential additions of 0.05 mM H_2O_2, 0.1mM AA, and 0.1 mM UA into 0.1 M PBS. Reproduced with permission from reference [108].

6.2.2.2. Graphene-noble metal nanostructures based non-enzymatic sensors

The enzymatic detection of analytes has limited its practical application because of the cost effectiveness of enzymes, sophisticated immobilization procedures [102]. Therefore, non-enzymatic detection has been receiving much importance. Several research groups reported the non-enzymatic detection of molecules and ions using graphene-noble metal hybrid nanomaterials. For example, Sun and co-workers reported the non-enzymatic detection of H_2O_2 using graphene-Ag nanoparticle composite synthesized under different conditions. They found that the high sensitivity (1.8 µM detection limit) of graphene-Ag NP synthesized under alkaline condition than that of graphene-Ag NP synthesized under acidic condition (3.6 µM detection limit) [107]. The enhancement in the electrochemical performance of the composites is attributed to the morphology of Ag NPs. Wang and co-workers reported graphene-Pt nanoparticles hybrid catalyst for non-enzymatic sensing of H_2O_2 with detection limit 80 nM which is sufficiently lower than other electrodes (Figure 9) [108].

Similarly, Wu et al. electrochemically deposited Pt nanoparticles on RGO and high sensitivity (0.6 µM) was obtained for glucose detection [109]. Du and co-workers developed RGO electrode–Au NPs by potentiostatic electro-deposition method to detect ssDNA [110]. Liu et al. reported the non-enzymatic detection of H_2O_2 by Pd nanoparticles supported on graphene nanosheets with high sensitivity and selectivity [111]. Hu et al. synthesized hollow

Pt-Ni nanostructures-graphene composite for non-enzymatic detection of glucose with high electrochemical performances [112].

CONCLUSION AND PERSPECTIVE

In this chapter, we discussed the various methods for synthesis of graphene and graphene-noble metal hybrid nanomaterials. The electrochemical applications towards fuel cell and sensor have been discussed briefly. The fundamental properties of graphene require exploration so that new and exciting properties can be observed. Although, many methods have been established for the synthesis of graphene, the new methodologies and techniques are highly desirable for massive production of high quality graphene.

ACKNOWLEDGEMENT

We acknowledge CSIR, New Delhi India (YSP-02, P-81-113) and Ministry of New and Renewable Energy, New Delhi, India (No.102/87/2011-NT) for financial support. SCS and AKS thank CSIR, India for fellowship. The authors are grateful to Dr. D. B. Ramesh Chief Librarian, Siksha 'O' Anusandhan University for helping in grammar and plagiarism checking.

REFERENCES

[1]   (a) Jain, PK; Huang, XI; El-Sayed, H; El-Sayed, MA. *Acc. Chem. Res.* 2008, 41, 1578-1586. (b) Guo, S; Wang, E. *Nano Today*, 2011, 6, 240-264.

[2]   Fahlman, BD. *Materials Chemistry*, Springer, Mount Pleasant, 2007, Vol.1, 282–283.

[3]   Xia, Y; Xiong, Y; Lim, B; Skrabalak, SE. *Angew. Chem. Int. Ed.*, 2009, 48, 60-103.

[4]   (a) Peng, Z; Yang, H. *Nano Today*, 2009, 4, 143-164. (b) Dunleavy, JK. *Platinum Met. Rev.*, 2006, 50, 110-110. (c) Williams, KR; Burstein, GT. *Catal. Today*, 1997, 38,401-410.

[5]   (a) Mohanty, A; Garg, N; Jin, R. *Angew. Chem., Int. Ed.*, 2010, 49, 4962-4966. (b) Langhammer, C; Zori, I; Kasemo, B; Clemens, BM. *Nano Lett.*, 2007, 7, 3122-3127. (c) Fernandez-Garcia, M; Martinez-Arias, A; Salamanca, LN; Coronado, JM; Anderson, JA; Conesa, JC; Soria, J. *J. Catal.*, 1999, 187, 474-485.

[6]   (a) Mazumder, V; Sun, SH. *J. Am. Chem. Soc.*, 2009, 131, 4588-4589. (b) Bianchini, C; Shen, PK. *Chem. Rev.*, 2009, 109, 4183-4206.

[7]   (a) Crooks, RM; Zhao, M; Sun, L; Chechik, V; Yeung, LK. *Acc. Chem. Res.*, 2001, 34, 181-190.(b) Cheng, Y; Samia, AC; Meyers, J. D; Panagopoulos, I; Fei, B; Burda, C. *J. Am. Chem. Soc.*, 2008, 130, 10643-10647. (c) Hu, MS; Chen, HL; Shen, CH; Hong, LS; Huang, BR; Chen, KH; Chen, LC. *Nat. Mater.*, 2006, 5, 102-106.

[8]   (a) Cao, YWC; Jin, R; Mirkin, CA. *Science*, 2002, 297, 1536-1540. (b) Jin, R. *Angew. Chem.*, 2008, 120, 6852-6855. *Angew. Chem. Int. Ed.*, 2008, 47, 6750-6753. (c)

Mohanty, A; Garg, N; Jin, R. *Angew. Chem.*, 2010, 122, 5082-5086. (d) Wu, Z; Chen, J; Jin, R. *Adv. Func. Mater.*, 2011, 21, 177-183.

[9]   (a) Nagy, A; Mestl, G. *Appl Catal A.*, 1999, 188, 337-353. (b) Frattini, A; Pellegri, N; Nicastro, D.; de Sanctis, O. *Mater Chem Phys.*, 2005, 94, 148-152. (c) Cao, YW; Jin, RC; Mirkin, CA. *Science*, 2002, 297, 1536-1540. (d) Rosi, NL; Mirkin, CA. *Chem. Rev.*, 2005, 105, 1547-1562.

[10]   Jariwala, D; Sangwan, VK; Lauhon, LJ; Marks, TJ; Hersam, MC. *Chem. Soc. Rev.*, 2013, 42, 2824-2860.

[11]   Scida, K; Stege, PW; Haby, G; Messina, GA; García, CD. *Anal. Chim. Acta*, 2011, 691, 6-17.

[12]   Falcao, EHL; Wudl, F. *J Chem Technol Biotechnol*, 2007, 82, 524–531.

[13]   (a) Geim, AK; Novoselov, KS. *Nat. Mater.*, 2007, 6, 183-191. (b) Nardecchia, S; Carriazo, D; Ferrer, ML; Gutiérreza, MC; Monte, F. *Chem. Soc. Rev.*, 2013, 42, 794-830.

[14]   (a) Gao, XF; Jang, J; Nagase, S. *J. Phy. Chem. C.*, 2010, 114, 832-842. (b) Stankovich, S; Dikin, DA; Piner, RD; Kohlhaas, KA; Kleinhammes, A; Jia, YY; Wu, Y; Nguyen, ST; Ruoff, RS. *Carbon.*, 2007, 45,1558-1565. (c) Shin, HJ; Kim, KK; Benayad, A; Yoon, SM; Park, HK; Jung, IS; Jin, MH; Jeong, HK; Kim, JM; Choi, JY; Lee, YH. *Adv. Funct. Mater.*, 2009, 19, 1987-1992. (d) Huang, X; Qi, X; Boey, F; Zhang, H. *Chem. Soc. Rev.*, 2012, 41, 666-686. (e) Bai, S; Shen, X. *RSC Adv.*, 2012, 2, 64-98. (f) Zhu, J; Chen, M; He, Q; Shao, L; Wei, S; Guo, Z. *RSC Adv.*, 2013, 3, 22790-22824.

[15]   Smalley, RE. *Angcw. Chem. Int. Ed. Engl.*, 1997, 36, 1594-1601.

[16]   Monthioux, M; Kuznetsov, VL. *Carbon.*, 2006, 44, 1621-1623.

[17]   Dresselhaus, MS; Araujo, PT. *ACS Nano*, 2010, 4 , 6297–6302.

[18]   Kroto, HW; Heath, JR; O'Brien, SC; Curl, RF; Smalley, RE; *Nature*, 1985, 318, 162-163.

[19]   Ganesh, EN. *Int. J. Inno. Technol. Expl. Eng.*, 2013, 2, 311-320.

[20]   (a) Wu, J; Pisula, W; Mu̇llen, K. *Chem. Rev.*, 2007, 107, 718-747. (b) Eda, G; Chhowalla, M. *Adv. Mater.*, 2010, 22, 2392–2415.

[21]   (a) Allen, MJ; Tung, VC; Kaner, RB. *Chem. Rev.*, 2010, 110, 132-145. (b) Dreyer, DR; Park, S; Bielawski, CW; Ruoff, RS. *Chem. Soc. Rev.*, 2010, 39, 228-240.

[22]   Burns, RC; Cvetkovic, V; Dodge, CV; Evans, DJF; Rooney, MLT; Spear, PM; Welbourn, CM. *J. Cryst. Growth*, 1990, 104, 257-279.

[23]   Zhu, Y; Murali, S; Cai, W; Li, X; Suk, JW; Potts, JR; Ruoff, RS. *Adv. Mater.*, 2010, 22, 3906-3924.

[24]   Peres, NMR. *Rev. Mod. Phys.*, 2010, 82, 2673-2673.

[25]   Mak, KF; Ju, L; Wang, F; Heinz, TF. *Solid State Commun.*, 2012, 152, 1341-1349.

[26]   Novoselov, KS; McCann, E; Morozov, SV; Fal'ko, VI; Katsnelson, MI; Zeitler, U; Jiang, D; Schedin, F; Geim, AK. *Nat. Phys.*, 2006, 2,177-180.

[27]   Lu, Q; Huang, R. *Phys. Rev. B*, 2010, 81, 155410-155410.

[28]   Novoselov, KS; Geim, AK; Morozov, SV; Jiang, D; Zhang, Y; Dubonos, SV; Grigorieva, IV; Firsov, AA. *Science.*, 2004, 306, 666-669.

[29]   (a) Hernandez, Y; Nicolosi, V; Lotya, M; Blighe, F; Sun, Z; De, S; McGovern, IT; Holland, B; Byrne, M; Gun'ko, YK; Boland, JJ; Niraj, P; Duesberg, G; Krishnamurthy, S; Goodhue, R; Hutchison, J; Scardaci, V; Errari, AC; Coleman, JN. *Nat. Nanotech.*,

2008, 3, 563-568. (b) Nuvoli, D; Alzari, V; Sanna, R; Scognamillo, S; Piccinini, M; Peponi, L; Kenny, JM; Mariani, A. *Nanoscale Res. Lett.*, 2012, 7, 674-684.

[30] (a) Zhao, W; Fang, M; Wu, F; Wu, H; Wang, L; Chen, G. *J. Mater. Chem.*, 2010, 20, 5817-5819. (b) Liu, L; Xiong, Z; Hu, D; Wu, G; Chen, P. *Chem. Commun.*, 2013, 49, 7890-5892. (c) Jeon, I-Y; Shin, Y-R; Sohn, G-J; Choi, H-J; Bae, S-Y; Mahmood, J; Jung, S-M; Seo, J-M; Kim, M-J; Chang, DW; Daia, L; Baek, J-B. *Proc. Natl. Acad. Sci. U.S.A*, 2012, 109, 5588-5593. (d) Jeon, I-Y; Choi, H-J; Jung, S-M; Seo, J-M; Kim, M-J; Dai, L; Baek, J-B. *J. Am. Chem. Soc.*, 2013, 135, 1386-1393.

[31] (a) Stankovich, S; Dikin, DA; Dommett, GHB; Kohlhaas, KM; Zimney, EJ; Stach, EA; Piner, RD; Nguyen, ST; Ruoff, RS. *Nature.*, 2006, 442, 282-286. (b) Wojtoniszak, M; Mijowska, E. *J. Nanopart. Res.*, 2012, 14, 1248-1254. (c) Loryuenyong, V; Totepvimarn, K; Eimburanapravat, P; Boonchompoo, W; Buasri, A. *Adv. Mater. Sci. Eng.*, 2013, 2013, 1-5.

[32] Sutter, P. *Nat. Mater.*, 2009, 8, 171-172.

[33] (a) Grandthyll, S; Gsell, S; Weinl, M; Schreck, M; Hüfner, S; Müller, F. *J. Phys. Condens. Matter.*, 2012, 24, 314204-314204. (b) Pletikosid, I; Kralj, M; Pervan, P; Brako, R; Coraux, J; N'Diaye, AT; Busse, C; Michely, T. *Phys. Rev. Lett.*, 2009, 102, 056808-056808. (c) Wassei, JK; Mecklenburg, M; Torres, JA; Fowler, JD; Regan, BC; Kaner, RB; Weiller, BH. *Small.*, 2012, 8, 1415-1422. (d) Lenski, DR; Fuhrer, MS. *J. Appl. Phys.*, 2011, 110, 013720-013720.

[34] Amini, S; Garay, J; Guanxiong, L; Balandin, AA; Abbaschian, R. *J. Appl. Phys.*, 2010, 108, 094321-094321.

[35] Chakrabarti, A; Lu, J; Skrabutenas, JC; Xu, T; Xiao, Z; Maguireb, JA; Hosmane, NS. *J. Mater. Chem.*, 2011, 21, 9491-9493.

[36] (a) Yang, XY; Dou, X; Rouhanipour, A; Zhi, LJ; Rader, HJ; Mullen, K. *J. Am. Chem. Soc.*, 2008, 130, 4216-4217. (b) Wu, J; Pisula, W; Müllen, K. *Chem. Rev.*, 2007, 107, 718-747. (c) Watson, MD; Fechtenkötter, A; Müllen, K. *Chem. Rev.*, 2001, 101, 1267-1300.

[37] (a) Kosynkin, DV; Higginbotham, AL; Sinitskii, A; Lomeda, JR; Dimiev, A; Price, BK; Tour, JM. *Nature.*, 2009, 458, 872-876. (b) Jiao, L; Zhang, L; Wang, X; Diankov, G; Dai, H. *Nature.*, 2009, 458, 877-880. (c) Shinde, DB; Debgupta, J; Kushwaha, A; Aslam, M; Pillai, VK. *J. Am. Chem. Soc.*, 2011, 133, 4168-4171.

[38] Brodie, BC. *Philos. Trans. R. Soc. London.*, 1859, 14, 249-259.

[39] Staudenmaier, L. *Ber. Dtsch. Chem. Ges.*, 1898, 31, 1481-1487.

[40] Hummers, WS; Offeman, RE. *J. Am. Chem. Soc.*, 1958, 80, 1339-1339.

[41] (a) Goncalves, G; Marques, PAAP; Granadeiro, CM; Nogueira, HIS; Singh, MK; Gracio, J. *Chem. Mater.*, 2009, 21, 4796-4802. (b) Marcano, DC; Kosynkin, DV; Berlin, JM; Sinitskii, A; Sun, Z; Slesarev, A; Alemany, LB; Lu, W; Tour, JM. *ACS Nano.*, 2010, 4, 4806-4814. (c) Zhang, L; Li, Y; Zhang, L; Li, D-W; Karpuzov, D; Long, Y-T. *Int. J. Electrochem. Sci.*, 2011, 6, 819-829. (d) Chen, J; Yao, B; Li, C; Shi, G. *Carbon.*, 2013, 64, 225-229.

[42] (a) Kempaiah, R; Salgado, S; Chunga, WL; Maheshwari, V. *Chem. Commun.*, 2011, 47, 11480-11482. (b) Gilje, S; Han, S; Wang, M; Wang, KL; Kaner, RB. *Nano Lett.*, 2007, 7, 3394-3398. (c) Kovtyukhova, NI; Ollivier, PJ; Martin, BR; Mallouk, TE; Chizhik, SA; Buzaneva, EV; Gorchinskiy, AD. *Chem. Mater.*, 1999, 11, 771-778.

[43] (a) Liao, K-H; Mittal, A; Bose, S; Leighton, C; Mkhoyan, KA; Macosko, CW. *ACS Nano.*, 2011, 5, 1253-1258. (b) Wong, CHA; Ambrosia, A; Pumera, M. *Nanoscale.*, 2012, 4, 4972-4977. (c) Larciprete, R; Fabris, S; Sun, T; Lacovig, P; Baraldi, A; Lizzit, S. *J. Am. Chem. Soc.*, 2011, 133, 17315-17321.

[44] (a) Stroyuk, AL; Andryushina, NS; Shcherban, ND; Il'in, VG; Efanov, VS; Yanchuk, IB; Kuchmii, SY; Pokhodenko, VD. *Theor. Exp. Chem.*, 2012, 48, 2-13. (b) Gong, P; Wang, Z; Li, Z; Mi, Y; Sun, J; Niu, L; Wang, H; Wang, J; Yang, S. *RSC Adv*, 2013, 3, 6327-6330. (c) Moon, G-H; Park, Y; Kim, W; Choi, W. *Carbon.*, 2011, 49, 3454-3462.

[45] (a) Shao, Y; Wang, J; Engelhard, M; Wang, C; Lin, Y. *J. Mater. Chem.*, 2010, 20, 743-748. (b) Xu, X; Huang, D; Cao, K; Wang, M; Zakeeruddin, SM; Grätzel, M. *Sci. Rep.*, 2013, 3, 1489-1489. (c) Viinikanoja, A; Wang, Z; Kauppila, J; Kvarnström, C. *Phys. Chem. Chem. Phys.*, 2012, 14, 14003-14009. (d) Kauppila, J; Kunnas, P; Damlin, P; Viinikanoja, A; Kvarnström, C. *Electrochimica Acta.*, 2013, 89, 84-89. (e) Guo, H-L; Wang, X-F; Qian, Q-Y; Wang, F-B; X-H. Xia, *ACS Nano.*, 2009, 3, 2653-2659.

[46] (a) Jasuja, K; Linn, J; Melton, S; Berry, V. *J. Phys. Chem. Lett.*, 2010, 1, 1853-1860. (b) Wang, S; Jiang, SP; Wang, X. *Electrochimica Acta.*, 2011, 56, 3338-3344. (c) Su, X; Chai, H; Jia, D; Bao, S; Zhou, W; Zhou, M. *New J. Chem.*, 2013, 37, 439-443. (d) Dai, Z; Wang, K; Li, L; Zhang, T. *Int. J. Electrochem. Sci.*, 2013, 8, 9384-9389.

[47] (a) Chen, H; Song, Z; Zhao, X; Li, X; Lin, H. *RSC Adv.*, 2013, 3, 2971-2978. (b) Zhou, Y; Bao, Q; Tang, LAL; Zhong, Y; Loh, KP. *Chem. Mater.*, 2009, 21, 2950-2956. (c) Krishnamoorthy, K; Veerapandian, M; Kim, G-S; Kim, SJ. *Curr. Nanosci.*, 2012, 8, 934-938. (d) Xu, Y; Sheng, K; Li, C; Shi, G. *ACS Nano*, 2010, 4, 4324-4330.

[48] (a) Chua, CK; Pumera, M. *Chem. Soc. Rev.*, 2013, 42, 3222-3233. (b) Chen, W; Yan, L; Bangal, PR. *J. Phys. Chem. C*, 2010, 114, 19885-19890. (c) Thomas, HR; Day, SP; Woodruff, WE; Vallés, C; Young, RJ; Kinloch, IA; Morley, GW; Hanna, JV; Wilson, NR; Rourke, JP. *Chem. Mater.*, 2013, 25, 3580-3588. (d) Park, S; Ruoff, RS. *Nat. Nanotechnol.*, 2009, 4, 217-224.

[49] Shin, H-J; Kim, KK; Benayad, A; Yoon, S-M; Park, HK; Jung, I-S; Jin, MH; Jeong, H-K; Kim, JM; Choi, J-Y; Lee, YH. *Adv. Funct. Mater.*, 2009, 19, 1987-1992.

[50] Stankovich, S; Dikin, DA; Piner, RD; Kohlhaas, KA; Kleinhammes, A; Jia, Y; Wu, Y; Nguyenb, ST; Ruoff, RS. *Carbon.*, 2007, 45, 1558-1565.

[51] Ren, P-G; Yan, D-X; Ji, X; Chen, T; Li, Z-M. *Nanotechnol.*, 2011, 22, 055705-055705.

[52] Wang, GX; Yang, J; Park, J; Gou, XL; Wang, B; Liu, H; Yao, J. *J. Phys. Chem. C.*, 2008, 112, 8192-8195.

[53] Lei, Z; Lu, L; Zhao, XS. *Energy Environ. Sci.*, 2012, 5, 6391-6399.

[54] Wakeland, S; Martinez, R; Grey, JK; Luhrs, CC. *Carbon.*, 2010, 48, 3463-6470.

[55] Fernández-Merino, MJ; Guardia, L; Paredes, JI; Villar-Rodil, S; Solís-Fernández, P; MartínezAlonso, A; Tascón, JMD. *J. Phys. Chem. C.*, 2010, 114, 6426-6432.

[56] Zhang, J; Yang, H; Shen, G; Cheng, P; Zhang, J; Guo, S. *Chem. Commun.*, 2010, 46, 1112-1114.

[57] Zhu, X; Liu, Q; Zhu, X; Li, C; Xu, M; Liang, Y. *Int. J. Electrochem. Sci.*, 2012, 7, 5172-5184.

[58] Some, S; Kim, Y; Yoon, Y; Yoo, H; Lee, S; Park, Y; Lee, H. *Sci. Rep.*, 2013, 3, 1929-1929.

[59] Fan, ZJ; Wang, K; Wei, T; Yan, J; Song, LP; Shao, B. *Carbon.*, 2010, 48, 1686-1689.

[60] Chen, D; Li, L; Guo, L. *Nanotechnol.*, 2011, 22, 325601-325601.

[61] Amarnatha, A; Hongb, CE; Kimc, NH; Kud, B-C; Kuilaa, T; Lee, JH. *Carbon.*, 2011, 49, 3497-3502.

[62] Chandra, V; Kim, KS. *Chem. Commun.*, 2011, 47, 3942-3944.

[63] Dreyer, DR; Murali, S; Zhu, Y; Ruoff, RS; Bielawski, CW. *J. Mater. Chem.*, 2011, 21, 3443-3447.

[64] Su, C-Y; Xu, Y; Zhang, W; Zhao, J; Liu, A; Tang, X; Tsai, C-H; Huang, Y; Li, L-J. *ACS Nano.*, 2010, 4, 5285-5292.

[65] Narayanan, B; Weeksa, SL; Jariwala, BN; Macco, B; Weber, J-W; Rathi, SJ; van de Sanden, MCM; Sutter, P; Agarwal, S; Ciobanu, CV. *J. Vac. Sci. Technol. A.*, 2013, 31, 040601-040601.

[66] Ambrosi, A; Chua, CK; Bonanni, A; Pumera, M. *Chem. Mater.*, 2012, 24, 2292-2298.

[67] Song, P; Zhang, X; Sun, M; Cui, X; Lin, Y. *RSC Adv.*, 2012, 2, 1168-1173.

[68] (a) Shang, N; Papakonstantinou, P; Wang, P; Silva, SRP. *J. Phys. Chem. C*, 2010, 114, 15837-15841. (b) Seger, B; Kamat, PV. *J. Phys. Chem. C*, 2009, 113, 7990-7995. (c) Liu, M; Zhang, R; Chen, W. *Chem. Rev.*, 2014, 114, 5117-5160. (d) Tan, C; Huang, X; Zhang, H. *Materials Today*, 2013, 16, 29-36.

[69] (a) Zhang, Z; Xu, F; Yang, W; Guo, M; Wang, X; Zhang, B; Tang, J. *Chem. Commun.*, 2011, 47, 6440-6442. (b) Jasuja, K; Linn, J; Melton, S; Berry, V. *J. Phys. Chem. Lett.*, 2010, 1, 1853-1860. (c) Li, Y; Fan, X; Qi, J; Ji, J; Wang, S; Zhang, G; Zhang, F. *Nano Res.*, 2010, 3, 429-437. (d) Qiu, J-D; Wang, G-C; Liang, R-P; Xia, X-H; Yu, H-W. *J. Phys. Chem. C.*, 2011, 115, 15639-15645. (e) Li, Y; Fan, X; Qi, J; Ji, J; Wang, S; Zhang, G; Zhang, F. *Materials Res. Bull.*, 2010, 45, 1413-1418.

[70] Liu, J; Fu, S; Yuan, B; Li, Y; Deng, Z. *J. Am. Chem. Soc.*, 2010, 132,7279-7281.

[71] Jin, T; Guo, S; Zuo, J-l; Sun, S. *Nanoscale.*, 2013, 5, 160-163.

[72] Huang, J; Zhang, L; Chen, B; Ji, N; Chen, F; Zhang, Y; Zhang, Z. *Nanoscale.*, 2010, 2, 2733-2738.

[73] Sudibya, HG; He, Q; Zhang, H; Chen, P. *ACS Nano.*, 2011, 5, 1990-1994.

[74] Li, FH; Yang, HF; Shan, CS; Zhang, QX; Han, DX; Ivaska, A; Niu, L. *J. Mater. Chem.*, 2009, 19, 4022-4025.

[75] (a) Kim, Y-K; Na, H-K; Min, D-H. *Langmuir.*, 2010, 26, 13065-13070. (b) Jasuja, K; Berry, V. *ACS Nano.*, 2009, 3, 2358-2366.

[76] Guo, S; Dong, S; Wang, E. *ACS Nano.*, 2009, 4, 547-555.

[77] Liu, M; Lu, Y; Chen, W. *Adv. Funct. Mater.*, 2013, 23, 1289-1296.

[78] Huang, X; Zhou, XZ; Wu, SX; Wei, YY; Qi, XY; Zhang, J; Boey, F; Zhang, H. *Small.*, 2010, 6, 513-516.

[79] (a) Sharma, S; Ganguly, A; Papakonstantinou, P; Miao, X; Li, M; Hutchison, JL; Delichatsios, M; Ukleja, S. *J. Phys. Chem. C.*, 2010, 114, 19459-19466. (b) Hassan, HMA; Abdelsayed, V; Khder, AERS; AbouZeid, KM; Terner, J; El-Shall, MS; AlResayes, SI; El-Azhary, AA. *J. Mater. Chem.*, 2009, 19, 3832-3837. (c) Zho, HQ; Qiu, CY; Liu, Z; Yang, HC; Hu, LJ; Liu, J; Yang, HF; Gu, CZ; Sun, LF. *J. Am. Chem. Soc.*, 2010, 132, 944-946.

[80] Hu, C; Cheng, H; Zhao, Y; Hu, Y; Liu, Y; Dai, L; Qu, L. *Adv. Mater.*, 2012, 24, 5493-5498.

[81] Giovanni, M; Poh, HL; Ambrosi, A; Zhao, G; Sofer, Z; Šaněk, F; Khezri, B; Webster, RD; Pumera, M. *Nanoscale.*, 2012, 4, 5002-5008.

[82] (a)Park, G; Lee, KG; Lee, SJ; Park, TJ; Wi, R; Kimdo, H. *J. Nanosci. Nanotechnol.*, 2011, 11, 6095-6101. (b) Chandra, S; Bag, S; Bhar, R; Pramanik, P. *J. Nanopart. Res.*, 2011, 13, 2769-2777. (c) Anandan, S; Manivel, A; Ashok kumar, M. *Fuel Cells.*, 2012, 12, 956-962.

[83] (a) Yao, Z; Zhu, M; Jiang, F; Du, Y; Wang, C; Yang, P. *J. Mater. Chem.*, 2012, 22, 13707-13713. (b) Zhou, Y-G; Chen, J-J; Wang, F; Sheng, Z-H; Xia, X-H. *Chem. Comm.*, 2010, 46, 5951-5953.

[84] (a) Guo, D; Luo, Y; Yu, X; Li, Q; Wang, T. *Nano Energy*, 2014, 8, 174-182. (b) Yang, P; Mai, W. *Nano Energy*, 2014, *8*, 274-290.

[85] (a) Arico, AS; Bruce, P; Scrosati, B; Tarascon, JM; Schalkwijk, WV; *Nat. Mater.*, 2005, 4, 366-377. (b) Demirci, UB; *J. Power Sources*, 2007, *169*, 239-246.

[86] (a) Leong, GJ; Schulze, MC; Strand, MB; Maloney, D; Frisco, SL; Dinh HN; Pivovar, B; Richards, RM. *Appl. Organometal. Chem.*, 2014, 28, 1–17. (b) Hong, JW; Kang, SW; Choi, BS; Kim, D; Lee, SB; Han, SW. *Acs Nano*, 2012, 6, 2410–2419. (c) Dai, L; Zhao, Y; Chi, Q; Liu, H; Li, J; Huang, Tao. *Nanoscale*, 2014, 6, 9944-9950.

[87] (a) Huang, X; Qi, X; Boey, F; Zhang, H. *Chem. Soc. Rev.*, 2012, 41, 666–686. (b) Liu, M; Zhang, R; Chen, Wei. *Chem. Rev.*, 2014, 114, 5117–5160.

[88] (a) Yoo, E; Okata, T; Akita, T; Kohyama, M; Nakamura, J; Honma, I. *Nano Lett*, 2009, 9, 2255-2259. (b) Li, Y; Gao, W; Ci, L; Wang, C; Ajayan, PM. *Carbon.* 2010, 48, 1124-1130. (c) Kou, R; Shao, YY; Wang, DH; Engelhard, MH; Kwak, JH; Wang, J; Viswanathan, VV; Wang, CM; Lin, YH; Wang, Y; Aksay, IA; Liu, J. *Electrochem. Commun.*, 2009, 11, 954-957.

[89] Xiao, Y-P; Wan, S; Zhang, X; Hu, J-S; Wei, Z-D; Wan, L-J. *Chem. Commun.*, 2012, 48, 10331-10333.

[90] Wang, R; Higgins, DC; Hoque, MA; Lee, DU; Hassan, F; Chen, Z. *Sci. Rep.*, 2013, 3, 2431-2431.

[91] Sahu, SC; Samantara, AK; Satpati, B; Bhattacharjee, S; Jena, BK. *Nanoscale*, 2013, *5*, 11265-11274.

[92] Sahu, SC; Samantara, AK; Dash, A; Juluri, RR; Sahu, RK; Mishra, BK; Jena, BK. *Nano Res.*, 2013, 6, 635-643.

[93] Zhao, H; Yang, J; Wang, L; Tian, C; Jiang, B; Fu, H. *Chem. Commun.*, 2011, 47, 2014-2016.

[94] Wang, Y; Liu, HL; Wang, L; Wang, HB; Du, X; Wang, F; Qi, T; Lee, J. M; Wang, X. *J. Mater. Chem. A.*, 2013, 1, 6839-6848.

[95] Gao, LN; Yue, WB; Tao, SS; Fan, LZ. *Langmuir*, 2013, 29, 957-964.

[96] Carrera-Cerritos, R; Baglio, V; Aricò, AS; Ledesma-García, J; Sgroi, MF; Pullini, D; Pruna, AJ; Mataix, DB; Fuentes-Ramíreza, R; Arriaga, LG. *Appl. Catal., B.*, 2014, 144, 554-560.

[97] (a) Yin, H; Tang, H; Wang, D; Gao, Y; Tang, Z. *ACS Nano*, 2012, 6, 8288-8297. (b) Davis, DJ; Raji, A-RO; Lambert, TN; Vigil, JA; Li, L; Nan, K; Tour, JM. *Electroanalysis*, 2014, 26, 164-170.

[98] (a) Zhang, S; Shao, Y; Liao, H; Liu, J; Aksay, IA; Yin, G; Lin, Y. *Chem. Mater.*, 2011, 23, 1079-1081. (b) Chen, X; Cai, Z; Chen, X; Oyama, M. *J. Mater. Chem. A.*, 2014, 2, 315-320. (c) Nethravathi, C; Anumol, EA; Rajamathi, M; Ravishankar, N. *Nanoscale.*, 2011, 3, 569-571. (d) Luo, B; Yan, X; Chen, J; Xu, S; Xu, Q. *Int. J. Hydrogen Energy.*, 2013, 38, 13011-13016.

[99]   Dong, LF; Gari, RRS; Li, Z; Craig, MM; Hou, SF. *Carbon.*, 2010, 48, 781-787.

[100] Xia, BY; Yan, Y; Wang, X; Lou, W. *Mater. Horiz.*, 2014, 1, 379-399.

[101] Chai, J; Li, F; Hu, Y; Zhang, Q; Han, D; Niu, L. *J. Mater. Chem.*, 2011, 21, 17922-17929.

[102] (a) Wu, S; He, Q; Tan, C; Wang, Y; Zhang, H. *Small.*, 2013, 9, 1160-1172. (b) Shao, Y; Wang, J; Wu, H; Liu, J; Aksay, IA; Lin, Y. *Electroanalysis.*, 2010, 22, 1027-1036. (c) Liu, Y; Dong, X; Chen, P. *Chem. Soc. Rev.*, 2012, 41, 2283-2307. (d) Fanga, Y; Wang, E. *Chem. Commun.*, 2013, 49, 9526-5939. (e) Hur, SH; Park, JN. *Asia-Pac. J. Chem. Eng.*, 2013, 8, 218-233. (f) Sun, Y; Wu, Q; Shi, G. *Energy Environ. Sci.*, 2011, 4, 1113-1132.

[103] Zhou, Y; Liu, S; Jiang, HJ; Yang, H; Chen, HY. *Electroanalysis.*, 2010, 22, 1323-1328.

[104] Dey, RS; Raj, CR. *J. Phys. Chem. C.*, 2010, 114, 21427-21433.

[105] Liang, J; Chen, Z; Guo, L; Li, L. *Chem. Commun.*, 2011, 47, 5476-5478.

[106] Yang, J; Deng, S; Lei, J; Ju, H; Gunasekaran, S. *Biosens. Bioelectron.*, 2011, 29, 159-166.

[107] (a) Li, QZ; Qin, XY; Luo, YL; Lu, WB; Chang, GH; Asiri, AM; Al-Youbi, AO; Sun, XP. *Electrochim. Acta.*, 2012, 83, 283-287. (b) Qin, X.Y.; Luo, Y.L.; Lu, W. B.; Chang, G.H.; Asiri, A.M.; Al-Youbi, AO; Sun, XP. *Electrochim. Acta.*, 2012, 79, 46-51.

[108] Guo, S; Wen, D; Zhai, Y; Dong, S; Wang, E. *ACS Nano*, 2010, 4, 3959-3968.

[109] Wu, H; Wang, J; Kang, XH; Wang, CM; Wang, DH; Liu, J; Aksay, IA; Lin, YH. *Talanta*, 2009, 80, 403-406.

[110] Du, M; Yang, T; Jiao, K. *J. Mater. Chem.*, 2010, 20, 9253-9260.

[111] Liu, H; Chen, X; Huang, L; Wang, J; Pan, H. *Electroanalysis*, 2014, 26, 556-564.

[112] Hu, Y; He, F; Ben, A; Chen, C. *J. Electroanal. Chem.*, 2014, 726, 55-61.

In: Innovations in Nanomaterials
Editors: Al-N. Chowdhury, J. Shapter, A. B. Imran

ISBN: 978-1-63483-548-0
© 2015 Nova Science Publishers, Inc.

Chapter 12

BIOCOMPATIBLE GRAPHENE: FROM SYNTHESIS TO BIOMEDICAL APPLICATIONS

*Dr. Mohyeddin Assali**

Department of Pharmacy, Faculty of Medicine and Health Sciences,
An Najah National University, Nablus, Palestine

ABSTRACT

The fabrication and biomedical application of nanomaterial-biomolecule nanohybrids is an exciting new field of research. Among the most popular bionanomaterial studied so far, are those based on the synthesis and bio-functionalization of graphene. Since its discovery in 2004, graphene has received a huge interest due to its unique structural, electrical, optical, and mechanical properties. The applications of graphene are diverse and include their use as reservoir for hydrogen storage, as additive in the fabrication of more advanced and resistant sport materials, or as nano-transporter for the internalization inside the cytoplasm of drugs unable to cross the cell membrane. The outstanding properties of graphene, such as its high surface area, make it excellent candidate for applications in cellular and molecular biology as well as in medicine. Indeed, to date, graphene has shown a tremendous advance in biomedical applications, including bioimaging, biosensing, and drug delivery especially in cancer treatment. The application of functionalized graphene as new nanovectors for the transport of biologically active material has emerged after the demonstration of their ability to penetrate inside the cell. Successful use of graphene as nanovehicle for the delivery of bionanocargoes includes the transport of small molecules and nucleotides for gene transfer. Among these systems, the most outstanding are those investigated for the treatment of devastating and still unsolved diseases such as cancer. Additionally, the capacity of graphene to accumulate heat when irradiated by near infrared allows it to act as nano-missile in hyperthermia therapy able to kill selectively cancer cells.

However, due to the low solubility of graphene in most organic solvents and especially in biological media, its biomedical applications have been hampered. Additionally, concerns about its potential toxicity have reduced much of the original enthusiasm about its promising medical use. Nevertheless, recent investigations have

* Email: m.d.assali@najah.edu; m.d.assali@hotmail.com.

concluded that conveniently functionalized water soluble graphene is completely cleared from the body via the renal pathway, and is non-toxic. Thus the set up of efficient and non-destructive methods for the synthesis of biocompatible water soluble graphene is an important preliminary condition for its clinical applications.

The present chapter will describe in depth the different approaches used for the synthesis, functionalization and water solubilization of graphene. A detailed overview of its biomedical applications will be discussed including bioimaging, drug & gene delivery, photothermal therapy, nervous and tissue engineering.

1. INTRODUCTION

Carbon as element presents various allotropic forms with a wide variety of dimensional structures such as diamond and graphite -which were discovered firstly- with 3D structures, fullerenes and carbon nanotubes that present 0D and 1D structures respectively; whereas the 2D structure allotrope was a challenge. More than 75 years ago, Peierls and Landau showed that the two dimensional carbon crystals were thermodynamically unstable and could not exist [1, 2]. This instability would be due to a divergent contribution of thermal fluctuations in low-dimensional crystals, leading to displacement of atoms that became comparable to the interatomic distances [3]. However, in the year of 2004, the group of Andre Geim and Konstantin Novoselov (Manchester University) were succeeded to isolate for the first time from graphite a unique 2D carbon sheet using micromechanical exfoliation, which was later renamed as "Graphene" [4].

Hence, graphene represents a new class of two-dimensional material of a carbon atom with thickness of 3.35 Å. It is formed by a mesh of hexagonal structures of sp^2 hybridized carbon and has a high specific surface area of 2630 m^2 g^{-1}. These characteristics confer its exceptional thermal, mechanical, optical and electrical properties.

1.1. Properties of Graphene

One of the interesting properties of graphene is its charge carriers behave as massless relativistic particles, or the Dirac fermions. It has been demonstrated that graphene is a zero-gap 2D semimetal with a tiny overlap between valence and conductance bands, and charge carriers move with little scattering under ambient conditions [4]. It also exhibits a strong ambipolar electric field effect with the concentration of charge carriers up to 10^{13} cm^{-2} and room-temperature mobilities of ~10 000 cm^2 V^{-1} s^{-1}, when the gate voltage is applied [5, 6]. This high mobility exceeds the current Si transistors and also other materials such as gold, gallium arsenide or carbon nanotubes [7].

Regarding the optical properties, graphene is a nearly transparent material as it absorbs about 2.3% of the intensity of the white light that reaches its surface [8, 9] and presents a dynamic conductivity in the visible range [10].

The thermal conductivity of graphene at room temperature is 3000-5000 W m^{-1} K^{-1} and is strongly dependent on the size of the graphene layer [11]. This value is 2 to 50 times greater than those of copper and silicon, respectively, common materials used in electronics today, which makes it ideal candidate for the fabrication of heat sinks and composite materials of high thermal conductivity.

The fracture strength of graphene is 42 N m^{-1} and its Young's modulus is 1.0 TPa, suggesting that graphene is the strongest material ever measured [12].

Due to its large surface area and high conductivity, a group of researchers from Harvard University and MIT have demonstrated that graphene sheets could make a big improvement over the currently used membranes for nanopore DNA sequencing- a technique that promises to accelerate and simplify the sequencing of long DNA chains [13].

1.2. Methods for the Synthesis of Graphene

As mentioned earlier, graphene was firstly obtained by micromechanical exfoliation of highly oriented pyrolytic graphite (HOPG) using adhesive tape and subsequently deposited on a silicon substrate, subsequently analyzed by atomic force microscopy (AFM) and transmission electron microscopy (TEM), figure 1 [14].

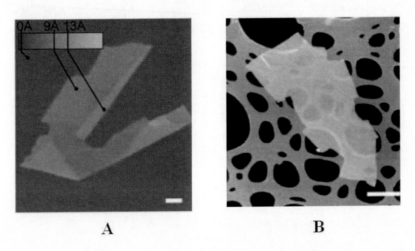

Figure 1. A) Graphene visualized by AFM. The folded region exhibits a relative height of ≈ 4 Å indicating that it is a single sheet. B) Single sheet of graphene observed by TEM. Reproduced with permission from ref. [14]. Copyright 2005, PNAS.

Although this methodology has allowed obtaining individual graphene sheets of highest quality and with dimensions of the order of 10 microns, the main problem lies in the extremely low yield. Therefore, alternative methods are developed for the preparation of graphene with low production costs while increasing the amount of the material obtained. Therefore, the methods that are currently most promising are:

a) Chemical vapor deposition (CVD) [15-18]: CVD processes generally use the transition metal surface for the growth of graphene using hydrocarbon gases as precursors at a deposition temperature of approximately 1000 °C. For example, the group of Kong [19] presented a scalable and low cost technique using ambient pressure CVD on polycrystalline Ni films, for which they prepared films with an elevated area of one or a few sheets of graphene and these prepared layers can be transferred to various nonspecific substrates.

b) Epitaxial growth of graphene on silicon carbide (SiC) [20-23]: It is a very promising method for the synthesis of graphene sheets with a uniform size in which a crystalline SiC substrate is heated in vacuum at high temperatures in the range of 1200-1600 °C. Since the sublimation rate of silicon is larger than that of carbon, excess carbon remains on the surface, being rearranged to form the layer of graphene. The necessity to work on ultra high vacuum and the required high temperature to produce the sublimation of silicon greatly limits their widespread application on large scale.

c) Chemical reduction of graphene oxide (GO): the oxidation and exfoliation of natural graphite by thermal technique or oxidation followed by the chemical reduction has been appreciating one of the most efficient methods for the production of large scale graphene and low cost. However, the major drawback of this method is the low structural quality of the sheets. With this method, strong oxidation of the graphite generates mixture (epoxy, hydroxyl, carbonyl and carboxyl ...) in both the edges and in the plane of the sheet which interrupting the electronic conjugation of the plane, figure 2 [24, 25]. These functional groups interrupt numerous van der Waals interactions between the layers of the graphene oxide giving them a hydrophilic character. However, due to its strong hydrophobicity, graphene obtained from the reduction of graphene oxide usually suffers from low solubility and irreversible agglomeration, thus limiting further processing and application.

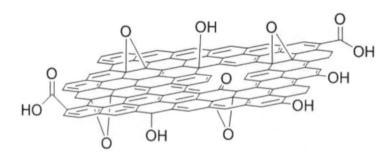

Figure 2. Structure of graphene oxide layers.

So far, many types of reduction methods have been developed to obtain a reduced graphene oxide sheets such as chemical reducing agents such as Hydrazine and NaBH$_4$ [26-28], thermal reduction [29, 30], photochemical reduction [31, 32], photothermal reduction [33], electrochemical reduction [34, 35] and microwave-assisted reduction [36]. Lee et al. have recently reviewed the effective techniques in the reduction of graphene oxide [37]. In this regards, it should be mentioned that the reducing reagents such as hydrazine and NaBH$_4$ are highly toxic. Since hydrazine is toxic and unstable, it is desirable to explore a new route within the green chemistry to reduce the graphene oxide. The group of Zhang has reported an environment friendly methodology using ascorbic acid as a reducing agent and an amino acid as a stabilizer [38]. Meanwhile, Loh's group reported a simple and clean hydrothermal dehydration route to convert graphene oxide to graphene using supercritical water (SC) which could act as a reducing agent under hydrothermal conditions and promote recovery of the π conjugation after dehydration [39]. More interestingly, the Tour's group demonstrated the use of environmental bacteria (*Shewanella oneidensis*) as an electron donor to reduce graphene. They have found that the conductivity and physical characteristics of graphene converted by the bacteria were comparable to other forms of chemically converted graphene [40]. These results are important in this regard not only of the use of environmental bacteria for the bioremediation but also in a more specific form in materials synthesis using green chemistry conditions.

d) Unzipping of carbon nanotubes (CNTs) for the synthesis of graphene nanoribbons (GNRs): the graphene nanoribbons with a width <10 nm exhibit interesting semiconductor properties that make them excellent candidates in electronic devices. There are several methodologies to obtain GNRs such as lithographic technique [41], chemical methods [42], sonochemical technique [43], CVD [44], or micromechanical cleavage [45]. However, these methods produce very limited quantities of GNRs. Recently; the unzipping of CNTs offers the possibility of producing large scale of narrow GNRs with well controlled width. Several methods were developed for the production of GNRs from the opening of CNTs such as oxidative treatment of CNTs [46, 47], plasma etching [48], cleavage caused by the tip of the scanning tunneling microscope (STM) through electrical current [49], and nanomanipulation [50], etc. An example in this matter, the group of Tour has prepared oxidized GNRs by suspending multiwalled carbon nanotubes (MWCNTs) in concentrated sulfuric acid followed by treatment with $KMNO_4$ (500 wt%) for 1 h at room temperature and then another hour at 55-70°C [51]. They have observed that $KMnO_4$ is a selective oxidant for the longitudinal splitting of MWCNTs. The opening of the nanotube is produced longitudinally or spirally, resulting ribbons with straight edges. As the process is oxidative, the obtained GNRs possess many carbonyl, carboxyl, hydroxyl and epoxy groups on the edges of opening. Therefore, it is necessary to transform them into GNRs with less oxygen content.

Simultaneously, Dai group subjected a film of poly (methyl methacrylate) (PMMA, etching mask)-MWCNT on argon plasma at different times, obtaining GNRs without the need for a subsequent reducing treatment [52]. Recently the same group has developed a method for large scale production of GNRs consisting of subjecting the pristine MWCNTs to calcination in air at 500°C, followed by dispersion in an organic solution of poly(p-phenylenevinylene-co-2,5-dioctyloxy-m-phenylenevinylene) (PMPV) in 1,2-dichloroethane (DCE), followed by sonication. With this method, it was observed that the GNRs were produced with high efficiency [53].

e) Total organic synthesis of graphene: the bottom-up strategy for the synthesis of advanced nano-materials with different sizes, shapes and composition has demonstrated its efficacy. Mention that graphene is composed of many interconnected polycyclic aromatic hydrocarbons (PAH), organic synthesis protocols can offer effective routes for the controlled synthesis of graphene with different sizes [54-58]. PAHs are attractive for their high versatility and can be modified by a range of aliphatic chains to modify their solubility.

The breakthrough was carried out by Müllen group in 2008, when they reported the synthesis of GNRs with lengths up to 12 nm using this methodology [59]. However, the extension of molecules of graphene becomes increasingly demanding despite the limited range sizes that can be reached because the increase in the molecular weight decreases the solubility and facilitates side reactions. Therefore, it remains a major challenge for organic chemists to synthesize large-scale molecules of well-defined graphene in shape, size and structure.

2. FUNCTIONALIZATION OF GRAPHENE

Graphene is substantially insoluble material, or it disperses with difficulty, in any type of solvent. Therefore, the potential application of graphene in the biological field implies the necessity to improve its solubility, especially in the water as it is the biological medium. For this purpose, several methodologies have been developed for its functionalization that can be classified into two main groups: covalent manner and non covalent routes.

2.1. Covalent Functionalization of Graphene

Covalent bonds can be established in two different ways, either by the use of the carboxyl groups of graphene oxide through amide or ester formation, or by direct reaction with the graphene sheet. The main advantage of this strategy is the great stability of the nanomaterial obtained after the functionalization; however its main drawback is the destruction of the π system of the graphene sheet which affects the electrical properties of the graphene.

2.1.1. Amidation and Esterification Reactions of Graphene Oxide

As have seen earlier, the production techniques of graphene generate different sizes of graphene sheets whose structure is furnished with oxygenated functional groups, mainly carbonyl and carboxylic groups which have been utilized for further functionalization using acid chloride as intermediates or directly using activating agents such as carbodiimide, figure 3.

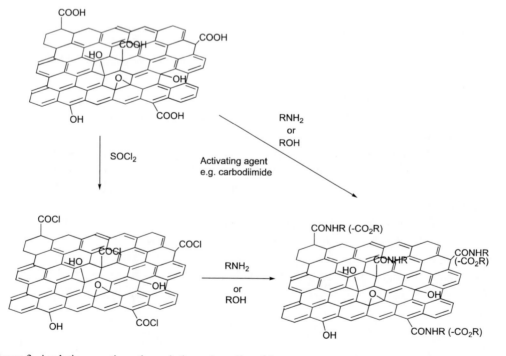

Figure 3. Acylation reactions through the carboxylic acid groups appeared on the graphene oxide sheet.

The group of Haddon has reported for the first time the grafting of octadecylamine (ODA) on the surface of graphene oxide after the activation of the carboxyl groups by thionyl chloride (SOCl$_2$). The functionalized graphene has a solubility of 0.5 mg/mL in THF and is also soluble in CCl$_4$ and 1,2-dichloroethane [60].

Dai group has decorated the graphene sheet with polyethylene glycol (PEG) using the EDCI as activating agent of the carboxyl groups on the surface of the graphene and the coupling with amine terminated PEG (PEG-NH$_2$) [61]. As known, PEG is a biocompatible hydrophilic polymer that extensively used to functionalize nanomaterials in order to enhance their biocompatibility, water solubility and pharmacokinetic profile. The obtained PEGylated graphene has shown excellent water solubility and great stability in all biological solutions. Then they tested the capability of the product to solubilize the water insoluble anticancer agent such as SN38, camptothecin analogue, through van der waals interactions. The resulting graphene-PEG-SN38 complex has shown high water solubility of SN38 and maintains its high anticancer activity similar to the drug alone in organic solvents. Other polymers like poly(vinyl alcohol) (PVA) [62], chitosan (CS) [63], polyethylenimine (PEI) [64] have been attached on the surface of the graphene following the same procedure.

2.1.2. Direct Formation of Covalent Bond with Graphene Sheet

This section will mainly cover the following chemical reactions: free radical addition, nucleophilic substitution and cycloaddition reactions.

2.1.2.1. Free Radical Addition Reaction

One of the most investigated lines within this type is the utilization of diazonium salts as functionalizing agents, figure 4. This methodology was developed in the group of Tour by electrophilic substitution of aryl diazonium salt on the surface of surfactant-wrapped graphene [65]. They demonstrated that the functionalization of the graphene oxide wrapped with the surfactant sodium dodecyl benzene sulfonate (SDBS) with the diazonium salt derivative can be achieved firstly by reducing the oxy groups with hydrazine monohydrate; followed by grafting of the aryl diazonium salt on the surface of SDBS wrapped graphene at room temperature. The degree of functionalization was estimated by TGA up to 1 functional group per 55 carbon atoms and the functionalized graphene is highly dispersable in N,N′-dimethylformamide (DMF), N,N′-dimethylacetamide (DMAc), and 1-methyl-2-pyrrolidinone (NMP).

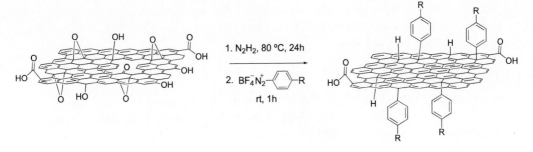

Figure 4. Reduction of the GO and functionalization with different derivatives of diazonium salts.

The same group has extended the functionalization to expanded graphite by using the diazonium salt of 4-bromo analine [66]. The obtained chemically-assisted exfoliated graphene (CEG) sheets are more soluble than pristine graphene in DMF after the ultrasonication and without the addition of surfactant and the microscopic data has shown that more than 70% of the CEG flakes have less than 5 layers.

2.1.2.2. Nucleophilic Substitution Reaction

The epoxy groups of graphene oxide are the main active sites for the nucleophilic substitution. The amine group in the organic compounds has lone pair of electrons that can attack the epoxy groups of the graphene oxide.

Niu and co-workers have reported the functionalization of graphene oxide using amine-terminated ionic liquid (IL-NH$_2$) [67]. The IL-NH$_2$ acts as a nucleophile and opens the ring of the epoxy group of the GO. The obtained functionalized graphene is totally dispersed in water, DMF and DMSO without the addition of any surfactant as stabilizer, figure 5.

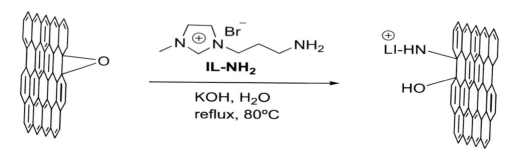

Figure 5. Scheme describes the preparation of polydisperse chemically-converted graphene (p-CCG) sheets.

Tagmatarchis and co-workers have used Bingel cyclopropanation reaction to introduce cylcopropane moiety onto the exfoliated graphene with the aid of microwave irradiation [68]. In a first step, the graphite flake has been exfoliated by sonication with benzylamine affording stable suspensions of few-layers graphene. In a second step, the exfoliated graphene has been functionalized with malonate moieties (diethyl malonate and extended tetrathiafulvalene (exTTF)) following bingel reaction using the microwave irradiation. This reaction provides high degree of functionalization with good solubility in various organic solvents such as DMF, dichloromethane, o-dichlorobenzene and toluene. Electrochemical examination of the hybrid exTTF-graphene allowed the determination of the redox potentials, and exhibited the formation of a radical ion pair that includes one-electron oxidation of exTTF and one electron reduction of graphene with a difference in energy gap of 1.23 eV.

2.1.2.3. Cycloaddition Reaction

In this section three major cycloaddition reactions will be mentioned, the [2 + 1] cycloaddition is one of the earliest methods that have been used to functionalize the graphene. One of the important examples of [2 + 1] cycloaddition is the introduction of aziridine adduct onto the graphene sheet by the formation of reactive intermediate nitrene, which can be achieved by thermal or photo decomposition of azide group. Various research groups have used this strategy to introduce different organic compounds onto graphene. Yan and co-workers have successfully functionalized exfoliated graphene by a variety of derivatives of perfluorophenylazide containing alkyl, ethylene oxide and perfluoroalkyl chains though their thermal and photochemical activation [69], figure 6. The functionalized graphene with alkyl and perfluoroalkyl chains were soluble in *o*-dichlorobenzene whereas the graphene functionalized with ethylene oxide chains was soluble in water.

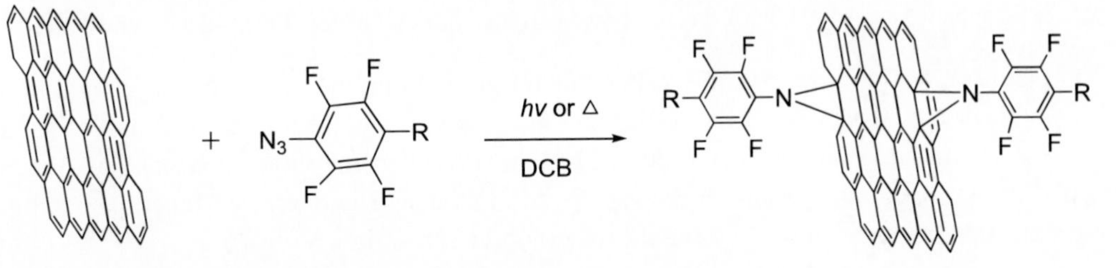

Figure 6. Functionalization of Pristine Graphene with azide derivatives.

Following the same strategy, graphene sheets have been functionalized with phenylalanine using nitrene addition to verify the solubility of the graphene sheets [70]. More interesting that the phenylalanine affords two other functional groups (the carboxyl and amine groups) that can be further functionalized. The product was determined to have 1 phenylalanine substituent per 13 carbons.

The other type of cycloaddition that has been applied on the chemistry of graphene is [2 + 2] cycloaddition. The reaction results in the formation of four-membered ring on the graphene sheet. Zhong et al. have demonstrated the addition of three derivatives of benzyne moiety which have been attached covalently to the graphene sheet [71], figure 7. Based on the TGA data, the degree of functionalization was estimated to be 1 functional group per 17 carbon atoms and the functionalized graphene demonstrated good solubility in various solvents such as DMF, ethanol, chloroform, 1,2-dichlorobenzene and water.

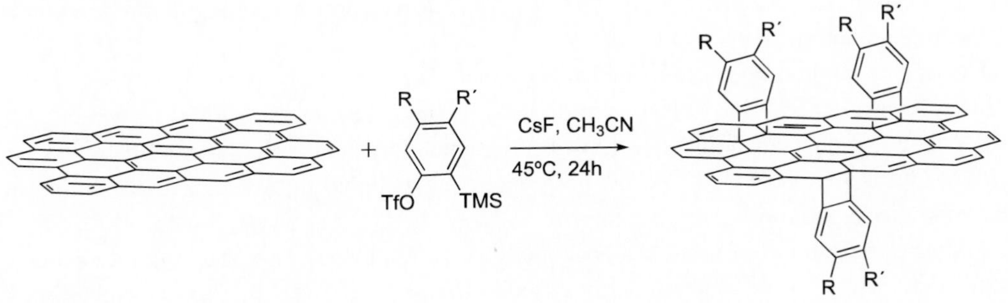

Figure 7. Illustration of the chemical modification of graphene sheets via aryne cycloaddition.

The third type is 1,3-dipolar or [3 + 2] cyloaddition reaction consists in the addition of azomethine ylides, generated *in situ* by thermal condensation between an α-amino acid and an aldehyde with a double bond of the graphene sheet resulting in pyrrolidone ring fused to the surface. Georgakilas et al. were the first who utilized the 1,3-dipolar cycloaddition to functionalize graphene π carbon network with high degree of functionalization reached 1 functional group per 40 carbon atoms as highlighted by TGA data [72]. The product demonstrated good solubility in polar organic solvents and water with high stability, figure 8.

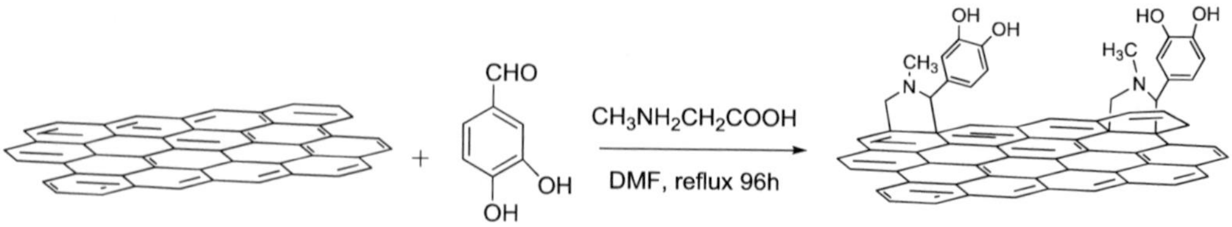

Figure 8. Schematic representation of the 1,3 dipolar cycloaddition of azomethine ylide on graphene.

On another work, Prato and co-workers have reported that the 1,3-dipolar cycloaddition occurs both in the edges as well as in the internal C = C bonds of the graphene sheet [73]. This observation has been confirmed by TEM analysis when the gold nanorods have been attached to the free amine group on the surface of graphene. The degree of functionalization is estimateed to be approximately 1 functional group in 128 carbon atoms.

2.2. Non-Covalent Functionalization of Graphene

The covalent functionalization of graphene has the drawback to modify the surface of the sheet and thus alter its physical and chemical properties, due to the conversion of the sp^2 carbons to sp^3 ones which decreases the mobility of the carrier, so the non-covalent approach, that does not alter the graphene sheet, is of greater interest.

Graphene sheets are difficult to disperse homogeneously in solution due to their tendency to form aggregates with tightly bounded packages. One approach that has been used to separate and obtain individual graphene sheets is to establish non-covalent interactions with the surface of the sheet with different species, such as polymers, aromatic compounds, detergents and biomolecules.

This type of functionalization is very attractive because it offers the ability to anchor molecules without affecting the electronic structure of the graphene. Non-covalent interactions are based in most cases on van der Waals forces or π-π stacking, hydrogen bonds and electrostatic interactions.

The first example of the non-covalent functionalization of graphene using poly(sodium 4-styrenesulfonate) (PSS) has been reported by Stankovich et al. [74], in which the functionalized graphene nanoplatelets have been obtained by the exfoliation and the reduction in situ of GO in the presence of PSS. These nanoplatelets have shown high dispersibility in water and other organic solvents.

Among the examples of non-covalent functionalization through π-π stacking interactions, pyrene derivatives have been mainly used. Thereby, Xu et al. have successfully functionalized a reduced graphene oxide non-covalently utilizing 1-pyrenebutyrate (PB⁻), as a stabilizer since the pyrene moiety has strong affinity with the basal plane of graphite via π-stacking. The obtained graphene is highly dispersed in aqueous media and has showed high magnitude (7 orders) of conductivity in comparison to the GO precursor [75].

Likewise, Liu et al. [76] have synthesized pyrene-terminated poly(N-isopropylacrylamide) (PNIPAAm) using reversible addition fragmentation chain transfer (RAFT) polymerization which is capable of the non-covalent functionalization of graphene sheet via π-π stacking between the pyrene moiety and π network of the graphene, Figure 9.

The obtained graphene-PNIPAAm composite is soluble and thermosensitive in water and has a reversible lower critical solution temperature (LCST)-induced dispersibility at 24°C.

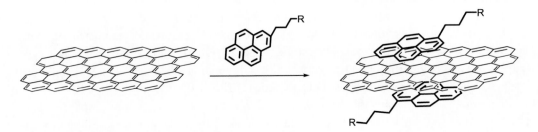

Figure 9. Scheme demonstrates the interaction between derivatives of pyrene and graphene.

In another study, Green and co-workers have utilized various derivatives of pyrene to functionalize and exfoliate graphite and to obtain one or few layers of graphene totally dispersed in water [77]. Among all investigated pyrene derivatives, i.e., pyrene (Py), 1-aminopyrene (Py–NH$_2$), 1-aminomethyl pyrene (Py–Me–NH$_2$), 1-pyrenecarboxylic acid (Py–CA), 1-pyrenebutyric acid (PyBA), 1-pyrenebutanol (PyBOH), 1-pyrenesulfonic acid hydrate (PySAH), 1-pyrenesulfonic acid sodium salt (Py–SO$_3$) and 1,3,6,8-pyrenetetrasulfonic tetra acid tetra sodium salt (Py–(SO$_3$)$_4$), the Py–SO$_3$ was found to be the most effective one, yielding graphene final concentrations as high as 0.8-1 mg/mL. The HRTEM images have confirmed the presence of one or few layers of graphene.

Other aromatic compounds that have been used in the exfoliation of graphite in water including perylene-based bolaamphiphiles [78, 79] or coronene tetracarboxylic acids (CTCA) [80]. Kaminska et al. have reported a one step method for the reduction of graphene oxide to reduced graphene taking advantage of the electron-donor properties of derivatives of tetrathiafulvalene (TTF) [81, 82]. Liu and co-workers have developed a nanosupermolecular structure consisting of the non-covalent functionalization of graphene oxide with adamantanyl porphyrin moiety via π-π stacking and the molecular recognition of the ademantane with folic acid-modified β-cyclodextrin [83]. The functionalized GO has been further functionalized by attaching the Doxorubicin (DOX, anticancer agent), so the whole nanostructure served as a targeted drug delivery system of doxorubicin with targeting agent (folic acid), the resulted nanosystem exhibited better anticancer activity with less toxicity than free doxorubicin *in vivo*. Malig et al. have prepared zinc phthalocyanines (ZnPc) complex for the non-covalent interaction with the surface of graphene through π-π interaction, the obtained hybrid material has showed an interesting material for solar cell applications [84].

Other type of non-covalent functionalization of graphene could be occurred by coating the surface of graphene with proteins, surfactants and polymers through van der Waals forces. The group of Suh has decorated graphene with ionic liquid polymer that exhibited as solution phase transferable graphene sheets [85]. The obtained nanomaterial is stable against chemical reduction and is highly dispersed in water. However, upon the anion exchange of the PIL on graphene sheets, the aqueous dispersed graphene sheets are readily transferred into the organic phase by changing their hydrophilic properties to hydrophobic.

In another study, the Suslick's group has utilized the sonochemical method to functionalize graphite starting from a reactive monomer of styrene obtaining polystyrene functionalized graphene. The ultrasonic irradiation of graphite in the presence of styrene exfoliates the graphite flakes to mono or few layers of graphene sheets. In addition, a radical polymerization of styrene has been initiated sonochemically forming polystyrene on the surface of the graphene sheets. The obtained graphene has showed high solubility in various organic solvents such as DMF, THF, toluene and chloroform [86].

Other range of conventional polymers have successfully exfoliated graphite in aqueous and organic solvents such as poly[styrene-b-(2-vinylpyridine)] and poly(isoprene-b-acrylic acid) block copolymers [87], ethyl cellulose [88], polyvinylpyrrolidone (PVP) [89], cellulose acetate (CA) [90], and hyperbranched polyethylene (HBPE) [91].

Regarding the surfactants that have been used in the functionalization of graphene, Zeng et al. have reported the self-assembly of sodium dodecyl benzene sulfonate (SDBS) on the surface of the graphene which has been further attached with horseradish peroxidase (HRP) by electrostatic attraction between the negative charges of SDBS and the positive charge of the HRP. Moreover, the HRP–graphene sheets electrode showed excellent electrocatalytic performance toward the reduction of H_2O_2 with fast response, wide linear range, high sensitivity and good stability [92].

The group of Coleman has functionalized the graphene with various ionic and non-ionic surfactants and has studied the effect of these surfactants on the dispersion of the graphene in water [93]. They have studied 8 ionic surfactants: SDBS, sodium dodecyl sulfate (SDS), lithium dodecyl sulfate (LDS), cetyltrimethyl ammonium bromide (CTAB), tetra-decyltrimethylammonium bromide (TTAB), sodium cholate (SC), sodium deoxycholate (DOC) and sodium taurodeoxycholate (TDOC), and 4 non-ionic surfactants which are IGEPAL CO-890, Triton X-100, Tween 20 and Tween 80. In all cases, the concentration of the surfactant was 0.1 mg/ml and the concentration of the graphite was 5 mg/ml, the dispersion was obtained by the ultrasonication method in aqueous solution. They have found that the degree of exfoliation, as characterized by flake length and thickness, was independent of the surfactant type; however the stability was dependent on the solvent type. The dispersed graphene was characterized to have 750 nm long and four layers thick on average.

3. BIOCOMPATIBILITY AND CYTOTOXICITY OF THE FUNCTIONALIZED GRAPHENE

For the graphene to reach the clinics and to have real clinical applications, it is necessary to study its biocompatibility and cytotoxicity *in vitro* and *in vivo*. Regarding the biodegradability, Star and co-workers have reported that low concentrations of hydrogen peroxide (~40 μM) and horseradish peroxidase (HRP) catalyzed the oxidation of graphene oxide (GO), therefore GO is susceptible to biodegradation, however the reduced graphene oxide failed to be oxidized by HRP [94]. This study will potentiate the development of safer biodegradable functionalized graphene.

In relation to the mechanism of graphene cell penetration and it is cross to the cell membrane. Various studies have proposed that the functionalized graphene penetrates the cells through clathrin-mediated endocytosis and phagocytosis [61, 95, 96]. Shi and co-workers have functionalized graphene with polyethylene glycol (PEG) shell attached via a disulfide linkage (GO-SS-mPEG) to develop a nanocarrier of anticancer agent (doxorubicin) that is soluble in water [97]. They also studied the mechanism of cell penetration and the release of the drug upon the response of glutathione (GSH), as they observed the functionalized graphene enters the cell through the endocytosis mechanism, Figure 10.

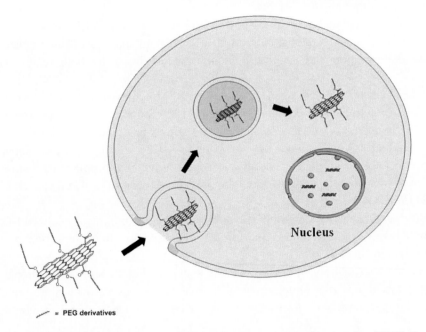

Figure 10. Schematic diagram showing the endocytosis process of the functionalized PEG-Graphene. Modified from ref [83].

Although, more studies are necessary to be conducted for better understanding of how graphene cross the cell membrane.

It has been found that the cytotoxicity of graphene is dose-dependent. The majority of *in vitro* studies have showed that the GO is safe at low concentration in various cell lines like HeLa cells [98], human fibroblasts [99, 100]. A549 human lung cancer cells [101], and human hepatoma HepG2 cells [102]. Wang et al. found that the GO with dose less than 20 mg/L did not exhibit toxicity to human fibroblasts; however the concentration above 50 mg/L exhibits obvious cytotoxicity such as decreasing cell adhesion, inducing cell apoptosis, entering into lysosomes, mitochondrion, endoplasm, and cell nucleus [100].

In vivo studies also showed concentration related toxicity of GO at high concentration. Zhang et al. used the radiotracer technique and a series of biological assays to study the distribution and biocompatibility of GO in mice [103]. They revealed that when the mice were exposed to 10 mg/kg body weight of GO a serious side effects were raised and a major accumulation of GO in lungs for long period of time, this could be due to the high hydrophobic interaction between the cells and the graphene oxide.

However, the functionalization of graphene, using both covalent and non-covalent approaches, plays a critical role in determining the biocompatibility of graphene. The functionalization of graphene with PEG, a hydrophilic biocompatible polymer, improves its biocompatibility and increases the solubility in water. Liu and collaborators studied the *in vivo* behaviors of nanographene sheets (NGS) functionalized with PEG coated by fluorescent label [104].

No obvious side effect of PEGylated NGS is noted for the injected mice by histology, blood chemistry, and complete blood panel analysis. The same group has studied the long-term *in vivo* biodistribution of ^{125}I-labeled NGS functionalized with PEG and examines the potential toxicity of graphene over time. They have observed that the PEGylated NGS mainly accumulate in the reticuloendothelial system (RES) including liver and spleen after intravenous administration and can be gradually cleared, likely by both renal and fecal excretion.

Blood biochemistry, hematological analysis, and histological examinations showed no toxicity to the treated mice in a period of 3 months [105]. Singh et al. have reported that the GO is highly thrombogenic in mouse and evoked strong aggregatory response in human platelets. Nevertheless, amine-modified graphene (G-NH$_2$) did not have absolutely no stimulatory effect on human platelets nor did it induce pulmonary thromboembolism in mice following intravenous administration, which contrast the observation obtained with GO and reduced GO (RGO) [106].

Other study conducted by the group of Liu revealed that dextran functionalized GO (Dex-GO) improved the stability and reduced the cell toxicity in comparison with the unfunctionalized graphene [107].

From the previous mentioned studies, we can conclude that the surface functionalization of graphene plays a great role in determining its toxicity. However, much more pre-clinical toxicity studies are necessary to translate this interesting nanomaterial in the clinical applications.

4. BIOMEDICAL APPLICATIONS OF THE FUNCTIONALIZED GRAPHENE

This section will cover the biomedical applications of the functionalized graphene including drug delivery and photo-thermal therapy of cancer, gene delivery, bio-imaging and tissue engineering.

4.1. Drug Delivery

There are many advantages of graphene over the various nanomaterials that promote it as an exceptional drug carrier. It has ultrahigh surface areas (2630 m^2/g) with delocalized π electrons which make it an efficient dug carrier with high loading capacity on the both sides of the single atom layer sheet. In addition, the GO is fabricated easily at low cost and more important it doesn't contain toxic metal particles [108]. Yang and co-workers have prepared a novel graphene oxide-doxorubicin hydrochloride nanohybrid (GO-DXR) through the π-π stacking between the aromatic rings of DXR and the π system of graphene sheet. In addition, the formation of hydrogen bond interactions between the carboxyl groups of the GO and amino groups of the DXR [109]. They obtained high loading capacity of DXR on GO that could reach 2.35 mg/mg. The release of DXR was pH-dependent due to the hydrogen bonding interactions and the efficient release was in acidic and basic conditions. Misra and collaborators have confirmed that the loading and the release of doxorubicin (DOX) is pH-dependent [110]. They firstly functionalized GO with DOX non-covalently then encapsulation of the functionalized graphene oxide with folic acid conjugated chitosan. They found that the release was the higher (35%) at acidic pH 5.3, this could be explained by the reduction of the hydrogen bond interactions between the amine group of the DOX and carboxyl groups of GO due to the protonation of the amino group. The group of Dai has reported the grafting of polyethylene glycol (PEG) star polymers onto the nano graphene oxide that conjugated with Rituxan (CD20 + antibody) to target specific cancer cells for selective cell killing [111]. The NGO sheets showed photoluminescence from visible to the near-infrared (NIR) range, which was used for cellular imaging. After that, doxorubicin was loaded onto the NGO sheets at a high capacity via π-π stacking. The drug release is pH-dependent as has been found by *in vitro* study. Kim et al. have prepared reduce graphene oxide covalently attached with branched polyethylenimine (BPEI) and polyethylene glycol (PEG) to form PEG-BPEI-rGO composite, the obtained nanocomposite was loaded with doxorubicin [112]. The nanocarrier was exposed to NIR irradiation to induce endosomal disruption and subsequent DOX release. The release was also enhanced by the presence of GSH producing greater cancer cell death efficacy with cells treated by NIR irradiation than those with no irradiation. Wang and co-workers modified the reduced graphene oxide (rGO) by noncovalent physisorption of gold nanoclusters (GNCs). It was found DOX loaded GNC-rGO inhibited HepG cell growth more strongly than DOX and GNC-rGO alone [113]. On another study, Yang and co-workers prepared superparamagnetic GO–Fe_3O_4 nanohybrid using chemical precipitation method [114]. The nanohybrid has on its surface amino groups that can be conjugated chemically with the folic acid as targeting agent, to give the multi-functionalized GO followed by DOX loading by π-π stacking, Figure 11.

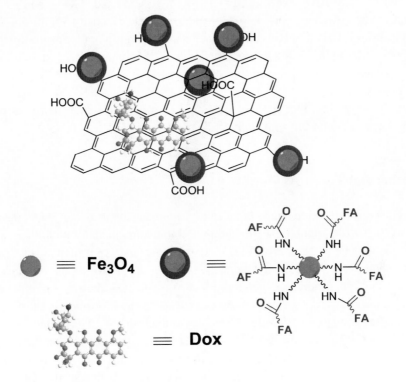

Figure 11. Scheme presents the multi-functionalized GO with anticancer agent Doxorubicin (DOX) and the targeting agent folic acid (FA). Modified from ref. [100].

The drug release is controlled by the pH conditions with maximum release in acidic conditions. *In vitro* study showed that GO-Fe$_3$O$_4$-FA-DOX has the highest cytotoxicity to Hela cells on comparison to the DOX loaded GO-Fe$_3$O$_4$ nanohybrid or DOX alone, which indicates that GO-Fe$_3$O$_4$-FA-DOX has the potential for selectively killing cancer cells *in vitro*.

Rana et al. have functionalized graphene oxide with chitosan in order to improve the water solubility of GO [115]. After that, they used the chitosan-GO hybrid for the delivery of ibuprofen (IBU) and 5-fluorouracil (5-FU). They observed that the 5-FU has lower loading capacity on the surface of the functionalized GO due to its hydrophilic character by the diamide group. They found that functionalized GO-chitosan is a promising new material for biological and medical applications. Pan et al. prepared graphene sheets (GS) covalently functionalized with thermo-responsive poly(N-isopropylacrylamide) (PNIPAM) via click chemistry [116]. PNIPAM-GS exhibited LCST at 33°C which is lower than the PINPAM alone because of the interaction between the graphene sheets and grafted PNIPAM. Moreover, PNIPAM-GS is able to load a water-insoluble anticancer drug, camptothecin (CPT), with a superior loading capacity of 15.6 wt%. *In vitro* test showed that PNIPAM-GS-CPT complex has high potency of killing A-5RT3 cancer cells compared with free CPT. Kakran et al. have functionalized GO covalently with various surfactants like Pluronic F38 (F38), Tween 80 (T80) and maltodextrin (MD) for loading and delivery of a poor water soluble antioxidant and anticancer drug, ellagic acid (EA). The release of EA is pH dependent and has a better cytotoxicity against human breast carcinoma cells (MCF7) and human colon adenocarcinoma cells (HT29) than that of free EA dissolved in DMSO [117].

Turcheniuk et al. have successfully prepared a GO and GO modified with 2-nitrodopamine coated magnetic particle (GO–MPdop) matrices loaded with insulin in order to protect insulin from the acidic pH in the stomach and to develop orally insulin formulations. They have obtained highly loading capacity at pH < 5.4 with a high stability at acidic conditions. The authors have observed the release profile of insulin from GO matrices were achieved at basic pH = 9.2.

These results open the door for the utilization of graphene oxide matrix as promising materials for the protection of insulin from acidic conditions [118].

4.2. Photothermal Therapy

Phototherapy is an interesting therapeutic approach that deals with the specific destruction of cancer cells upon light irradiation. Phototherapy includes photothermal therapy (PTT) and photodynamic therapy (PDT) in treating and control the disease [119]. Photothermal therapy is based on the optical-absorbing capability of the nanomaterials to generate heat upon light irradiation, which promotes the selective destruction of the abnormal cells. Graphene has received a great attention due to the strong absorption in the near-infrared (NIR) region. The group of Guo has developed doxorubicin-loaded PEGylated nanographene oxide (NGO-PEG-DOX), which can have dual effect (synergistic effect) on the cancer cells, the delivery of anticancer agent (DOX) and the destruction of the cancer cells by heat, in one system [120]. *In vitro* and *in vivo* studies showed that the designed NGO-PEG-DOX nanoparticle was superior to chemotherapy or photothermal treatment alone, Figure 12.

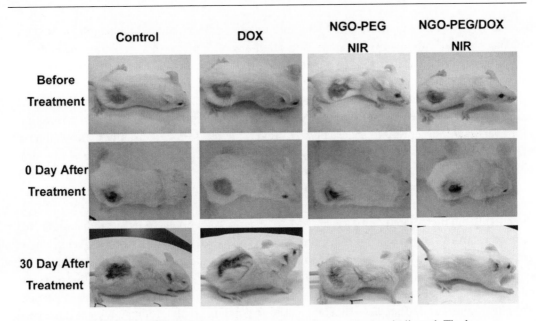

Figure 12. Representative photos of tumors on mice after various treatments indicated. The laser irradiated tumor on NGO-PEG-DOX injected mouse was completely destructed. Reproduced with permission from ref. [105]. Copyright 2011, Elsevier.

The group of Dai has developed a nano-rGO sheets with high near-infrared (NIR) light absorbance for potential photothermal therapy. They increased the selectivity of cellular uptake in U87MG cancer cells by attaching a targeting peptide bearing the Arg-Gly-Asp (RGD) moiety. They observed a total destruction of U87MG cells incubated with nano-rGO-RGD and irradiated with 15 W/cm^2 NIR light for 8 min [121]. Liu and co-workers have functionalized ultra-small reduced graphene oxide (nRGO) noncovalently with PEG and have studied how size and surface chemistry affect the *in vivo* behaviors of graphene, and remarkably improve the performance of graphene-based *in vivo* photothermal cancer treatment. They found also excellent *in vivo* treatment efficacy with 100% of tumor elimination after intravenous injection of nRGO-PEG and the followed 808 nm laser irradiation, and more importantly is that the power of the laser was 0.15 W/cm^2 for 5 min of which is an order of magnitude lower than that usually applied for *in vivo* tumor ablation using many other nanomaterials [122]. Other work of the same group, concerns the preparation of RGO–iron oxide nanoparticle (IONP)-PEG as a novel probe for *in vivo* multimodal tumor imaging and MRI guidance for photothermal therapy. They achieved ultra-efficient tumor ablation using low laser power at 0.5 W cm^{-2}, with no observation of any *in vitro* or *in vivo* toxicities [123]. Hu et al. have developed multifunctional quantum dot rGO nanocomposite that works as imaging agent and for photothermal treatment of cancer [124]. They used a suitable spacer between the QDs and rGO to reduce their fluorescent quenching. The nanocomposite is also functionalized with folic acid to selectively target the MCF-7 cells. After 9 min of irradiation at 808 nm, the cell viability decreased to less than 5% when QD-rGO was used, which was marked by the decrease of the QD brightness and serves as indicator of the progress of the photothermal therapy. Very recently, Szunerits and co-workers have reported the development of gold nanorods (Au NRs) coated with rGO-PEG (rGO-PEG-Au NRs) for the selective killing of uropathogenic E. coli UTI89. They got an advantage of the rGO-PEG-Au NRs hybrid in its photothermal properties and also the usage of relative low concentrations (20-40 µg/ml) to reach the ablation of bacteria present in aqueous solution with low cytotoxicity. These findings offer a new biocompatible material for the killing of pathogen in a selective manner for clinical treatments of patients with urinary infections [125].

On the other hand, the photodynamic therapy (PDT) depends on the generation of reactive oxygen species (ROS) upon the light irradiation of photosensitiziers (PSs) resulting in irreversible destruction of cancer cells. However, the used PSs are usually hydrophobic in nature, so it is necessary to develop new approaches to increase their water solubility and biocompatibility. Dong et al. functionalized non-covalently methoxy-poly(ethylene glycol) modified nano-graphene oxide (NGO-mPEG) with Zinc phthalocyanine (ZnPc), a photosensitizer for photodynamic therapy, thorugh π-π stacking. The obtained GO-mPEG-ZnPc exhibited high cytotoxicity towards MCF7 cancer cells upon Xe light irradiation [126]. In another work, Hu and co-workers prepared graphene oxide/TiO$_2$ hybrid (GOT), where graphene oxide acted as electron sink to enhance the photodynamic activities of the TiO$_2$. Following exposure of cells to GOT and light visible irradiation, a marked decrease in mitochondrial membrane potential, cell viability, activities of superoxide dismutase, catalase and glutathione peroxidase, as well as increased malondialdehyde production were observed. Moreover, GOT caused significant elevation in caspase-3 activity, and induced apoptotic death [127]. Tian et al. functionalized graphene oxide with PEG and then loaded Chlorin e6 (Ce6), photosensitizer molecule, on its surface via supramolecular π-π stacking to develop a nanosystem that combined PDT and PTT [128]. The obtained material is highly soluble in water, and more interesting, it was found that the activity and the delivery of Ce6 molecules were promoted by the photothermal activity of the graphene after its laser irradiation at 808 nm with low power density. This work highlights the promise of using graphene for potential multifunctional cancer therapies.

4.3. Gene Delivery

Gene therapy is the science that deals with the delivery of specific gene to treat certain diseases. However, the main challenge is the development of a suitable delivery system that can protect the gene from the degradation by nucleases. Various non-viral vectors used in this application, one of these vectors which has been extensively investigated in gene delivery is polyethyleneimine (PEI), a cationic polymer, that has showed a strong interaction by electrostatic forces between its positive charge and the negatively charged phosphate groups of RNA and DNA. However, its use for clinical applications has been limited due to low biocompatibility and high toxicity of PEI [129]. On the other hand, the nanohybrid GO-PEI exhibited lower cytotoxicity and better transfection efficacy than PEI alone [130]. Chen et al. have functionalized graphene oxide covalently though amide bond with branched PEI. The prepared PEI-GO has lower cytotoxicity than PEI alone and has demonstrated better transfection of plasmid DNA than the PEI 25 kDa alone and can effectively deliver plasmid DNA into cells and be localized in the nucleus [131]. Zhang and co-workers prepared PEI-grafted GO nanocarrier for delivery of siRNA and anticancer agent, doxorubicin (DOX). This study also confirmed the lower toxicity of the PEI-GO comparing with the PEI alone and it is also showed the higher transfection efficacy of siRNA, at optimal N/P ratio, than PEI 25 K. Moreover, the sequential delivery of siRNA and DOX by the PEI-GO nanocarrier enhanced the anticancer activity of DOX [64]. Bao et al. have successfully improved the water solubility and biocompatibility of graphene oxide by functionalized it covalently with chitosan (CS). This nanocarrier is capable of loading the anticancer agent camptothecin on its surface via $\pi - \pi$ stacking and hydrophobic interactions. GO–CS is also able to condense plasmid DNA into stable, nanosized complexes, and the resulting GO–CS/pDNA nanoparticles exhibit reasonable transfection efficiency in HeLa cells at certain nitrogen/phosphate ratios [132]. Zhi et al. prepared multifunctional nanocomplex, composed of polyethylenimine (PEI)/ poly(sodium 4-styrenesulfonates) (PSS)/ graphene oxide (GO) and termed PPG for the co-delivery of novel multi drug resistance-reversing agent [MicroRNA-21(miR-21)] and anticancer drug [adriamycin (ADR)] to MCF-7/ADR cancer cells. They observed that PPG significantly enhanced the accumulation of ADR in MCF-7/ADR cells and exhibited much higher cytotoxicity than free ADR [133]. Moreover, the enhanced activity could be correlated with effective silencing of miR-21 and with increased accumulation of ADR in drug-resistant tumor cells. The mechanism of the penetration of the PPG into the cells was confirmed to be via the caveolae and clathrin-mediated endocytosis pathways, Figure 13.

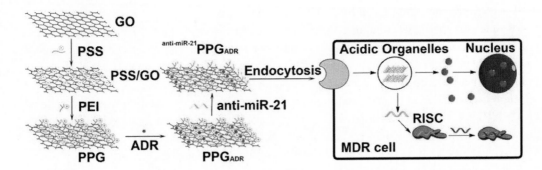

Figure 13. Schematic of the PPG fabrication and MDR reversion. Reproduced with permission from ref. [117]. Copyright 2013, PLOS.

As shown above, applying graphene as nanovector for gene delivery is still in its first steps, although there are a lot of promising results in this regard. So, it is necessary to improve the functionalization of the surface of the graphene to fulfill the requirement for gene therapy and explore more *in vitro* and *in vivo* studies to demonstrate its potential and determine its safety and biodistribution profiles.

4.4. Bio-Imaging

As mentioned in the introduction, graphene has interesting optical and electrical properties that make it a good candidate for bio-imaging application. Liu and co-workers have studied for the first time the *in vivo* behavior of nanographene sheets (NGS) functionalized covalently with polyethylene glycol (PEG) by a fluorescent labeling method [104]. They labeled NGS-PEG with Cy7, NIR fluorescent dye, by conjugating Cy7 with the amino terminated groups of PEG. They have observed high tumor uptake of NGS in various xenograft tumor mouse models like xenograft 4T1 murine breast cancer tumor, KB human epidermoid carcinoma tumor and U87MG human glioblastoma tumor mouse models. In the same approach, Wate et al. have developed a multifunctional graphene oxide using covalent method that consists of three components; Fe_3O_4 (Fe) nanoparticles, PAMAM-G4-NH2 (G4) dendrimer and Cy5 (Cy) as fluorescent group. GO-G4-Fe-Cy nanosystem was successfully uptaken by MCF7 cancer cells exhibiting bright and stable fluorescence for imaging [134].

In another investigation, Huang et al. have studied the cellular uptake of graphene oxide functionalized covalently with gold nanoparticle (AuNP) using surface enhanced Raman scattering (SERS) technique. After measuring the SERS spectra of GO at different spots in the cell, they have found that the GO loaded with Au NP cargos is distributed inhomogeneously inside the Ca Ski cell. Moreover, they confirmed that the mechanism of internalization of the functionalized graphene oxide is through clathrin-mediated endocytosis [135].

Hong et al. have studied the pharmacokinetic and tumor targeting efficacy of the covalently functionalized GO with tumor targeting agent (TRC105 that recognizes specifically CD105) labeled with ^{64}Cu using positron emission tomography (PET) technique [136]. Serial *in vivo* PET imaging demonstrated the radiolabeled GO conjugates accumulated in 4T1 murine breast tumor quickly, which demonstrated that TRC105 conjugated GO can be used as the *in vivo* imaging agent for CD105 targeting.

Beside the previous imaging techniques, magnetic resonance imaging (MRI) is a powerful technique used frequently in diagnostic medicine. Superparamagnetic Fe_3O_4 nanoparticles (NPs) are widely used in MRI, it induces localized inhomogeneity of magnetic field, and therefore causes a decrease in regional signal intensity due to shortening of T2 relaxation time. Chen et al. have reported for the first time the conjugation of functionalized Fe_3O_4 NPs with GO for cellular MRI. They have found that the Fe_3O_4-GO composites possess good physiological stability, low cytotoxicity and internalized by Hela cells. Moreover, the Fe_3O_4-GO composites showed significantly enhanced cellular MRI comparing with the isolated Fe_3O_4 NPs [137]. In another work, Ma et al. have prepared multifunctional magnetic nanosystem in a multistep process: formation of graphene oxide–iron oxide hybrid nanocomposite (GO–IONP), followed by the biofunctionalization with PEG polymer and finally loaded with doxorubicin (DOX) to form a GO–IONP–PEG–DOX complex, which enabled magnetically targeted drug delivery [138]. They have successfully used GO–IONP–PEG as T2 contrast agent for MRI in tumor bearing mice *in vivo*. Moreover, they have exploited the strong optical absorbance from the visible to the near-infrared (NIR) region of GO–IONP–PEG for magnetic field-directed localized photothermal ablation of cancer cells. Therefore, this work demonstrates great potential for applications in cancer theranostics.

4.5. Tissue Engineering

Graphene is an excellent candidate for tissue engineering, an interdisciplinary field that deals with developing biological alternatives that restore, maintain, repair or improve function of a tissue or whole organ. Graphene has interesting properties such as the mechanical (strength, high elasticity, flexibility), electrical and high conductivity which make it a physical analogue of extracellular matrix (ECM) components. Kalbacova et al. have produced large single layer of graphene by chemical vapor deposition (CVD) which has showed a better substrate for the adhesion and proliferation of human osteoblasts and mesenchymal stem cells in comparison with SiO$_2$ substrate [139], Figure 14. Due to the electrical property of graphene, it has a great potential for neural tissue engineering.

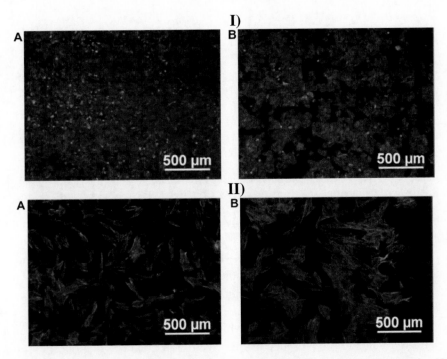

Figure 14. I) Fluorescent images of human cells incubated on graphene and SiO$_2$ substrate for 48 h: (A) human osteoblasts on graphene, (B) human osteoblasts on SiO$_2$ substrate (actin filaments are stained green and nuclei are stained blue). II) Fluorescent images of human MSCs incubated on graphene and on a SiO$_2$ substrate for 48 h: (A) MSCs on graphene, (B) MSCs on a SiO$_2$ substrate (actin filaments are stained green). Reproduced with permission from ref. [123]. Copyright 2012, Elsevier.

Li et al. have found that the graphene substrates exhibited excellent biocompatibility without affecting the cell viability and morphology of the development of neurites, one of the key structures for neural functions until its maturation in a mouse hippocampal culture model [140]. In addition, the expression of growth associate in protein-43 (GAP-43) was greatly enhanced in graphene group compared to tissue culture polystyrene (TCPS) group, which might result in the boost of neurite sprouting and outgrowth, Figure 15.

3D scaffolds based graphene have been successfully prepared recently to mimic the 3D micro-enviroment of most types of cells. The group of Shi has constructed for the first time a self-assembled graphene hydrogel (SGH) through hydrothermal method [141]. The SGH exhibited excellent mechanical, electrical, and thermal properties.

Shi and co-workers have studied the formation of three-dimensional self assembly of graphene oxide in aqueous solution to form hydrogels. The 3D network formation is related to the balance between the static repulsion and bonding interaction [142]. The adding of cross linkers (polymers, small ammonium salts, and metal ions) or acidizing the GO solution will increase the bonding force or decrease the repulsion force between GO sheets in solution which reinforce the 3D GO network and induce GO gelation or participation. The formed GO hydrogels showed low critical gelation concentrations.

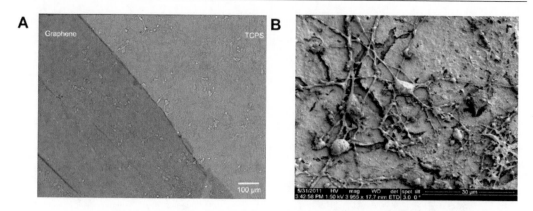

Figure 15. Neurons cultured on different substrates. (A) An optical image of neurons cultured on the border of graphene (left) and TCPS (right), (B) scanning electron microscopy image of neurons on graphene. Reproduced with permission from ref. [124]. Copyright 2011, Elsevier.

Graphene oxide has been added to polyvinyl alcohol (PVA) to produce GO/PVA composite hydrogels. PVA hydrogel suffers from poor mechanical and water-retention properties, but the GO/PVA hydrogels showed 132% increase in tensile strength and a 36% improvement of compressive strength were achieved [143]. Lu et al. prepared chitosan–PVA nanofibers containing graphene using electrospinning technique [144]. The formed composite showed better wound healing with a faster rate in both mice and rabbit comparing with the chitosan-PVA nanofibers without graphene. Moreover, graphene in the composite exhibited antibacterial activity due to the presence of free electron in graphene that does not affect the multiplication of eukaryotic cells but inhibits the prokaryotic cell multiplication therefore prevents the growth of microbes.

The group of Park has synthesized graphene–$CaCO_3$ hybrid film, which then undergo a transformation into graphene-incorporated hydroxyapaptite [$Ca_{10}(PO_4)_6(OH)_2$; HAp] [145]. The formed GO/graphene–HAp composites supported high viability of osteoblast cells with elongated morphology, which considered a potential candidate for bone tissue engineering.

CONCLUSION

In conclusion, graphene and its derivatives have been explored in many scientific fields due to their interesting mechanical, electrical and optical properties. Various studies have been developed in order to synthesize graphene in large scale, with high reproducibility, low cost and high purity. However, the biomedical applications of graphene have been hampered due to their low solubility in water and low biocompatibility, therefore, it is necessary to functionalize and modify the surface of graphene sheets to increase its biocompatibility and the most common polymer used in this purpose is polyethylene glycol (PEG) that improved the water solubility of graphene and mask it from phagocytosis. For this purpose, there are two strategies in the functionalization of graphene the covalent and non-covalent ones. The covalent approach depends in the formation of covalent bond between the graphene sheet and the ligand moieties. This approach has the advantage to give the formed composite high stability but interferes with the electrical properties of the graphene sheets. On the other hand, the non-covalent approach doesn't disrupt the π network of the graphene and is based on the formation of non-covalent interactions such as hydrophobic, van der Waals interactions and π-π stacking. Therefore, the functionalization of graphene sheets is needed to improve their biocompatibility, as recent studies showed that the functionalized graphene can internalize the cells through endocytosis and can be totally eliminated from the body without demonstrating side effects with reasonable therapeutic dose. Moreover, the functionalized graphene has been applied in various biomedical fields such as drug delivery, photothermal therapy of cancer, gene delivery, bio-imaging and tissue engineering.

Although the tremendous applications of graphene, it is still in its infancy and many *in vivo* studies are necessary to determine its pharmacokientic profile and its long term toxicity. Further investigations of preclinical studies also will be essential in order to move this interesting material toward clinical studies.

REFERENCES

[1]   Peierls, R. E. *Ann Inst Henri Poincare* 1935, *5*, 177.

[2]   Landau, L. D. *Phys Z Sowjetunion* 1937, *11*, 26.

[3]   Mermin, N. D. *Phys Rev* 1968, *176*, 250-254.

[4]   Novoselov, K. S.; Geim, A. K.; Morozov, S. V.; Jiang, D.; Zhang, Y.; Dubonos, S. V.; Grigorieva, I. V.; Firsov, A. A. *Science* 2004, *306*, 666-669.

[5]   Morozov, S. V.; Novoselov, K. S.; Katsnelson, M. I.; Schedin, F.; Elias, D. C.; Jaszczak, J. A.; Geim, A. K. *Phy. Rev. Lett.* 2008, *100*, 016602.

[6]   Chen, J. H.; Jang, C.; Xiao, S.; Ishigami, M.; Fuhrer, M. S. *Nat. Nanotechnol.* 2008, *3*, 206-209.

[7]   http://phys.org/news119030362.html

[8]   Nair, R. R.; Blake, P.; Grigorenko, A. N.; Novoselov, K. S.; Booth, T. J.; Stauber, T.; Peres N. M. R.; Geim, A. K. *Science* 2008, *320*, 1308.

[9]   Kuzmenko, A. B.; van Heumen, E.; Carbone F.; van der Marel, D. *Phys. Rev. Lett.* 2008, *100*, 117401.

[10] Gusynin, V. P.; Sharapov, S. G.; Carbotte, J. P. *Phys. Rev. Lett.* 2006, *96*, 256802.

[11] Balandin, A. A.; Ghosh, S.; Bao, W.; Calizo, I.; Teweldebrhan, D.; Miao F.; Lau, C. N. *Nano Lett.* 2008, *8*, 902-907.

[12] Lee, C.; Wei, X. D.; Kysar, J. W.; Hone, J. *Science* 2008, *321*, 385-388.

[13] Garaj, S.; Hubbard, W.; Reina, A.; Kong, J.; Branton, D.; Golovchenko, J. A. *Nature* 2010, *467*, 190-193.

[14] Novoselov, K. S.; Jiang, D.; Schedin, F.; Booth, T. J.; Khotkevich, V. V.; Morozov, S. V.; Geim, A. K. *Proc. Natl. Acad. Sci. USA* 2005, *102*, 10451-10453.

[15] Vang, R.T.; Honkala, K.; Dhal, S., Vestergaard, E. K.; Schadt, J.; Laegsgaard, E.; Clausen, B. S.; Norskov, J. K.; Besenbacher, F. *Nat. Mater.* 2005, *4*, 160-162.

[16] Sutter, P. W.; Flege, J.-I.; Sutter, E. A. *Nat. Mater.* 2008, *7*, 406-411.

[17] Li, X.; Cai, W.; An, J.; Kim, S.; Nah, J.; Yang, D.; Piner, R.; Velamakanni, A.; Jung, I.; Tutuc, E.; Banerjee, S. K.; Colombo, L.; Ruoff, R. S. *Science* 2009, *324*, 1312-1314.

[18] Dervishi, E.; Li, Z.; Watanabe, F.; Biswas, A.; Xu, Y.; Biris, A. R.; Saini, V.; Biris, A. S. *Chem. Commun.* 2009, 4061-4063.

[19] Reina, A.; Jia, X.; Ho, J.; Nezich, D.; Son, H.; Bulovic, V.; Dresselhaus, M. S.; Kong, J. *Nano Lett.* 2009, *9*, 30-35.

[20] Shivaraman, S.; Barton, R. A.; Yu, X.; Alden, J.; Herman, L.; Chandrashekhar, M. V. S., Park, J.; McEuen, P. L.; Parpia, J. M.; Craighead, H. G.; Spencer, M. G. *Nano Lett.* 2009, *9*, 3100-3105.

[21] Aristov, V. Y.; Urbanik, G.; Kummer, K.; Vyalikh, D. V.; Molodtsova, O. V.; Preobrajenski, A. B.; Zakharov, A. A.; Hess, C.; Hänke, T.; Büchner, B.; Vobornik, I.; Fujii, J.; Panaccione, G.; Ossipyan, Y. A.; Knupfer, M. *Nano Lett.* 2010, *10*, 992-995.

[22] Emtsev, K. V.; Bostwick, A.; Horn, K.; Jobst, J.; Kellogg, G. L.; Ley, L.; McChesney, J. L.; Ohta, T.; Reshanov, S. A.; Röhrl, J.; Rotenberg, E.; Schmid, A. K.; Waldmann, D.; Weber, H. B.; Seyller, T. *Nat. Mater.* 2009, *8*, 203-207.

[23] Deng, D.; Pan, X.; Zhang, H.; Fu, Q.; Tan, D.; Bao, X. *Adv. Mater.* 2010, *22*, 2168-2171.

[24] Rao, C. N. R.; Sood, A. K.; Subrahmanyam, K. S.; Govindaraj, A. *Angew. Chem. Int. Ed.* 2009, *48*, 7752-7777.

[25] Compton, O. C.; Nguyen, S. T. *Small* 2010, *6*, 711-723.

[26] Li, D.; Müller, M. B.; Gilje, S.; Kaner, R. B.; Wallace, G. G. *Nat. Nanotechnol.* 2008, *3*, 101-105.

[27] Liu, J.; Lin, Z.; Liu, T.; Yin, Z.; Zhou, X.; Chen, S.; Xie, L.; Boey, F.; Zhang, H.; Huang, W. *Small.* 2010, *6*, 1536-1542.

[28] He, Q.; Wu, S.; Gao, S.; Cao, X.; Yin, Z.; Li, H.; Chen, P.; Zhang, H. *ACS Nano.* 2011, *5*, 5038-5044.

[29] Mao, S.; Lu, G.; Yu, K.; Bo, Z.; Chen, J. *Adv. Mater.* 2010, *22*, 3521-3526.

[30] Zhu, Y.; Stoller, M. D.; Cai, W.; Velamakanni, A.; Piner, R. D.; Chen, D.; Ruoff, R. S. *ACS Nano.* 2010, *4*, 1227-1233.

[31] Williams, G.; Seger, B.; Kamat, P. V. *ACS Nano.* 2008, *2*, 1487-1491.

[32] Huang, X.; Zhou, X.; Wu, S.; Wei, Y.; Qi, X.; Zhang, J.; Boey, F.; Zhang, H. *Small.* 2010, *6*, 513-516.

[33] Cote, L. J.; Cruz-Silva, R.; Huang, J. *J. Am. Chem. Soc.* 2009, *131*, 11027-11032.

[34] Hosu, I. S.; Wang, Q.; Vasilescu, A.; Peteu, S. F.; Raditoiu, V.; Railian, S.; Zaitsev, V.; Turcheniuk, K.; Wang, Q.; Li, M.; Boukherroub, R.; Szunerits, S. *RSC Adv.* 2015, *5*, 1474-1484.

[35] Oprea, R.; Peteu, S. F.; Subramanian, P.; Qi, W.; Pichonat, E.; Happy, H.; Bayachou, M.; Boukherroub, R.; Szunerits, S. *The Analyst.* 2013, *138*, 4345-4352.

[36] Hassan, H. M. A.; Abdelsayed, V.; Khder, A. E. R. S.; AbouZeid, K. M.; Terner, J.; El-Shall, M. S.; Al-Resayes, S. I.; El-Azhary, A. A. *J. Mater. Chem.* 2009, *19*, 3832-3837.

[37] Kuila, T.; Mishra, A. K.; Khanra, P.; Kim, N. H.; Lee, J. H. *Nanoscale.* 2013, *5*, 52-71.

[38] Gao, J.; Liu, F.; Liu, Y.; Ma, N.; Wang, Z.; Zhang, X. *Chem. Mater.* 2010, *22*, 2213-2218.

[39] Zhou, Y.; Bao, Q.; Tang, L. A. L.; Zhong, Y.; Loh, K. P. *Chem. Mater.* 2009, *21*, 2950-2956.

[40] Salas, E. C.; Sun, Z.; Lüttge, A.; Tour, J. M. *ACS Nano* 2010, 4, 4852-4856.

[41] Han, M. Y.; Ozyilmaz, B.; Zhang, Y. B.; Kim, P. *Phys. Rev. Lett.* 2007, *98*, 206805.

[42] Datta, S. S.; Strachan, D. R.; Khamis, S. M.; Johnson, A. T. C. *Nano Lett.* 2008, *8*, 1912-1915.

[43] Li, X. L.; Wang, X.; Zhang, L.; Lee, S.; Dai, H. *Science* 2008, *319*, 1229-1232.

[44] Campos-Delgado, J.; Romo-Herrera, J. M.; Jia, X.; Cullen, D. A.; Muramatsu, H.; Kim, Y. A.; Hayashi, T.; Ren, Z.; Smith, D. J.; Okuno, Y. *Nano Lett.* 2008, *8*, 2773-2778.

[45] Moreno-Moreno, M.; Castellanos-Gomez, A.; Rubio-Bollinger, G.; Gomez-Herrero, J.; Agrait, N. *Small* 2009, *5*, 924-927.

[46] Hirsch, A. *Angew. Chem. Int. Ed.* 2009, *48*, 6594-6596.

[47] Higginbotham, A. L.; Kosynkin, D. V.; Sinitskii, A.; Sun, Z.; Tour, J. M. *ACS Nano* 2010, *4*, 2059-2069.

[48] Jiao, L. Zhang, L.; Ding, L.; Liu, J.; Dai, H. *Nano Res.* 2010, *3*, 387-294.

[49] Paiva, M. C.; Xu, W.; Proenca, M. F.; Novais, R. M.; Lagsgaard, E.; Besenbacher, F. *Nano Lett.* 2010, *10*, 1764-1768.

[50] Kim, K.; Sussman, A.; Zettl, A. *ACS Nano* 2010, *4*, 1362-1366.

[51] Kosynkin, D. V., Higginbotham, A. L., Sinitskii, A.; Lomeda, J. R., Dimiev, A.; Price, B. K.; Tour, J. M. *Nature* 2009, *458*, 872-876.

[52] Jiao, L.; Zhang, L.; Wang, X.; Diankov, G.; Dai, H. *Nature* 2009, *458*, 877-880.

[53] Jiao, L.; Wang, X.; Diankov, G.; Wang, H.; Dai, H. *Nat. Nanotechnol.* 2010, *5*, 321-325.

[54] Wu, J.; Pisula, W.; Müllen, K. *Chem. Rev.* 2007, *107*, 718-747.

[55] Sakamoto, J.; van Heijst, J.; Lukin, O.; Schlüter, A. D. *Angew. Chem. Int. Ed.* 2009, *48*, 1030-1069.

[56] Qian, H.; Wang, Z.; Yue, W.; Zhu, D. *J. Am. Chem. Soc.* 2007, *129*, 10664-10665.

[57] Qian, H.; Negri, F.; Wang, C.; Wang, Z. *J. Am. Chem. Soc.* 2008, *130*, 17970-17976.

[58] Shi, Y.; Qian, H.; Li, Y.; Yue, W.; Wang, Z. *Org. Lett.* 2008, *10*, 2337-2340.

[59] Yang, X. Y.; Dou, X.; Rouhanipour, A.; Zhi, L. J.; Rader, H. J.; Müllen, K. *J. Am. Chem. Soc.* 2008, *130*, 4216-4217.

[60] Niyogi, S.; Bekyarova, E.; Itkis, M.E.; McWilliams, J.L.; Hamon, M.A.; Haddon, R.C. *J. Am. Chem. Soc.* 2006, *128*, 7720–7721.

[61] Liu, Z.; Robinson, J.T.; Sun, X.; Dai, H. *J. Am. Chem. Soc.* 2008,*130*,10876–10877.

[62] Salavagione, H.J.; Gomez, M.A.; Martinez, G. *Macromolecules* 2009, *42*, 6331–6334.

[63] Hu, H.; Wang, X.; Wang, J.; Liu, F.; Zhang, M.; Xu, C. *Appl. Surf. Sci.* 2011, *257*, 2637–2642.

[64] Zhang, L.M.; Lu, Z.X.; Zhao, Q.H.; Huang, J.; Shen, H.; Zhang, Z.J. *Small* 2011, *7*, 460–464.

[65] Lomeda, J. R.; Doyle, C. D.; Kosynkin, D. V.; Hwang W. F.; Tour, J. M. *J. Am. Chem. Soc.* 2008, *130*, 16201–16206.

[66] Sun, Z.; Kohama, S.I.; Zhang, Z.; Lomeda, J.R.; Tour, J.M. *Nano Res* 2010, *3*, 117–125.

[67] Yang, H.; Li, F.; Shan, C.; Han, D.; Zhang, Q.; Niu, L. *J Mater. Chem.* 2009, *19*, 4632–4638.

[68] Economopoulos, S. P.; Rotas, G.; Miyata, Y.; Shinohara H.; Tagmatarchis, N. *ACS Nano* 2010, *4*, 7499–7507.

[69] Liu, L.-H.; Lerner, M.M.; Yan, M. *Nano Letters* 2010, *10*, 3754-3756.

[70] Strom, T. A.; Dillon, E. P.; Hamilton, C. E.; Barron, A. R. *Chem. Commun.* 2010, *46*, 4097-4099.

[71] Zhong, X.; Jin, J.; Li, S.; Niu, Z.; Hu, W.; Li, R.; Ma, J. *Chem. Commun.* 2010, *46*, 7340-7342.

[72] Georgakilas, V.; Bourlinos, A. B.; Zboril, R.; Steriotis, T. A.; Dallas, P.; Stubos A. K.; Trapalis, C. *Chem. Commun.* 2010, *46*, 1766–1768.

[73] Quintana, M.; Spyrou, K.; Grzelczak, M.; Browne, W. R.; Rudolf, P.; Prato, M. *ACS Nano* 2010, *4*, 3527–3533.

[74] Stankovich, S.; Piner, R.D.; Chen, X.; Wu, N.; Nguyen, S.T.; Ruoff, R.S. *J. Mater. Chem.* 2006, *16*, 155–158.

[75] Xu, Y.; Bai, H.; Lu, G.; Li, C.; Shi, G. *J. Am. Chem. Soc.* 2008, *130*, 5856–7585.

[76] Liu, J.; Yang, W.; Tao, L.; Li, D.; Boyer, C.; Davis, T.P. *J. Polym. Sci. Part A Polym. Chem.* 2010, *48*, 425–433.

[77] Parviz, D.; Das, S.; Ahmed, H. S. T.; Irin, F.; Bhattacharia, S.; Green, M. J. *ACS Nano* 2012, *6*, 8857–8867.

[78] Schmidt, C. D.; Boettcher C.; Hirsch, A. *Eur. J. Org. Chem.* 2007, 5497–5505.

[79] Englert, J. M.; Rohrl, J.; Schmidt, C. D.; Graupner, R.; Hundhausen, M.; Hauke F.; Hirsch, A. *Adv. Mater.* 2009, *21*, 4265–4269.

[80] Ghosh, A.; Rao, K. V.; George S. J.; Rao, C. N. R. *Chem. Eur. J.* 2010, *16*, 2700–2704.

[81] Kaminska, I.; Barras, A.; Coffinier, Y.; Lisowski, W.; Roy, S.; Niedziolka-Jonsson, J.; Woisel, P.; Lyskawa, J.; Opallo, M.; Siriwardena, A.; Boukherroub, R.; Szunerits, S. *ACS Appl. Mater. Interfaces.* 2012, *4*, 5386-5393.

[82] Kaminska, I.; Das, M. R.; Coffinier, Y.; Niedziolka-Jonsson, J.; Woisel, P.; Opallo, M.; Szunerits, S.; Boukherroub, R. *Chem. Commun.* 2012, *48*, 1221-1223.

[83] Yang, Y.; Zhang, Y.M.; Chen, Y.; Zhao, D.; Chen, J.T.; Liu, Y. *Chem. Eur. J.* 2012, *18*, 4208–4215.

[84] Malig, J.; Jux, N.; Kiessling, D.; Cid, J.J.; Vazquez, P.; Torres, T.; Guldi, D.M. *Angew. Chem. Int. Ed.* 2011, *50*, 3561–3565.

[85] Kim, T.Y.; Lee, H.; Kim, J.E.; Suh, K.S. *ACS Nano* 2010, *4*, 1612–1618.

[86] Xu, X.; Suslick, K.S. *J. Am. Chem. Soc.* 2011, *133*, 9148–9151.

[87] Skaltsas, T.; Karousis, N.; Yan, H.; Wang, C.-R.; Pispas S.; Tagmatarchis, N. *J. Mater. Chem.* 2012, 21507–21512.

[88] Liang Y. T.; Hersam, M. C. *J. Am. Chem. Soc.* 2010, *132*, 17661–17663.

[89] Bourlinos, A. B.; Georgakilas, V.; Zboril, R.; Steriotis, T. A.; Stubos A. K.; Trapalis, C. *Solid State Commun.* 2009, *149*, 2172–2176.

[90] May, P.; Khan, U.; Hughes J. M.; Coleman, J. N. *J. Phys. Chem. C* 2012, *116*, 11393–11400.

[91] Xu, L.; McGraw, J.-W.; Gao, F.; Grundy, M.; Ye, Z.; Gu Z.; Shepherd, J. L. *J. Phys. Chem. C* 2013, *117*, 10730–10742.

[92] Zeng, Q.; Cheng, J.; Tang, L.; Liu, X.; Liu, Y.; Li, J.; Jiang, J. *Adv. Funct. Mater.* 2010, *20*, 3366–3372.

[93] Smith, R.J.; Lotya, M.; Coleman, J.N. *New J. Phys.* 2010, *12*, 125008-125019.

[94] Kotchey, G.P.; Allen, B.L.; Vedala, H.; Yanamala, N.; Kapralov, A.A.; Tyurina, Y.Y.; Klein-Seetharaman, J.; Kagan, V.E.; Star, A. *ACS Nano* 2011, *5*, 2098–2108.

[95] Zhang, L.M.; Xia, J.G.; Zhao, Q.H.; Liu, L.; Zhang, Z. *Small* 2010, *6*, 537-544.

[96] Mu, Q.; Su, G.; Li, L.; Gilbertson, B.O.; Yu, L.H., Zhang, Q.; Sun, Y.-P.; Yan, B. *ACS Appl. Mater. Interfaces* 2012, *4*, 2259–2266.

[97] Wen, H.Y.; Dong, C.Y.; Dong, H.Q.; Shen, A.; Xia, W.; Cai, X.; Song, Y.; Li, X.; Li, Y.; Shi, D. *Small* 2012, *8*, 760-769

[98] Lu, C.H.; Zhu, C.L.; Li, J.; Chen, X.; Yang, H.H. *Chem. Commun.* 2010, *46*, 3116-3118.

[99] Liao, K.H.; Lin, Y.S.; Macosko, C.W.; Haynes, C.L. *ACS Appl. Mater. Interfaces* 2011, *3*, 2607-2615.

[100] Wang, K.; Ruan, J.; Song, H.; Zhang, J.; Wo, Y.; Guo, S.; Cui, D. *Nanoscale Res. Lett.* 2011, *6*, 1-8.

[101] Chang, Y.; Yang, S-.T.; Liu, J-.H.; Dong, E.; Wang, Y.; Cao, A.; Liu, Y.; Wang, H. *Toxicol. Lett.* 2011, *200*, 201-210.

[102] Yuan, J.F.; Gao, H.C.; Sui, J.J.; Duan, H.; Chen, W.N.; Ching, C.B. *Toxicol. Sci.* 2012, *126*, 149-161.

[103] Zhang, X.Y.; Yin, J.L.; Peng, C.; Hu, W.Q.; Zhu, Z.Y.; Li, W.X.; Fan, C.; Huang, Q. *Carbon* 2011, *49*, 986–995.

[104] Yang, K.; Zhang, S.A.; Zhang, G.X.; Sun, X.M.; Lee, S.T.; Liu, Z.A. *Nano Lett.* 2010, *10*, 3318–3323.

[105] Yang, K.; Wan, J.M.; Zhang, S.A.; Zhang, Y.J.; Lee, S.T.; Liu, Z.A. *ACS Nano* 2011, *5*, 516–522.

[106] Singh, S.K.; Singh, M.K.; Kulkarni, P.P.; Sonkar, V.K.; Gracio, J.J.A.; Dash, D. *ACS Nano* 2012, *6*, 2731–2740.

[107] Zhang, S.A.; Yang, K.; Feng, L.Z.; Liu, Z. *Carbon* 2011, *49*, 4040–4049.

[108] Banks, C.E.; Crossley, A.; Salter, C.; Wilkins, S.J.; Compton, R.G. *Angew. Chem. Int. Ed.* 2006, *45*, 2533-2537.

[109] Yang, X.Y.; Zhang, X.Y.; Liu, Z.F.; Ma, Y.F.; Huang, Y.; Chen, Y. *J. Phys. Chem. C* 2008, *112*, 17554–17558.

[110] Depan, D.; Shah, J.; Misra, R.D.K. *Mater. Sci. Eng. C* 2011, *31*, 1305-1312.

[111] Sun, X.; Liu, Z.; Welsher, K.; Robinson, J.; Goodwin, A.; Zaric, S.; Dai, H. *Nano Res.* 2008, *1*, 203–212.

[112] Kim, H.; Lee, D.; Kim, J.; Kim, T.-I.; Kim, W.J. *ACS Nano* 2013, *7*, 6735–6746.

[113] Wang, C.S.; Li, J.Y.; Amatore, C.; Chen, Y.; Jiang, H.; Wang, X.M. *Angew. Chem. Int. Ed.* 2011, *50*, 11644–11648.

[114] Yang, X.Y.; Wang, Y.S.; Huang, X.; Ma, Y.F.; Huang, Y.; Yang, R.C.; Duan, H.Q.; Chen, Y.S. *J. Mater. Chem.* 2010, *21*, 3448–3454.

[115] Rana, V.K.; Choi, M.-C.; Kong, J.-Y.; Kim, G.Y.; Kim, M.J.; Kim, S.-H.; Mishra, S.; Singh, R.P.; Ha, C.-S. *Macromol. Mater. Eng.* 2011, *296*, 131–140.

[116] Pan, Y.Z.; Bao, H.Q.; Sahoo, N.G.; Wu, T.; Li, L. *Adv. Funct. Mater.* 2011, *21*, 2754-2763.

[117] Kakran, M.; Sahoo, N.G.; Bao, H. Pan, Y.; Li, L. *Curr. Med. Chem.* 2011, *18*, 4503-4512.

[118] Turcheniuk, K.; Khanal, M.; Motorina, A.; Subramanian, P.; Barras, A.; Zaitsev, V.; Kuncser, V.; Leca, A.; Martoriati, A.; Cailliau, K.; Bodart, J.-F.; Boukherroub, R.; Szunerits, S. *RSC Adv.* 2014, *4*, 865-875.

[119] Yang, K.; Feng, L.; Shia, X.; Liu, Z. *Chem. Soc. Rev.* 2013, *42*, 530-547.

[120] Zhang, W.; Guo, Z.; Huang, D.; Liu, Z.; Guo, X.; Zhong, H. *Biomaterials* 2011, *32*, 8555-8561.

[121] Robinson, J.T.; Tabakman, S.M.; Liang, Y.Y.; Wang, H.L.; Casalongue, H.S.; Vinh, D.; Dai, H. *J. Am. Chem. Soc.* 2011, *133*, 6825–6831.

[122] Yang, K.; Wan, J.M.; Zhang, S.; Tian, B.; Zhang, Y.J; Liu, Z. *Biomaterials* 2012, *33*, 2206–2214.

[123] Yang, K.; Hu, L.; Ma, X.; Ye, S.; Cheng, L.; Shi, X.; Li, C.; Li, Y.; Liu, Z. *Adv. Mater.* 2012, *24*, 1868-1872.

[124] Hu, S.H.; Chen, Y-.W.; Hung, W-.T.; Chen, I-.W.; Chen, S-Y. *Adv. Mater.* 2012, *24*, 1748-1754.

[125] Turcheniuk, K.; Hage, C.-H.; Spadavecchia, J.; Serrano, A. Y.; Larroulet, I.; Pesquera, A.; Zurutuza, A.; Pisfil, M. G.; Héliot, L.; Boukaert, J.; Boukherroub, R.; Szunerits, S. *J. Mater. Chem. B.* 2015, *3*, 375-386.

[126] Dong, H.Q.; Zhao, Z.L.; Wen, H.Y.; Li, Y.Y.; Guo, F.F.; Shen, A.J.; Pilger, F.; Lin, C.; Shi, D.L. *Sci. China Chem.* 2010, *53*, 2265–2271.

[127] Hu, Z.; Huang, Y.; Sun, S.; Guan, W.; Yao, Y.; Tang, P.; *Carbon* 2012, *50*, 994–1004.

[128] Tian, B.; Wang, C.; Zhang, S.; Feng, L.; Liu, Z. *ACS Nano* 2011, *5*, 7000-7009.

[129] Jager, M.; Schubert, S.; Ochrimenko, S.; Fischer, D.; Schubert, U.S. *Chem. Soc. Rev.* 2012, *41*, 4755–4767.

[130] Feng, L.; Zhang, S.; Liu, Z. *Nanoscale* 2011, *3*, 1252–1257.

[131] Chen, B.A.; Liu, M.; Zhang, L.M.; Huang, J.; Yao, J.L.; Zhang, Z.J. *J Mater Chem* 2011, *21*, 7736–7741.

[132] Bao, H.Q.; Pan, Y.Z.; Ping, Y.; Sahoo, N.G.; Wu, T.F.; Li, L.;Gan, L.H. *Small* 2011, *7*, 1569–1578.

[133] Zhi, F.; Dong, H.; Jia, X.; Guo, W.; Lu, H.; Yang, Y.; Ju, H.; Zhang, X.; Hu, Y. *PLoS ONE* 2013, *8*, e60034.

[134] Wate, P.S.; Banerjee, S.S.; Jalota-Badhwar, A.; Mascarenhas, R.R.; Zope, K.R.; Khandare, J.; Misra, R.D.K. *Nanotechnology* 2012, *23*, 415101.

[135] [135] Huang, J.; Zong, C.; Shen, H.; Liu, M.; Chen, B.; Ren, B.; Zhang, Z. *Small* 2012, *8*, 2577–2584.

[136] Hong, H.; Yang, K.; Zhang, Y; Engle, J. W.; Feng, L; Yang, Y.; Nayak, T.R.; Goel, S.; Bean, J.; Theuer, C. P.; Barnhart, T. E.; Liu, Z.; Cai, W. *ACS Nano* 2012, *6*, 2361-2370.

[137] Chen, W.; Yi, P.; Zhang, Y.; Zhang, L.; Deng, Z.; Zhang, Z. *ACS Appl. Mater. Interfaces* 2011, *3*, 4085-4091.

[138] Ma, X.; Tao, H.; Yang, K.; Feng, L.; Cheng, L.; Shi, X.; Li, Y.; Guo, L.; Liu, Z. *Nano Res.* 2012, *5*, 199–212.

[139] Kalbacova, M.; Broz, A.; Kong, J.; Kalbac, M. *Carbon* 2010, *48*, 4323-4329.

[140] Li, N.; Zhang, X.; Song, Q.; Su, R.; Zhang, Q.; Kong, T.; Liu, L.; Jin, G.; Tang, M.; Cheng, G. *Biomaterials* 2011, *32*, 9374-9382.

[141] Xu, Y.; Sheng, K.; Li, C.; Shi, G. *ACS Nano* 2010, *4*, 4324-4330.

[142] Bai, H.; Li, C.; Wang, X.; Shi, G. *J. Phys. Chem. C* 2011, *115*, 5545-5551.

[143] Zhang, L.; Wang, Z.; Xu, C.; Li, Y.; Gao, J.; Wang, W.; Liu, Y. *J. Mater. Chem.* 2011, *21*, 10399–10406.

[144] Lu, B.; Li, T.; Zhao, H.; Li, X.; Gao, C.; Zhang, S.; Xie, E. *Nanoscale* 2012, *4*, 2978–2982.

[145] Kim, S.; Ku, S. H.; Lim, S. Y.; Kim, J. H.; Park, C. B. *Adv. Mater.* 2011, *23*, 2009-2014.

In: Innovations in Nanomaterials
Editors: Al-N. Chowdhury, J. Shapter, A. B. Imran

ISBN: 978-1-63483-548-0
© 2015 Nova Science Publishers, Inc.

Chapter 13

GRAPHENE-BASED NANOCOMPOSITE MATERIALS FOR THE PHOTOREDUCTION OF CARBON DIOXIDE INTO VALUABLE ORGANIC COMPOUNDS

Pawan Kumar[1], Suman L. Jain[1,*]
and Rabah Boukherroub[2,†]

[1]Chemical Sciences Division, CSIR-Indian
Institute of Petroleum, Dehradun, India
[2]Institut d'Electronique, de Microélectronique et de Nanotechnologie (IEMN),
Villeneuve d'Ascq, France

ABSTRACT

Photocatalytic reduction of carbon dioxide (CO_2) into hydrocarbon fuels such as methane, methanol, etc is an attractive strategy for simultaneously harvesting solar energy and capturing CO_2, which is a major cause for global warming. Recently, graphene-based nanocomposites due to their large surface area, high conductivity, ease of functionalization and low cost have emerged as efficient photocatalysts for the reduction of CO_2 to hydrocarbons. Furthermore, graphene oxide (or reduced graphene oxide) can be coupled with various semiconductors to form graphene oxide (graphene)-semiconductor nanocomposites, which act as high-performance visible light active photoredox catalysts owing to their enhanced quantum efficiency. In this chapter, we will summarize the various graphene/semiconductor/homogeneous catalyst nanocomposites used as high performance visible light active photocatalysts for the reduction of CO_2 to high value chemicals such as methane, methanol, etc.

---

* Suman L. Jain: Chemical Sciences Division, CSIR-Indian Institute of Petroleum, Dehradun-248005, India. E-mail: suman@iip.res.in.

† Rabah Boukherroub: Institut d'Electronique, de Microélectronique et de Nanotechnologie (IEMN), UMR CNRS 8520, Avenue Poincaré – BP 60069, 59652 Villeneuve d'Ascq, France. E-mail: rabah.boukherroub@iri.univ-lille1.fr.

INTRODUCTION

Manmade emissions of carbon dioxide (CO_2) have increased its atmospheric concentration from around 258 ppm in 1958 to 398 ppm in 2013 [1-6]. This CO_2 can be used as carbon feedstock for the production of valuable chemicals such as dimethyl carbonate, polycarbonates, carbamates, etc. [7-12]. But synthesis of such compounds is not viable because of conversion of thermodynamically more stable CO_2 molecule into less stable molecule is an energy intensive step [13-16]. Thus, the development of visible light—particularly solar assisted—conversion of CO_2 to high energy materials, has attracted increasing attention from the scientific community due to its great potential of fulfilling our future energy demands and getting rid from increasing global warming. Photochemical conversion of CO_2 to value added products is a most challenging task, because of the high reduction potential of CO_2, multiple electron transfer pathway and need of a highly efficient robust catalyst that can derive electrons for CO_2 reduction by water splitting under visible light irradiation [17-21]. Many types of semiconductor photocatalysts such as TiO_2 [22, 23], ZnO [24, 25], $BiVO_4$ [26], $InVO_4$ [27, 28] and combinations [29-32] thereof have been widely studied for this purpose. By far the most researched photocatalytic material is anatase TiO_2 because of its low cost, high thermodynamic stability, strong oxidizing power and relative nontoxicity. However, the rapid recombination of electrons and holes is one of the main limitations for using TiO_2 as a high performance photoactive catalyst for practical applications [33]. Moreover, its wide band gap of 3.2 eV limits its application to the ultraviolet (UV) region. Or only a small fraction ($\approx$ 5%) of the total solar spectrum reaches the earth's surface. In order to utilize irradiation from sunlight or from artificial room light sources, the development of visible light-active TiO_2 is essentially necessary.

In order to improve the photocatalytic performance of semiconductor-based photocatalysts such as TiO_2, a variety of strategies such as doping with various metals like Cu [34-39], Ag [40, 41], Ru [42], etc. and nonmetal N [43, 44], C [45], iodine [46], etc. have been investigated. Semiconductor doping with these elements not only slows down the rate of electron and hole recombination, but also lowers the band gap position by orbital overlapping. However, the conversion efficiencies are quite low far from the acceptable levels. Another approach to increase the absorption pattern of semiconductor type photocatalyst is the dye sensitization [47-51]. In this approach, any organic or inorganic dye (in solution or immobilized) is used as a visible light absorber which transfers the excited electrons to the semiconductor. This strategy, although advantageous as it provides visible light promoted CO_2 reduction, has major limitations which include the degradation of the dye by photogenerated oxidizing holes and leaching of photosensitizer. Recently, the coupling of semiconductor photocatalysts with carbonaceous nanomaterials has gained considerable interest due to their unique and controllable structural and electrical properties, which has proven to be beneficial for enhancing the photocatalytic activities. Among the carbon nanostructures (e.g., C_{60}, carbon nanotubes, and graphene), graphene offers new opportunities in photovoltaic conversion and photocatalysis through formation of hybrid structures with a variety of nanomaterials [52-56]. For example, when combined with TiO_2 graphene acts as an electron trap, which promotes electron–hole separation and facilitates interfacial electron transfer (Figure 1) [57, 58]. Graphene, a one-atom-thick sp^2-hybridized carbon sheet, has attracted a great deal of scientific interest since its discovery by Geim and co-workers [59-64].

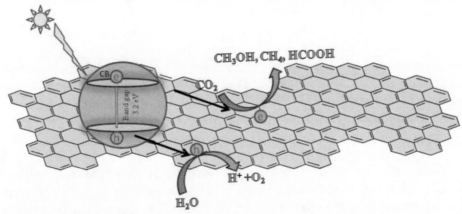

CH$_3$OH, CH$_4$, HCOOH

CO$_2$

H$^+$+O$_2$

H$_2$O

Figure 1. Mechanistic pathway of CO$_2$ reduction on graphene-TiO$_2$ nanocomposite.

Owing to its unique properties such as high mobility of charge carriers (250, 000 cm^2 V^{-1} s^{-1}) [65-67], thermal conductivity (5000 W m^{-1} K^{-1}) [68, 69], mechanical strength (1060 GPa) [70] and high specific surface area (2630 m^2 g^{-1}) [71, 72], graphene grasps a huge potential for various applications including electronic devices, solar energy conversion, supercapacitors, sensors, catalysis and particularly for the preparation of composite materials [73, 74]. Furthermore, graphene has an extensive two-dimensional π-π conjugation structure, which endows it with excellent conductivity of electrons. Several methods for the synthesis of graphene sheets are available namely micromechanical exfoliation, chemical vapour deposition (CVD), epitaxial growth, longitudinal unzipping of carbon nanotubes, direct sonication, chemical and thermal reduction/exfoliation as well as solution-based approaches [75, 76]. However, graphene synthesis faces a problem that is common to many novel materials: the absence of process for production in high yields (gram scale) [77, 78].

The exceptional properties of graphene have generated huge interest for the synthesis of graphene-semiconductor nanocomposite materials. The principal driving force for the growing research trends in the field of graphene-semiconductor nanohybrids is the formation of nanocomposite materials of superior properties, resulting from the combination of the properties of graphene and semiconductors. In this context, extensive work has been carried out on the preparation of graphene–TiO$_2$ composite and its application in various fields such as photocatalytic bactericide [79, 80], hydrogen evolution [81-83], dye-sensitized solar cell [84, 85]. The nanohybrids of graphene/graphene oxide with TiO$_2$ and other doped/nondoped semiconductors were also reported for the photocatalytic dye degradation: Cu-TiO$_2$/grapheme [86, 87], P-25/grapheme [88, 89], functionalized graphene sheets/ZnO nanocomposites [90], N-doped graphene/ZnSe naocomposite [91], graphene oxide enwrapped Ag/AgX nanocomposites [92], graphene/Bi$_2$WO$_6$ nanocomposite [93], Nd/TiO$_2$ modified with grapheme [94], for water splitting: ternary polyaniline–graphene–TiO$_2$ [95], N-doped Sr$_2$Ta$_2$O$_7$ coupled with graphene sheets [96], and for organic pollutant degradation: TiO$_2$/ graphene nanocomposites [97, 98]. Nevertheless, there is a limited number of reports on the use of graphene- based TiO$_2$ (and other semiconductors) nanocomposites for photocatalytic CO$_2$ conversion into high value chemicals (Figure 1). Tu et al., [99] reported on robust hollow spheres consisting of molecular-scale alternating titania ($Ti_{0.91}O_2$) nanosheets and graphene (G) nanosheets prepared by a layer-by-layer assembly technique with polymer beads as sacrificial templates under microwave irradiation.

The use of microwave technique helps in removing the template and simultaneously reduces graphene oxide into graphene.

The developed photocatalyst was subsequently used for photocatalytic conversion of CO_2 into renewable fuels (CO and CH_4) in the presence of water vapour. The formation rates of CO and CH_4 over the as-prepared G–$Ti_{0.91}O_2$ hollow spheres were found to be 8.91 and 1.14 mmol g^{-1} h^{-1}, respectively. The total conversion of CO_2 over G–$Ti_{0.91}O_2$ hollow spheres was found to be five-times higher than blank $Ti_{0.91}O_2$ hollow spheres and nine-times higher than commercial P-25. The superior photoactivity of G–$Ti_{0.91}O_2$ hollow spheres can be attributed to the ultrathin nature of $Ti_{0.91}O_2$ nanosheets, allowing charge carriers to move rapidly onto the surface to participate in the photoreduction process. In addition, the sufficiently compact stacking of ultrathin $Ti_{0.91}O_2$ nanosheets with G nanosheets allowed the photo-generated electron to transfer fast from the titania nanosheets to G to enhance lifetime of the charge carriers; the hollow structure probably acted as a photon trap-well to allow the multiscattering of incident light for the enhancement of light absorption.

Very recently, Zhou et al., [100] described another TiO_2–graphene (G) hybrid nanocomposite, prepared by an *in situ* reduction–hydrolysis technique in a binary ethylenediamine (En)–H_2O solvent. The reduction of graphene oxide (GO) into graphene (G) by En and the formation of TiO_2 nanoparticles loaded onto graphene through chemical bonds (Ti–O–C bond) was achieved simultaneously. Due to the reducing role of En, abundant Ti^{3+} is formed on the surface of the TiO_2 in the obtained G–TiO_2 nanohybrids. The synthesized G–TiO_2 nanocomposite exhibited higher photoactivity (8 mmol g^{-1} h^{-1} CH_4 and 16.8 mmol g^{-1} h^{-1} C_2H_6) in comparison with the blank TiO_2 sample (10.1 mmol g^{-1} h^{-1} CH_4 and 7.2 mmol g^{-1} h^{-1} C_2H_6) and commercial P-25 (0.69 mmol g^{-1} h^{-1} CH_4, and minor CO 0.16 mmol g^{-1} h^{-1}, C_2H_6 is absent). The enhanced photocatalytic performance of the G–TiO_2 nanocomposite was probably due to the synergistic effect of the introduction of graphene and the presence of surface Ti^{3+} sites. Particularly, the yield of C_2H_6 was found to be increased with the content of incorporated graphene.

Graphene not only facilitates the transportation of electrons, but in some studies it was found that graphene can elevate the position of the semiconductor's conduction band. For example, tungsten trioxide (WO_3) is a visible light active semiconductor due to its narrow band gap of 2.7 eV. However, the low position of its conduction band (lower than −0.1 V *vs.* NHE, pH = 7) has limited its application for reduction of CO_2 to hydrocarbons. Wang et al., [101] have shown that nanocomposite of WO_3 with graphene synthesized using *in situ* hydrothermal method can elevate the position of conduction band of WO_3. By using this hybrid system, 0.89 µmol g^{-1} cat CH_4 was obtained after 8 h visible light illumination. XPS VB spectra recorded on WO_3 and WO_3-graphene nanocomposite allowed to determining the position of the VB band edge. It was found that WO_3 has a deeper VB maximum of 0.37 eV than WO_3-graphene nanocomposite, suggesting a stronger oxidation power of holes photogenerated in the WO_3 VB. A correlation of this value with the CB minimum indicated its higher value by -0.24 V than CO_2/CH_4 redox potential.

Apart from TiO_2/graphene and WO_3/graphene composites, graphene oxide (GO) was also found to be a promising photocatalyst for the catalytic conversion of CO_2 to methanol (MeOH). GO can be easily synthesized through extensive oxidation of natural graphite [102].

GO contains functional oxygen groups (hydroxyl, epoxy, carbonyl and carboxyl) and sp^3 carbon, responsible for its variable properties from pristine graphene. Although the chemistry is still under debate, these oxygen-containing groups provide graphene with hydrophilic character and chemical reactivity [103, 104]. In contrast to graphene, GO is insulating in nature due to presence of oxygen carrying functionalities on its surface.

The presence of these oxy-containing groups generates lots of sp^3 hybridized carbons and simultaneously disrupts the conjugated sp^2 carbons π-π network. However, the controlled reduction of GO by removing most of the oxygen-containing groups can transform it into semiconductor of appropriate band gap [105, 106].

Recently, Hsu et al., [107] have conducted a systematic investigation of the photocatalytic CO_2 reduction on various GO samples synthesized under different conditions. They found that the GO, obtained by the modified Hummer's method [108] in the presence of excess $KMnO_4$ and excess H_3PO_4 to raise the level of oxidation, exhibited the highest photocatalytic efficiency among the studied samples. The photocatalytic conversion rate of CO_2 to methanol was found to be 0.172 mmol g^{-1} cat h^{-1} under simulated solar-light source irradiation for 4 h, which is six-fold higher than pure TiO_2. They proposed a possible explanation that in the modified GO, the oxygenated functional groups provide a 2D network of sp^2 and sp^3 bonded atoms, leading to the presence of a finite band gap depending on the isolated sp^2 domains. During the photocatalytic reduction process, the modified GO with surplus oxygenated components is photoexcited to generate electron–hole pairs (e^-–h^+), which then migrate to the GO surface to react with adsorbed reactants [109]. The reduction potential of an e^- in the GO conduction band (ca. - 0.79 V $vs.$ NHE) is lower than the potential of CO_2/CH_3OH (- 0.38 V $vs.$ NHE), while the oxidation potential of the h^+ in the GO valance band (ca. 4 V $vs.$ NHE) is higher than the potential of H_2O/O_2, H^+ (E = 0.82 V $vs.$ NHE). Therefore, the photogenerated electrons and holes on the irradiated GO can react with adsorbed CO_2 and H_2O to produce CH_3OH via a six-electron reaction.

Yadav et al., [110] reported the use of a novel graphene-based visible light active photocatalyst consisting of a covalently attached chromophore, such as multianthraquinone substituted porphyrin to GO coupled with the activity of an enzyme, formate dehydrogenase. In this system, under visible light irradiation, the chromophore is excited and transfers its electrons through GO to a rhodium complex in solution, which becomes reduced. The reduced rhodium complex then reduces the co-enzyme NAD^+ to NADH thus forming the photocatalysis cycle. Finally, NADH is consumed by the CO_2 substrate for its enzymatic (formate dehydrogenase) conversion to formic acid. The NAD^+ released from the enzyme can undergo photocatalysis cycle in the same way, leading to the photoregeneration of NADH. The higher conversion rate of CO_2 using this system indicates the potential use of enzyme incorporation into a photocatalytic system to enhance the catalytic activity of the graphene-based materials in CO_2 photoreduction systems.

An et al., [111] synthesized Cu_2O/reduced graphene oxide (rGO) composite using a facile one-step microwave-assisted chemical method and investigated its performance for the photoreduction of CO_2. In this system, rGO acted not only as an ideal electron trap to hinder fast charge recombination, but also as a stabilizer to improve the stability of Cu_2O. The rGO coating dramatically increases Cu_2O activity for CO_2 photoreduction and provided a nearly six times higher activity than the optimized Cu_2O and 50 times higher activity than Cu_2O/RuO_x junction in the 20th hour. Furthermore, an apparent initial quantum yield of approximately 0.34% at 400 nm has been achieved by the Cu_2O/rGO junction for CO_2

photoreduction. The photocurrent of the junction is nearly double that of the blank Cu_2O photocathode. The improved activity together with the enhanced stability of Cu_2O is attributed to the efficient charge separation and transfer to rGO as well as the protection function of rGO. By coupling with rGO, the photoreduction activity of Cu_2O was enhanced by two times, with CO as the only reduction product.

The incorporation of rGO into Cu_2O further improves the photocatalyst stability remarkably, which shows a great potential for CO_2 conversion in a sustainable manner.

Tan et al., [112] investigated the photocatalytic activity of rGO-TiO_2 hybrid nanocrystals, synthesized through a simple solvothermal route, for the reduction of CO_2 to hydrocarbons. The prepared rGO/TiO_2 nanocomposites exhibited superior photocatalytic activity (0.135 $\mu mol\ g^{-1}\ cat\ h^{-1}$) in the reduction of CO_2 over graphite oxide and pure anatase. The intimate contact between TiO_2 and rGO was proposed to accelerate the transfer of photogenerated electrons on TiO_2 to rGO, leading to an effective charge anti-recombination and thus enhancing the photocatalytic activity. Furthermore, rGO/TiO_2 nanocomposite photocatalysts were found to be active even under the irradiation of low-power energy saving light bulbs, which renders the entire process economically and practically feasible. In another work [113], a composite of solvent exfoliated graphene (SEG) synthesized by ultrasonication in N,N-dimethyl formamide (DMF) and TiO_2 (P-25) in the form of film made by using ethyl cellulose as a stabilizing and film forming polymer was used for the photoreduction of CO_2 to methane.

The as obtained SEG (0.27%)-TiO_2 gave 8.3 $\mu mol\ h^{-1}\ m^{-2}$ methane under UV light irradiation that was 4.5 times higher than bare TiO_2. Further it was demonstrated that less defective SEG composite with TiO_2 possessed higher photocatalytic activity than more defective rGO composite with TiO_2, for the reduction of CO_2 to CH_4, with up to a 7-fold improvement as compared to pure TiO_2 under visible illumination.

As described above a number of improved semiconductor-graphene (graphene oxide) nanocomposites have been developed as potential catalysts for the photoreduction of CO_2, but their quantum yields and selectivities of products are still far from the acceptable level for practical applications. Thus, considerable efforts are being paid towards using transition metal complexes as visible light active photoredox catalysts for CO_2 reduction [114-118]. Such catalysts can have multiple and accessible redox states that have been shown to promote multiple electron transfer (MET) reactivity.

Furthermore, the formal reduction potentials can be systematically tuned through ligand modification to better match the potential required for CO_2 reduction. A number of transition metal based molecular complexes such as ruthenium(II) polypyridine carbonyl complex [119-122], cobalt(II) trisbipyridine [123], cobalt(III) macrocycles [124] rhenium [125, 126] and iridium complex [127, 128] with a photosensitizer have been developed as efficient photocatalytic systems to reduce CO_2 with relatively high quantum yield and high selectivity of products. Rhenium(I) complex and $Re(bpy)(CO)_2X_2$ (X = Cl, Br, I) were found to be superior catalysts for CO_2 reduction reaction. It was demonstrated that if the halogens were replaced by phosphene-based ligands then the yield of product was increased manifold [129]. This was attributed to the overlap between phosphorous and metal orbital. However, in some instances, a weak absorption in the visible region makes the utility of these complexes limited. This limitation can be overcome by using high nuclearity transition metal complexes having more than one metal unit joint together in a single complex [130-133]. Furthermore, supporting a soluble metal complex onto an inorganic solid matrix is one of the promising approaches to combine the advantages of both homogeneous catalyst such as high reactivity, selectivity with heterogeneous one such ease of separation from products, lack of corrosiveness, and robustness for operation at high temperatures. In this context, a number of homogeneous photoredox complexes have been immobilized onto various supports such as TiO_2, organic polymers, ion exchange resins, etc. [134-136].

However, in some instances, the non-covalent attachment of the metal complex to support materials leads to the leaching of the active species during the reaction. Graphene/graphene oxide have been acknowledged to be ideal supporting materials for the attachment of homogenous redox metal complexes not only due to their high surface area and presence of various oxygen-containing functional groups, but work synergistically with metal complexes to provide better photocatalytic activity [137-140].

In a study by Zhu et al., [141] Ru(dcbpy)$_3$ complex was immobilized to rGO *via* non covalent interaction and then platinum nanoparticles (co-catalyst) were deposited on the surface of rGO. This hybrid system was found to be much more active for hydrogen evolution reaction in comparison to Ru(dcbpy)$_3$-sensitized Pt composites and rGO/Pt nanocomposites under visible light. Here, it is important to mention that photoexcited electrons flow from Ru(dcbpy)$_3$ to rGO to Pt because of low lying Fermi level of rGO and lower work function of Pt nanoparticles. Although, a number of homogeneous complexes such as zinc phthalocyanine, iron phthalocyanine, porphyrine, ruthenium complexes and organic dyes have been immobilized to GO/rGO for various applications, a very few reports are known on the photocatalytic CO_2 reduction.

Very recently, Kumar et al., [142] reported a novel GO tethered Co(II) phthalocyanine complex (CoPc–GO), prepared by a stepwise procedure, as an efficient, cost effective and recyclable photocatalyst for the reduction of CO_2. The covalent attachment of the CoPc to the GO support not only enhances the stability of the developed heterogeneous catalyst, but also prevents the leaching of the active metal from the support. The beauty of the method was that CoPc was also decorated on the basal plane of GO so a higher loading of CoPc was achieved in comparison to conventional methods where only edge groups were used (Figure 2) [143]. The prepared catalyst was used for the photocatalytic reduction of CO_2 by using water as a solvent and triethylamine as sacrificial donor.

Methanol was obtained as the major reaction product along with the formation of minor amount of CO (0.82%). It was found that GO-grafted CoPc exhibited higher photocatalytic activity than homogeneous CoPc as well as GO and showed good recoverability without significant leaching during the reaction. The yield of methanol after 48 h of reaction by using GO-CoPc catalyst in the presence of sacrificial donor triethylamine was found to be 3781.8881 μmol g^{-1} cat and the conversion rate was found to be 78.7893 μmol g^{-1} cat h^{-1}.

After the photoreduction experiment, the catalyst was easily recovered by filtration and reused for the subsequent recycling experiment without significant change in the catalytic efficiency.

In another report by Kumar et al., [144] ruthenium trinuclear polyazine complex was immobilized onto GO support through complexation using phenanthroline ligands (GO-phen) as coupling moieties (Ru-phen-GO). The three ruthenium units were joined through bipyrimidine bridging ligand and works like antenna [145, 146].

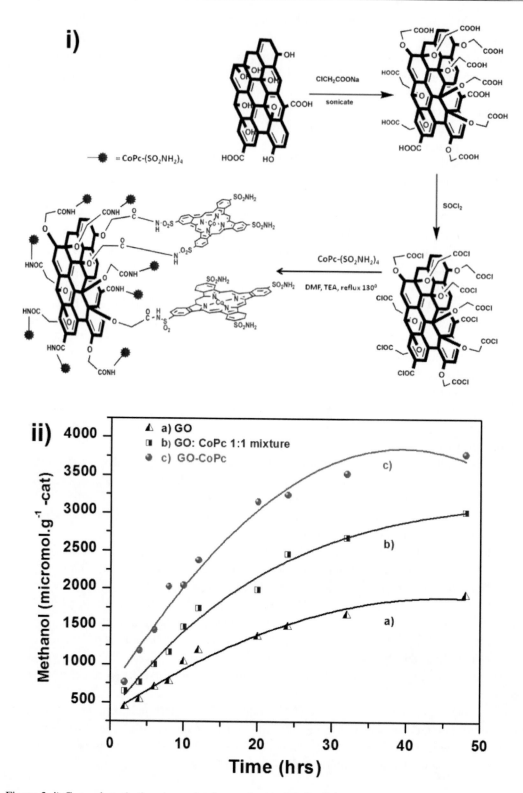

Figure 2. i) General synthetic scheme for the synthesis of CoPc-GO; ii) Methanol formation rate by using CoPc-GO catalyst.

The developed photocatalyst was used for the photocatalytic reduction of CO_2 to methanol by using 20 watt white cold LED flood light in dimethyl formamide/water mixture containing triethylamine as a reductive quencher. After 48 h illumination, the yield of methanol was found to be 3977.57 ± 5.60 µmol g^{-1} cat (Figure 3). The developed photocatalyst exhibited higher photocatalytic activity than GO: 2201.40 ± 8.76 µmol g^{-1} cat (Figure 3). After the reaction, the catalyst was easily recovered and reused for subsequent four runs without significant loss of catalytic activity and no leaching of the metal/ligand was detected during the reaction.

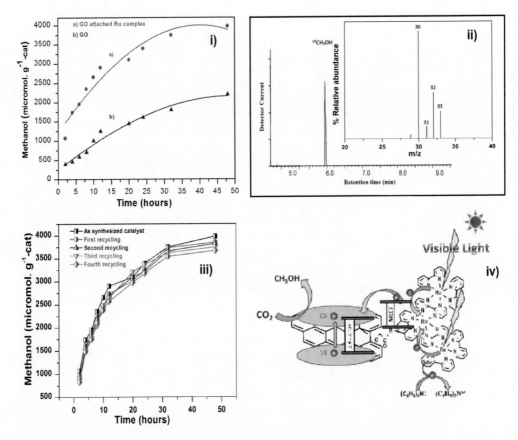

Figure 3. i) Methanol formation rate by using Ru-phen-GO catalyst; ii) GC-MS of product formed by using $^{13}CO_2$; iii) Recycling experiments; iv) Plausible reaction mechanism of the reaction.

CONCLUSION

In this chapter, we have attempted to review the prior art in the field of photoreduction of CO_2 using graphene-based nanocomposite materials as photoredox catalysts under visible light irradiation. We also covered its conversion to molecules that are precursors to or are directly usable as fuels.

In most studies, the enhanced photocatalytic performance of the nanocomposites was attributed to graphene playing the role as a medium for electron capturing/transporting. In case of graphene oxide/TiO_2 nanocomposites, the photocatalytic enhancement can be brought up by: 1) increasing the amount of surface-adsorbed reactant species, 2) lowering the recombination rate of photogenerated electron–hole pairs, and 3) extending the light absorption range.

At this stage, it can be concluded from past work that graphene-based nanocomposites can be established as efficient heterogeneous catalysts to be applied as high performance photoactive catalytic materials for reduction of CO_2 to valuable products.

REFERENCES

[1]   Halmann, M. M., Steinberg, M. *Greenhouse Gas Carbon Dioxide Mitigation Science and Technology*, Lewis Publishers, Boca Raton, Florida. 1999.

[2]   Song, C. *Catal. Today* 2006, 115, 2–32.

[3]   Olah, G. A. *Angew. Chem. Int. Ed.* 2013, 52, 104–107.

[4]   Maginn, E. J. *J. Phys. Chem. Lett.* 2010, 1, 3478–3479.

[5]   Keeling, C. D.; Whorf, T. P. Atmospheric CO_2 records from sites in the SiO air sampling network, In: *Trends: A Compendium of Data on Global Change. Carbon Dioxide Information Analysis Center*, Oak Ridge National Laboratory, US Department of Energy, Oak Ridge, TN, US, 2005.

[6]   Solomon, S.; Quin, D.; Manning, M.; Marquis, M.; Averyt, K.; Tignor, M. M. B.; Miller, Jr. H. L.; Chen, Z. *The Physical Science Basis, Climate Change, 4th Assessment Report of the IPCC*; Cambridge University Press: Cambridge, UK, 2007.

[7]   Aresta, M.; Dibenedetto, A. *Dalton Trans.* 2007, 2975.

[8]   Aresta, M. *Carbon dioxide reduction and use as a chemical feedstock,* Tolman, W. B., Ed.; Wiley-VCH: Weinheim, Germany, 2006, 1–41.

[9]   Mikkelsen, M.; Jorgensen, M.; Krebs, F. C. *Energy Environ. Sci.* 2010, 3, 43–81.

[10]   Omae, I. *Catal. Today* 2006, 115, 33–52.

[11]   Choi, J. C.; He, L. N.; Sakakura, T. *Green Chem.* 2002, 4, 230.

[12]   Xiaoding, X.; Moulijn, J. A. *Energy Fuels* 1996, 10, 305.

[13]   Zeverhoven, R.; Eloneva, S.; Teir, S. *Catal. Today* 2006, 115, 73.

[14]   Ferguson, J. E. *Carbon Dioxide. MacMillan Encyclopedia of Chemistry*, 1997, MacMillan References, New York, Vol. 1, 302–308.

[15]   Morris, A. J.; Meyer, G. J.; Fujita, E. *Acc. Chem. Res.* 2009, 12, 1983-1994.

[16]   Indrakanti, V. P.; Kubicki, J. D.; Schobert, H. H. *Energy Environ. Sci.* 2009, 2, 745–758.

[17]   Fujita, E. *Coord. Chem. Rev.* 1999, 185-186, 373–384.

[18]   Benson, E. E.; Kubiak, C. P.; Sathrum, A. J.; Smieja, J. M. *Chem. Soc. Rev.* 2009, 38, 89–99.

[19]   Barber, J.; Tran, P. D. *J. R. Soc. Interface* 2013, 10, 20120984.

[20]   Roy, S. C.; Varghese, O. K.; Paulose, M.; Grimes, G. A. *ACS Nano* 2010, 4, 1259–1278.

[21]   Michl, J. *Nature Chem.*, 2011, 3, 268–269.

[22]   Habisreutinger, S. N.; Mende, L. S.; Stolarczyk, J. K. *Angew. Chem. Int. Ed.* 2013, 52, 7372–7408.

[23]   Tahir, M.; Amin, N. S. *Energy Conversion Management* 2013, 76, 194–214.

[24] Sakthivel, S.; Neppolian, B.; Shankar, M. V.; Arabindoo, B.; Palanichamy, M.; Murugesan, V. *Solar Energy Mater Solar Cells*, 2003, 77, 65–82.

[25] Li, X.; Wang, Q.; Zhao, Y.; Wu, W.; Chen, J.; Meng, H. *J. Colloid Interface Sci.* 2013, 411, 69–75.

[26] Liu, Y.; Huang, B.; Dai, Y.; Zhang, X.; Qin, X.; Jiang, M.; Whangbo, M.-H. *Catal. Commun.* 2009, 11, 210–213.

[27] Wang, Z. Y.; Chou, H. C.; Wu, J. C. S.; Tsai, D. P.; Mul, G. *Appl. Catal. A* 2010, 380, 172–177.

[28] Pan, P. W.; Chen, Y. W. *Catal. Commun.* 2007, 8, 1546–1549.

[29] Naval, S.; Dhakshinamoorthy, A.; lvaro, M.; Garcia, H. *ChemSusChem* 2013, 6, 562–577.

[30] Cowan, A. J.; Durrant, J. R. *Chem. Soc. Rev.* 2013, 42, 2281-2293.

[31] Benson, E. E.; Kubiak, C. P.; Sathruma, A. J.; Smieja, J. M. *Chem. Soc. Rev.* 2009, 38, 89-99.

[32] Tran, P. D.; Wong, L. H.; Barber, J.; Loo, J. S. C. *Energy Environ. Sci.* 2012, 5, 5902–5918.

[33] Furube, A.; Asahi, T.; Masuhara, H.; Yamashita, H.; Anpo, M. *J. Phys. Chem. B* 1999, 103, 3120–3127.

[34] Liu, L.; Gao, F.; Zhao, H.; Li, Y.; *Appl. Catal. B* 2013, 134–135, 349-358.

[35] Slamet, H. W.; Nasution, E.; Purnama, S.; Kosela, J.; Gunlazuardi, J. *Catal. Commun.* 2005, 6, 313–319.

[36] Li, Y.; Wang, W.-N.; Zhan, Z.; Woo, M.-H.; Wu, C.-Y.; Biswas, P. *Appl. Catal. B* 2010, 100, 386–392.

[37] Izumi, Y. *Coord. Chem. Rev.* 2013, 257, 171-186.

[38] Liu, S.; Guo, E.; Yin, L. *J. Mater. Chem.* 2012, 22, 5031–5041.

[39] Yu, J.; Ran, J. *Energy Environ. Sci.* 2011, 4, 1364-1371.

[40] Yanyuan, W.; Hanming, D. *Chin. J. Catal.* 2011, 32, 36–45.

[41] Liu, L.; Pitts, D. T.; Zhao, H.; Zhao, C.; Li, Y. *Appl. Catal. A* 2013, 467, 474–482.

[42] Sasirekha, N.; Basha, S. J. S.; Shanthi, K. *Appl. Catal. B* 2006, 62, 169–180.

[43] Asahi, R.; Morikawa, T.; Ohwaki, T.; Aoki, K.; Taga, Y. *Science* 2001, 293, 169-293.

[44] Zhang, Q. W.; Wang, J.; Yin, S.; Sato, T.; Saito, F. *J. Am. Chem. Soc.* 2004, 126, 1161–1163.

[45] Khan, S. U. M.; Al-Shahry, M.; Ingler, Jr. W. B. *Science* 2002, 297, 2243–2245.

[46] Zhanga, Q.; Li, Y.; Ackermanb, E. A.; Josifovskac, M. G. Li, H. *Appl. Catal. A* 2011, 400, 195–202.

[47] Sato, S.; Koike, K.; Inoue, H.; Ishitani, O. *Photochem. Photobiol. Sci.* 2007, 6, 454-461.

[48] Fujita, E.; Milder, S. J.; Brunschwig, B. S. *Inorg. Chem.* 1992, 31, 2079-2085.

[49] Lang, X.; Chen, X.; Zhao, J. *Chem. Soc. Rev.* 2014, 43, 473-486.

[50] Suzuki, T. M.; Tanaka, H.; Morikawa, T.; Iwaki, M.; Sato, S.; Saeki, S.; Inoue, M.; Kajino, T.; Motohiro, T. *Chem. Commun.* 2011, 47, 8673-8675.

[51] Sato, S.; Morikawa, T.; Saeki, S.; Kajino, T.; Motohiro, T. *Angew Chem. Int. Ed.*, 2010, 49, 5101-5105.

[52] Leary, R.; Westwood, A. *Carbon* 2011, 49, 741-772.

[53] Huang, X.; Qi, X.; Boey, F.; Zhang, H. *Chem. Soc. Rev.* 2012, 41, 666–686.

[54] Dai, L.; Chang, D. W.; Baek, J. B.; Lu, W. *Small* 2012, 8, 1130–1166.

[55] Yu, Y.; Yu, J. C.; Yu, J. G.; Kwok, Y. C.; Che, Y. K.; Zhao, J. C.; Ding, L.; Ge, W. K.; Wong, P. K. *Appl. Catal. A* 2005, 289, 186-196.

[56] An, X.; Yu, J. C. *RSC Adv.* 2011, 1, 1426–1434.

[57] Park, Y.; Kang, S. H.; Choi, W. *Phys. Chem. Chem. Phys.* 2011, 13, 9425–9431.

[58] Gao, Y.; Pu, X.; Zhang, D.; Ding, G.; Shao, X.; Ma, J. *Carbon* 2012, 50, 4093-4101.

[59] Geim, A. K.; Novoselov, K. S. *Nature Mater.* 2007, 6, 183-191.

[60] Novoselov, K.; Geim, A.; Morozov, S.; Jiang, D.; Zhang, Y.; Dubonos, S.; Grigorieva, I.; Firsov, A. *Science* 2004, 306, 666-669.

[61] Meyer, J. C.; Geim, A.; Katsnelson, M. Novoselov, K.; Booth, T.; Roth, S. *Nature* 2007, 446, 60-63.

[62] Castro Neto, A. H.; Guinea, F.; Peres, N. M. R.; Novoselov, K. S.; Geim, A. K. *Rev. Mod. Phys.* 2009, 81, 109-162.

[63] Singh, V.; Joung, D.; Zhai, L.; Das, S.; Khondaker, S. I.; Seal, S. *Prog. Mater. Sci.* 2011, 56, 1178–1271.

[64] Rao, C. N. R.; Sood, A. K.; Subrahmanyam, K. S.; Govindaraj, A. *Angew. Chem. Int. Ed.* 2009, 48, 7752-7777.

[65] Novoselov, K. S.; Geim, A. K.; Morozov, S. V.; Jiang, D. Katsnelson, M. I.; Grigorieva, I. V.; Dubonos, S. V.; Firsov, A. A. *Nature* 2005, 438, 197-200.

[66] Novoselov, K. S.; Geim, A. K.; Morozov, S. V.; Jiang, D.; Zhang, Y.; Dubonos, S. V.; Grigorieva, I. V.; Firsov, A. A. *Science* 2004, 306, 666-669.

[67] Du, X.; Skachko, I.; Barker, A.; Andrei, E. Y. *Nat. Nanotechnol.* 2008, 3, 491-495.

[68] Balandin, A. A.; Ghosh, S.; Bao, W.; Calizo, I.; Teweldebrhan, D.; Miao, F.; Lau, C. N. *Nano Lett.* 2008, 8, 902-907.

[69] Ghosh, S.; Calizo, I.; Teweldebrhan, D.; Pokatilov, E. P.; Nika, D. L.; Balandin, A. A.; Bao, W.; Miao, F.; Lau, C. N. *Appl. Phys. Lett.* 2008, 92, 151911-151913.

[70] Lee, C.; Wei, X.; Kysar, J. W.; Hone, J. *Science* 2008, 321, 385.

[71] Chae, H. K.; Siberio-Perez, D. Y.; Kim, J.; Go, Y.; Eddaoudi, M.; Matzger, A. J.; Keeffe, M. O'; Yaghi, O. M. *Nature* 2004, 427, 523-527.

[72] Stoller, M. D.; Park, S.; Zhu, Y.; An, J.; Ruoff, R. S. *Nano Lett.* 2008, 8, 3498-3502.

[73] Mas, A. A.; Wei, D. *Nanomaterials* 2013, 3, 325-356.

[74] Biro, L. P.; Incze, P. N.; Lambin, P. *Nanoscale* 2012, 4, 1824–1839.

[75] Eda, G.; Fanchini, G.; Chhowalla, M. *Nature Nanotechnol.* 2008, 3, 270-274.

[76] Juang, Z. Y.; Wu, C. Y.; Lu, A. Y.; Su, C. Y.; Leou, K. C.; Chen, F. R.; Tsai, C. H. *Carbon* 2010, 48, 3169-3174.

[77] Sharma, P.; Hussain, N.; Das, M. R.; Deshmukh, A. B.; Shelke, M. V.; Szunerits, S.; Boukherroub, R. *Handbook of Research on Nanoscience, Nanotechnology and Advanced Materials*; Ed. Mohamed Bououdina, M.; Davim, J. P., IGI Global; US, 2014, Chapter 10, pp. 196-225.

[78] Das, M. R.; Sharma, P.; Borah, S. C.; Szunerits, S.; Boukherroub, R. *Innovative Graphene Technologies: Developments and Characterization*; Tiwari A. Rapra Publishers, US, 2013.

[79] Akhavan, O.; Ghaderi, E. *J. Phys. Chem. C* 2009, 113, 20214-20220.

[80] Akhavan, O.; Ghaderi, E. *Carbon* 2012, 50, 1853-1860.

[81] Zhang, X.; Sun, Y.; Cui, X.; Jiang, Z. *Int. J. Hydrogen Energy* 2012, 37, 811-815.

[82] Cheng, P.; Yang, Z.; Wang, H.; Cheng, W.; Chen, M.; Shangguan, W.; Ding, G. *Int. J. Hydrogen Energy* 2012, 37, 2224-2230.

[83] Ni, M.; Leung, M. K. H.; Leung, D. Y. C.; Sumathy, K. *Renew. Sust. Energy Rev.* 2007, 11, 401–425.

[84] Yang, N. L.; Zhai, J.; Wang, D.; Chen, Y. S.; Jiang, L. *ACS Nano* 2010, 4, 887–894.

[85] Sun, S. R.; Gao, L.; Liu, Y. Q. *Appl. Phys. Lett.* 2010, 96, 083113.

[86] Xiang, Q.; Yu, J.; Jaroniec, M. *Chem. Soc. Rev.* 2012, 41, 782–796.

[87] Liu, S.; Tian, J.; Wang, L.; Luo, Y.; Sun, X. *Catal. Sci. Technol.* 2012, 2, 339–344.

[88] Zhang, H.; Lv, X.; Li, Y.; Wang, Y.; Li, J. *ACS Nano* 2010, 4, 380–386.

[89] Zhang, Y.; Tang, Z. R.; Fu, X.; Xu, Y. J. *ACS Nano* 2010, 4, 7303–7314.

[90] Yang, Y.; Ren, L.; Zhang, C.; Huang, S.; Liu, T. *ACS Appl. Mater. Interfaces* 2011, 3, 2779–2785.

[91] Chen, P.; Xiao, T. Y.; Li, H. H.; Yang, J. J.; Wang, Z.; Yao, H. B.; Yu, S. H. *ACS Nano* 2012, 6, 712-719.

[92] Zhu, M.; Chen, P.; Liu, M. *ACS Nano* 2011, 5, 4529-4536.

[93] Gao, E.; Wang, W.; Shang, M.; Xu, J. *Phys. Chem. Chem. Phys.* 2011, 13, 2887–2893.

[94] Khalida, N. R.; Ahmed, E.; Hong, Z.; Zhang, Y.; Ullah, M.; Ahmed, M. *Ceram. Int.* 2013, 39, 3569–3575.

[95] Jing, L.; Yang, Z. Y.; Zhao, Y. F.; Zhang, Y. X.; Guo, X.; Yan, Y. M.; Sun, K. N. *J. Mater. Chem. A* 2014, 2, 1068-1075.

[96] Mukherji, A.; Seger, B.; Lu, G. Q.; Wang, L. *ACS Nano* 2011, 5, 3483-3492.

[97] Stengl, V.; Popelkova, D.; Vlacil, P. *J. Phys. Chem. C* 2011, 115, 25209–25218.

[98] Jiang, G.; Lin, Z.; Chen, C.; Zhu, L.; Chang, Q.; Wang, N.; Wei, W.; Tang, H. *Carbon* 2011, 49, 2693-2701.

[99] Tu, W.; Zhou, Y.; Liu, Q.; Tian, Z.; Gao, J.; Chen, X.; Zhang, H.; Liu, J.; Zou, Z. *Adv. Funct. Mater.* 2012, 22, 1215–1221.

[100] Zhou, Y.; Bao, Q. L.; Tang, L. A. L.; Zhong, Y. L.; Loh, K. P. *Chem. Mater.* 2009, 21, 2950-2956.

[101] Wang, P. Q.; Bai, Y.; Luo, P. Y.; Liu, J. Y. *Cat. Commun.* 2013, 38, 82–85.

[102] Brodie, B. C. *Phil. Trans.* 1859, 149, 249-259.

[103] Loh, K. P.; Bao, Q.; Ang, P. K.; Yang, J. *J. Mater. Chem.* 2010, 20, 2277–2289.

[104] Georgakilas, V.; Otyepka, M.; Bourlinos, A. B.; Chandra, V.; Kim, N.; Kemp, K. C.; Hobza, P.; Zboril, R.; Kim, K. S. *Chem. Rev.* 2012, 112, 6156–6214.

[105] Loh, K. P.; Bao, Q.; Eda, G.; Chhowalla, M. *Nature Chem.* 2010, 2, 1015-1024.

[106] Eda, G.; Mattevi, C.; Yamaguchi, H.; Kim, H.; Chhowalla, M. *J. Phys. Chem. C* 2009, 113, 15768–15771.

[107] Hsu, H. C.; Shown, I.; Wei, H. Y.; Chang, Y. C.; Du, H. Y.; Lin, Y. G.; Tseng, C. A.; Wang, C. H.; Chen, L. C.; Lind, Y. C.; Chen, K. H. *Nanoscale* 2013, 5, 262-268.

[108] Hummers, W. S.; Offeman, R. E. *J. Am. Chem. Soc.* 1958, 80, 1339.

[109] Eda, G.; Chhowalla, M. *Adv. Mater.* 2010, 22, 2392.

[110] Yadav, R. K.; Baeg, J. O.; Oh, G. H.; Park, N.-J.; Kong, K.-J.; Kim, J.; Hwang, D. W.; Biswas, S. K. *J. Am. Chem. Soc.* 2012, 134, 11455-11461.

[111] An, X.; Li, K.; Tang, J.; *ChemSusChem.* 2014, 7, 1086-1093.

[112] Tan, L. L.; Ong, W. J.; Chai, S. P.; Mohamed, A. R. *Nanoscale Res. Lett.* 2013, 8, 465-474.

[113] Liang, Y. T.; Vijayan, B. K.; Gray, K. A.; Hersam, M. C. *Nano Lett.* 2011, 11, 2865-2870.

[114] Takeda, H.; Koizumi, H.; Okamotoa, K.; Ishitani, O. *Chem. Commun.* 2014, 50, 1491-1493.

[115] Inagakia, A.; Akita, M. *Coord. Chem. Rev.* 2010, 254, 1220–1239.

[116] Reithmeier, R.; Bruckmeier, C.; Rieger, B. *Catalysts* 2012, 2, 544-571.

[117] Takeda, H.; Ishitani, O. *Coord. Chem. Rev.* 2010, 254, 346–354.

[118] Cokoja, M.; Bruckmeier, C.; Rieger, B.; Kuhn, F.; Hermann, W. A. *Angew. Chem. Int. Ed.* 2011, 50, 8510–8537.

[119] Planas, N.; Ono, T.; Vaquer, L.; Miro, P.; Buchholz, J. B.; Gagliardi, L.; Cramer, C. J.; Llobet, A. *Phys. Chem. Chem. Phys.* 2011, 13, 19480–19484.

[120] Voyame, P.; Toghill, K. E.; Mendez, M. A.; Girault, H. H. *Inorg. Chem.* 2013, 52, 10949–10957.

[121] Chen, Z.; Concepcion, J. J.; Brennaman, M. K.; Kang, P.; Norris, M. R.; Hoertz, P. G.; Meyer, T. J. *Proc. Nat. Acad. Sci.* 2012, 109, 15606–15611.

[122] Tamaki, Y.; Morimoto, T.; Koike, K.; Ishitani, O. *Proc. Nat. Acad. Sci.* 2012, 2012, 1-6.

[123] Grodkowski, J.; Neta, P.; Fujita, E.; Mahammed, A.; Simkhovich, L.; Gross, Z. *J. Phys. Chem. A* 2002, 106, 4772–4778.

[124] Ogata, T.; Yamamoto, Y.; Wada, Y.; Murakoshi, K.; Kusaba, M.; Nakashima, N.; Ishida, A.; Takamuku, S.; Yanagida, S. *J. Phys. Chem.* 1995, 99, 11916–11922.

[125] Gholamkhass, B.; Mametsuka, H.; Koike, K.; Tanabe, T.; Furue, M.; Ishitani, O. *Inorg. Chem.* 2005, 44, 2326–2336.

[126] Hori, H.; Johnson, F. P. A.; Koike, K.; Ishitani, O.; Ibusuki, T. *Photochem. J. Photobiol. A* 1996, 96, 171–174.

[127] Sato, S.; Morikawa, T.; Kajino, T.; Ishitani, O. *Angew. Chem. Int. Ed.* 2013, 52, 988-992.

[128] Hull, J. F.; Balcells, D.; Blakemore, J. D.; Incarvito, C. D.; Eisenstein, O.; Brudvig, G. W.; Crabtree, R. H. *J. Am. Chem. Soc.* 2009, 131, 8730-8731.

[129] Sato, S.; Koike, K.; Inoue, H.; Ishitani, O. *Photochem. Photobiol. Sci.* 2007, 6, 454–461.

[130] Fujita, E.; Milder, S. J.; Brunschwig, B. S. *Inorg. Chem.* 1992, 31, 2079–2085.

[131] Miao, R.; Mongelli, M. T.; Zigler, D. F.; Winkel, B. S. J.; Brewer, K. J. *Inorg. Chem.* 2006, 45, 10413-10415.

[132] Miao, R.; Brewer, K. J. *Inorg. Chem. Commun.* 2007, 10, 307-312.

[133] Windle, C. D.; Campian, M. V.; Klair, A. K. D.; Gibson, E. A.; Perutz, R. N.; Schneider, J. *Chem. Commun.* 2012, 48, 8189–8191.

[134] Zhou, X. T.; Ji, H. B.; Huang, X. J. *Molecules,* 2012, 17, 1149-1158.

[135] Sato, S.; Morikawa, T.; Saeki, S.; Kajino, T.; Motohiro, T. *Angew. Chem. Int. Ed.* 2010, 49, 5101–5105.

[136] Hirose, T.; Maeno, Y.; Himed, Y. *J. Mol. Catal. A* 2003, 193, 27–32.

[137] Ragoussi, M. E.; Malig, J.; Katsukis, G.; Butz, B.; Spiecker, E.; Torre, G. de la; Torres, T.; Guldi, D. M. *Angew. Chem. Int. Ed.* 2012, 51, 6421–6425.

[138] Zhu, J.; Li, Y.; Chen, Y.; Wang, J.; Zhang, B.; Zhang, J.; Blau, W. J. *Carbon* 2011, 49, 1900-1905.

[139] Mahyari, M.; Shaabani, A., *Appl. Catal. A* 2014, 469, 524–531.

[140] Yu, Y.; Zhou, M.; Shen, W.; Zhang, H.; Cao, Q.; Cui, H. *Carbon* 2012, 50, 2539-2545.

[141] Zhu, M.; Dong, Y.; Xiao, B.; Du, Y.; Yang, P.; Wang, X. *J. Mater. Chem.* 2012, 22, 23773-23779.

[142] Kumar, P.; Kumar, A.; Sreedhar, B.; Sain, B.; Ray, S. S.; Jain, S. L. *Chem. Eur. J.* 2014, 20, 6154-6161.

[143] Sun, X.; Liu, Z.; Welsher, K.; Robinson, J. T.; Goodwin, A.; Zaric, S.; Dai, H. *Nano Res.* 2008, 1, 203-212.

[144] Kumar, P.; Sain, B.; Jain, S. L. *J. Mater. Chem. A* 2014, 2, 11246-11253.

[145] Fleming, C. N.; Maxwell, K. A.; Simone, J. M. D.; Meyer, T. J.; Papanikolas, J. M. *J. Am. Chem. Soc.* 2001, 123, 10336-10347.

[146] Rybtchinski, B.; Sinks, L. E; Wasielewski, M. R. *J. Am. Chem. Soc.* 2004, 126, 12268-12269.

In: Innovations in Nanomaterials
Editors: Al-N. Chowdhury, J. Shapter, A. B. Imran

ISBN: 978-1-63483-548-0
© 2015 Nova Science Publishers, Inc.

Chapter 14

HIERARCHICAL MICROSTRUCTURES FOR SUPERCAPACITOR AND BATTERY ELECTRODES: "NANO ENVIES MICRO"

S. Roshny, R. Ranjusha, Ajay Amrutha, Joseph Jickson, Rejinold Sanoj, Nair V. Shantikumar, R. Jayakumar and A. Balakrishnan[*]

Amrita Center for Nanosciences, Kochi, India

ABSTRACT

Energy storage has been revolutionized by the inception of nanotechnology. Energy storage components such as supercapacitors and batteries exploit the advantages of nanomaterials such as the reduced size and high surface area which enhance the performance in terms of efficiency and compactness. Nanomaterials essentially provide a high surface area which ensures higher participation of ions involved in the working of a storage device. They have been employed as electrode materials as well as current collectors. Materials when scaled down to the nano-regime pose stability issues which can be solved by using surfactants, but this will in turn compromise the performance of the system due to the insulating properties of the surfactants. It is suggested that full potential of nanotechnology can be explored through the manipulation of the material morphology thereby coming up with superior electrode architectures. Recently, hierarchical nanostructures have gained tremendous attention as electrode materials. Various hierarchical structures have been reported, such as urchin shapes, flower shapes, spiked balls etc which have shown better efficiency while answering the toxicity and stability issues. These micro/nano hybrid structures have been shown to tremendously increase the charge transport, better diffusion and large electrochemically active surface area. The unique architecture they provide retains the advantages of nanostructures while remaining a microstructure as a whole. Hierarchical nanostructures are also believed to be able to accommodate strains that result from reversible redox reactions as well as the intercalation-deintercalation process associated with supercapacitors and batteries. The stability of the device is affected when the strains result in structural damage.

[*] Corresponding author email: avinash.balakrishnan@gmail.com.

Transition metal oxides and sulphides have gained interest as storage electrode materials due to their high stability and high energy density, typically 2–3 times higher than those of the carbon/graphite-based materials. Among the various metal oxides, MnO_2, NiO and NiS have been given special attention due to their high theoretical capacitance and improved cycling stability. Various structures have been reported in the literature which utilizes hierarchical shapes for better performance. They can be synthesized using simple methods and by defining the parameters which could influence the particle morphology, their performance can be predicted and this could aid in the design of high performing devices. This chapter deals with the syntheses and properties of metal oxides, sulphides and their composites possessing novel architectures. A further toxicity issue in reference to particle size has also been discussed.

1. INTRODUCTION

In the coming years it is projected that the world energy consumption will increase significantly with a dire need for clean and environment friendly energy sources. Electrical energy generated from renewable sources has attracted attention and has been pursued with great vigilance [1-3]. However, electricity generated from such resources calls for competent and efficient energy storage devices. This means that the generated electricity should be made available to all throughout the day and night without any fluctuations, which otherwise can be an expensive affair when it comes to commercial and residential grid applications [4, 5]. Thus, large scale energy generation such as solar, wind, hydro-thermal and other renewable sources will be applicable only through the development of energy storage devices which can meet the continuous energy demands. In order to ensure smooth operation, safety of these devices is critical to prevent any supply failure. The available storage technology focuses on devices comprising of different types of batteries and supercapacitors. The working of these devices is governed by the principles of electrochemistry. Batteries store electrical energy via chemical reactions and supercapacitors generally store energy through charge separation mechanism [6, 7].

Unfortunately, the current technology for storage devices falls short of meeting the demands generated by commercial, residential and transportation applications. This is mainly because batteries lack the required volumetric power density and supercapacitors show poor volumetric energy density. Thus, there is a strong need for high energy and power density devices coupled with stability and life which can replace the conventional storage devices flourishing in the market. It is quite well-known that the particle size has tremendous effect on the properties when it is scaled down to the nano-regime [8]. Recent developments in nanostructured materials for electrode applications showed promising results where the capacity and power densities of the storage devices improved significantly [9-11]. This means that, a thorough fundamental understanding of the correlation between the performance of nano-electrode materials and the electrolyte could pave the way for utilizing nanotechnology for fabricating high performing storage devices.

Different nanomaterials have been employed for electrode development in both energy storage and generation devices [9-11]. These include different nanostructures of metal oxides, nanocarbons, chalcogenides and their composites [11-18]. An increase in the number of particles per volume (by an order of nine) is observed when the particle size is reduced from micron to nano regime [8]. This effectively increases the surface area, which is highly beneficial for storage applications as it implies more active sites can be utilized during the redox reactions/double layer mechanism which leads to improved specific capacitance of the electrode overlay. However, the advantages displayed by the nanomaterials are often negated by the issues of environmental toxicity and cost hindering its entry into the commercial market. The disadvantages are illustrated by the fact that nanoparticles in their free form or as a part of a final product can accumulate in the soil, water, or plant life either during the manufacturing, processing or disposal thereby contributing to the environmental pollution. This harmfully affects the ecosystem, disturbing the sensitive balance [19-21]. The interaction of nanoparticles with the human body and its adverse effects has been widely investigated as well. These particles, on entering the human blood stream, show high mobility and has been reported to cross the blood-brain barrier resulting in permanent damage [22-24].

This problem can be addressed by retaining micron size morphology and finding innovative ways to increase the surface area [25]. There are various high surface area nanostructures reported in the literature which include nanorods, nanobelts, nanowires etc for energy storage applications [26, 27]. If these nano-morphologies can be engineered on to the surface of a micron size particle it can effectively increase the surface area while remaining a microstructure as a whole. Such hierarchical microstructures comprising collective nanostructures are hereby referred as micro/nano hybrids. These micro/nano hybrids can be beneficial during the electrode fabrication as these will enhance the active sites participating in the redox reactions during the charging-discharging process. The synthesis of micro/nano hybrids is closely governed by the process parameters like time, temperature, precursors, concentration etc. The present chapter discusses the details of three such micro/hybrid materials namely NiS, NiO and MnO_2 and their impact on the storage properties in batteries and supercapacitors. Reports showing the level of toxicities these materials can induce on the human body has also been discussed in detail.

2. ADVANTAGES OF 3-D HIERARCHICAL STRUCTURES OVER 1D AND 2D STRUCTURES

The dawn of the 20th century saw an increase in the popularity of portable electronic products coupled with an urgent need for high energy density storage devices in the sectors of aviation, space exploration, electric vehicles and other fields [28-30]. This necessity brought about the revolution of nickel hydrogen and lithium ion batteries which seemed to be the solution to the energy storage problem [28, 31]. However, even in the present scenario, these devices are still not fully functional due to the lower capacities making them inadequate to meet the general requirements of the electric vehicles and grid applications. These applications require high energy density of 10 ~100 Wh/kg, high power density of 500~1500 W/kg, low cost of ~50 \$/kWh, long cycle life, environmental compatibility etc. [32, 33]. It is quite anticipated that micro/nano hybrid systems when translated into a high surface area cathode overlays can achieve capacities near to the theoretical limit. The advantages of using 1-D nanomaterials in lithium ion batteries are:

a) Improvement in the life cycle of the batteries by successfully absorbing the strain introduced into the overlay by the insert ion and removal of lithium ions.

b) Increased contact area between the electrode and the electrolyte and the consequent increase in the number of active sites resulting in a higher charge discharge rates and high power.

c) A shorter diffusion path distance for electron and lithium ions which makes possible the operation of batteries at higher power [7, 35-37].

However, they suffer from issues like aggregation/agglomeration [38-40] which limits their applicability in thin film coatings thereby losing the essence of nano. With a view to maximizing the potential uses of incorporating nanotechnology in the system, the nanostructures of electrochemical active materials should be optimized in terms of not only the size but also morphology, texture, and overall cell architecture [37].

To circumvent these obstacles of agglomeration and instability, high surface 3-D materials received significant attention. It can be said that the 1-D nanostructures like nanobelts, nanowires, nanoribbons etc are more often not self-supported, requiring them to be translated into an electrode film [37]. Moreover while translating these materials into a thin film as electrode overlays, they must be reinforced with additional support to maintain the diffusion length as well as continuously provide electrical and mechanical contact, withstanding the stress and the subsequent strain brought about by the electrode reactions. Graphitic additives are used for this purpose which can improve the performance of the nano-electrodes. An alternative is to use self-supported 3D architectures, which can improve the current collecting properties by integrating electrodes, thereby providing better interconnection between them [37, 40-45]

A battery system is characterized by the nature of the electrodes used which essentially determines the capacity and the lifetime of a battery is determined by the quality of the electrode electrolyte interface. The Li-ion battery system works on insertion compounds whose capacity is determined by the number of Li ions the framework structure can uptake. This number is a characteristic of the crystal and electronic structure of the material of the electrode [7, 46]. Traditional intercalation compounds (For instance $LiCoO_2$) can provide at most one Li ion per 3d metal. When such intercalation compounds were replaced with oxides, sulfides, phosphides, nitrides etc, it was surprisingly revealed that these compounds even with structures unsuitable for intercalation chemistry were shown to exhibit incredible capacities as high as 1000 $mAhg^{-1}$ for Co_3O_4 [7, 46, 47]. These new conversion reactions that provided two or even more electrons per 3d metal paved the way for the creation of a new class of electrodes exhibiting incredible capacity gains over various voltage ranges which was determined by the anion [47].

These novel conversion reactions could be better explored for improvements in reaction kinetics by either

1. employing conductive coatings on the particle surfaces [48]
2. moving from bulk to thin film material [49] or
3. playing with the new electrode design.

The present chapter focuses on electrode designs for 3D hierarchical architecture in sulfide systems for Li-ion battery technology and their unique properties.

3. SULFIDES IN LITHIUM ION BATTERIES

Recent advances and research in lithium ion batteries are primarily focused on $LiCoO_2$, $LiNiO_2$, $LiMn_2O_4$, $LiFePO_4$ and other compounds [50-53]. However these materials still suffer from low capacity and lack of environmental benignity. Among the alternatives, lithium sulfur batteries have been hailed as promising system for high energy density application [54, 55]. Although these systems have reportedly shown a high first discharge capacity i.e ~ 1320 mAh/g [56] which is close to the theoretical capacity (1672 mAh/g) [57], they suffer from low cycling stability (few tens of cycles) resulting from electrode dissolution into the electrolyte.

As a solution to this problem, research have been focused on employing sulfide based compounds like chalcogenides as cathode materials for Li ion batteries because they are cheap, relatively less toxic and readily available in abundance while showing a high theoretical capacity of ~590-650 mAh/g [32]. The added advantage of these systems is that they are conducting in nature thereby improve the rate capability of the battery. The reaction mechanism of lithium/metal sulfide couples can be described as follows [58]:

$$Mn+ X + ne^- + nLi^+ \rightarrow M^0 + nLiX, (X = S,O) \tag{1}$$

It has been observed that metal sulfide cathode materials that react with lithium based on this conversion mechanism show high reversible capacities and high energy densities which can be attributed to the large numbers of ion/electrons exchange involved in conversion reactions. Among the different metal sulfides, nickel sulfide is being extensively studied upon because of its high capacity. Jaisinski and Burrows in 1969 first reported that nickel sulfide can be used as a cathode material in lithium ion batteries [59]. Various sulfides exist in the forms of Ni_3S_2, Ni_6S_5, Ni_7S_6, NiS, Ni_3S_4, and NiS_2. [60] Among these materials, NiS is considered to be lucrative because of its good rate capability and stability over a wide range of temperatures. It exhibits good electrical conductivity and has a high theoretical capacity of ~590 mAh/kg [50, 61].

3.1. Fabrication of 3-D NiS Flower Like Structure

A lot of studies involving different bottom up and top down approaches have been reported for the synthesis of nickel sulfide nanostructures which includes colloidal micro-emulsion [62], ball milling [63], co-polymer approach [64] hydrothermal process [65], UV irradiation [66], sol–gel [67], laser ablation [68], thermal degradation [69], reduction/organic synthesis [70, 71] etc. The performance of NiS in the redox systems are controlled by two parameters: a) interfacial properties associated to its morphology and b) the ionic/electronic conduction. The formation of poly-sulphides in the electrolyte medium which occurs at high current densities causes active mass loss which affects the specific capacity. To enhance the specific capacity, one method suggested is to provide shorter diffusion pathways for ions employing high surface area electrodes possessing good current collection properties. One way of doing this is by the incorporation of secondary phases like different forms of carbon onto the cathode materials. Nano-carbonaceous phases like activated carbon (AC) [72], carbon nanotubes (CNTs) [73] and graphene with metal oxides/sulfides [74, 75] have been shown to improve the power and energy density values. The effect of incorporating carbonaceous phases so as to improve the current collection properties has not been exploited widely. This is attributed to the fact that such functionalization using carbon and its different phases for the cathode material lead to a reduction in the contribution of NiS in the performance. [76, 77]. Moreover, the absence of a surfactant will lead to agglomeration in the system, resulting in the reduction in conductivity. [78, 79]. This has also shown imposed restriction in the pore volume available for electrolyte percolation, reducing the specific capacity [80, 81]. These problems signify the importance of research in hierarchical structures where the overall structure remains micron sized but comprising of nanostructures arrayed in peculiar fashion. Such structures can prevent agglomeration while retaining the surface area. These structures when functionalized with different carbon forms can enhance the rate capability performance of the cathode and/or anode.

Graphene is widely researched as a secondary phase cathode material mainly for current collection function. However retaining the graphene form during electrode processing is difficult due to the agglomeration issues [82]. Further synthesizing graphene in bulk is an expensive affair with the existing technology. Thus, researchers have begun focusing on obtaining nanocarbons from other natural sources. The viability of mesoporous carbon synthesised from an unmodified commercial carbon precursor such as coal tar was exploited by Tong et al. [83]. He et al., have synthesized hierarchical porous carbon from coal tar pitch by with nano-Fe_2O_3 as template and activation agent coupled with KOH activation with microwave heating [84]. Similarly other natural carbon sources such as rice husk [85, 86], bamboo [87, 88], coconut shell [89, 90], sugar [91] etc have been exploited widely. Camphor is another carbon source which has been used for functionalizing cathode/anode materials [92, 93]. The ease in obtaining carbon nanobeads by simple pyrolysis of camphor has directed the recent research to use camphor as carbon source. The following section discusses some of the results where camphor carbon was used as secondary phase in cathode materials like NiS. Recently [61] hydrothermal technique was used to develop NiS hierarchical hybrid structures as bulk powder. Figure 1 shows the architecture of these powders.

Figure 1. SEM image of NiS nano/micro hybrids.

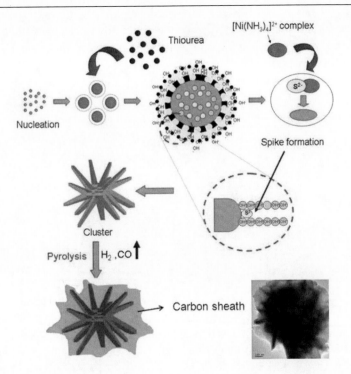

Figure 2. Schematic Diagram of the mechanism of formation of carbon sheathed NiS heirarchical structures. [inset: TEM image of the carbon sheathed NiS structure] (Adapted with modifications [49]).

The structure obtained was a nano/micro hybrid structure which was treated with camphor through pyrolysis, resulting in a unique structure of NiS clusters anchored with porous sheaths of carbon (Figure. 2). These clusters showed nanospikes whose diameter varied from 50-80 nm. During pyrolysis, the hexagonal ring like structure of camphor breaks into C-H, C = O and C-C bonds of pentagonal rings. These carbon atoms are twisted in such a way that three of these atoms lie in one plane and the other three on another plane, which facilitates nucleation of the carbon nanoparticles possessing a graphitic outer layer which then wraps around the hybrid nanospiked structure of NiS clusters.

The amount of carbon content deposited on each batch of pyrolyzed NiS powders were determined using CHN analysis based on which the samples were designated as NC-1, NC-2 and NC-5. Here N and C represent nickel sulfide and carbon respectively and number indicates the carbon content.

3.2. CV and Cycling Studies

The influence of carbon content was found to be crucial in determining the performance of the electrodes fabricated from these nano/micro hybrids. This was elucidated based on CV studies where a lower mass loading of carbon could not contribute to the conductivity because of an inter-aggregate gap and a higher loading of carbon masked the effect of NiS with carbon predomination (Figure. 3).

From CV analysis these structures were found to exhibit higher current densities than pristine NiS clusters, which were attributed to the combined influence of increased surface area offered by the carbon sheaths, increased conductivity due to electron tunneling as well as graphitic conduction. A fully functional coin cell fabricated from these electrodes showed a first discharge capacity of 500 mAhg^{-1} and reasonably a good cycling stability of 100 cycles with 20% fading (Figure. 3 (b)).

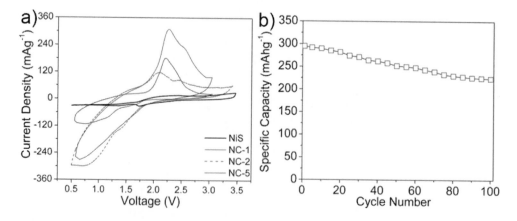

Figure 3. a) Cyclic voltammograms of graphs of pure NiS and carbon sheathed NiS (NC) b) Cycling study of coin cell (Adapted with modifications from [49]). Here N, C and number stands for nickel, carbon and weight percentage of carbon, respectively.

3.3. Impedance Spectroscopy Measurements and Equivalent Circuit Modelling

To further study the role of carbon in influencing the electronic conductivity of the NiS system, impedence measurements were performed on the coin cells fabricated using NiS and NC-2 electrodes. A sine-wave of 300 mV amplitude was applied over a frequency range of 100 kHz to 0.1 Hz. The Nyquist plots obtained are given in Figure 4. The two graphs exhibited two semicircular regions at high and medium frequency regions followed by a linear region tilted at 45° to the x- axis in the low frequency region. The spectra can be modelled into an electronic circuit using a Frequency Response Analyser (FRA) circuit fitting tool, corresponding to a second order transient response behavior.

The plot intercepts the x-axis in the high frequency range at the point which corresponds to the ohmic resistance of the electrolyte, membrane and electrode. The first semicircular region represents the the reaction between the electrolyte and the surface of the electrode. Consequently, C_{SEI} and R_{SEI} are the capacitance and the interfacial resistance of the solid–electrolyte interface layer. This high frequency semicircle is assigned to Li$^+$ diffusion through the SEI layer of the active material. The second semicircle in the low frequency region can be corresponded to charge transfer kinetics and C_{dl} and R_{ct} are the double layer capacitance and the charge transfer resistance. The impedance connected with Li$^+$ transport in the solid phase appears in the low frequency range. Similar results have been reported in literature previously. [94-96]. The linear region at ~45° to the imaginary Z axis at low frequency represents the Warburg impedance (W) and represents the diffusion of Li$^+$ within the bulk anode/electrode. [97] From the Nyquist plots, it is observed that the semicircle diameter of NC-2 is lesser than that of NiS coin cells, implying the charge transfer resistance (R_{ct}) of NC-

2 (~2 M) is smaller than that of NiS coin cells (~5.5 M). This reveals that the transfer rate of electrons in NC-2 is higher than NiS, and carbon anchoring indeed improves the surface electronic conductivity of NiS.

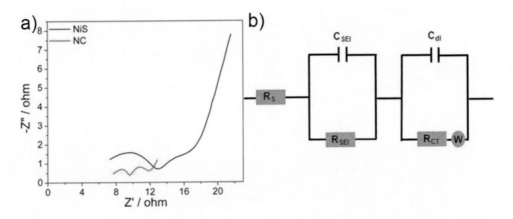

Figure 4. a) Nyquist plot of the NiS and NC-2 coin cells b) Equivalent circuit model of the Nyquist Plot.

The performance of such carbon sheathed NiS systems is promising for future electrode materials in lithium ion batteries considering their low cost and facile fabrication process. But, with the demand for fast charging energy storage devices for portable electronic devices and electric vehicle systems has been increasing over the last few years, a great deal of effort has been made to develop and design high surface area supercapacitor electrodes with good physico-chemical properties as a replacement to batteries Various materials belonging to transition metal oxides have been explored for fabricating supercapacitors [94]. Among these materials, NiO is considered to be lucrative because of its good rate capability and stability over a wide range of temperature. The next section will deal with metal oxide structures for supercapacitor electrodes.

4. TRANSITION METAL OXIDES IN BATTERIES

Poizot et al., was the first to demonstrate the use of nanoparticles of transition metal oxides for enhancing the surface electrochemical reactivity of lithium ion batteries [98]. Lithium alloying agglomeration [99] and the growth of passivation layers [100] are said to be the factors that prevent the effective reversible insertion of Lithium ions into the cathodes. Poizot reported electrodes made of nanoparticles of transition metal oxides such as Cobalt, Nickel, Copper and Iron oxides which showed capacity retention up to 100 cycles and with electrochemical capacity of 700 mAhg^{-1}. The high reactivity of these cathode materials towards lithium was attributed to the nano size. They further argued that the capacity retention properties of these materials were closely associated with the precursor morphology. This was validated using studies on Cu_2O, which established that a change in the particle morphology could influence the capacity retention. Further it was stated that there existed an optimum particle size and morphology for each metal oxide system which determines the best electrochemical performance.

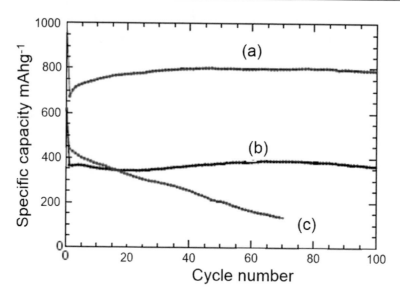

Figure 5. Capacity fading of a) 2μm CoO b) 1μm Cu$_2$O 0.15 μm Cu$_2$Oas a function of particle size at a cycling rate of C/5(Adapted with modifications from [98]).

The mechanism of the reactivity of lithium towards these metal oxides is different from conventional processes such as reversible lithium insertion/de-insertion into the host structures as well as lithium alloying. As mentioned earlier, the electrochemical performance of these metal oxides and sulphides in lithium ion batteries came as a surprise as most of these materials showed a rock salt structure which was unsuitable for lithium intercalation/de-intercalation chemistry due to the absence of empty sites for lithium ions. Further, these materials did not exhibit tendency to form alloys with lithium. The reversible product formation of LiO$_2$ which was electrically inactive has lead researchers to believe that such metal oxide systems were not feasible. The unusual enhanced electrochemical activity of these metal particles towards the reversible product LiO$_2$ is attributed to the size confinement of the particles.

4.1. Nio Heirarchical Structures for Battery Applications

A wide variety of transition metal oxides have been used for energy storage as well as energy generation. [101-109]. NiO is one of the most researched transition metal oxide for electrode materials in storage devices [110-112]. As an anode in Lithium ion batteries, it exhibits a theoretical capacity of 718 mAh^{-1} [113, 114]. But it is a poor conductor at room temperature ($<10^{-13}$ Ω^{-1} cm^{-1}) and is classified as Mott-Hubbard insulator [115-117]. Several methods have been used for the synthesis of NiO such as spray pyrolysis, chemical precipitation etc.

Over the decade, NiO nanostructures exhibiting different morphologies, such as microspheres [118], films [119], nanocolumns [120], flower-like microspheres [121], nanoplatelets [122] etc have been successfully synthesized applied to LIBs. These materials showed good electrochemical performance because of their special structure. Controllable synthesis of NiO precursors under low temperatures which on thermolysis resulted in the formation of tremella like NiO hierarchical structures was reported by Lingling et al. [123].

The paper reported a morphology of tremella-like NiO with porous ultrathin nano-flakes prepared by hydrothermal method at low temperature (Figure 6). The electrode showed high capacity when used as an anode for LIBs, which could be attributed to the specific 3D structure which can accommodate the strain induced by the volume change during fast charge/discharge while offering a large surface area. In addition to the 3D structure, the porosity can also provide a short diffusion path for the lithium ions.

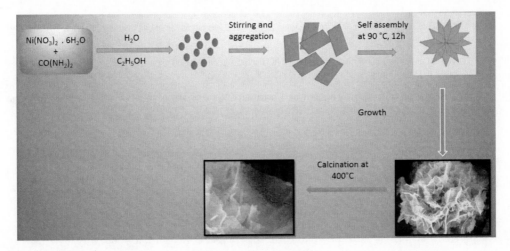

Figure 6. Formation mechanism of tremella like NiO heirarchical structures (Adapted with modifications from [123]).

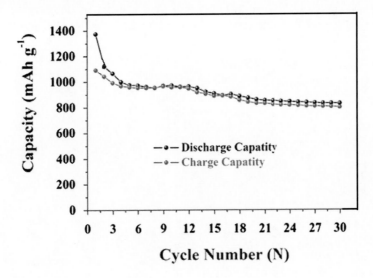

Figure 7. Cycling studies of NiO structures at 0.1 C (Adapted with modifications from [123]).

But such structures when employed as battery electrodes showed capacity reduction (Figure 7) during cycling, which is one of the primary areas where batteries fail to meet the requirements of the technological applications. Even though, all the systems discussed so far exhibited promising energy densities, all of them suffered from poor cycling capacities and rate capabilities. A simple capacitor can overcome this but they suffer from low energy densities, which again render them insufficient for today's applications which require a continuous supply of energy delivered at a faster rate. This is where the need for supercapacitors got highlighted, with higher energy density than a conventional capacitor and a higher power density than available battery systems.

5. SUPERCAPACITORS

Supercapacitor is an electrical energy storage device similar in fabrication to a battery. It consists of two electrodes immersed in an electrolyte and a separator between the electrodes. The electrodes are usually fabricated from a porous material exhibiting a very large surface area. The pores have a diameter of the order of nanometers. These porous electrodes exhibit high surface area, greater than that of a battery. The electric charges are stored in these micropores at or near the electode-electrolyte interface. In contrast to batteries they exhibit long shelf and cycle life. This has been exhibited mainly by those supercapacitors fabricated using carbon electrodes. While most rechargeable batteries show corrosion effects and self-discharge if rendered inactive over longer periods of time, supercapacitors were observed to retain their capacitance thereby recharge to their initial condition. The remarkable fact is that supercapacitors can be deep cycled at high rates for 500,000–1,000,000 cycles with only a relatively small change in their properties, accounting to 10–20% degradation in capacitance and resistance. The primary disadvantage of supercapacitors is their relatively low energy density compared to batteries limiting their use to applications in which relatively small quantities of energy are required before the supercapacitors can be recharged. Supercapacitors can, however, be recharged in very short interval of time (usually seconds) in the presence of a source of energy able to deliver the level of power required for the application.[124]

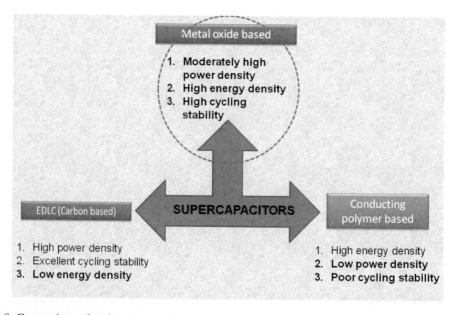

Figure 8. Comparison of various types of supercapacitors.

Supercapacitors falls under three categories- electric double layer capacitors employing carbon (Ultra-capacitors), pseudo-capacitors employing transition metal oxides and polymer based supercapacitors. In an EDLC, capacitance storage action takes place through charge separation at the electrode-electrolyte interface, giving rise to an electrochemical double layer, hence the name. Current is generated on the application of a voltage, resulting from the rearrangement of the charges. The mechanism of charge storage in pseudo-capacitors, on the other hand is based on fast redox reactions occurring at the interface brought about by the change in oxidation state of the material. Pseudocapacitors are known to have a higher capacitance than the EDLCs. [125-130]. The advantages and disadvantages of the three types of supercapacitors are highlighted in Figure 8.

6. NICKEL OXIDE AS SUPERCAPACITOR ELECTRODE MATERIAL

Transition metal oxides such as ruthenium oxide, manganese oxide, cobalt oxide, and nickel oxide are used as electrode materials for supercapacitors. [131]. Among these materials, nanostructured nickel oxide (NiO) is of great significance since it exhibits a very high theoretical capacitance of 2573 Fg^{-1} within 0.5 V, which is dependent on the size and morphology of the nanostructured material [132-134]. There have been extensive research to utilize the potential of nickel oxide for supercapacitors, due to its good pseudocapacitive behavior, ready availability, environmental compatibility and low cost compared to the high performing supercapacitor material RuO_2.[135, 136].

Various morphologies have been reported for NiO nanostructures which include porous nano/microspheres [137], nanoflowers [138], nanosheets [139], and nanofibers [140]. Studies have proved that the electrochemical behavior of NiO is dependent on various parameters like its microstructure, surface area, and the presence of dopants [141-146, 135]. This highlights the importance of a means of controlled synthesis of such nanostructures with desirable properties such as enhanced electronic conductivity, lower resistance to the diffusion process of ions as well as a very large electrochemically active surface area [135]. To ensure effective contact with a large number of electrochemically active sites even at large current densities, it is desirable for the materials to possess a large surface area. This interaction between electrode and the ions in the electrolyte also depends on the morphology of the nanostructures as well as the porosity. [147-149, 105]. The ion diffusion kinetics as well the electronic conductivity governs the charge discharge performance of the supercapacitor. [150, 131]. Many techniques have been reported for the synthesis of porous structures such as surfactant template [150], sol–gel [151-153], anodization [154], and hard template [155-157] solvothermal/ hydrothermal techniques [158, 141]. Further detailed research enthusiasm in exploiting such nanostructures has been stunted by the disadvantages such as toxicity as well as the mechanical stability [159, 160].

Reports have adequately warned about the harmful effects of inhalation exposure to nickel oxide nanoparticles [161]. Studies have shown significant lung toxicity and inflammation being induced in rats following intra-tracheal instillation resulting from exposure to nickel oxide nanoparticles [162]. In the same study, size was found to be a major factor in determining the toxicity. Ultrafine nickel oxide nanoparticles with an average size of 800 nm were found to be more toxic than micron sized nickel oxide (4.8 μm) particles.

Another study revealed that a shift to the nano-size range in nickel-containing particles induces toxicity which could be attributed to Ni (II) carcinogenesis [163, 164]. Taking these factors into consideration, waste materials from devices containing nanoparticles can be an environmental hazard. Porosity optimization is another problem associated with electrode fabrication. An increase in the density of micro pores may enhance the overall surface area leading to a higher energy density but it limits the ion diffusion and increases the resistance as a result of which the power density will be compromised [165]. Hence it becomes imperative that in order to design high performing porous nanostructured electrodes as well as reducing the toxicity, the morphology of the electrode materials should be suitably tailored. In these contexts nano/micro hierarchical porous structures are best suited as electrode materials in energy storage devices.

6.1. Nano/Micro Hybrid Hierarchical Structures of Nickel Oxide

Hierarchical nano/micro hybrids of NiO structures was reported by Anjali et al., [166]. They described a soft template synthesis of bulk nanostructured NiO with microporous flake like structures which provided easily accessible diffusion paths enhancing the rate capability and capacitance.

6.1.1. Fabrication

The schematic illustration elucidating the stages in the formation of these hierarchical structures is shown in Figure 9. The nanoflakes formation can be explained by an assembly-by-assemble mechanism and a subsequent mechanism of folding.

The reactions involved are explained as follows.

$$Ni^{2+} + 6\,NH_3 \leftrightarrow [Ni\,(NH_3)_6]^{2+} \tag{i}$$

$$NH_3 + H_2O \rightarrow NH_4^+ + OH^- \tag{ii}$$

$$Ni^{2+} + 2OH^- \rightarrow Ni(OH)_2 \tag{iii}$$

Ni^{2+} with its strong affinity with NH_3 results in the formation of a stable complex, $[Ni(NH_3)_6]^{2+}$. This complex gets decomposed as temperature increases, thereby providing OH^- ions for the formation of $Ni(OH)_2$. The complex decreases the free Ni^{2+} concentration in the solution, which results in a relatively low reaction rate of Ni^{2+} ions with OH^- ions. The role of PVP is to provide a soft template which separates the different $Ni(OH)_2$ nanosheets formed. $Ni(OH)_2$ layers intercalated with PVP shows weaker hydrogen bonding as compared to the hydrogen bonding between the hydroxyl groups of the adjacent sheets, yielding thinner and more flexible sheets. The formation of the hierarchical structure can be ascribed to the surface charges along with the induced elastic strain, causing the sheets to fold when there is the formation of many short range forces, to give a final micro structure. The samples retained their hierarchical structure on annealing resulting in the formation of pure NiO. The final structure showed a highly microporous structure which could be due to the removal of water molecules from $Ni(OH)_2$ (Figure 10).

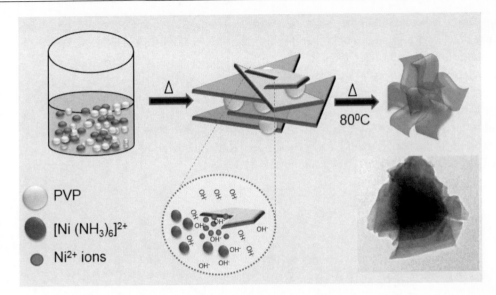

Figure 9. Schematic illustration of the formation of heirarchical structures (Adapted with modifications from [166]).

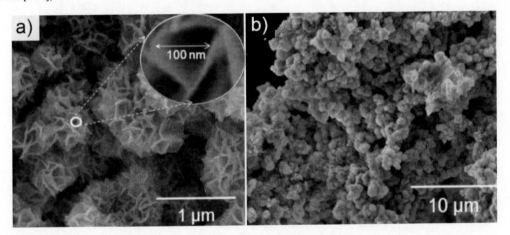

Figure 10. SEM images of the NiO structures formed (Adapted with modifications from [166]).

6.1.2. Electrochemical Performance

The electrochemical performance of the microspheres was evaluated by employing them as electrodes and subjecting them to CV analysis. The result is shown in Figure 11.

CV curves showed the presence of prominent redox patterns which can be represented by the following reaction.

$$NiO + yOH^- \leftrightarrow NiOOH + ye^- \tag{iv}$$

The specific capacitance values obtained were 261, 473, and 1004 Fg^{-1} for the scan rates of 100, 50 and 10 mVs^{-1}. It was observed that at lower scan rates the specific mass capacitance tend to increase, indicating that the diffusion of OH$^-$ that are primarily responsible for the redox behavior depends on the CV scan rates. It is expected that at lower scan rates, both the outer and the inner pores on the electrode overlay are effectively utilized for OH$^-$

propagation, whereas at higher scan rates this interaction of the OH⁻ could be limited to mainly outer regions of the pores thereby limiting the capacitance values [167, 168]. This is because at lower sweep rates more amount of time is provided for the OH⁻ to access the bulk of the electrode.

Specific mass capacitance values as high as 1950 Fg⁻¹ was calculated at scan rates of 1 mVs⁻¹ with less than 10% fading at the end of 1000 cycles. This high capacitance values can be attributed to the porous nature of the electrode overlay which not only increases the effective surface area but could also provide accessible diffusion pathways for the ions. BET surface area of these spheres was found to be ~110 m²g⁻¹.

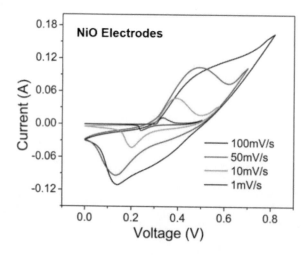

Figure 11. CV curves of the NiO electrodes at different scan rates of 1, 10, 50 and 100 mV/s. (Adapted with modifications from [166]).

6.1.3. Coin Cell Fabrication

The application of these porous hierarchical structures was explored in a fully functional coin-cell supercapacitor unit.

Figure 12 a) depicts the discharging behavior at different current densities. The discharge curves showed a deviation from linearity which is uncharacteristic of a typical electrochemical double layer capacitor (EDLC). This anomaly was explained in the study as arising due to the pseudocapacitance resulting from the redox reaction in the given voltage range. As the current density increases the voltage drop at the onset of the discharge profile becomes more prominent which can be attributed to the electrolyte resistance. Figure 12(b) displays the discharge profile at a lower current density of 0.05 mA cm⁻². The electrical parameters such as specific capacitance power density (P) and energy density (E) were calculated according to DIN IEC 62391-1 [169, 170].

The specific capacitance value was calculated to be 17.6 Fg⁻¹. Such low values of capacitance for the coin cells when compared to the single electrode studies were attributed to the fact that since the NiO electrode was in series with the mesoporous carbon, the capacitance is merely of EDLC with no redox reaction happening. The high ionic resistance inside the porous structure also adds to the decreased capacitance values. The maximum values of the power density and energy density were found to be 24 kWkg⁻¹ and 17 Whkg⁻¹, respectively.

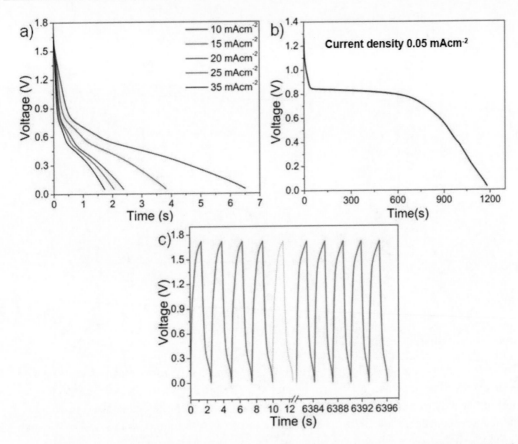

Figure 12. Discharge behavior of the coin cell (a) at high current densities (b) at low current density of 0.05 mAcm^{-2} and (c) for a continuous 1000 cycles. (Adapted with modifications from [166]).

Figure 12 (c) represents the charge discharge profile of the coin cell performed for a continuous 1000 cycles. At the end of 1000 cycles the capacitance fade was observed to be <10%. This fading can be attributed to the dissolution to the electrode in the electrolyte. This was confirmed by ICP-AES tests which showed that the NiO dissolution in the electrolyte increased from 1 ppm to 8 ppm at the end of 1000th cycle.

6.1.4. Electrochemical Impedence Studies

The EIS of as-prepared coin cells and the cells after 1000th cycles was measured in the frequency range of 100-0.1 kHz (Figure 13). Two major characteristic features in the high and low frequency regions are attributed to various resistance phenomena during different interfacial processes in faradaic reactions. Figure 13 (inset) represents the equivalent circuit diagram R_s and R_{ct} are solution and charge-transfer resistances, respectively. C_{dl} and C_p in the circuit represents double layer and pseudocapacitance, respectively. The slope of the 45° portion of the curve representing Warburg resistance (W) is a result of the frequency dependence of ion diffusion/transport in the electrolyte [171-173].

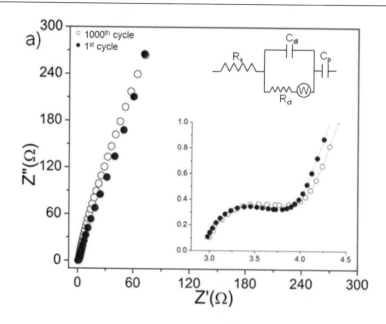

Figure 13. Nyquist plot representation of the coin cell unit (inset) the equivalent circuit diagram (Adapted with modifications from [166]).

The R_{ct} values were found to be 1 and 1.2 Ω respectively for the samples at 1st and 1000th cycles. The increase in the R_{ct} values can be attributed to the structural changes happening at the electrode overlay. EIS tests showed these electrodes when used in coin cell exhibited a response time of 1.8 ms. The toxicity issues were found to be less severe for such nano/micro hybrids as shown by Roshny et al., in MnO_2 nano/micro hybrids for supercapacitor applications which will be discussed in the next section.

7. MANGANESE DIOXIDE NANO/MICRO HYBRIDS FOR SUPERCAPACITORS

Manganese dioxide (MnO_2) as a cathode material for supercapacitors and batteries has been widely explored because of its low-cost, ease in availability, non-toxicity and high theoretical specific capacitance of 1370 Fg^{-1}[174-177]. However, even though MnO_2 is a biocompatible material, when it is reduced to the nano regime, it shows significantly high cellular uptake than a micro particle, which results in cell inflammation at dangerous levels. Hence these nanoparticles of even benign materials will pose a threat to physiological function, even causing permanent damage. This is one of the factors hindering the commercialization of nanotechnology. These nanoparticles can enter market only when it has been proven that there is no threat to health and the environment by regulatory bodies like Food and Drug Administration (FDA).

Roshny et al., showed a simple a low temperature molten salt route which can generate unique architecture of MnO_2 nanospikes arrayed in an atypical fashion to form micron sized ball morphology. [178]. These hybrid nano/micro structures showed superior electrochemical performance as compared to the values reported in the literature for pristine nanostructures of

MnO_2. In addition, the paper reported cell uptake studies which established that compared to a nano particle, these nano/micro hybrids showed considerably lower cellular uptake, which meant these structures posed no threat to humans if exposed during functioning as well as the disposal.

7.1. Morphological Analysis

Figure 14 (a-d) shows the TEM images displaying the evolution of MnO_2 morphology synthesized under different processing times of 1, 2, 5 and 10 h.

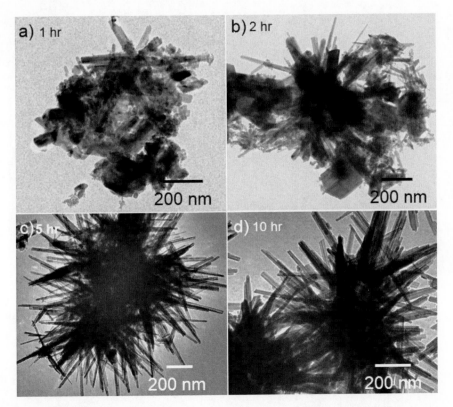

Figure 14. Evolution of MnO_2 morphology as a function of time (Adapted with modifications from [178]).

Microspheres with clear nano spikes on the surface was seen which became stable structure at a processing time of 5 hours and retained that structure at further processing times. These microspheres showed diameter varying from 1-2 μm. The spike showed thickness ranging from 30-70 nm.

This time dependent morphological evolution is schematically represented in Figure 15.

The molten salt synthesis employs a suitable salt, $LiNO_3$ in this case, in which the precursors get diffused and dissolved during the preliminary stage. This stage is followed by the formation of nuclei which can be expressed by the reaction (1) shown below:

$$MnSO_4 \rightarrow MnO_2 + SO_2 \qquad (v)$$

By means of Ostwald ripening, the MnO_2 nuclei formed aggregate into nanosized spheroidal particles which act as crystal growth sites, absorbing the newly formed nuclei onto its surfaces.

The degree of adsorption is determined by the surface energies of the different facets of the base crystal. Surface energy minimisation plays an important role when it comes to the structure formation. The nuclei gets attached itself to the high surface energy facet as per the criteria for surface energy minimisation and continues to grow along the impelled direction.

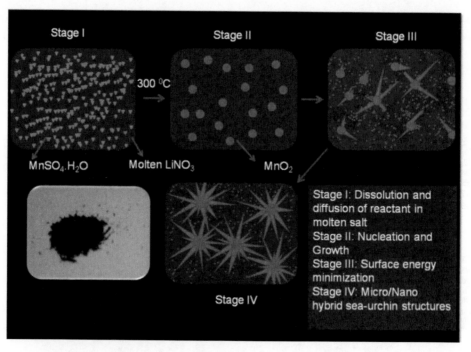

Figure 15. Schematic representation of the time dependent morphological evolution of the nano/micro structure. (Adapted with modifications from [178]).

7.2. Cytotoxicity Studies

Cytotoxicity experiments were carried out on human dermal fibroblasts (HDF) by MTT assay. MTT [3-(4, 5-dimethylthiazole-2-yl)-2, 5-diphenyl tetrazolium] assay for cytotoxic evaluation is a colorimetric test based on the selective ability of viable cells to reduce the tetrazolium component of MTT in to purple colored formazan crystals. The results are shown in Figure 16.

From Figure 16, it was observed that nano MnO_2 showed higher cellular uptake, which resulted in the modulation of the cytoskeletal arrangements of HDF cells which is indicated in the figure with purple spots representing actin dye penetration.

Cellular uptake in this level can be dangerous because of the increased probability for blood brain barrier penetration. Further studies showed distorted cell nucleus with expanded cell morphology confirming enhanced cellular uptake of the nano MnO_2 in comparison to the nano/micro hybrid structures. Such a cell inflammatory response was also an indication of its toxicity for humans at the topical level. The role that the morphology plays in controlling the cell uptake and thereby the danger of cell interaction was clearly depicted indicating a basic micron structure posed no threat when compared to a pristine nanoparticle, but at the same time showing superior performance as well.

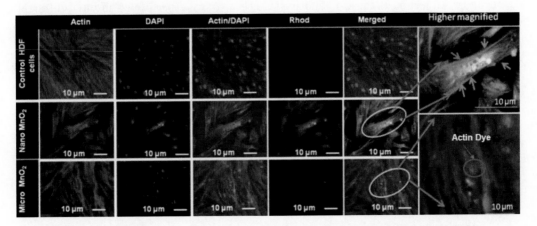

Figure 16. HDF cell uptake study of nano MnO_2 and nano/micro hybrid MnO_2 (Adapted with modifications from [178]).

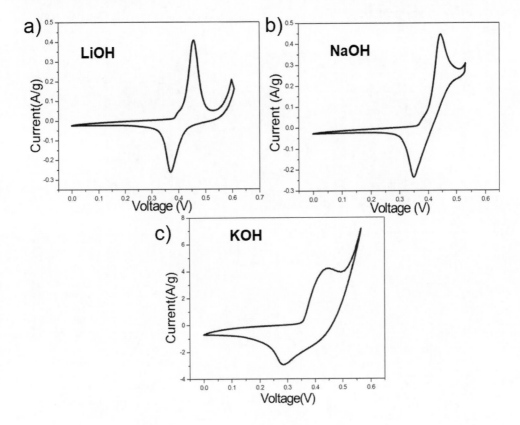

Figure 17. CV analysis of the electrode in 1M a) LiOH b) NaOH and c) KOH electrolyte systems (Adapted with modifications from [178]).

7.3. Electrochemical Studies

Electrochemical performance was studied by employing these nano/micro structures as electrodes and subjecting them to CV analysis. The tests were performed in three different electrolyte systems viz KOH, LiOH and NaOH, out of which the best system was determined for further evaluation. (Figure 17).

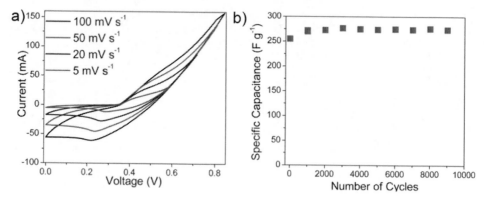

Figure 18. (a) CV curves at different scan rates, and (b) cycling studies at 200 mV/s. (Adapted with modifications from [178]).

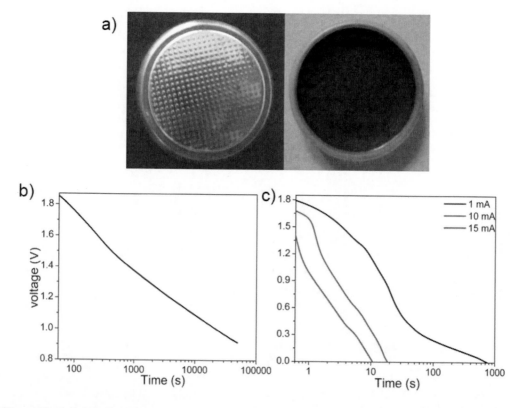

Figure 19. a) Coin cell with the cathode coating b) self-discharge behaviour of M-5 coin cells and c) constant current discharge performance at different discharging currents (Adapted with modifications from [178]).

These electrodes were subjected to cycling studies at 100 mVs^{-1} scan rate. The electrodes showed no fading even after the end of 10,000 cycles. (Figure 18)

A coin cell fabricated using these electrodes showed capacitance values of 450, 12 and 11 mF at discharge currents of 1 mA, 10 mA and 15 mA respectively. (Figure 19).

7.4. Electrochemical Impedence Studies

The EIS of as-prepared coin cells was measured in the frequency range of 100000-0.1 Hz.

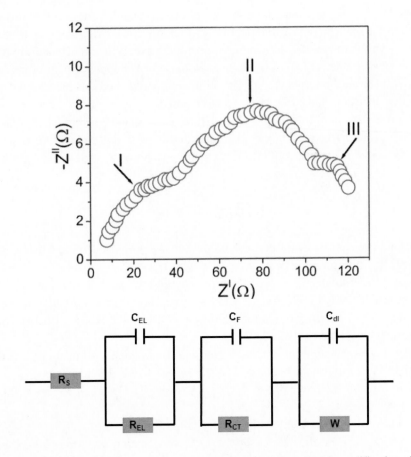

Figure 20. Nyquist plot of M-5 coin cell and Equivalent circuit (Adapted with modifications from [178]).

The Nyquist plot showed three distinct regions characteristic of impedance spectrum of an electrochemical system. (Figure 20). Each arc could be related to the distinct phase boundary. The high frequency arc corresponds to the electrolyte. The point of intersection with the real axis gives the electrolyte/solution resistance (Rs). This solution resistance is the current limiting factor in a supercapacitor. The arc at medium frequencies reflects the electrode reactions, namely the charge transfer (R_{ct}) at the electrode/electrolyte interface. Diffusion of ions (mass transport) can also give rise to impedance called Warburg impedance (R_w), which is represented by the third arc in the Nyquist plot. This impedance depends on the frequency of the potential perturbation, prominent at lower frequencies, as the ions are expected to diffuse further. The Warburg resistance can also be correlated to the microporous structure of the interface between electrode and electrolyte. These coin cells further exhibited energy and power densities of of 4.5 Whkg^{-1} and 14 kW kg^{-1}, respectively, with a low self-discharge.

CONCLUSION

Tailoring electrode architecture at the microscale with implementing hierarchical structures is largely an unexplored research direction mainly due to the difficulty in processing and parameter control. Fabrication of hierarchically structured electrodes which can answer the problems related to synthesis, cost and toxicity is expected to revolutionize every field employing the principles of nanotechnology for the enhancement of performance. These high surface area micro/nano hybrid structures could increase the charge transport, facilitate better diffusion and offer large electrochemically active sites. Further research is needed to explore the extent to which such structures can enhance the advantages and reduce the toxicity issues. From the preliminary examples discussed in this chapter it is expected that such structures could be the future of efficient energy generation as well as energy storage.

REFERENCES

[1]   Mclarnon F. R.; Cairns E. J. Energy storage. *Ann. Re. Energy.* 1989, 14, 241–247.
[2]   Baker, J. N; Collinson, A. *Power. Eng J.*1999, 6, 107–112.
[3]   Ahearne, J. *International Union of Pure and Applied Physics*, 2004,76–86.
[4]   Basic research needs for electrical energy storage, *Report of The Basic Energy Science Workshop For Electrical Energy Storage*, Office Of Basic Energy Sciences, Department Of Energy, July 2007
[5]   Hussain, T. *Energy Storage - Technologies and Applications;* Ahmed F.Z. , ISBN 978-953-51-0951-8, Published: January 23, 2013 under CC BY 3.0 license
[6]   Whittingham, M. S.; Savinelli, R. F.; Zawodzinski, T. *Chemical Reviews.* 2004, 104, 4243–4886.
[7]   Tarascon, J. M.; Armand, M. *Nature.* 2001, 414, 359.
[8]   Schodek, D. L.; Ferreira, P.; Ashby, M. F. *Nanomaterials, Nanotechnologies and Design: An Introduction for Engineers and Architects,* Butterworth-Heinemann Ltd, U.K, 2009, 187.
[9]   Mehrens, M. W.; Schenk, J.; Wilde, P. M.; Abdelmula, E.; Axmann, P.; Garche, J. *J. Power Sources,*2002, 105, 182–188.
[10]   Conway, B. E.; Birss, V.; Wojtowicz, J. *J. Power Sources*, 1997, 66, 1-14.
[11]   Wang, G.; Zhang, L.; Zhang, J. *Chem. Soc. Rev*, 2012,41, 797-828.
[12]   Jiang, J.; Li, Y.; Liu, J.; Huang, X.; Yuan, C.; Lou, X. W. *Adv. Mater.* 2012, 24, 5166–5180.
[13]   Nazar, L.F.; Goward, G.; Leroux, F.; Duncan, M.; Huang, H.; Kerr, T.; Gaubicher, J. *Int. J. Inorg. Mater.* 2001, 3, 191–200.

[14] Xia, X.; Tu, J.; Zhang,Y.; Wang, X.; Gu, C.; Zhao, X.; Fan, H. J. *ACS Nano*. 2012, 6, 5531–5538.

[15] Wang, D.; Kou, R.; Choi, D.; Yang, Z.; Nie, Z.; Li, J.; Saraf, L. V.; Hu, D.; Zhang, J.; Graff, G. L.; Liu, J.; Pope, M. A.; Aksay, I. A. *ACS Nano*. 2010, 4, 1587–1595.

[16] Gao, M.; Xu, Y.; Jianga, J.; Yu, S. Chem. Soc. Rev. 2013,42, 2986-3017.

[17] Frackowiak, E.; Beguin, F. *Carbon*. 2001, 39, 937–950.

[18] Candelaria, S. L.; Shao, Y.; Zhou, W.; Li, X.; Xiao, J.; Zhang, J.; Liu, Y. W. J.; Li, J.; Cao, G. *Nano Energy*. 2012,1, 195–220.

[19] Sharma, H. S.; Sharma, A. *Progress in Brain Research*. 2007, 162, 245–273.

[20] Nel, A.; Xia, T Ma¨dlerL.; Li, N. *Science,* 2006, 311, 622.

[21] Vishwakarma, V.; Samal, S. S.; Manoharan, N. *Journal of Minerals & Materials haracterization & Engineering*. 2010, 9,455-459.

[22] Chau, C.; Wu, S.; Yen, G. *Trends Food Sci Tech*. 2007,18, 269-280.

[23] MyrtillSimkó; Mats-Olof Mattsson; *Part. Fibre. Toxicol*. 2010, 7:42.

[24] Balakrishnan, A.; Subramanian, K. R. V. *Nanostructured Ceramic Oxides for Supercapacitor Applications*, CRC Press. 2014.

[25] FDA's Approach to Regulation of Nanotechnology Products, http://www.fda.gov/ ScienceResearch/SpecialTopics/Nanotechnology/ucm301114.htm

[26] Khoo, E.; Wang, J.; Ma, J.; Lee, P. S. *J. Mater. Chem*. 2010, 20, 8368-8374.

[27] Liu, R.; Duaya, J.; Lee, S. B. *Chem. Commun*. 2011, 47, 1384–1404.

[28] Chen, H.; Cong, T. N.; Yang, W.; Tan, C.; Li, Y.; Ding, Y. *Progress in Natural Science*. 2009, 19, 291–312.

[29] Baker, J. N.; Collinson, A. *Power Eng. J*. 1999, 6, 107–112.

[30] Koot, M.; Kessels, J.; Jager B. Heemels, ; W.; Bosch,P.;Steinbuch, M.; *IEEE T VehTechnol*. 2005, 54, 771–782.

[31] Palacín, M. R. Chem. Soc. Rev. 2009,38, 2565-2575.

[32] Liu, X. J.; Kang, S. D.; Kim, J. S.; Ahn, I. S.; Ahn, H. J. *Rev. Adv. Mater. Sci*. 2011, 28, 98-102.

[33] Wu, F.; Yang, H. X. *Green Seconday Batteries*, Science Press, Beijing, 2009, 16.

[34] Arico, A. S.; Bruce, P.; Scrosati, B.; Tarascon, J. M. ; Van Schalkwijk, W. *Nat. Mater.* 2005, 4, 366.

[35] J. Maier, *Nat. Mater*. 2005, 4, 805.

[36] Wang, Y.; Cao, G. Z.; *Adv. Mater*. 2008, 20, 2251.

[37] Liu, J.; Cao, G.; Yang, Z.; Wang, D.; Dubois, D.; Zhou, X.; Graff, G. L.; Pederson, L. R.; Zhang, *J. Chem Sus Chem*. 2008, 1, 676 – 697.

[38] Balakrishnan, A.; Pizette, P.; Martin, C. L.; Joshi, S. V.; Saha, B. P. *Acta Materialia*. 2010, 58, 802–812.

[39] Balakrishnan, A.; Martin, C. L.; Saha, B. P.; Joshi, S. *J. Am. Ceram. Soc*. 2010, 94, 1046-1052.

[40] Chen, X.; Du, Y.; Zhang, N. Q.; Sun, K. N. J. *Nano. Mat*. 2012, 2012, Article ID 905157.

[41] Sreelakshmi, K. V.; Sasi, S.; Balakrishnan, A.; Sivakumar, N.; Nair, A. S.; Nair, S. V. Subrmanian, K. R. V. *Energy Technology*. 2014, 2, 257-262.

[42] Praveen, P.; Ravi, S.; Soumya, M. S.; Binitha, G.; Balakrishnan, A. *Sci. Adv. Mater*. 5, 2021-2026.

[43] Jacob, D.; Mini, P. A.; Balakrishnan, A.; Nair, S. V.; Subramanian, K. R. V. *Bull. Mater. Sci.* 2014, 37, 61-69.

[44] Mini, P. A.; Balakrishnan, A.; Nair, S. V.; Subramanian, K. R. V. *Chem. Commun.* 2011, 47, 5753-5755.

[45] Kalluri, S.; Madhavan, A. A.; Bhupathi, P. A.; Vani, R.; Paravannoor, A.; Nair, A. S. *Sci. Adv. Mater.* 2012, 4, 1220-1225.

[46] Grugeon, S.; Laruelle, S.; Dupont, L.; Tarascon, J. M. *Solid. State. Sci.* 2003, 5, 895.

[47] Gillot, F.; Boyanov, S.; Dupont, L.; Doublet, M. L.; Morcrette, M.; Monconduit, L.; Tarascon, J. M. *Chem. Mater.*, 2005, 17, 6327–6337.

[48] Hu, J.; Li, H.; Huang, X. Electrochem. Solid-State Lett. 2005, 8, A66.

[49] Pralong, V.; Leriche, J. B.; Beaudoin, B.; Naudin, E.; Morcrette, M.; Tarascon, J. M. *Solid State Ionics.* 2004, 166, 295.

[50] Zhu, L. Z.; Han, E. S.; Cao, J. L. *Adv. Mat. Res.* 2011, 236-238, 694.

[51] Priya Nair, S.; Jyothsna, U.; Praveen, P.; Balakrishnan, A.; Subramanian, K. R. V. *ISRN Nanotechnology.* 2013. Volume 2013, Article ID 653237.

[52] Brutti, S.; Panero, S. *Nanotechnology for Sustainable Energy*, 2013, Chapter 4, pp 67–99.

[53] Parvathy, S.; Ranjusha, R.; Sujith, K.; Subramanian, K. R. V.; Sivakumar, N. *J. Nano Mat.* 2012,41.

[54] Sonia, T. S.; Sivakumar, N.; Nair, S.; Balakrishnan, A.; Nair, S. V. *Sci. Adv. Mater.* 5, 1828-1836.

[55] Schuster, J.; He, G.; Mandlmeier, B.; Yim, T.; Lee, K. T.; Bein, T.; Nazar, L. F. *Angew. Chem. Int. Ed.* 2012, 51, 3591 –3595.

[56] Ji, X.; Lee, K. T.; Nazar, L. F. *Nat Mater.* 2009, 8,500-506.

[57] Mikhayalik, Y. V.; Akridge, J. R. *J. Electrochem. Soc.* 2004, 11, A1969.

[58] Poizot, P. et al.; *J. Electrochem. Soc.* 2002, 149(9), A1212-A1217.

[59] Jasinski, R.; Burrows B. *J. Electrochem. Soc.* 1969, 116, 422-424.

[60] Ramasamy, K.; Malik, M. A.; O'Brien, P.; Raftery, J.; Helliwell, M. *Chem. Mater.* 2010, 22, 6328–6340.

[61] Sebastian, S. T.; Jagan, R. S.; Rajagopalan, R.; Paravannoor, A.; Menon, L.V.; Nair, S. V.; Balakrishnan, A. V. *RSC Advances.* 2014, 4, 11673-11679.

[62] Khiew, P. S.; Huang, N. M.; Radiman, S; Md. Soot *Ahmad. Matter Lett.* 2004, 58, 762-767.

[63] Han, S. C.; Kim, H. S.; Song, M. S.; Kim, J. H.; Ahn, H. J.; Lee, J. Y. J. *Alloys Compd.* 2003, 351, 273–278.

[64] Türkera, Y.; Dag, O. *J. Mater. Chem.* 2008, 18, 3467–3473.

[65] Salavati, M.; Davar, F.; Emadi, H. *Chalcogenide Lett.* 2010, 7, 647–655.

[66] Ni, Y.; Ma, X.; Hong, J.; Xu, Z. *Mater. Lett.* 2004, 58, 2754–2756.

[67] Kovtyukhova, N. I.; Buzaneva, E. V.; Waraksa, C. C.; Mallouk, T. E. *Mater. Sci. Eng.* 2000, B69–70, 411–417.

[68] Niu, K.Y.; Yang, J.; Kulinich, S. A.; Sun, J.; Du, X. W. *Langmuir.* 2010, 26, 16652–16657.

[69] Ghezelbash, A.; Sigman, M. B.; Korgel, B. A. *Nano Lett.* 2004, 4, 537–542.

[70] Shen, G.; Chen, D.; Tang, K.; An, C.; Yang, Q.; Qian, Y. *J. Solid State Chem.* 2003, 173, 227–231.

[71] Chen, N.; Zhang, W.; Yu, W.; Qian, Y. *Mater. Lett.* 2002, 55, 230–233.

[72] Wang, D.; Zeng, Q.; Zhou, G.; Yin, L.; Li, F.; Cheng, H.; Gentle, I. R.; Lu, G. Q. M. *J. Mater. Chem. A*. 2013, 1, 9382.

[73] Shi, Y.; Wang, Y.; Wong, J. I.; Tan, A. Y. S.; Hsu, C.L.; Li, L.; lu, Y.; Yang, H. Y. *Sci Rep*. 2013, 3, 2169.

[74] Zhu, J.; Wang, D.; Wang, L.; You, W.; Wang, Q. *Int. J. Electrochem. Sci*. 2012, 7, 9732.

[75] Mahmood, N.; Zhang, C.; Hou, Y. *Small,* 2013, 9, 1321.

[76] Ho, X.; Wei, J.; *Materials*. 2013, 6, 2155.

[77] Singh, V.; Joung, D.; Zhai, L.; Das, S.; Khondaker, S. I.; Seal, S. *Prog. Mater Sci*. 2011, 56, 1178.

[78] Rodríguez-Pérez, L.; Herranza M. A.; Martín, N. *Chem. Commun*. 2013, 49, 3721.

[79] Luo, J.; Kim, J.; Huang, J. *Acc. Chem. Res*, 2013, 46, 2225.

[80] Lui, G.; Liao, J.; Duan, A.; Zhang, Z.; Fowler , M.; Yu, A. J.; *Mater. Chem. A*. 2013, 1, 12255.

[81] Park, M.; Yu, J.; Kim, K.; Jeong, G.; Kim, J.; Jo, Y.; Hwang, U. K.; Kang, S.; Woo, T.; Kim, Y. Phys. *Chem. Chem. Phys*.2012,14, 6796.

[82] Aparna, R.; Sivakumar, N.; Balakrishnan, A.; Nair, A.S.; Nair, S.V. *Journal of Renewable and Sustainable Energy*. 2013, 5, 033123.

[83] Tong, S.; Mao, L.; Zhang, X.; Jia, C. Q. *Ind. Eng. Chem. Res*. 2011, 50, 13825.

[84] He, X.; Zhao, N.; Qiu, J.; Xiao, N.; Yu, M.; Yu, C.; Zhang, X.; Zheng, M. J. *Mater. Chem. A*. 2013, 1, 9440.

[85] Cao, Z.; Cao, X.; Sun, L.; He, Y. *Adv. Mater. Res*. 2011,239, 2101.

[86] Ghosh, R.; Bhattacherjee, S. J. *Chem. Eng Process Technol*. 2013, 4, doi:10.4172/2157-7048.1000156.

[87] Rajbhandari, R.; Shrestha, L. K.; Pokharel, B. P.; Pradhananga. R. R. J. *Nanosci. Nanotechnol*. 2013, 13, 2613.

[88] Boonpoke, A.; Chiarakorn, S.; Laosiripojana1, N.; Towprayoon, S.; Chidthaisong, A. *J. Sustain. Energ. Enviro*. 2011, 2, 77.

[89] Sun, L.; Tian, C.; Li, M.; Meng, X.; Wang, L.; Wang, R.; Yin, J.; Fu, H. *J. Mater. Chem. A*, 2013, 1, 6462.

[90] Rosi, M.; Muhamad, P.; Ekaputra; Ferry, I.; Mikrajuddin, A.; Khairurrijal. *AIP Conf. Proc*.2010, 86 1325.

[91] Praveen, P.; Jyothsna, U.; Nair, P.; Ravi, S.; Balakrishnan, A. *J. Nanosci Nanotech*. 2013 ,8, 5607-5612.

[92] Awasthi, K.; Kumar, R.; Raghubanshi, H.; Awasthi, S.; Pandey, R.; Singh, D.; Yadav. T. P.; Srivastava, O. N.; *Bull. Mater. Sci*. 2011, 34,607.

[93] Mukul, K.; Yoshinori, A. *Dam. Relat. Mater*. 2003, 12, 1845.

[94] Rahman, M. M.; Wang, J.; Wexler, D.; Zhang, Y.; Li, X. J.; Chou, S. L.; Liu, H. K. *J. Solid State Electrochem*. 2010,14,571.

[95] He, B. L.; Dong, B.; Li, H. L. *Electrochem Commun*.2007, 9,425.

[96] Quan-Chao. Z; Xiang Yun QiuShou-Dong.X; Ying. H, Qiang ;Shi Gang. S; *Lithium Ion Batteries – New Developments*, Chapter 8, ISBN 978-953-51-0077-5, InTech, 2012. 189.

[97] Kim, I. S.; Blomgren, G. E.; Kumta, P. N.; *J. Electrochem.Soc*. 2005, 152.

[98] Poizot, P.; Laruelle, S.; Grugeon, S.; Dupont, L.; Tarascon, J. M. *Nat*. 2000, 407, 496.

[99] Ian, A.; Courtney, A.; McKinnonb, W. R.; Dahnadoi.J.R. *J. Electrochem. Soc.* 1999, 1 46, 1, 59-68.

[100] Denis; Baudrin, E.; Touboul, M.; Tarascon, J. M. *J. Electrochem. Soc.* 1997, 144, 12,4099-4109.

[101] Anjusree, G. S.; Asha, A. M.; Subramanian, K. R. V.; Sivakumar, N; Nair, A. S. *Inter. J. of Mater. Res.*104 (6), 573-577.

[102] Binitha, G.; Soumya, M. S.; Madhavan, A. A.; Praveen, P.; Balakrishnan, A. J. *Mater. Chem. A.* 2014,1, 11698-11704.

[103] Madhavan, A. A.; Mohandas, A.; Licciulli, A.; Sanosh, K. P.; Praveen, P.; Jayakumar, R.; Nair, S. V.; Nair, A. S.; Balakrishnan, A. *RSC Adv.* 3, 25312-25316.

[104] Ranjusha, R.; Sreeja, R.; Mini, P. A.; Subramanian, K. R. V.; Nair, S. V. *Mater. Res. Bull.*47, 1887-1891.

[105] Ranjusha, R.; Lekha, P.; Subramanian K. R. V.; Shantikumar. V. N. *J. Mater. Sci. Tech.* 2011, 27, 961-966.

[106] Madhavan, A. A.; Ranjusha, R.; Daya, K. C.; Arun, T. A.; Praveen, P.; Sanosh, K. P. *Sci. Adv. Mater.* 2014, 6, 828-834.

[107] Desilvestro, J.; Haas, O. J. *Electrochem. Soc.* 1990,137, 5C-22C.

[108] Lekha, P; Balakrishnan, A.; Subramanian, K. R. V.; Nair, S. V. *Mater. Chem. Phy.* 2013,141, 216-222.

[109] Nair, S. V.; Balakrishnan, A.; Subramanian, K. R. V.; Anu, A.; Asha, A. M. *Bul. Mater. Sci.* 2012, 35, 489-493.

[110] Anjali, P.; Vani, R.; Sonia, T. S.; Nair, A. S; Ramakrishna, S.; Ranjusha, R. *Sci. Adv. Mater.* 2013, 6, 94-101.

[111] Anjali, P; Nair, A. S.; Ranjusha, R.; Praveen, P.; Subramanian, K. R. V. *Chem Plus Chem.*2013,78, 1258-1265.

[112] Anjali, P.; Ranjusha, R.; Asha, A. M.; Vani, R.; Kalluri, S. *Chem. Eng. J.*2013, 220, 360-36.

[113] Mamak, M.; Coombs, N.; Ozin, G. A. *Chem Mater.* 2001, 13, 3564–3570.

[114] Wang, X.; Song, J.; Gao, L.; Jin, J.; Zheng, H.; Zhang, Z.; *Nanotech,* 2005, 16,37–39.

[115] Morin, F. *J. Phys. Rev.* 1954, 93, 1199.

[116] Lunkenheimer, P.; Loidl, A.; Ottermann, C. R.; Bange, K.; *Phys. Rev.* 1991, B 44, 5927.

[117] 117. Idris; Nurul Hayati; Advanced materials for lithium rechargeable battery, Doctor of Philosophy thesis, Faculty of Engineering, University of Wollongong, 2011. http://ro.uow.edu.au/theses/3492

[118] Liu, L.; Li, Y.; Yuan, S. M.; Ge, M.; Ren, M. M.; Sun, C. S. J. *Phy. Chem. C.* 2010, 114, 251–255.

[119] Wu, M. S.; Lin, Y. P. *Electrochem Acta.* 2011, 56, 2068–73.

[120] Wang, X. H.; Li, X. W.; Sun, X. L.; Li, F; Liu, Q. M.; Wang, Q. *J Mater. Chem.* 2011, 21, 3571–3.

[121] Shang, S. Q.; Xue, K. Y.; Chen, D. R.; Jiao, X. L. *Cryst. Eng. Comm.* 2011, 13, 5094–9.

[122] Wang, X.; Li, L.; Zhang, Y. G.; Wang, S. T.; Zhang, Z. D.; Fei, L. F. *Crys. Grow. Des.* 2006, 6, 2163–2165.

[123] Hu, L.; Qu, B.; Chen, L.; Li, Q. *Mater. Let T.* 2013, 108, 92–95.

[124] Burke, A.; J. *Power Sources.* 2000, 91, 37–50.

[125] Lota, G.; Centeno, T. A.; Frackowiak, E.; Stoeckli, F. *Electrochim Acta.* 2008, 53, 2210–2216.

[126] Fang, B.; Binder, L. J. *Power Sources.* 2006, 163, 616–622.

[127] Conway, B. E. *J Electrochem Soc.* 1991, 138, 1539–1548.

[128] Sarangapani, S.; Tilak, B. V.; Chen, C. P. *J Electrochem Soc.* 1996,143, 3791–3799.

[129] Conway, B. E. Electrochemical Supercapacitors: *Scientific Fundamentals and Technological Applications.* New York: Plenum; 1999.

[130] Zheng, J. P.; Cygan, P. J.; Jow, T. R. *J Electrochem Soc.* 1995,142, 2699–2703.

[131] Farrukh, I. D.; Kevin, R. M.; Mohammed, E. S. *Nanoscale. Res. Lett.* 2013, 8, 363.

[132] Zhang, B.; Jimin, D.; Jing, C.; LiaLiu, J. Z. S. *RSC Advances.* 2012, 2, 2257–2261.

[133] Liu, K. C.; Anderson, M. A. *Electrochem. Soc.*1996, 143, 124–130.

[134] Nam, K. W.; Kim, K. B. *J. Electrochem. Soc.* 2002, 149, 346–354.

[135] Xiaojun, Z.; Wenhui, S.; Jixin, Z.; Weiyun, Z.; Jan, M.; Subodh, M.; Tuti, L. M.; Yanhui, Y.; Hua, Z.; Huey, H. H.;Qingyu,Y. *Nano Res.* 2010, 3,643–652.

[136] Chang, K. H.; Hu, C. C.; Chou, C. Y. *Chem. Mater.* 2007, 19, 2112–2119.

[137] Yuan, C. Z.; Zhang, X. G.; Su, L. H.; Gao, B.; Shen, L. F. *J. Mater. Chem.* 2009, 19, 5772–5777.

[138] Lang, J. W.; Kong, L. B.; Wu, W. J.; Luo, Y. C.; Kang, L. *Chem. Commun.* 2008, 4213–4215.

[139] Zhu, J. X.; Gui, Z. *Mater. Chem. Phys.* 2009, 118, 243–248.

[140] Qiu, Y. J.; Yu, J.; Zhou, X. S.; Tan, C. L.; Yin, J. *Nanoscale Res. Lett.* 2009, 4, 173–177.

[141] Yuan, C. Z.; Chen, L.; Gao, B.; Su, L. H.; Zhang, X. G.J. *Mater. Chem.* 2009, 19, 246–252.

[142] Yuan, C. Z.; Gao, B.; Zhang, X. G. J. *Power Sources.* 2007, 173, 606–612.

[143] Xing, W.; Li, F.; Yan, Z. F.; Lu, G. Q. J. *Power Sources* 2004, 134, 324–330.

[144] Jiao, F.; Hill, A. H.; Harrison, A.; Berko Jiao, F.; Hill, A. H.; Harrison, A.; Berko, A.; Chadwick, A. V.; Bruce, P. G. *J. Am. Chem. Soc.* 2008, 130, 5262–5266.

[145] Yan, H. W.; Blanford, C. F.; Holland, B. T.; Parent, M.; Smyrl, W. H.; Stein, A. *Adv. Mater.* 1999, 11, 1003–1006.

[146] Wei, T. Y.; Chen, C. H.; Chien, H. C.; Lu, S. Y.; Hu, C. *Adv. Mater.* 2010, 22, 347–351.

[147] Xiong, S.; Yuan, C.; Zhang, X.; Qian, Y. Cryst. *Eng. Comm.* 2011, 13:626–632.

[148] Wang, D.W.; Li, F.; Cheng, H.M. *J Power Sources.* 2008, 185, 1563–1568.

[149] Hou, Y.; Cheng, Y.W.; Hobson, T.; Liu, J. *Nano Lett.* 2010, 10, 2727–2733.

[150] Brezesinski, K.; Wang, J.; Haetge, J.; Reitz, C.; Steinmueller, S. O.; Tolbert, S. H.; Smarsly, B. M.; Dunn, B.; Brezesinski, T. *J. Am. Chem. Soc.* 2010, 132, 6982–6990.

[151] Han, Y.; Zhang, D. L.; Chang, L. L.; Sun, J. L.; Zhao, L.; Zou, X. D.; Ying, J. Y. *Nature Chem.* 2009, 1, 123–127.

[152] Chen, C.; Cai, W. M.; Long, M. C.; Zhang, J. Y.; Zhou, B. X.; Wu, Y. H.; Wu, D. Y. *J Hazard. Mater.* 2010, 178, 560–565.

[153] Hu, X. L.; Li, G. S.; Yu, J. C. *Langmuir.* 2010, 26, 3031–3039.

[154] Frey, S.; Keipert, S.; Chazalviel, J. N.; Ozanam, F.; Carstensen, J.; Foll, H. Phys. *Status. Solidi A.* 2007, 204, 1250–1254.

[155] Kresge, C. T.; Leonowicz, M. E.; Roth, W. J.; Vartuli, J. C.; Beck, J. S. *Nature* 1992, 359, 710–712.

[156] Zhang, Z. Y.; Zuo, F.; Feng, P. Y. J. *Mater. Chem.* 2010, 20, 2206–2212.

[157] Peterson, A. K.; Morgan, D. G.; Skrabalak, S. E. *Langmuir* 2010, 26, 8804–8809.

[158] Lou, X. W.; Deng, D.; Lee, J. Y.; Feng, J.; Archer, L. A. *Adv. Mater.* 2008, 20, 258–262.

[159] Xiong, S.; Yuan, C.; Zhang, X.; Qian, Y.; *Cryst Eng Comm.*2011, 13, 626-632.

[160] Pietruska, J. R.; Liu, X.; Smith, A.; McNeil, K.; Weston, P.; Zhitkovich, A.; Hurt, R.; Kane, A. B. *Toxicol Sci.* 2011, 124, 138-148.

[161] Ogami, Y.; Morimoto, T.; Myojo, T.; Oyabu, M.; Murakami, M.; Todoroki, K.; Nishi, C.; Kadoya, M.; Yamamoto, I.; Tanaka, *Inhal. Toxicol.* 2009, 21, 812-818.

[162] Zhong, H.; DeMarzo, A. M.; Laughner, E.; Lim, M.; Hilton, D. A.; Zagzag, D.; Buechler, P.; Isaacs, W. B.; Semenza, G. L.; Simons, J. W. *Cancer Res.*1999, 59, 5830-5835.

[163] Pietruska, J. R.; Johnston, T. J.; Zhikovich, A.; Kane, A. B. Environ. *Health Perspect.*2010, 118, 1707-1713.

[164] Pietruska, J. R.; Kane A. B. *Cancer Res.* 2007, 67, 3637-3645.

[165] Han, D.; Xu, P.; Jing, X.; Wang, J.; Yang, P.; Shen, Q.; Liu, J.; Song, D.; Gao, Z.; Zhang, M. J. *Power Sources.* 2013, 235, 45-53.

[166] Anjali, P.; Sonia, T. S.; Brahme, S.; Kim, T. N.; Nair, S. V.; Subramanian, K. R. V.; Balakrishnan, A. 2015, 618, 396-402.

[167] Sun, X. M.; Liu, J. F.; Li, Y. D. *Chem. Eur. J.* 2006, 12, 2039-2047.

[168] Varghese, M. V.; Reddy, Y. W.; Zhu, S.; Chang, T. C.; Hoong, G. V. ; SubbaRao, B. V. R.; Chowdari, A. T. S.; Wee, C. T.; Sow, C. H. *Chem. Mater.*2008, 20, 3360-3367.

[169] Wang, Y.; Zhu, Q. S.; Zhang, H. G. *Chem. Commun.* 2005, 20, 5231-5233.

[170] Wang, B.; Song, C. X.; Hu, Z. S.; Fu, X. J. *Phys. Chem. B* 2005, 109, 1125-1129.

[171] Zheng, J. P.; Cygan, P. J.; Zow, T. R. J. *Electrochem. Soc.* 1995, 142, 2699-2703.

[172] Liu, K. C.; Anderson, M. A. J. *Electrochem. Soc.*1996, 143, 124-130.

[173] Srinivasan, V.; Weidner, J. W. J. *Electrochem. Soc.*1997, 144, L210- L213.

[174] Ranjusha, R.; Ramakrishna, S.; Nair, A. S.; Anjali, P.; Vineeth, S.; Sonia, T. S.; *RSC Adv.* 2013, 3, 17492-17499.

[175] Ranjusha, R.; Prathibha, V.; Ramakrishna, S.; Nair, A. S.; Anjali, P.; Nair, S. V.; Avinash, B. *Scripta Materialia.* 2013, 68, 881-884.

[176] Ranjusha, R.; Sajesh, K. M.; Roshny, S.; Lakshmi, V.; Anjali, P.; Sonia, T. S.; Nair, S. V.; Avinash, B. *Microporous Mesoporous Mater.* 2014, 186, 30-36.

[177] Ranjusha, R.; Nair, A. S.; Ramakrishna, S.; Anjali, P.; Sujith, K. J. *Mater. Chem.* 2012, 22, 20465-2047.

[178] Roshny, S.; Ranjusha, R.; Deepak, M. S.; Rejinold, N. S.; Jayakumar, R.; Nair, S. V.; Avinash, B. *RSC Adv.* 2014, 4, 15863-15869.

In: Innovations in Nanomaterials
Editors: Al-N. Chowdhury, J. Shapter, A. B. Imran

ISBN: 978-1-63483-548-0
© 2015 Nova Science Publishers, Inc.

Chapter 15

ZnO Nanowires: An Excellent Nanomaterial for Next Generation Sensing Devices

Niranjan S. Ramgir*, S. Kumar, N. Datta, M. Kaur,
A. K. Debnath, D. K. Aswal and S. K. Gupta
Thin Films Devices Section, Technical Physics Division,
Bhabha Atomic Research Center, Mumbai, India

Abstract

In recent years, a great deal of research has been focused on the synthesis of metal oxides based nanomaterials owing to their superior and enhanced functional properties for realizing functional nanodevices. Among different nanostructures, nanowires (NWs) in particular, are looked upon as a promising candidate for realizing improved sensing response. They offer various advantages including high surface area-to-volume ratio, effective pathway for electron transfer, dimensions comparable to the extension of the surface charge region, enhanced and tunable surface reactivity, faster response kinetics, relatively simple preparation methods allowing large-scale production, ease of fabrication and manipulation, high integration density, and low power consumption. In order to harness the complete advantage of nano-dimension it is desirable to use single nanowire. However, associated problems namely sensor to sensor variation, complexity of the sensor fabrication approach and the in-built issue of randomness raise a major concern over the important parameters like reproducibility and repeatability. Use of NWs in thin film form wherein the average properties of multiple NWs is measured circumvents the above measured problems to a greater extend. Herein NWs can be selectively grown between the predefined electrodes or electrical contacts can be provided by depositing the electrodes with known dimensions on the NWs network itself. Accordingly, NWs based sensors in thin film form have been investigated widely for possible sensor device applications.

Among these, ZnO NWs in particular provide the advantages of ease of synthesis using physical/chemical processes, high thermal stability, excellent biocompatibility, and tunable morphology and surface reactivity. For sensor applications ZnO offers the advantages of flexibility of cost effective synthesis using various precursors, ease of

---

* E-mail: niranjanpr@yahoo.com; ramgirns@barc.gov.in. Tel: +91-022-2559 5839.

synthesis in conventional (thin and thick films) as well as nano forms and possibility of morphology control. Besides, the ability to manipulate the wide band gap provides the opportunity to tailor the response of ZnO NWs towards different gases. Accordingly, it is often modified with sensitizers like In, Pd, Pt, Au, and Cu etc, and has demonstrated sensors with improved response kinetics towards gases like C_2H_5OH, CO and H_2S. In the present chapter the ability to tailor the response of ZnO NWs towards different gases will be demonstrated citing some of our recent findings. As a case study for H_2S detection, the optimum temperature of maximum sensitivity was systematically brought down to room temperature using Au and CuO as sensitizers.

Modification with CuO resulted in a H_2S sensor with an enhanced response at 200°C in comparison to that of pure ZnO NWs that exhibited a maximum response at 300°C. Additionally, modification with Au resulted in a H_2S sensor with maximum response at room temperature. The enhanced and selective response in these cases were attributed to the formation of p-n junction (between p-CuO and n-type ZnO) and Schottky contacts (between Au and ZnO), respectively. Work function measurements using Kelvin probe method in the presence and absence of test gases further corroborated the formation of nano-Schottky type barriers and nano p-n junctions.

Keywords: ZnO, NWs, hydrothermal growth, vapor phase growth, chemiresistive gas sensors

1. INTRODUCTION

Recent advances in nanoscience and nanotechnology have resulted in a widespread investigation of various nanostructures like nanoparticles, nanorods, nanowires (NWs), nanobelts and nanoheterostructures. This has been attributed mainly to the improved functional properties coupled with the ease of synthesis. Of the different nanoforms being investigated NWs, in particular have been looked upon as a potential material for realizing sensors with improved response and response kinetics [1, 2, 3, 4]. They offer the advantage of

- high surface area-to-volume ratio,
- effective pathway for electron transfer,
- faster response kinetics,
- dimensions comparable to the extension of the surface charge region,
- enhanced and tunable surface reactivity,
- superior thermal stability,
- crystalline dislocation-defect free structure with precise chemical composition, surface and terminations,
- relatively simple preparation methods allowing large-scale production,
- ease of fabrication and manipulation,
- high integration density owing to smaller size,
- low power consumption, and
- low cost

Possible room temperature application and ultrahigh sensor response values are the main factors governing the research direction based on NWs. Accordingly, various metal and metal oxide NWs have been investigated for possible sensing application. Now in order to realize a complete commercially viable sensor based on NWs demands certain criteria to be met:

- Should satisfy the '4S' criteria-sensitivity, selectivity, stability and suitability
- Synthesis- process should be easy, low temperature based, reproducible, repeatable, economically viable and should have the scope for batch production
- Sensor fabrication process- fast, accurate, compatible with CMOS technology and ready for batch production
- Integration to microelectronics-both synthesis and sensor fabrication process should be compatible with the CMOS technology
- Reproducibility and repeatability- a challenge often faced by nanotechnologists for realizing commercial viability. The sensor developed should exhibit repeatable and reproducible response
- Long shelf life

In accordance to these, various organic and inorganic NWs of Cu, Pd, Pt, In_2O_3, CuO, SnO_2, ZnO, Fe_2O_3, TiO_2, polypyrrole, polyaniline, etc, have been explored. Among them ZnO, in particular has been looked upon as a promising material with all the above mentioned properties to realize an ultimate next generation gas sensors. In particular, it exhibits a wide band gap of 3.37 eV at 300 K, an exciton binding energy of 60 meV, large piezoelectronic constant, high thermal stability, tunable surface reactivity, biocompatibility and excellent optoelectronic properties [5, 6]. For sensor applications it offers the advantages of

- flexibility of cost effective synthesis using various precursors,
- ease of synthesis using physical/chemical processes,
- precise control over size and aspect ration,
- tunable surface reactivity,
- possible room temperature application [7]

All the above mentioned properties definitely make ZnO an excellent material for next generation sensing application. In the present chapter, reasons behind making ZnO an ideal sensing material has been demonstrated citing references from the literature and our recent works. The important outcome of the present chapter is to demonstrate that the ZnO NWs indeed form an excellent nanomaterials class with a tremendous potential for commercial sensing application.

2. GROWTH OF ZnO NWs

ZnO NWs have been synthesized generally using both physical (vapor phase) as well as chemical (hydrothermal) processes [8, 9]. Both the techniques have their own merits and demerits. However, the former has been more appealing due to the high crystal quality of the resulting NWs. On the other hand the latter has been employed as it involves low growth temperatures and has the high potential for the scale up coupled with the advantage of low cost [10, 11].

2.1. Vapor Phase Growth

The vapor phase growth is generally performed in a tubular furnace under controlled conditions of temperature, pressure, reactants, and carrier gas as predicted by respective phase diagrams. The source material either pure Zn metal or zinc oxide powder once evaporated is transported by a carrier gas (usually Ar + O_2 mixture) towards the growth site where it nucleates. The deposition technique is very simple and cheap, and the size and shape can easily be controlled. The resulting NWs are highly pure, structurally uniform, and single crystalline. The growth often involves the use of catalyst to control size and aspect ratio of resulting NWs [12]. And depending on the presence of a catalyst, there are different growth mechanisms, i.e., vapor–liquid–solid (VLS), solution–liquid–solid (SLS), vapor-solid-solid (VSS) or vapor–solid (VS) process [13, 6]. The controlled pressure of the inert atmosphere and the gradient of temperature within the furnace allow condensation and nucleation of the nanostructures downstream the gas flow. Such a peculiar thermodynamic condition promotes formation of NWs.

2.1.1. Vapor–Liquid–Solid Mechanism

The controlled catalytic growth of whiskers, and NWs, was discovered and demonstrated by Wagner and Ellis in 1964 [14]. They have named the mechanism as VLS mechanism for the three phases that are involved: the vapor phase precursor, the liquid catalyst droplet, and the solid crystalline product. In most of the catalytic growths, NWs have uniform diameters. Although it is commonly believed that in the VLS process, the size of the catalyst particles determines the NWs width this is not true for all the deposition conditions. Experimental studies on ZnO NWs growth on $Al_{0.5}Ga_{0.5}N$ substrate confirm that this rule only applies when the catalyst particles are reasonably small (<40 nm) [15]. NWs grown using this method are characterized by the presence of catalyst at the tip.

As shown in Figure 1 (a, b) ZnO NWs grown using VLS are characterized by the presence of Au on the tip of the NWs. And accordingly, formation of liquid Zn and liquid Au-Zn alloy are mandatory for the VLS process. NWs production by simple oxidation of the metal composing the metal oxide has also been reported [16, 17, 18]. The VLS growth in general produces NWs of the best crystalline quality.

2.1.2. Vapor–Solid Mechanism

The VS growth takes place when the NW crystallization originates from the direct condensation from the vapor phase without the use of a catalyst (Figure 1 c, d).

This is the most commonly used method for the growth of NWs generally referred to as 'carbothermal –reduction' method wherein the ZnO NWs are grown as per VS mechanism. At the beginning the growth was attributed to the presence of lattice defects, but when defects-free NWs were observed this explanation could not be accepted any longer. Another peculiar effect registered was NW growth rate was higher than the calculated condensation rate from the vapor phase.

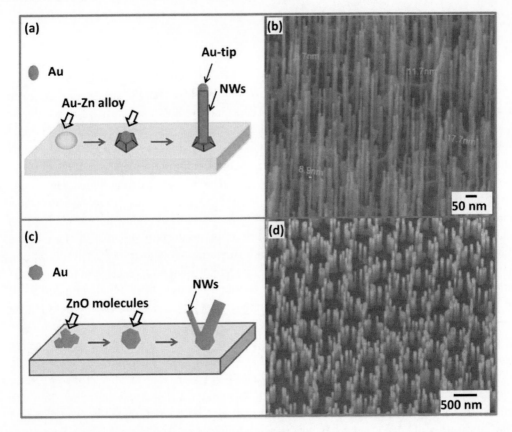

Figure 1. Shows the schematic representation and corresponding SEM images of ZnO NWs grown using (a, b) VLS and (c, d) VS mechanism..

A possible interpretation is that all the faces of the NW adsorb the molecules that afterwards diffuse on the principal growth surface of the wire. VS process occurs in many catalyst-free growth processes. Quite a few experimental and theoretical works have proposed that the minimization of surface free energy primarily governs the VS process. Under high temperature condition, source materials are vaporized and then directly condensed on the substrate placed in the low temperature region. Once the condensation process happens, the initially condensed molecules form seed crystals serving as the nucleation sites. As a result, they facilitate directional growth to minimize the surface energy.

The ZnO NWs growth using vapor phase technique is highly sensitive towards the process parameters. These include starting material, catalyst, substrates used, tube geometry, deposition temperature, duration, carrier gas and its flow rate. It was recently demonstrated that controlling the gas and reactant flow kinetics inside the furnace using deposition temperature, profile of the furnace and the flow rate of gases it is possible to get NWs growth at a desired position inside the furnace. Moreover, controlling the starting materials it was also shown to control the quality of resulting NWs. In particular, employing ionic liquid as a carbon source the growth mode was reversibly switched from VLS to VS mechanism [6].

2.2. Hydrothermal Growth of ZnO NWs

Besides, vapour phase deposition the hydrothermal approach has been investigated, and employed widely for the growth of ZnO NWs [8]. The simple chemistry involved, the minimum cost and the instrument required is the major reason for its widespread usage. In this method, ZnO NWs are grown in two steps [19]. First the ZnO nanoparticles as seed layer were deposited on to the substrate and later the substrate is subjected to the hydrothermal growth. In brief, ZnO NPs were synthesized via chemical route by adding 10 mM NaOH solution in methanol dropwise to the 30 mM zinc acetate solution in methanol at 60°C under rigorous stirring. The stirring was continued further for ~120 minutes. This led to the formation of NPs ranging from 5 - 15 nm. The resulting nanoparticles in solution are stable upto two weeks. After two weeks nanoparticles were found to agglomerate, grow bigger in dimensions (< 100-500 nm) and accordingly, were observed to settle down at the base of the container. Nanoparticles were subsequently spin coated onto the substrates Si/SiO$_2$ (100 nm) for seed layer deposition. Chemical route provides the advantage of precise control over the particle sizes. For example, particles of size 5, 10 and 15 nm could be obtained using 5, 10 and 15 mM solution of zinc acetate dihydrate along with 30, 30 and 45 mM solution of NaOH at growth temperatures of 60, 65 and 75°C, respectively.

Hydrothermal growth was achieved by suspending the seed layer coated wafers upside down over an aqueous equimolar (25 mM) solution of zinc nitrate hexahydrate, Zn(NO$_3$)$_2$.6H$_2$O, and hexamethylenetetramine (HMTA), (CH$_2$)$_6$N$_4$, at 90°C for 7 h. After growth, the wafers were rinsed thoroughly with de-ionized water, dried under Ar flow and used for gas sensing studies and further characterization. NWs were found to be well adherent to the substrate. ZnO NWs were having diameter from 50 to 200 nm and length between 6 and 10 μm. Length of the NWs was found to increase with growth time. Additionally, growth of NWs onto the substrate is highly reproducible and repeatable under identical conditions of temperature and solution concentrations. The results of NWs grown on substrates without seed particles and with seed particles for different growth time are shown in Figure 2. With seed layer NWs grow uniformly over the substrates and have lower diameter of ~50-100 nm as compared to that of NWs grown using seed layer which exhibited ~500 nm diameter. Length of NWs was found to depend strongly on the growth duration with length increasing with time.

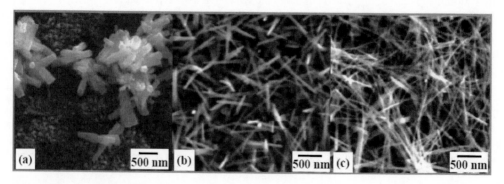

Figure 2. SEM images of NW-films grown on silicon substrates: (a) without seed particles and with seed particles for (b) 6 h and (c) 21 h growth times. Reprinted from publication Sens. Actuators B 2011, 156, 875-880, Copyright (2011), with permission from Elsevier.

Growth mechanism of NWs in hydrothermal method can be understood considering the salvation of Zn ions in water, pH of the solution and the growth temperatures. In aqueous solution, salvation of zinc(II) by water gives rise to various aquo ions including $ZnOH^+_{(aq)}$, $Zn\text{-}(OH)_{2(aq)}$, $Zn(OH)_{2(s)}$, $Zn(OH)_{3\text{-}(aq)}$, and $Zn(OH)_4^{2-}{}_{(aq)}$ [20]. For a given concentration, the stability of these complexes is governed by the pH and the growth temperature. Dehydration of these hydroxyl species results in the formation of solid ZnO nuclei. And the crystal grows continuously by the condensation of the surface hydroxyl groups with the zinc-hydroxyl complexes. These hydrolysis and condensation reactions of zinc salts result in one-dimensional crystals. Hexamine (HMTA) which is a nonionic cyclic tertiary amine that can act as a Lewis base to metal ions and also as a bidentate ligand capable of bridging two zinc(II) ions in solution is used very often to promote one-dimensional ZnO precipitation [21]. In particular, it functions in part, by decomposing during the reaction and increasing the pH to above ~9 at the crystal surface. It is also known to hydrolyze, producing formaldehyde and ammonia in the pH and temperature range of the ZnO NW reaction. It acts as a pH buffer by slowly decomposing to provide a gradual and controlled supply of ammonia, which can form ammonium hydroxide as well as complex zinc(II) to form $Zn(NH_3)_4^{2+}$ [22]. Because dehydration of the zinc hydroxide intermediates controls the growth of ZnO, the slow release of hydroxide may have a profound effect on the kinetics of the reaction. Additionally, ligands such as HMTA and ammonia can kinetically control species in solution by coordinating to zinc(II) and keeping the free zinc ion concentration low. Additionally, they can also coordinate to the ZnO crystal, hindering the growth of certain surfaces. However, the exact role of HMTA in NW growth is still under a debate and demands a detailed investigation [23].

2.3. Other Growth Methods

2.2.1. Template-Assisted Synthesis

Anodization growth technique is a well-established process to grow porous material, i.e., anodic aluminium oxide (AAO), used for the deposition of 1-D NWs into the pores [24, 25, 26, 27] as shown in Figure 3. The advantages of anodization and electroplating processes for 1D nanostructures production are low costs, repeatability and potential compatibility with silicon technologies which make these nanostructure synthesis procedures interesting.

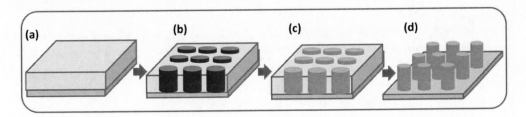

Figure 3. Schematic representation of template-assisted growth of NWs using AAO template. (a) Aluminium (Al) film coating on substrate, (b) anodization of Al resulting in AAO template with uniform distribution of pores, (c) NWs growth in the pores and (d) free standing NWs upon template removal.

Control in NW dimensions and the morphology of the ordered arrays can be achieved. Because the diameter of these nano-channels and the inter-channel distance are easily controlled by the anodization voltage, it provides a convenient way to manipulate the aspect ratio and the area density of nanostructures.

Other periodic structured templates, such as molecular sieves, and polymer membranes, have also been found significant for the growth of NWs [28]. NWs can be deposited into the pores using electro-deposition or sol–gel deposition methods [29]. The pore dimensions govern the diameter of the NWs while the time duration of the deposition governs the length. Subsequent to fabrication, the resultant NWs can be released from the templates by selective removal of the host matrix. The template methods are generally low-cost, scalable to mass production, and environmentally benign. However, the polycrystalline nature of most NWs obtained by this route somehow restricted their applications to sensors. Besides, after the removal of template, the embedded arrays of NWs with a high aspect ratio are found to normally collapse into an entangled mess due to the surface tension force exerted on NWs during the evaporation of the liquid [30]. Recently, graphene with a unique 2-D structure composed of planes of honeycomb carbon lattice (ZnO and RuO_2) [31], channel diffused plasma-etched self-assembled monolayer templates, [32] have also been investigated as a novel template material for NWs growth.

2.4. Control over Size, Aspect Ratio and Distribution

Using vapour phase deposition it was demonstrated to have a control over the size and aspect ratio of resulting NWs. In VLS method, the size and distribution of Au over the substrate is crucial to have a control over the size and position control. Using template deposition approach it was demonstrated to have a control over the size of Au seed layer. Prof. Zacharias group demonstrated that using nano sphere lithography (NSL) i.e., polystyrene sphere (PSS) (single or double layer) it is possible to have a hexagonal or triangular distribution of Au seeds over the surface of the substrate [33]. Varying the deposition time of Au it is possible to have a control over the size of the Au particle. Similarly, for VS grown the Au layer acts a defect site for ZnO adsorption and thus is demonstrated to control the site of NWs growth. All the above mentioned points clearly indicates that using vapour phase deposition and controlling the seed layer distribution over the substrate it is possible to have a precise control of ZnO NWs size, aspect ratio and position on the substrate.

Now, hydrothermal growth has been demonstrated to result in uniform and dense growth of NWs over the substrates. Using conventional photolithography it is possible to confine or define the region for seed layer deposition. This can be effectively used to get a control over the distribution of NWs over the substrate. Yuon et al. demonstrated control over the ZnO distribution over the substrate using topographical confinement and preferential chemisorptions as shown schematically in the Figure 4, [34].

In particular, ZnO seed layer was confined in the channel govern by the photoresist pattern on the substrate and later the substrate was subjected to the NWs growth. For this both vapour phase as well as hydrothermal growth methods can be utilized for the NWs growth.

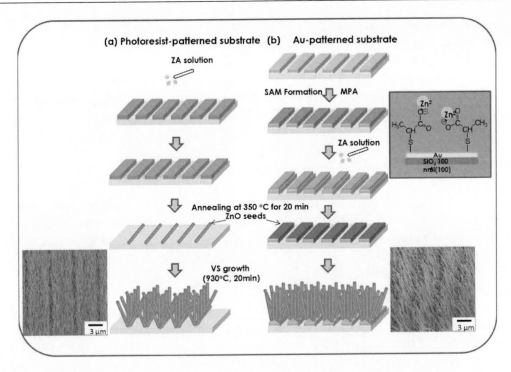

Figure 4. Schematic of the growth of ZnO NWs using (a) topographical confinement and (b) preferential chemisorptions. Reprinted from publication J. Phys. Chem. C 2010, 114, 10092–10100, Copyright (2010), with permission from American Chemical Society.

2.5. Industrial Scale Production Compatible with Complementary Metal Oxide Semiconductor (CMOS) Technology

Of the different techniques employed for the ZnO NW growth, hydrothermal in particular is looked upon as a potential technique for industrial scale production. This is owing to the advantages of low growth temperatures, simple chemistry involved, no requirement of complex or Hi-tech instruments and the low cost of precursors or the process. Besides, it has been demonstrated to result in uniform and dense growth of NWs over the Si wafers as big as 2 or 4 inch in diameter. This is the dimensional scale which is generally employed for the industrial CMOS processes. Additionally, the Si substrate and the other chemicals required for the hydrothermal growth pose no harm to the CMOS processes or are the one commonly employed in fabrication processes. And once the NWs are grown they can be rinsed with the common solvents like water, ethanol, isopropanal commonly used for CMOS processes and the whole wafer can be subjected to the CMOS processes required to realise various device applications.

3. SENSOR FABRICATION METHODS

This forms an important step from the commercial view point. The sensor fabrication process involves various steps; NW deposition, providing electrical contacts, heater arrangement, and integration into the measurement circuits. ZnO NWs are used as either chemiresistive sensors or in field effect transistor (FET) configuration for the sensing purpose. Of the two methods chemiresistive sensors working on the principle of change in resistance upon chemical interaction with the test gas is widely exploited. For this NWs are generally being studied or employed in three different configurations; single, multiple or thin film form. To realise a sensor based on the NWs following approaches are used.

3.1. Pick And Place Approach: Single-or Multiple NWs

In this approach the NWs are generally stripped off from the substrate. NWs in the powder form are used as the starting material. They are dispersed in the suitable solvents and then spin coated or drop casted on the substrates. Electrode contacts are then realised using different means.

3.1.1. Dielectrophoresis

In this method NWs are aligned between the predefined electrodes under the application of an electric field of ~15 Vpp at 10 MHz between the electrodes [35]. Consequently, NWs are trapped and aligned along the electric field lines bridging the electrode gap where the electric field becomes higher. Using this approach it is possible to have a sensor device based on single and multiple NWs. Although, the approach is widely used owing to its simplicity it suffers from the inherent drawback of reproducibility. It is very difficult and unpredictable to have a similar contact every time. It is also difficult to control the number density of the NWs.

3.1.2. Direct Deposition over the Single or Multiple NWs Using Focused Ion Beam Technique

Focused ion beam technique provides the advantage of precise control over the deposition process contact area and the number density [36]. In this method NW suspension is prepared first and then a small amount is drop-casted over the substrate so as to have a scarce distribution over the surface. NWs are then located under the microscope (SEM) and contact to the ends of the NWs is provided with the help of focused ion beam. The samples are then subjected to annealing for better adherence. Although, the method results in good Ohmic contact between the NWs and the electrode material, it however suffers from various drawbacks. These include high processing time, requires complex and hi-tech instruments, limited reproducibility attributed to the random distribution of NWs.

Both single and multiple NWs based sensors have not pick-up the interest as there are still issues that need to be answered. One of the important limitations is the sample to sample variation arising due to poor control over the NWs aspect ratio and density.

A small change in aspect ratio results in a drastic variation in the electronic properties and accordingly demands a proper averaging of the performance between different sensor devices. Besides, the complexity and the randomness of the device fabrication approach raise the important concern over the reproducibility and repeatability of the sensors.

Table 1. Chemiresistive sensors based on ZnO NWs

Sensor Material	NW case	Target	OT (°C)	Lowest conc. (ppm)	Sensor response (S or S%)	D (nm)	Ref
ZnO	Single	H$_2$	RT	500	55%	200	[38]
ZnO	Single	H$_2$	RT	100	34%	100	[39]
ZnO	Branched	NO$_2$	300	1	26	<100	[40]
ZnO	TF	CO	RT	0.5	126%	100-200	[41]
ZnO	TF	C$_2$H$_5$OH	400	859	2	70	[42]
ZnO	TF	C$_2$H$_5$OH	250	25	10	40-60	[43]
ZnO	TF	O$_2$	50	300 ml/min	10	70-80	[44]
ZnO	TF	Ethanol	RT	40	2	150	[45]
ZnO-Pt	Single	H$_2$	RT	500	20%	30	[46]
ZnO-He$^+$	Multiple	H$_2$S	RT	100	7	400	[47]
ZnO	Multiple	H$_2$S	RT	4	25%	100–200	[48]
ZnO	Multiple	NO$_2$	225	0.5	12	-	[49]
ZnO-In	Multiple	C$_2$H$_5$OH	5 V	1	3	60–150	[50]
ZnO-Pd	Multiple	C$_2$H$_5$OH	170	5	8%	50	[51]
ZnO-Pd	TF	CO	RT	0.1	1.02	20-50	[52]
ZnO-Au	Multiple	CO	350	5	30%	100	[53]
ZnO-Au	Thick Film	C$_2$H$_5$OH	380	2	<10	30-50	[54]
ZnO-Pt	TF	NH$_3$	300	5	2.88%	-	[55]
ZnO	TF	NO$_2$	300	10	263%	20-100	[56]
ZnO	TF	C$_2$H$_5$OH	300	1	1.9	25	[57]
ZnO-CuO	TF	H$_2$S	200	0.5	1.4	50-200	[54]
ZnO	TF	methanol	RT	0.5%	17	100	[58]
ZnO-Au	Pellet	C$_2$H$_5$OH	240	100	5.5	60-180	[59]
ZnSnO$_3$	Single	O$_2$	RT	1.0 x 10^{-4} Pa	6 orders	60	[60]
ZnSnO$_3$	Multiple	C$_2$H$_5$OH	RT	1	2.7	20–90	[61]
ZnO	TF	NO$_2$	250	0.5	105%	-	[62]
ZnO-LaOCl	Thick film	CO	400	10	**1.6**	50-150	[63]
ZnO-Co	TF	p-Xylene	400	5	**19.55**	-	[64]
ZnO-Graphene	Single	Ethanol	RT	50 sccm	**8**	<200	[65]

*TF: Thin film, S% = 100 × (Rair − Rgas)/Rgas or [(Igas−Io)/Io]×100%, S = Rg/Ra.

NWs grown on the predefined electrodes provide a good adherence and the conductance properties. They have also been shown to demonstrate the improved response characteristics as compared to the NWs films on which contacts are realized later [34].

3.2. NW Thin Films As Sensing Material

Use of NW thin film circumvents the problems associated with the pick and place approach. There is no need to have a position and site control as the average properties of multiple NWs is measured. Herein NWs can be selectively grown between the predefined electrodes or electrical contacts can be provided on top of the NWs network.

This method removes the complexity involved towards a longer extend [37], is easy and assures reproducible and repeatable measurements. Owing to the advantages offered and the simplicity involved large number of sensors has been realized using the present approach as also mentioned in table 1.

4. IMPROVEMENT OF SENSING PROPERTIES OF SENSORS BASED ON ZNO NWS

Different intrinsic ZnO NWs (all of them n-type) have been synthesized and demonstrated for application as sensors for H_2,[66] C_2H_5OH, [67] NO_2 [68] and H_2S gases [69, 70]. In particular, for H_2S gas, Wang et al. have reported a room temperature sensitivity of 1.7 towards 0.05 ppm H_2S [18]. However, the sensors have large response time (~25 min). This could be attributed to the use of binder and the room temperature operation. Liao et al. have also reported a sensitivity ($S = R_a/R_g$) of nearly 2.5 for detection of 100 ppm of H_2S gas at 100°C [19]. The lower sensitivity could be due to a sandwich type configuration, where NWs were sandwiched between two electrodes namely Si and Cu that may result in slow diffusion of gas. For gas sensors it is desirable that no binders are used to permit fast diffusion of gas and adherence of NWs to the substrate should be good. In order to improve the sensor response and response kinetics towards a particular gas different methods have been utilized, these include;

4.1. Surface Modification with Sensitizers

Zinc oxide (ZnO) NWs have been extensively researched for the gas sensing activity owing to their excellent optoelectronic, thermal and chemical properties [71, 72, 73, 74]. Similar to other semiconducting oxides, it also suffers from the inherent drawback of poor selectivity or cross sensor response. In order to improve selectivity various strategies like incorporation of dopants, use of binary compounds and surface modification using different sensitizers like Au, Pd, Pt and metal oxides (Cu, Fe, Co) have been employed [75, 76]. Sensitizers, in particular furnishes additional absorption sites and surface electronic states by providing catalytic sites for enhanced surface chemical reactions like chemisorptions, oxidation and reduction, which mediate electronic transfer processes [77]. Here, the size, nature, oxidation state and surface distribution of the sensitizers are determining factors in governing the sensor response [78]. For instance, McAleer et al. found that the optimal responses occur when the cluster sizes are between 1 and 5 nm [79]. An optimal distance between clusters can be estimated as equal to an oxygen surface diffusion length at operating temperatures [80].

4.1.1. Modification with CuO Forming Random Distribution of Nano P-N Junctions

CuO is intrinsically p-type in nature which is attributed to the presence of Cu^+ vacancies. ZnO on the other hand shows both p-type as well as n-type nature and is a current matter of debate. The n-type nature is usually attributed to the oxygen vacancies and/or native defects like H [81, 82] while the p-type nature is attributed to the zinc vacancy, surface acceptor levels created by the adsorbed oxygen and/or the unintentional carbon doping [83, 84, 85]. In the case of carbon doping, carbon immobilizes the oxygen in the interstitial site forming a carbon-oxygen cluster defect that acts as a shallow acceptor [86]. However, it is difficult to predict under what conditions ZnO would exhibit n-type or p-type nature. ZnO nanoparticles used as a seed layer for the NWs growth have exhibited a p-type nature attributed to the zinc acetate dehydrate, a starting material, that induces Zn vacancies [29, 87]. NWs, on the other hand, exhibited n-type nature owing to oxygen vacancies probably arising from the precursor zinc nitrate hexahydrate.

Thus, modification of n-type ZnO with p-type CuO would result in the formation of randomly distributed nano p-n junctions over the surface as also observed experimentally [88, 89]. CuO modification was carried out by depositing an ultrathin layer of Cu (10 nm) using Cu wire (99.9% purity). Samples were subjected to thermal oxidation at 400°C for 1 h thereby allowing conversion of Cu into p-type CuO phase. Thermal oxidation of Cu at temperatures >270°C leads to the formation of pure CuO phase [90, 91]. In order to confirm that the Cu is indeed present in the CuO form XPS measurements over the sensor films were performed. Figure 5 (a, b) shows the O1s spectra for both pure and CuO modified ZnO NWs. For pure ZnO, two peaks at BE values of 530.5 and 531.6 eV corresponding to O 1s and adsorbed oxygen species, respectively were observed. Upon CuO modification, O1s peaks shifted to lower BE values of 529.9 and 531 eV. The peak at 529.9 is mainly attributed to the O 1s corresponding to CuO [92].

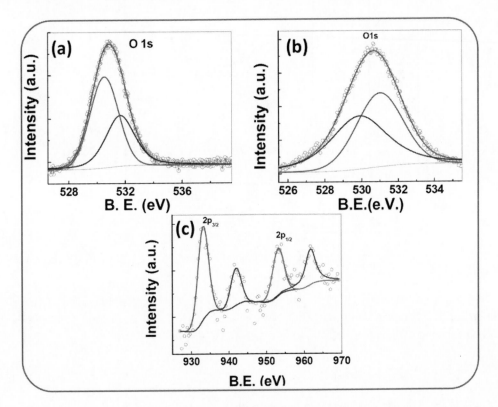

Figure 5. O1s spectra for both (a) pure and (b) CuO modified ZnO NWs, and (c) Cu 2p peaks for CuO modified ZnO NWs. Part of the figure reprinted from publication Sens. Actuators B 2014, 202, 1270-1280, Copyright (2014), with permission from Elsevier.

Additionally, for CuO modified ZnO NWs the Cu 2p peaks can be fitted and deconvoluted into individual four peaks at binding energies of 933.2, 941.9, 953.2 and 962 eV (Figure 5 (c, d)). The two peaks located at 933.2 and 953.2 eV correspond to the Cu $2p_{3/2}$ and $2p_{1/2}$ core levels, respectively [93]. Shake up features at 941.9 and 961.7 eV for the Cu

$2p_{3/2}$ and $2p_{1/2}$ core levels are evident and diagnostic of an open $3d^9$ shell corresponding to Cu^{2+} state [94, 95]. The presence of satellite peaks confirms the formation of CuO phase [96]. These studies clearly indicate that Cu indeed exists in Cu^{2+} state in the form of CuO over ZnO random NWs network.

Now, CuO interacts with H_2S selectively forming CuS, a degenerate semiconductor with metallic type conductance behavior. Accordingly, CuO modified ZnO NWs have exhibited an excellent sensitivity and selectivity towards H_2S at an operating temperature of 200°C [8].

Figure 6 shows the typical response curve recorded for CuO modified ZnO random NWs network towards 5 ppm of H_2S. The sensor films exhibited a typical response and recovery times of 6 and 30 min, respectively. The response and recovery times (t_{90}) were defined as the time required by the sensor to reach to 10% and 90% of its saturation and original values upon exposure to test gas and fresh air, respectively.

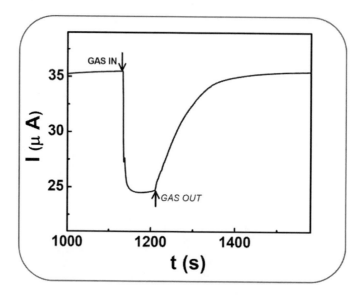

Figure 6. A typical response curve recorded for CuO modified ZnO random NWs network towards 5 ppm of H_2S.

The improved sensing characteristic has been attributed to the response of the p-n junction formed between CuO and ZnO-NWs towards H_2S. Upon exposure, CuO interacts with H_2S forming CuS as per the following reaction: [97]

$$CuO + H_2S \rightarrow CuS + H_2O \qquad (1)$$

CuS is a degenerated semiconductor with a metallic conductance behavior and accordingly, destroys the n-p-n junction as well as the potential barrier. This causes release of large number of electrons to the host matrix. In other words the destruction of potential barrier results in a sharp decrease in the electrical resistance and consequently in sensitivity. During recovery, CuS reacts with the oxygen forming CuO rejuvenating the surface via the following reaction:

$$2CuS + 3O_2 \rightarrow 2CuO + 2SO_2 \qquad (2)$$

Since CuS converts back to CuO, potential barrier reappears due to the formation of n-p-n heterojunctions, and original high resistance is regained. This model explains well the H_2S sensing mechanism CuO modified ZnO-NWs random networks. The chemical reaction between CuO and H_2S also explains why CuO modified ZnO-NWs random networks are very specific towards H_2S.

As has been described previously, the sensing layer consists of interconnected random NWs network. The sensor film comprises of NWs, NWs junction (NWs-CuO-NWs) and the metal-NWs contact resistance. Impedance measurement clearly indicated that the bulk contribution (R_b) and NWs-electrode interface contribution (R_c and Q_j) remain unaffected by the ambient atmosphere. And upon exposure to H_2S the dominating factor governing the decrease in resistance could be attributed to the decrease in band bending as a result of CuS formation.

Modification with CuO results in the formation of nano p-n junction between p-type CuO and n-type ZnO distribution randomly over the ZnO NWs surface. Because of this the NWs are electron depleted and hence are expected to have a higher work function values. The complex surface morphology of NWs renders the conductivity changes at the surface difficult to measure or interpret. Gas sensing is a surface phenomenon and so is the work function, accordingly, the change in work function could also provide the evidence or information about the interaction between the surface and the gas [98, 99]. Work function (ϕ) is defined as the energy difference between the Fermi level (E_F) and the vacuum level (E_{vac}). Figure 7 shows the schematic of the set-up used for the Kelvin-probe measurements [100]. ZnO NWs deposited onto a conducting Au coated (Cr/Au: 10/120 nm) glass substrates were used for this purpose.

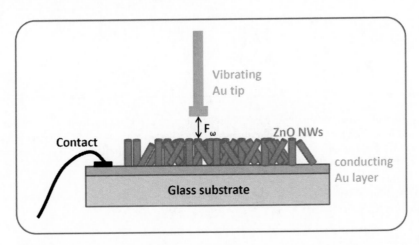

Figure 7. Schematic of the set-up used for the Kelvin-probe measurements. Reprinted from publication Sens. Actuators B 2013, 186, 718-726, Copyright (2013), with permission from Elsevier.

Figure 8 shows the work function area map recorded for both pure and CuO modified sensor films. Interestingly, the average work function values were found to be 5.06 and 5.36 eV for pure and CuO modified ZnO, respectively. This increase in work function by ~0.2 eV upon CuO modification further supports the formation of p–n junctions over the surface of the sensor films and thus leads to the upward band bending. Figure 8 shows the change in work function upon exposure to reducing (H_2S: 50 ppm) and oxidising gases (Cl_2: 5 ppm) for pure

(Figure 8 (b)) and CuO modified ZnO sensor films, (Figure 8 (c)) respectively. The average work function values before and after the exposures have been tabulated in Table 2. As expected, exposure to H_2S and Cl_2 causes the work function to decrease and increase, respectively. In case of H_2S exposure, the change in work function value is attributed mainly to the interaction of CuO with H_2S leading to CuS formation.

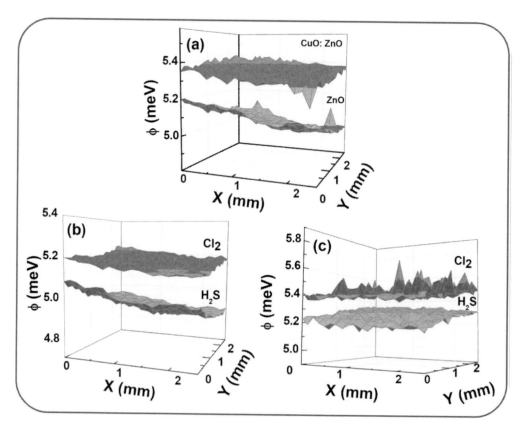

Figure 8. (a) The work function area map recorded for both pure and CuO modified sensor films. Change in work function upon exposure to reducing (H_2S: 50 ppm) and oxidizing gases (Cl_2: 5 ppm) for (b) pure and (c) CuO modified ZnO sensor films. Reprinted from publication Sens. Actuators B 2014, 202, 1270-1280, Copyright (2014), with permission from Elsevier.

The work function of the material, has two components that can be influenced by the surface reactions; (relation (13))

$$\phi = qVs + \chi \tag{3}$$

$$\Delta\phi = \pm q\Delta Vs \pm \Delta\chi \tag{4}$$

where, qVs is the surface band bending and χ represents the electron affinity which is the measure of surface dipole moment. The $q\Delta Vs$ can be directly calculated for n-type semiconductors using:

$$q\Delta Vs = -kT \ln \left(\frac{G_{gas}}{G_{air}}\right) \qquad (5)$$

where, G_{gas} and G_{air} are the values of conductance in the presence of gas and ambient air, respectively, which can be experimentally determined. Band bending contribution to the work function calculated using equation (5) is -75 and -135 meV (for 20 ppm H_2S) for pure and CuO modified sensor films, respectively. The negative sign indicates the decrease in upward band bending after H_2S exposure. After subtracting the band bending (equation (4)) we arrive at a dipole induced increase in work function of 195 meV and 225 meV, respectively. This indicates a surface dipole layer with positive side towards the vacuum. The possible factors affecting the surface dipole layer are charge transfer across the interface, rearrangement of electron cloud at the semiconductor surface in presence of adsorbate and also existence of interface state serving as a buffer of charge carriers [101]. ZnS formation on ZnO nanostructures upon reaction with H_2S has been reported, [102] which might result in charge transfer from ZnS to ZnO due to varying electronegativities of O and S and hence increasing the dipole contribution to the work function.

4.1.2. Modification with Au Forming Random Nanoschottky Barriers

A randomly distributed p-n junction formed at the interface between p-type CuO and n-type ZnO along with the elite interaction between CuO and H_2S forming metallic CuS were the key towards achieving high sensor response. Analogously, modification with metal nanoparticles is anticipated to result in the formation of Schottky type barrier junctions between metal Au and semiconductor ZnO. And could be effectively exploited to tailor the sensors response or realize improved sensing characteristics. For metal-semiconductor (n-type) junction, a Schottky barrier is formed when the work function of metal (ϕ_M) is greater than the electron affinity (χ) of semiconductor. The barrier (ϕ_B) height thus formed is given as

$$\phi_B = \phi_M - \chi \qquad (6)$$

As Au has ϕ_M of 5.1 eV and ZnO has an affinity of ~ 2.08 eV, modification with Au will result in randomly distributed Schottky junction at the interface. Recently, an improvement in ethanol response of ZnO NWs ($\chi = 2.08\ eV$) upon Au modification attributed to the nano-Schottky type barrier junctions has been reported [103, 104]. Modification with Au has been demonstrated to result in a sensor with improved sensor characteristics towards CO (sensor response S = 53% [$(R_a-R_b)/R_a$] × 100%], 100 ppm, 350°C) [105] and ethanol (sensor response S = 89.5 [R_a/R_g], 100 ppm, 300°C) [81]. Improvement in sensing properties towards CO has been attributed mainly to the interaction of CO gas with surface adsorbed oxygen species. While for ethanol it has been assigned to the surface depletion controlled sensing mechanism. More specifically, modification with Au nanoparticles resulted in nano-Schottky barriers like junctions on the surface of ZnO, that can be modulated by adsorption and desorption of target gases. The sensor in this case has been realized using drop casting method and thus probably would suffer from the drawback of sample to sample variation. NWs grown on a substrate would be advantageous to overcome this drawback. NWs films modified with thin layer of Au have been found to exhibit enhanced sensing characteristics towards ethanol with maximum sensor response at 325°C. The response and recovery times for Au modified (10 nm) sensor films towards 50 ppm of ethanol at 325°C were 5 and 20 s, respectively.

Similar to other semiconducting oxides, ZnO surface is also characterized by the presence of adsorbed oxygen species (O_2^-, O^-, O^{2-}). At 325°C, the adsorbed oxygen species present on the surface are mainly O^{2-}. These species interact with the ethanol gas as per the reaction:

$$CH_3CH_2OH + O^{2-}_{(ads)} \rightarrow C_2H_4O + H_2O + 2e^- \tag{7}$$

thereby causing release of electrons into the host ZnO matrix. The net result is the decrease in the resistance of the sensor film. The faster reaction kinetics is attributed to the catalytic role of Au in imparting the sensor response mainly via electronic sensitization mechanism. In such mechanism the additives enhances the rate of reaction by providing the additional active sites. Besides, the interaction with the test gas occurs mainly on the Au surface and the corresponding changes are transferred immediately to the host matrix reflected as a fast drop in the resistance of the sensor film. Deposition of thin Au layer (10 nm) would result in a discontinuous film or island formation on the NWs surface. These islands are randomly distributed along the length of NWs and could possibly lead to the formation of nano-Schottky type barriers or junction between metallic Au and semiconducting ZnO [13]. Presence of Au promotes the catalytic dissociation of molecular oxygen species. The potential barrier thus formed would result in the electron depleted NWs surface. This is the reason why the resistance of Au modified NWs is higher than the pristine NWs. Now, the barrier properties can easily be altered by the adsorption or desorption of adsorbed species or target gas molecules.

Besides improvement in the ethanol sensing properties sensor modified with Au also exhibited a remarkable enhancement in the response towards H_2S and that too at room temperature. Figure 9 shows the room temperature (25°C) response curves for pure and Au modified (1.2 at.%) ZnO NWs towards 1 and 5 ppm of H_2S, respectively. The sensor response of Au modified sample towards 1 and 5 ppm, were 38 and 79.4, respectively as compared to that of pure ZnO NWs for which the values were 2 and 5, respectively. Modification with Au also resulted in the improvement of the recovery time. It being 860 and 170 s for pure and Au modified samples towards 5 ppm H_2S, respectively.

Further, in order to confirm the resistance increase or the nano-Schottky junction barrier formation the work function measurements were performed using Kelvin probe method also known as Kelvin probe force microscopy (KPFM). The interaction between the conductive tip and sensors surface allows mapping of local electric charge and surface potential distribution. Figure 8 shows the work function area scan recorded for 4 mm^2 of sensor samples. The color variation indicates the localized variation in the CPD.

The final work function values of the sensor samples calculated taking the average of the relative variation in the CPD are 5.085 and 5.305 eV, for pure and Au modified ZnO sensor samples, respectively. As is clearly evident the incorporation of Au in the ZnO resulted in a work function increase by 0.22 eV. Since work function of the material is defined as the energy required to set Fermi electron free, an increase in work function further corroborates the observation of increase in resistance for Au modified ZnO NWs. The difference between the work function of Au (5.1) and ZnO (5.3 eV) is ~0.2 eV. The barrier height thus formed will contribute at lower thickness of the material (ZnO). The ZnO NWs grown over the surface of Au were having the length in the 1-2 μm range. At this length scale there is no contribution from the underlying Au layer for work function measurements. Therefore, we expect the similar behavior if different contacts (Ohmic or Schottky type) were used for the measurements. It is important to note that CPDs are highly material-dependent and related to the work functions of pure material and to additional surface dipole moments. The electrical potential ϕ can be separated into two components, namely, ψ, which is due to the presence of an electrostatic charge on the surface and is called the Volta potential and χ, which is due to the presence of a dipole-charge distribution at the surface and is called the surface potential. These changes occur over the top layer of the material (~ 100 nm).

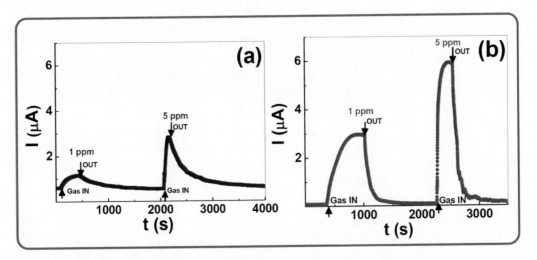

Figure 9. Room temperature (25°C) response curves for pure and Au modified (1.2 at.%) ZnO NWs towards 1 and 5 ppm of H_2S, respectively. Reprinted from publication Sens. Actuators B 2013, 186, 718-726, Copyright (2013), with permission from Elsevier.

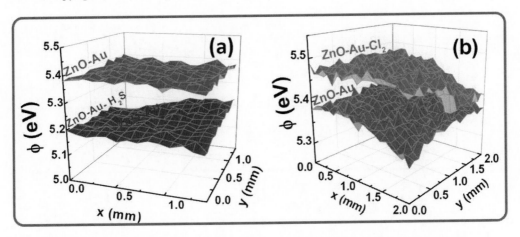

Figure 10. (a) Work function area map for pure and Au modified ZnO NWs and the raster area scans for Au modified ZnO NWs upon exposure to (b) reducing (H_2S) and (c) oxidizing gases (Cl_2), respectively. Reprinted from publication Sens. Actuators B 2013, 186, 718-726, Copyright (2013), with permission from Elsevier.

As shown in Figure 10 (a), Au modified ZnO NWs exhibited a higher work function of 5.39 eV as compared to that of pure NWs which exhibited ϕ of 5.09 eV. This higher value of ϕ can be attributed to the modulation of the Schottky barrier height at ZnO-Au interface. Work function of the material is strongly dependent upon the surface effects, grain boundary effects, and defect chemistry that are sensitive to O_2 process pressure [106]. Besides, work function is highly sensitive towards the charge distribution on the surface of the sample [107]. Hence any change in the structural or chemical character modifies the overall charge distribution and correspondingly the work function [108]. In the present case, modification with Au facilitates the dissociation of molecular oxygen species which forms oxygen ions by capturing the free electrons from ZnO. The localization of charges in these defect states leads to a band bending and result in the formation of nano Schottky-type barrier junction influencing the work function [109, 110]. Modification with Au results in randomly distributed nano Schottky barrier type junctions over the NWs surface. An increase of work function by 0.2 eV is thus attributed to the nano Schottky-type barrier junction. Such type of defect states are absent in both pure ZnO and Au films and accordingly, the work function value for Au modified ZnO NW is higher. Figure 9 (b) and (c) shows the raster area scans for Au modified ZnO NWs upon exposure to reducing (H_2S) and oxidizing gases (Cl_2), respectively. Work function is found to decrease and increase correspondingly. In both the cases area scan map upon exposure exhibited a lowering trend attributed to the recovery of the sensor film during measurements.

Figure 11 represents schematically the changes in the work function upon exposure to reducing (H_2S) and oxidizing (Cl_2) gases, respectively. At room temperature the adsorbed oxygen species on the surface of ZnO NWs is mainly $O_2^-{}_{(ads)}$. H_2S interacts with the adsorbed oxygen species and losses a large number of electrons. Herein the presence of Au nanoparticles on the surface facilitates the H_2S oxidation process by providing large number of adsorbed oxygen species.

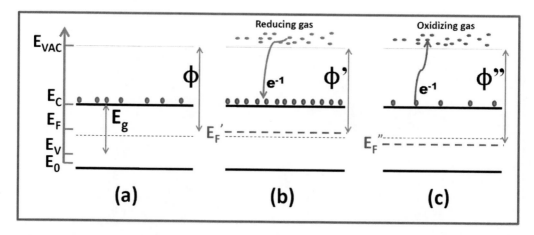

Figure 11. Schematic representation of the changes in the work function upon exposure to reducing (H_2S) and oxidizing (Cl_2) gases, respectively. Reprinted from publication Sens. Actuators B 2013, 186, 718-726, Copyright (2013), with permission from Elsevier.

The donation of electrons by reducing gas at the sensor surface results in the shift of Fermi level (E_F) towards CB depicted as lower ϕ (Figure 11 (b)). On the other hand exposure to Cl_2, which is a strong oxidizing gas leads to the further extraction of electrons (trap) from the host matrix causing resistance to increase and accordingly ϕ to increase (Figure 11 (c)) [111]. In the case of Schottky barrier junction devices, gas permeation toward the metal/semiconductor interface (electrode contact) works only for hydrogen or hydrogen producing compounds and causes the formation of a dipole layer. Hence contribution to the work function change from the electrode material is expected to be negligible [112]. Thus, the nano Schottky-barrier junction formed at the Au and ZnO interface can easily be altered in the presence of test gases and contributes effectively for the enhanced sensor response.

The contact provided to the ZnO NWs films is by Au layer which also results in a symmetrical Schottky-type barrier or the contact resistance. For the flow of electrons at the two electrodes there exist two junctions one at entering Au/ZnO interface and the other leaving the ZnO/Au interface. The room temperature resistance of pure and Au-modified ZnO NWs is 0.16 and 1.75 $M\Omega$, respectively. The contribution of symmetrical contact resistance to the total resistance is minimum compared to its bulk value and can be neglected. The deposition of thin Au layer (10 nm) results in a discontinuous film or island formation that is randomly distributed along the length of NWs. This leads to the formation of nano-Schottky type barriers or junction between metallic Au and semiconducting ZnO. In other words, a continuum of gap states induced by the Au islands locally pins the ZnO Fermi level [113]. The Fermi levels are forced to coincide and electrons pass from the ZnO to Au. The result is an excess of negative charge on Au and the formation of a positive charge depletion zone in the ZnO near its surface. The gap between Au and ZnO vanishes and the electric field now corresponds to a gradient of the electron potential in the depletion region. The Schottky barrier height (SBH) at the interface is controlled by the interaction between the metal adatoms and the semiconductor dangling bonds [26]. Also, metal lattices could introduce different interface geometries, associated with the various adsorption sites of the metal adatoms. The potential barrier thus formed would result in the electron depleted NWs surface. This is the reason why the resistance of Au modified NWs is higher than the pristine NWs.

Another crucial parameter in terms of commercial viability is the long term stability of the sensors. Figure 12 shows the stability measurements performed over a period of 5 weeks for the Au modified ZnO NWs films towards 10 ppm of H_2S at room temperature. The resistance of the sensor film (R_a) was found to increase with time with a corresponding increase in the sensor response. Throughout the measurement period the sensor response was found to vary between from 75 to 95.

The change in the resistance of the sensor film could be attributed to the surface poisoning, catalyst poisoning, migration and/or segregation of additives, degradation of contacts (diffusion), fluctuations of temperatures in the surrounding atmosphere and humidity related effects [114]. It has been shown that the adsorption of water is a dominant factor in the formation of surface characteristics, both with respect to the adsorption of other species and in surface catalysis. The electronic properties of ZnO NWs have been shown to be affected by the presence of water vapor [115, 116]. More specifically, with increase in the relative humidity values the resistance is observed to decrease attributed to the chemisorbed hydroxyl groups (OH^-) onto the surface following the Grotthuss mechanism [24, 117]. Further at constant humidity values the resistance variation was found to be 2% measured over a period of 3 months.

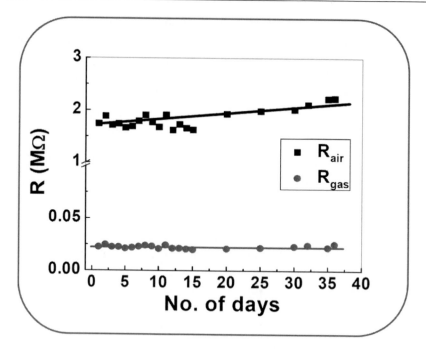

Figure 12. Long term stability measurements performed over a period of 5 weeks for the Au modified ZnO NWs films towards 10 ppm of H_2S at room temperature. Reprinted from publication Sens. Actuators B 2013, 186, 718-726, Copyright (2013), with permission from Elsevier.

5. COMPATIBILITY WITH CMOS TECHNOLOGY

To realize an ultimate commercially viable sensor based on NWs processes like NWs growth or deposition, surface modification, isolation and sensor fabrication should be made compatible with the CMOS technology. ZnO NWs provide the advantage of the growth processes compatible with CMOS technology. Besides, surface modification and sensor fabrication can also be easily carried out. Recently, Subannajui et al. demonstrated an upscalable process combining conventional micromachining with phase shift lithography as a suitable tool for NW device technology [118]. In particular, using phase shift lithography as a large scale patterning tool position of Au was controlled precisely over the substrate as big as 4 inch in diameter. At these positions, ZnO NWs were successfully grown using VS growth. More importantly, they demonstrated the fabrication of Si/ZnO NW heterojunction diodes as the prototype device and an example of device fabrication in large arrays.

Other techniques based on ZnO NWs that also demonstrated a potential for realizing commercial viability includes laser interference lithography with hydrothermal growth [119], light stamping lithography [120], localized joule heating [121], electron beam lithography [122, chemically anchored Au particles [123] and focused ion-beam (FIB) technique [124]. Thus, all the above methods can be used effectively to realize programmable positioning, size and aspect ratio control with scale up potential.

CONCLUSION

All the above mentioned points definitely make ZnO NWs a prospective candidate for the realization of an ultimate next generation sensors. It fulfils both the requirement from the sensor point of view and the commercial aspect. To summarize, Table 2 illustrates the points that justify ZnO NWs as next generation sensing material.

Table 2. Properties of ZnO NWs demonstrating its potential as next generation sensor

Sr. No.	Sensor requirements	Solution offered by ZnO NWs
1.	Ease of synthesis	Can be grown using both physical and chemical routes
2.	Flexibility of synthesis	Various starting material can be employed
3.	Control over dimensions	Aspect ration of NWs depends on the growth duration and can be controlled easily. Besides, diameter of the NWs could easily be controlled using VLS mechanism
4.	Mass production	Hydrothermal method yields uniform array of NWs over 2-4 inch substrates
5.	Compatibility with CMOS technology	Si wafers can be used for the synthesis and can easily be extended for further CMOS processes
6.	Tailored response	The response of ZnO NWs could be tailored towards different gases by a proper choice of sensitizer (Au, CuO) and control over its amount and distribution over the NWs surface [70, 82]
7.	Understanding of sensing mechanism	Using p-n junction (between p-CuO and n-type ZnO) and Schottky contacts (between Au and ZnO) a highly sensitive and selective sensor towards H_2S was achieved
8.	Low operating temperature	Can easily be tuned as demonstrated for H_2S wherein the optimum temperature of maximum sensitivity was systematically brought down to room temperature using CuO and Au as sensitizers
9.	Reproducibility	NWs growth using both physical and chemical routes is highly reversible and yields NWs with identical aspect ratios. Besides, sensor fabrication processes are simple.
10.	Repeatability	Modification of NWs network by a thin layer of sensitizer or electrode deposition using a mask can easily be repeated or reproduced thereby making the complete fabrication process highly repeatable.
11.	Low cost	Hydrothermal process provides a very low cost mean to generate an array of NWs over desired substrates. Besides simple patterning (masking/photolithography) can yield precise control over position and aspect ratio.

Thus, the 4 'S' factors namely sensitivity, selectivity, stability and suitability required for making the sensor commercially viable can easily be achieved using ZnO NWs. All the above mentioned definitely makes it the most promising material for realizing next generation sensing devices.

REFERENCES

[1] Ramgir, N. S.; Yang, Y.; Zacharias, M.; *Small* 2010, *6*, 1705– 1722.

[2] Hong, W.-K.; Sohn, J. I.; Hwang, D.-K.; Kwon, S.-S.; Jo, G.; Song, S.; Kim, S.-M.; Ko, H.-J.; Park, S.-J.; Welland, M. E.; Lee, T. *Nano Lett.* 2008, *8*, 950-956.

[3] Rout, C. S.; Hari Krishna, S.; Vivekchand, S.R.C.; Govindaraj, A. ; Rao, C.N.R. *Chem. Phys. Lett.* 2006, *418*, 586-590.

[4] Pradhan, B.; Batabyal, S. K.; Pal, A. J. *Sol. Energy Mater. Sol. Cells*, 2007, *91*, 769-773.

[5] Subannajui, K.; Ramgir, N. S.; Grimm, R.; Michiels, R.; Yang, Y.; Muller, S.; Zacharias, M. *Crys. Grow. Des.* 2010, *10*, 1585–1589.

[6] Ramgir, N. S.; Subannajui, K.; Yang, Y.; Grimm, R.; Michiels, R.; Zacharias, M. *J. Phys. Chem. C* 2010, *114*, 10323-10329.

[7] Menzel, A;. Goldberg, R.; Burshtein, G.; Lumelsky, V.; Subannajui, K.; Zacharias M. ; Lifshitz, Y. *J. Phys. Chem. C* 2012, *116 (9)*, 5524–5530.

[8] Greene, L. E.; Law, M.; Goldberger, F.; Kim, F.; Johnson, J. C.; Zhang, Y.; Saykally, R. J.; Yang, P. *Angew. Chem., Int. Ed.* 2003, *42*, 3031-3034.

[9] Pacholski, C.; Kornowski, A.; Weller, H. *Angew. Chem. Int. Ed.* 2002, *41*, 1188-1191.

[10] Tam, K. H.; Cheung, C. K.; Leung, Y. H.; Djurišić, A. B.; Ling, C. C.; Beling, C. D.; Fung, S.; Kwok, W. M.; Chan, W. K.; Phillips, D. L.; Ding, L.; Ge, W. K.; *J. Phys. Chem. B* 2006, *110*, 20865–20871.

[11] Liu, B.; Zeng, H. C.; *J. Am. Chem. Soc.* 2003, *125*, 4430-4431.

[12] Kang, H. W.; Yeo, J.; Hwang, J. O.; Hong, S.; Lee, P.; Han, S. Y.; Lee, J. H.; Rho, Y. S.; Kim, S. O.; Ko, S. H.; Sung, H. J. *J. Phys. Chem. C* 2011, *115*, 11435–11441.

[13] Subannajui, K.; Wongchoosuk, C.; Ramgir, N. S.; Wang, C.; Yang, Y.; Hartel, A.; Cimalla, V.; Zacharias, M. *J. Appl. Phys.* 2012, *112*, 034311.

[14] Wagner, R. S.; Ellis, W.C. *Appl. Phys. Lett.* 1964, *4*, 89–90.

[15] Wang, X.; Song, J.; Summers, C.J.; Ryou, J.H.; Li, P.; Dupuis, R.D.; *J. Phys. Chem. B* 2006, *110*, 7720–7724.

[16] Zhi, C.Y.; Bai, X.D.; Wang. E.G. *Appl. Phys. Lett.* 2004, *85*, 1802–1804.

[17] Ng, H.T.; Chen, B.; Li, J.; Han, J.; Meyyappan, M.; Wu, J. *Appl. Phys. Lett.* 2003, *82*, 2023–2.255.

[18] Zhao, Q.X.; Willander, M.; Morjan, R.R.; Hu, Q.H.; Campbell, E.E.B. *Appl. Phys. Lett.* 2003, *83*,165–167.

[19] Ramgir, N.S.; Ghosh, M.; Veerender, P.; Datta, N.; Kaur, M.; Aswal, D.K.; Gupta, S.K. *Sens. Actuators B* 2011, *156*, 875-880.

[20] Greene, L. E.; Yuhas, B. D.; Law, M.; Zitoun, D.; Yang, P. *Inorg. Chem.* 2006, *45*, 7535-7543.

[21] Govender, K.; Boyle, D. S.; Kenway, P. B.; O'Brien, P. *J. Mater. Chem.* 2004, *14*, 2575-2591.

[22] Wang, Z.; Qian, X. F.; Yin, J.; Zhu, Z. K. *Langmuir* 2004, *20*, 3441-3448.

[23] Vayssieres, L. *Adv. Mater.* 2003, *15*, 464-466.

[24] Foss, C.A.; Hornyak, G.L.; Stockert, M. C. R.; *J. Phys. Chem.* 1992, *96*, 7497–7499.

[25] Cepak, V.M.; Hulteen, J.C.; Che, G.; Jirage, KBJA; Lakshmi, B.B.; Fisher, E.R. *Chem. Mater.* 1997, *9*, 1065.

[26] Jessensky, O.; Muller, F.; Gosele, U. *Appl. Phys. Lett.* 1998, *72*, 1173–1175.

[27] Hulteen, J.C.; Martin, C.R. *J. Mater. Chem.* 1997, *7*, 1075–1087.

[28] Zhou, H.; Wong, S. S. *ACS Nano* 2008, *2*, 944–958.

[29] Wang, D.; Jakobson, H. P.; Kou, R.; Tang, J.; Fineman, R. Z.; Yu, D.; Lu, Y. *Chem. Mater.* 2006, *18*, 4231–4237.

[30] Liang, Y.; Zhen, C.; Zou, D.; Xu, D. *J. Am. Chem. Soc.* 2004, *126*, 16338–16339.

[31] Park, J.; Lee, Y.; Lee, J.; Jang, H. S.; Shin, H.-Y.; Yoon, S.; Baik, J. M.; Kim, M. H.; Kim, S.-J. *Crys. Grow. Des.* 2012, *12*, 3829-3833.

[32] George, A.; Maijenburg, A. W.; Maas, M. G.; Blank, D. H. A.; ten Elshof, J. E. *Langmuir* 2011, *27*, 12235–12242.

[33] Fan, H. J.; Werner, P.; Zacharias, M. *Small* 2006, *2*, 700-717.

[34] Youn, S. K.; Ramgir, N. S.; Wang, C.; Subannajui, K.; Cimalla, V.; Zacharias, M. *J. Phys. Chem. C* 2010, *114*, 10092–10100.

[35] Guo, L.; Zhang, H.; Zhao, D.; Li, B.; Zhang, Z.; Jiang, M.; Shen, D.; Sens. Actuators B 2012, *166–167*, 12-16.

[36] Hernández-Ramírez, F.; Rodríguez, J.; Casals, O.; Russinyol, E.; Vilà, A.; Romano-Rodríguez, A.; Morante, J.R.; Abid, M. *Sens. Actuators B* 2006, *118*, 198-203.

[37] Ahn, M.-W.; Park, K.-S.; Heo, J.-H.; Kim, D.-W.; Choi, K.J.; Park, J.-G. *Sens. Actuators B* 2009, *138*, 168-173.

[38] Das, S. N.; Kar, J. P.; Choi, J.-H.; Lee, T. Il; Moon, K.-J.; Myoung, J.-M. *J. Phys. Chem. C* 2010, *114*, 1689–1693.

[39] Lupan, O.; Ursaki, V.V.; Chai, G.; Chow, L.; Emelchenko, G.A.; Tiginyanu, I.M.; Gruzintsev, A.N.; Redkin, A.N. *Sens. Actuators B* 2010, *144*, 56-66.

[40] An, S.; Park, S.; Ko, H.; Jin, C.; Lee, W. I.; Lee, C. *Thin Solid Films* 2013, *547*, 241-245.

[41] Park, J.Y.; Park, Y. K.; Kim S. S. *Mater. Lett.* 2011, *65*, 2755-2757.

[42] Santra, S.; Guha, P.K.; Ali, S.Z.; Hiralal, P.; Unalan, H.E.; Covington, J.A.; Amaratunga, G.A.J.; Milne, W.I.; Gardner, J.W.; Udrea, F. *Sens. Actuators B* 2010,*146*, 559-565.

[43] Rai, P.; Khan, R.; Ahmad, R.; Hahn, Y.-B.; Lee, I.-H.; Yu, Y.-T. *Curr. Appl. Phys.* 2013, *13*, 1769-1773.

[44] Minaee, H.; Mousavi, S.H.; Haratizadeh, H.; de Oliveira P.W. *Thin Solid Films* 2013, *545*, 8-12.

[45] Hsu, N.-F.; Chung, T.-K. *Appl. Phys. A* 2014, *116*, 1261-1269.

[46] Tien, L. C.; Wang, H. T.; Kang, B. S.; Ren, F.; Sadik, P. W.; Norton, D. P.; Pearton, S. J.; Lin, J. *Electrochem. Solid State Lett.* 2005, *8*, G230.

[47] Liao, L.; Lu, H. B.; Li, J. C.; Liu, C.; Fu, D. J.; Liu, Y. L. *Appl. Phys. Lett.* 2007, *91*, 173110.

[48] Kaur, M.; Bhattacharya, S.; Roy, M.; Deshpande, S. K.; Sharma, P.; Gupta, S. K.; Yakhmi, J. V. *Appl. Phys. A* 2007, *87*, 91-96.

[49] Ahn, M.-W.; Park, K.-S.; Heo, J.-H.; Park, J.-G.; Kim, D.-W.; Choi, K. J.; Lee, J.-H.; Hong, S.-H. *Appl. Phys. Lett.* 2008, *93*, 263103.

[50] Li, L. M.; Li, C. C.; Zhang, J.; Du, Z. F.; Zou, B. S.; Yu, H. C.; Wang, Y. G.; Wang, T. H. *Nanotechnology* 2007, *18*, 22.

[51] Hsueh, T.-J.; Chang, S.-J.; Hsu, C.-L.; Lin, Y.-R.; Chen, I.-C. *Appl. Phys. Lett.* 2007, *91*, 053111.

[52] Choi, S.-W.; Kim, S. S. *Sens. Actuators B* 2012, *168*, 8-13.

[53] Chang, S.-J.; Hsueh, T.-J.; Chen, I.-C.; Huang, B.-R. *Nanotechnology* 2008, *19*, 175502.

[54] Guo, J.; Zhang, J.; Zhu, M.; Ju, D.; Xu, H.; Cao, B. *Sens. Actuators B* 2014, *199*, 339-345.

[55] Chang, S. J.; Weng, W. Y.; Hsu, C. L.; Hsueh, T. J. *Nano Comm. Networks*, 2010, *1*, 283-288.

[56] Park, S.; An, S.; Ko, H.; Jin, C.; Lee, C. *ACS Appl. Mater. Interfaces* 2012, *4*, 3650–3656.

[57] Wan, Q.; Li, Q. H.; Chen, Y. J.; Wang, T. H.; He, X. L.; Li, J. P.; Lin, C. L. *Appl. Phys. Lett.* 2004, *84*, 3654.

[58] Tiong, T. Y.; Dee, C. F.; Hamzah, A. A.; Majlis, B. Y.; Rahman, S. A. *Sens. Actuators B* 2014, *202*, 1322-1332.

[59] Hongsith, N.; Viriyaworasakul, C.; Mangkorntong, P.; Mangkorntong, N.; Choopan, S. *Cer. International*, 2008, *34*, 823-826.

[60] Xue, X. Y.; Feng, P.; Wang, Y. G.; Wang, T. H. *Appl. Phys. Lett.* 2007, *91*, 022111.

[61] Xue, X. Y.; Chen, Y. J.; Wang, Y. G.; Wang, T. H. *Appl. Phys. Lett.* 2005, *86*, 233101.

[62] Nguyen, H.; Quy, C. T.; Hoa, N. D.; Lam, N. T.; Duy, N. V.; Quang, V. V.; Hieu, N. V. *Sens. Actuators B* 2014, *193*, 888-894.

[63] Hieu, N. V.; Khoang, N. D.; Trung, D. D.; Toan, L. D.; Duy, N. V.; Hoa, N. D. *J. Haz. Mater.* 2013, *244–245*, 209-216.

[64] Woo, H.-S.; Kwak, C.-H.; Chung, J.-H.; Lee, J.-H. *ACS Appl. Mater. Inter.* 2014, *6*, 22553-22560.

[65] Jebreiil Khadema, S.M.; Abdia, Y.; Darbarib, S.; Ostovari, F. Curr. Appl. Phys. 2014, 14, 1498-1503.

[66] Das, S. N.; Kar, J. P.; Choi, J-H.; Lee, T.; Moon, K.-J.; Myoung, J.-M. *J. Phys. Chem. C* 2010, *114*, 1689-1693.

[67] Wan, Q.; Li, Q. H.; Chen, Y. J.; Wang, T. H.; He, X. L.; Li, J. P.; Lin, C. L. *Appl. Phys. Lett.* 2004, *84*, 3654-3656.

[68] Ahn, M.-W.; Park, K.-S.; Heo, J.-H.; Park, J.-G.; Kim, D.-W.; Choi, K. J.; Lee, J.-H.; Hong, S.-H. *Appl. Phys. Lett.* 2008, *93*, 263103.

[69] Wang, C.; Chu, X.; Wu, M. *Sens. Actuators B* 2006, *113*, 320–323.

[70] Liao, L.; Lu, H. B.; Li, J. C.; He, H.; Wang, D. F.; Fu, D. J.; Liu, C.; Zhang, W. F. *J. Phys. Chem. C* 2007, *111*, 1900-1903.

[71] Tam, K. H.; Cheung, C. K.; Leung, Y. H.; Djurisic, A. B.; Luig, C. C.; Beling, C. D.; Fung, S.; Kwok, W.; Chan, W. K.; Philips, D. L.; Diug, L.; Ge, W. K. *J. Phys. Chem. B* 2006, *110*, 20865-20871.

[72] Menzel, A.; Subannajui, K.; Güder, F.; Moser, D.; Paul, O.; Zacharias, M. *Adv. Funct. Mater.* 2011, *21*, 4342-4348.

[73] Rout, C. S.; Krishna, S. H.; Vivekchand, S. R. C.; Govindaraj, A.; Rao, C. N. R. *Chem. Phys. Lett.* 2006, *418*, 586-590.

[74] Klingshirn, C. *Chem. Phys. Chem.* 2007, *8*, 782-803.

[75] Aswal, D. K.; Gupta S. K. (Eds.), *Science and Technology of Chemiresistive Gas Sensors*, Nova Science Publisher, NY, USA, (2007) ISBN-13: 978-1-60021-514-8.

[76] Li, Y. X.; Trinchi, A.; Wlodarski, W.; Galatsis, K.; Kalantarzadeh, K. *Sens. Actuators B* 2003, *93*, 431-434.

[77] Barsan, N.; Schweizer-Berberich, M.; Gopel, W. *J. Anal. Chem.* 1999, *365*, 287-304.

[78] Korotcenkov, G. *Sens. Actuators B* 2005, *107*, 209-232.

[79] McAleer, J. F.; Moseley, P. T.; Norris, J. O. W.; Williams, D. E.; Tofield, B. C. *J. Chem. Soc.-Faraday Trans.* 1988, *184*, 441-457.

[80] Korotcenkov, G.; Brinzari, V.; Boris, Y.; Ivanova, M.; Schwank, J.; Morante, J. *Thin Solid Films* 2003, *436*, 119-126.

[81] Xiang, B.; Wang, P.; Zhang, X.; Dayeh, S. A.; Aplin, D. P. R.; Soci, C.; Yu, D.; Wang, D.; *Nano Lett.* 2007, *7*, 323-328.

[82] Look, D. C.; Farlow, G. C.; Reunchan, P.; Limpijumnong, S.; Zhang, S. B.; Nordlund, K. *Phys. Rev. Lett.* 2005, *95*, 225502.

[83] Tan, S. T.; Chen, B. J.; Sun, X. W.; Yu, M. B.; Zhang, X. H.; Chu, S. J. *J. Electron. Mater.* 2005, *34*, 1172-1176.

[84] Wang, B.; Min, J.; Zhao, Y.; Sang, W.; Wang, C. *Appl. Phys. Lett.* 2009, *94*, 192101.

[85] Hsu, Y. F.; Xi, Y. Y.; Tam, K. H.; Djurisic, A. B.; Luo, J.; Ling, C. C.; Cheung, C. K.; Ching Ng, A. M.; Chan, W. K.; Deng, X.; Beling, C. D.; Fung, S.; Cheah, K. W.; Fong, P. W. K.; Surya, C. C. *Adv. Funct. Mater.* 2008, *18*, 1020–1030.

[86] Tan, S. T.; Sun, X. W.; Yu, Z. G.; Wu, P.; Lo, G. Q.; Kwong, D. L. *Appl. Phys. Lett.* 2007, *91*, 072101.

[87] Chaudhuri, S. K.; Ghosh, M.; Das, D.; Raychaudhuri, A. K. *J. Appl. Phys.* 2010, *108*, 064319.

[88] Datta, N.; Ramgir, N. S.; Kaur, M.; Kailasaganapathi, S.; Debnath, A. K.; Aswal, D. K.; Gupta, S. K. *Sens. Actuators B* 2012, *166– 167*, 394– 401.

[89] Datta, N.; Ramgir, N. S.; Kumar, S.; Veerender, P.; Kaur, M.; Kailasaganapathi, S.; Debnath, A.K.; Aswal, D.K.; Gupta, S.K. *Sens. Actuators B* 2014, *202*, 1270-1280.

[90] Ramgir, N.S.; Goyal, C.P.; Sharma, P.K.; Goutam, U.K.; Bhattacharya, S.; Datta, N.; Kaur, M.; Debnath, A. K.; Aswal, D. K.; Gupta, S.K. *Sens. Actuators B* 2013, *188*, 525-532.

[91] Platzman, I.; Brener, R.; Haick, H.; Tannnenbaum, R. *J. Phys. Chem. C* 2008, *112*, 1101–1108.

[92] Tahir, D.; Tougaard, S. *J. Phys.: Condens. Matter* 2012, *24*, 175002.

[93] Yin, M.; Wu, C.-K.; Lou, Y.; Burda, C.; Koberstein, J. T.; Zhu, Y.; O'Brien, S. *J. Am. Chem. Soc.* 2005, *127*, 9506.

[94] Kim, J.; Kim, W.; Yong, K. *J. Phys. Chem. C* 2012, *116*, 15682–15691.

[95] Moulder, J.F.; Stickle, W.F.; Sobol, P.E.; Bombson, K.D. Handbook of X-ray Photoelectron Spectroscopy, Physical Electronics Inc., Minnesota, (1995).

[96] Zhang, Z.; Wang, P.; *J. Mater. Chem.* 2012, *22*, 2456-2464.

[97] Ramgir, N.S.; Kailasa Ganapathi, S.; Kaur, M.; Datta, N.; Muthe, K.P.; Aswal, D.K.; Gupta, S.K.; Yakhmi, J.V. *Sens. Actuators B* 2010, *151*, 90-96.

[98] Anothainart, K.; Burgmair, M.; Karthigeyan, A.; Zimmer, M.; Eisele, I. *Sens. Actuators B* 2003, *93*, 580–584.

[99] Nonnenmacher, M.; O'Boyle, M. P.; Wickramasinghe, H. K. *Appl. Phys. Lett.* 1991, *58* 2921.

[100] Ramgir, N. S.; Sharma, P. K.; Datta, N.; Kaur, M.; Debnath, A.K.; Aswal, D.K.; Gupta, S.K. *Sens. Actuators B* 2013, *186*, 718-726.

[101] Ishii, H.; Sugiyama, K.; Ito, E. *Adv. Mater.* 1999, *11*, 605 - 625.

[102] Lahiri, J.; Batzill, M. *J. Phys. Chem. C* 2008, *112*, 4304- 4307.

[103] Li, C.; Li, L.; Du, Z.; Yu, H.; Xiang, Y.; Li, Y.; Cai, Y.; Wang, T. *Nanotechnology* 2008, *19*, 035501.

[104] Ramgir, N. S.; Kaur, M.; Sharma, P. K.; Datta, N.; Kailasaganapathi, S.; Bhattacharya, S.; Debnath, A. K.; Aswal, D. K.; Gupta, S. K. *Sens. Actuators B* 2013, *187*, 313-318.

[105] Chang, S.-J.; Hsueh, T.-J.; Chen, I.-C.; Huang, B.-R. *Nanotechnology* 2008, *19*, 175502.

[106] Chen, S.-H.; Yu, C.-F.; Lin, Y.-S.; Xie, W.-J.; Hsu,T.-W.; Tsai, D. P. *J. Appl. Phys.* 2008, *104*, 114314.

[107] Brillson, L. J.; Lu, Y. *J. Appl. Phys.* 2011, *109*, 121301.

[108] Hwang, J. D.; Lin, Y. L.; Kung, C. Y. *Nanotechnology* 2013, *24*, 115709.

[109] Murdoch, G.B.; Hinds, S.; Sargent, E.H.; Tsang, S.W.; Mordoukhovski, L.; Lu, Z.H. *Appl. Phys. Lett.* 2009, *94*, 213301.

[110] Glatzel, T; Sadewasser, S.; Shikler, R.; Rosenwaks, Y.; Lux-Steiner, M.C.; *Mater. Sci. Eng. B*, 2003, *102*, 138-142.

[111] Miyata, T.; Hikosaka, T.; Minami, T. *Sens. Actuators B* 2000, *69*, 16-21.

[112] Potje-Kamloth, K. *Chem. Rev.* 2008, *108*, 367-399.

[113] Flores, F.; Miranda, R. *Adv. Mater.* 1994, *6*, 540–548.

[114] Korotcenkov, G.; Cho, B.K. *Sens. Actuators B* 2011, *156*, 527-538.

[115] Chang, S.-P.; Chang, S.-J.; Lu, C.-Y.; Li, M.-J.; Hsu, C.-L.; Chiou, Y.-Z.; Hsueh, T.-J.; Chen, I-C. *Superlatt. Microstructures* 2010, *47*, 772-778.

[116] Asar, N.; Erol, A.; Okur, S.; Arikan, M.C. *Sens. Actuators A* 2012, *187*, 37-42.

[117] Barsan, N.; Weimar, U. *J. Electroceramics*, 2001, *7*, 143–167.

[118] Subannajui, K.; Güder, F.; Zacharias, M. *Nano Lett.* 2011, *11 (9)*, 3513–3518.

[119] Wei, Y.; Wu, W.; Guo, R.; Yuan, D.; Das, S.; Wang, Z. L. *Nano Lett.* 2010, *10*, 3414–3419.

[120] Hwang, J. K.; Cho, S.; Seo, E. K.; Myoung, J. M.; Sung, M. M. *ACS Appl. Mater. Interfaces*, 2009, *1*, 2843–2847.

[121] Chen, C. C.; Lin, Y. S.; Sang, C. H.; Sheu, J.-T. *Nano Lett.* 2011, *11*, 4736–4741.

[122] Consonni, V.; Sarigiannidou, E.; Appert, E.; Bocheux, A.; Guillemin, S.; Donatini, F.; Robin, I.-C.; Kioseoglou, J.; Robaut, F. *ACS Nano* 2014, *8*, 4761–4770.

[123] Ito, D.; Jespersen, M. L.; Hutchison, J. E. *ACS Nano* 2008, *2*, 2001–2006.

[124] Wang, X.; Xie, S.; Liu, J.; Kucheyev, S. O.; Wang, Y. M.; *Chem. Mater.* 2013, *25*, 2819–2827.

EDITORS CONTACT INFORMATION

Dr. Al-Nakib Chowdhury
Professor
Department of Chemistry
Bangladesh University of Engineering and Technology,
Dhaka-1000, Bangladesh
nakib@chem.buet.ac.bd

Dr. Joe Shapter
Professor & Research Leader,
Flinders Centre for NanoScale Science and Technology,
School of Chemical and Physical Sciences,
Flinders University of South Australia,
Bedford Park, South Australia,
5042 Australia
joe.shapter@flinders.edu.au

Dr. Abu Bin Imran
Assistant Professor
Department of Chemistry
Bangladesh University of Engineering and Technology
Dhaka-1000, Bangladesh
abimran@chem.buet.ac.bd

INDEX

D

E

F

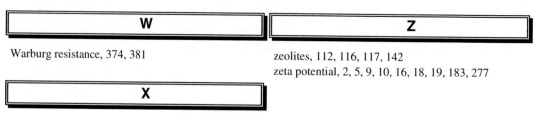